Conserving Forest Biodiversity

A Comprehensive
Multiscaled Approach

ABOUT ISLAND PRESS

Island Press is the only nonprofit organization in the United States whose principal purpose is the publication of books on environmental issues and natural resource management. We provide solutions-oriented information to professionals, public officials, business and community leaders, and concerned citizens who are shaping responses to environmental problems.

In 2002, Island Press celebrates its eighteenth anniversary as the leading provider of timely and practical books that take a multidisciplinary approach to critical environmental concerns. Our growing list of titles reflects our commitment to bringing the best of an expanding body of literature to the environmental community throughout North America and the world.

Support for Island Press is provided by The Nathan Cummings Foundation, Geraldine R. Dodge Foundation, Doris Duke Charitable Foundation, Educational Foundation of America, The Charles Engelhard Foundation, The Ford Foundation, The George Gund Foundation, The Vira I. Heinz Endowment, The William and Flora Hewlett Foundation, Henry Luce Foundation, The John D. and Catherine T. MacArthur Foundation, The Andrew W. Mellon Foundation, The Moriah Fund, The Curtis and Edith Munson Foundation, National Fish and Wildlife Foundation, The New-Land Foundation, Oak Foundation, The Overbrook Foundation, The David and Lucile Packard Foundation, The Pew Charitable Trusts, The Rockefeller Foundation, The Winslow Foundation, and other generous donors.

The opinions expressed in this book are those of the author(s) and do not necessarily reflect the views of these foundations.

Conserving Forest Biodiversity

A Comprehensive Multiscaled Approach

DAVID B. LINDENMAYER
JERRY F. FRANKLIN

ISLAND PRESS

Washington • Covelo • London

Library of Congress Cataloging-in-Publication Data

Lindenmayer, David.
Conserving Forest Biodiversity : a comprehensive multiscaled approach
/ David B. Lindenmayer and Jerry F. Franklin
    p. cm.
Includes bibliographical references (p.).
ISBN 1-55963-934-2 (hardcover : alk. paper) — ISBN 1-55963-935-0
(pbk. : alk. paper)
1. Forest conservation. 2. Plant diversity conservation. I. Franklin,
Jerry F. II. Title.
  SD411 .L56 2002
  333.95'16—dc21

                                                    2002005948

British Cataloguing-in-Publication Data available.

Printed on recycled, acid-free paper ♲

Manufactured in the United States of America
09 08 07 06 05 04 03 02   8 7 6 5 4 3 2 1

This book is dedicated to

the late Edmundo Fahrenkrog (1947–2001).

This outstanding Chilean conservationist and forester was a leader in advocating and implementing the kind of integrated natural resource management promoted in this text.

CONTENTS

PREFACE

Our deeply held belief that "the matrix matters" was the primary stimulus to initiate and complete this volume. Many participants in the debates over conservation of biodiversity have been willing to view biodiversity as a "set-aside" issue best dealt with by designating a system of large ecological reserves while largely ignoring conservation values on the rest of the landscape. By and large, the stakeholders in these debates over forest resource management—who include both environmental- and commodity-oriented individuals and organizations—seem to favor this position, differing only over reserve location and size and total reserved area.

Conservation of biodiversity is not a set-aside issue but rather one that requires significant effort throughout the forest estate. It is our conviction that a strategy to maintain forest biodiversity based primarily upon ecological reserves will fail. Most of the world's temperate and boreal forests—including all of the most productive and diverse areas—are already being utilized by human societies, or soon will be. The management practices used on these lands—the unreserved forest areas, or the *matrix*—will largely determine how successful human society is at conserving forest biodiversity and maintaining forest health.

The ecological literature contains many papers and books on reserve systems, the value of habitat patches and vegetation remnants, and the design of wildlife corridors. Few publications focus on the ecological value of the unreserved lands that contain and surround these reserves and corridors. Too often, forests are viewed as either "habitat" (the reserves) or "not-habitat" (everything else) in the literature of conservation biology and in the programs and promotional literature of environmental organizations.

Our premise is that the conservation of a significant proportion of forest biodiversity requires a comprehensive and multiscaled approach that includes both reserves and the matrix; we attempt to lay the foundations for such a comprehensive strategy in this book. Although we do discuss the value of large ecological reserves—they are a fundamental part of any credible conservation strategy—the book emphasizes the management of the matrix. We believe that the importance of the matrix has been overlooked or at least underemphasized by many conservation biologists, environmental organizations, and resource management entities. Our objectives are to make the critical role of the matrix more apparent to stakeholders, managers, and decision makers involved in efforts to sustain biodiversity and ecosystem processes in forest landscapes and to suggest issues and approaches to sustained management of the forest matrix.

Readers expecting generic "recipes" for conserving forest biodiversity will be disappointed. There are no universal recipes that can be applied uniformly and uncritically to all landscapes and stands. Rather, this book is designed to stimulate readers to identify for themselves the best strategies with which to achieve conservation objectives in particular stands and landscapes.

In this book we address approaches to forest management that enhance the conservation of biodiversity. We are, however, acutely aware that ecologically sustainable forest management involves a lot more than the conservation of biodiversity. For example, we have largely ignored the importance of the matrix for the provision of goods and services. However, the need to develop comprehensive multiscaled strategies that span large ecological reserves and matrix lands applies equally well to the maintenance of key ecosystem processes as it does to biodiversity conservation. Similarly, there are many issues associated with social and economic aspects of sustainable forest management that we do not discuss in this book. Covering all topics would take many more volumes.

Our primary focus is on temperate forests, largely because these are the ecosystems with which we have had the most professional experience. However, the general themes and principles have application in other landscapes, such as tropical forests, as well as in landscapes used for agriculture and grazing (see Lefroy and Hobbs [2000] for example).

We are aiming for a wide audience—undergraduate and postgraduate university students, academics and teachers, foresters and other natural resource managers, conservation biologists, ecologists, and decision makers in natural resource management. We have assumed readers will have a reasonable understanding of basic ecology, conservation biology, and forestry concepts.

Finally, when the idea for this book first surfaced (in Patagonia in 1997), the aim was to produce a short text on matrix management for biodiversity conservation. It quickly grew to be a substantially more difficult and larger task than was initially envisaged. Nevertheless, in the treatment of a topic of this size (and the ever-increasing body of literature associated with it), there can be no doubt that we have missed some key issues and not done justice to others. Given this, we welcome criticism of the book, as it will help improve a future edition.

ACKNOWLEDGMENTS

The themes addressed in a book like this can only evolve from a prolonged period of collaboration with many colleagues. DBL would like to thank a wide range of friends and outstanding scientists who have made major contributions to other studies, particularly in the Central Highlands of Victoria and at Tumut (both in southeastern Australia). These studies have helped develop the concepts and ideas explored in this book. A number of people deserve special mention in this regard: Mark Burgman, Ross Cunningham, Christine Donnelly, Joern Fischer, Sandy Gilmore, Phil Gibbons, Malcolm Hunter, Ryan Incoll, Bob Lacy, Rob Lesslie, Sue McIntyre, Mike McCarthy, Bill McComb, Chris MacGregor, Dan McKenney, Brendan Mackey, Adrian Manning, Henry Nix, David Norton, Ian Oliver, Rod Peakall, Matthew Pope, Hugh Possingham, Harry Recher, Peter L. Smith, Andrea Taylor, Ann Svendrup-Thygeson, and Karen Viggers.

Several research teams and management programs have been critical contributors to JFF's work with regard to the conceptualization of many of the ideas presented in this book. Foremost among these are the H. J. Andrews Ecosystem Team, led by Fred Swanson and Mark Harmon, a uniquely open and collegial group that has been the collective source of many key concepts, empirical data, and syntheses. Another is the Rio Condor Project, including the academic scientists involved in the original Independent Scientific Commission (chaired by Mary Kalin-Arroyo) and subsequent monitoring activities and the many dedicated employees of Trillium Corporation who have struggled to create a new model of sustainable forestry. Among the many noteworthy colleagues and contributors are Fred Swanson, Mark Harmon, Jim Sedell, Tom Spies, Juan Armesto, Mary Kalin-Arroyo, Dean Berg, Andy Carey, John Gordon, Stan Gregory, Art McKee, Robert Mitchell, Ken Lertzman, Jim MacMahon, Dick Miller, Will Moir, Jack Ward Thomas, Bob Van Pelt, Dick Waring, and Tim Brown. K. Norman Johnson has been a continuing mentor in the area of policy analysis and social science. Steve Brinn and the late Edmundo Fahrenkrog of Trillium Corporation were principled associates in some practical lessons on implementing ecosystem and adaptive management that are reflected in this book. Participation with literally hundreds of colleagues in such efforts as the Gang of Four (Scientific Panel for Late Successional Forest Ecosystems), Forest Ecosystem Management Assessment Team, Sierra Nevada Ecosystem Project, and Clayoquot Sound Scientific Panel has provided very practical experiences for JFF in the application of environmental science to the development of regional forest policies; the true frontiers of conservation science are revealed in such efforts.

Joern Fischer, Ryan Incoll, and Rosie Smith helped admirably with figures and tables and assisted in the collection of the enormous volume of literature used to write this book. Lynne Hendrix provided valuable assistance in digitization and improvement of photographic images and also with communications between the two continents. Ray Brereton, John Hickey, Sandy Gilmore, Mick Brown, Sue McIntyre, Sarah Munks, Adrian Manning, David Norton, Bob Pressey, Tomasz Wesolowski, Peter Kanowski, and Adrian Wayne kindly provided additional literature. Esther Beaton, and John Hickey kindly provided access to some wonderful photographic material. Anne Findlay did a fantastic editing job on the text—and coped with a large quantity of often quite densely written material. Bob Van Pelt allowed us to use two of his beautifully drawn stand profiles.

Many workshops and projects have provided us with stimulation, ideas, data, and images (although they may occasionally be stunned by what we did with their input!). Workshops on biological legacies (Wind River, Oregon, 1997), natural disturbance regimes as templates for logging regimes (Orono, Maine, 1999), and annual reviews of the BC Coastal Forests Project in British Columbia (Parksville, Vancouver Island) made important contributions to the concepts presented in this book. A invitation from Denis Saunders and John Craig to write a review chapter for a conference on the matrix at Lake Taupo, New Zealand, in

1997 provided a start on this book; we are grateful to them and to Neil Mitchell for their support of that initial effort.

We are most grateful to Jerry Alexander, Mick Brown, Joern Fischer, Sandy Gilmore, Ryan Incoll, Adrian Manning, and Peter L. Smith, who read and commented on the entire manuscript; their suggestions greatly improved earlier versions of the book. Fred Swanson provided invaluable and exhaustive reviews of Chapters 6, 7, and 8 and, with Mark Harmon, coauthored the paper (with JFF) that is extensively utilized in the discussion of monitoring. These reviewers generously and unselfishly suggested many useful ideas and insights from their own extensive experience.

Finally, we are most grateful to our partners, Karen Viggers and Phyllis Franklin, who tolerated the long hours and hard work that went into writing this book.

Conserving Forest Biodiversity

A Comprehensive
Multiscaled Approach

INTRODUCTION

Nearly everyone is familiar with poet (and cleric) John Donne's quotation, "No man is an Island . . . ," with its message that we are all connected—a continent will be affected by the loss of even a tiny part of itself. This notion of interconnectedness reinforces one premise underlying this book—that the small network of existing ecosystem reserves is crucial for the health of ecosystems extending far beyond their borders—and turns it upside down: if the matrix can be affected by what happens in reserves, how much greater is the effect of the matrix on reserves? From this perspective, we can see that stands, landscapes, regions—and all their parts—are intertwined. For this reason, the conservation of biodiversity requires a comprehensive strategy across multiple spatial and temporal scales.

> No man is an Island, entire of it self; every man is a piece of the Continent, a part of the main; if a clod be washed away by the sea, Europe is the less, as well as if a promontory were, as well as if a manor of thy friends or of thine own were; any man's death diminishes me, because I am involved in Mankind; And therefore never send to know for whom the bell tolls; it tolls for thee. (John Donne, *Meditation 17* [1624])

This book consists of four parts, each containing several logically linked chapters. The focus is on the use of matrix management—management of lands not currently protected in reserves—because until very recently the emphasis has been on the creation of large ecological reserves, with conservation management outside these protected-area networks receiving only limited attention.

Part I consists of four chapters that outline general themes and principles for developing comprehensive plans for forest biodiversity conservation through matrix management. This section describes the critical roles the matrix plays in biodiversity conservation and explores its importance in key areas of ecology such as metapopulation dynamics, habitat fragmentation, and landscape connectivity. It also suggests how knowledge about natural disturbance regimes can be used to lessen the impacts of human-caused disturbance.

Part II presents the essential elements of a comprehensive approach to forest biodiversity conservation. These include large ecological reserves, landscape-level management strategies, and stand-level management strategies. Reserves are currently the cornerstone of the ecosystem conservation effort but problematic when used as the sole biodiversity conservation strategy. This section argues that comprehensive plans for biodiversity conservation must rely not only on the use of ecological reserves, but also on matrix management applied in both near-natural forests and plantations at multiple spatial and temporal scales.

Five case studies compose Part III, illustrating aspects and elements of applied matrix management in forests. These studies build on the general principles for maintaining habitat across a full range of spatial scales and the landscape- and stand-level strategies outlined in Part II, and they illustrate the need for multiscaled strategies as part of a comprehensive approach to forest biodiversity conservation. These case studies cover conservation planning and matrix management issues from North America, South America, and Australia, providing examples that range from relatively intact forest ecosystems to intensively managed plantations.

Part IV covers additional aspects of matrix management in forest landscapes, such as the role of adaptive management and monitoring, ideas for the ongoing refinement of matrix management, and observations about the social dimensions and tensions in implementing matrix-based forest management.

Because so many aspects of matrix management are intimately interrelated, there is inevitably some repetition in themes and ideas between the different chapters and parts of this book. Nevertheless, we have tried to make the text accessible to as many readers from different backgrounds as possible. We anticipate that different readers will use this book in different ways. Some will dip in and out according to their interests and requirements. Others, such as field practitioners responsible for implementing on-ground matrix management, may wish to move directly to Chapter 3 and subsequently focus most on Chapters 6, 7, and 8 (landscape- and stand-level strategies).

However, earlier chapters of the book give a theoretical grounding for matrix management, and later chapters give a social and economic context for comprehensive approaches to forest biodiversity conservation.

PART I Principles for Biodiversity Conservation in the Matrix

Part I primarily explores the topic of matrix management. This is because much of the focus of conservation biologists has been on reserve allocation, with conservation management outside these protected-area networks receiving only limited attention.

In Chapter 1 we define what we mean by "the matrix"—landscape areas not designated primarily for conservation purposes. We also define what we consider to be ecologically sustainable forest management given its critical importance for conserving biodiversity in the matrix. Most of Chapter 1 is given over to a discussion of the critical roles of the matrix for biodiversity conservation, including supporting populations of species, regulating the movement of organisms, buffering sensitive areas and reserves, and maintaining the integrity of aquatic ecosystems. Finally, we briefly highlight the limitations of reserve systems and why these roles for the matrix are critical for biodiversity conservation, a theme that is revisited in considerable detail in Part II (Chapter 5).

Because the role of the matrix for biodiversity conservation has largely been ignored in much of ecology and conservation biology, Chapter 2 is dedicated to an exploration of the importance of the matrix in key topics such as metapopulation dynamics, habitat fragmentation, and landscape connectivity. This sets a theoretical and applied framework for identifying a set of general principles to guide matrix management in Chapter 3. We argue that the overarching principle for matrix management is the maintenance of suitable habitat at multiple spatial scales. Underpinning this is the maintenance of stand structural complexity, the maintenance of connectivity, the maintenance of land-

scape heterogeneity, and the maintenance of aquatic ecosystem integrity. Because of the varying needs of different species at different spatial and temporal scales coupled with the uncertainty of the effectiveness of any given single strategy in its own right, a fifth guiding principle—risk-spreading, or the application of multiple conservation strategies—is also discussed in Chapter 3. A sixth principle—using knowledge and inferences from natural disturbance regimes—is such a large and important topic in informed matrix management for biodiversity conservation that an entire chapter (Chapter 4) is dedicated to it. The fundamental premise of this chapter is that the impacts of human disturbance on forest biodiversity can be reduced if those impacts are within the bounds of natural disturbance regimes such as fires, floods, and windstorms.

The four chapters in Part I set a practical and theoretical foundation for the detailed discussion in Part II of a multiscaled set of approaches to conserving forest biodiversity ranging from large ecological reserves to individual trees within managed stands. How these approaches are implemented will vary between stands, landscapes, and regions. No generic "cookbook" can be applied uncritically everywhere. This is clearly demonstrated in the series of case studies that are featured in Part III, which also illustrate many of the critical roles of the matrix and reemphasize the general principles for matrix management that are the core of Part I. These case studies also highlight many of the social and political realities of matrix management in the real world, which Parts IV and V discuss in greater detail.

Critical Roles for the Matrix

The days are over when the forest may be viewed only as trees and the trees viewed only as timber.

—U.S. SENATOR HUBERT HUMPHREY (IN PATTON 1992)

The conservation of biodiversity is one of the fundamental guiding principles for ecologically sustainable forest management. Many existing conservation programs are limited to a primary or exclusive focus on lands contained in reserves for biodiversity conservation. Yet, most forest will be in off-reserve, or *matrix*, lands in the vast majority of forest regions and forest types. Comprehensive strategies for the conservation of forest biodiversity must include *both* reserves and matrix-based strategies. The importance of the matrix for the conservation of biodiversity in forests reflects its dominance in both temperate and tropical regions—most forest landscapes have been, or will be, actively used and managed. Therefore, many forest-dependent species will occur primarily in matrix lands—or not at all.

How the matrix is managed will influence the size and viability of populations of many forest taxa and thus biodiversity per se. Matrix conditions also greatly influence connectivity between reserves and the movement of organisms. In addition, by acting as buffers, matrix conditions strongly control reserve effectiveness. The matrix must sustain functionally viable populations of organisms that are fundamental to the maintenance of essential ecosystem processes such as nutrient cycling, seed dispersal, and plant pollination—processes that underpin the long-term productivity of ecosystems and their ability to produce goods and services for human use.

The conservation of biodiversity has become a major concern for resource managers and conservationists worldwide, and it is one of the foundation principles of ecologically sustainable forestry (Carey and Curtis 1996; Hunter 1999). This represents a major challenge for forest management because forests support approximately 65 percent of the world's terrestrial taxa (World Commission on Forests and Sustainable Development 1999). They are the most species-rich environments on the planet, not only for vertebrates, such as birds (Gill 1995), but also for invertebrates (Erwin 1982; Majer et al. 1994) and microbes (Torsvik et al. 1990).

Setting aside networks of dedicated reserves has been the traditional approach advocated by many conservation biologists to conserve the extraordinary biodiversity that characterizes forest ecosystems. Many books and vast numbers of scientific articles have been written on reserve design and selection (Shafer 1990; Noss and Cooperrider 1994; Margules et al. 1995; Anonymous 1996; Pigram and Sundell 1997). In this book, we argue that the conservation of a significant proportion of the world's forest biodiversity will require a far more comprehensive and multiscaled approach than simply partitioning forest lands into reserves and production areas, which we term the *matrix*. This book attempts to lay the foundations for such a comprehensive strategy. Although large ecological reserves are discussed (see Chapter 5), most of this book addresses management of the matrix.

Most temperate and subtropical forest landscapes are composed primarily (or even exclusively) of off-reserve forests, or matrix lands. It has been estimated that between 90 and 95 percent of the world's forests have no formal protection (Sugal 1997). This is particularly true in temperate regions where the most productive (and species-diverse) forested lands have already been extensively modified by humans (Franklin 1988; Virkkala et al. 1994). Therefore, forests outside reserves are extremely important for the conservation of biodiversity—how they are managed will ultimately determine the fate of much biodiversity.

Our primary objective in this book is to illustrate the importance of the matrix for biodiversity conservation and to propose strategies for enhanced matrix management that can be the basis for a comprehensive approach to maintaining forest biodiversity. We begin in this first chapter by providing our definitions of biodiversity and the matrix. We then illustrate the importance of the matrix for conserving forest biodiversity.

Defining Biodiversity and Ecologically Sustainable Forest Management

There are many definitions of biodiversity. Ours is relatively simple:

Biodiversity encompasses genes, individuals, demes, metapopulations, populations, species, communities, ecosystems, and the interactions between these entities.

There are also many interpretations of ecologically sustainable forest management (Amaranthus 1997). Ours follows Lindenmayer and Recher (1998):

Ecologically sustainable forest management perpetuates ecosystem integrity while continuing to provide wood and non-wood values; where ecosystem integrity means the maintenance of forest structure, species composition, and the rate of ecological processes and functions with the bounds of normal disturbance regimes.

Two other terms widely used in this book are *stands* and *landscapes*. We define a stand as "a patch of forest distinct in composition or structure or both from adjacent areas."

This definition is often inadequate, such as when modified cutting practices like retention at the time of harvest are employed (see Chapter 8); this means that

stands can actually be composed of structural mosaics (Franklin et al. 2002). However, the simple definition is widely used and understood (see Helms 1998) and, except where noted, we use it in this book.

Given that the focus of this book is on forests, we crudely define a landscape as "many sets of stands," or patches, that cover an area ranging from many hundreds to tens of thousands of hectares. Drainage basins are a good landscape unit, but it often is necessary to consider much smaller areas or very large regional landscape units.

Defining the Matrix from a Conservation Biology and Landscape Ecology Perspective

In the technical language of landscape ecology, the *matrix* is defined as the dominant and most extensive "patch type" (Forman 1995; Crow and Gustafson 1997). Other criteria used in its definition include the portion of the landscape that is best connected and that has a controlling influence over key ecosystem processes such as water and energy flows (Forman 1995).

In conservation biology and forest planning literature, the "matrix" often refers to areas that are not devoted primarily to nature conservation. In temperate regions in particular, these areas are generally available for resource extraction and use, including the production of commodities, as well as for many other human uses. The definitions of "matrix" from both landscape ecology and conservation biology perspectives are congruent in many temperate regions where reserved lands are clearly in the minority. Conversely, in undeveloped regions, the matrix sensu landscape ecology (the dominant patch type) may not be equivalent to the matrix sensu conservation biology because the majority of the forested land is in a "natural" condition. For this book, we have adopted a very broad definition of the matrix:

The matrix comprises landscape areas that are *not* designated primarily for conservation of natural ecosystems, ecological processes, and biodiversity regardless of their current condition (i.e., whether natural or developed).

Much of our focus is on biodiversity conservation in wood production areas outside the dedicated reserve system because land allocation in many jurisdictions around the world has created a distinction between reserves and commodity landscapes. The term *matrix management* is used frequently throughout the book, and it refers to approaches to conserve biodiversity in forests outside the reserve system.

Critical Roles for the Matrix

There are four critical roles the matrix plays that relate specifically to biodiversity conservation: (1) supporting populations of species, (2) regulating the movement of organisms, (3) buffering sensitive areas and reserves, and (4) maintaining the integrity of aquatic systems.

Conditions in the matrix will determine the degree to which it contributes positively or negatively to these roles.

Conserving biodiversity for its own sake is only one of many possible goals of matrix management. Another is the production of commodities, such as wood, and services, such as well-regulated flows of high-quality water. Management practices in the matrix will determine whether these goods and services can be sustained, because such practices also influence whether elements of biodiversity critical to long-term sustainability, such as mycorrhizal-forming fungi, are maintained (Perry 1994). Such organisms need to be conserved at functionally effective levels to maintain ecosystem processes (Conner 1988). Hence, conservation of biodiversity in the matrix is fundamental to achieving intrinsic goals (e.g., sustainable production of wood products) and extrinsic goals (e.g., maintenance of regional biodiversity and regulation of streamflow).

Supporting Populations of Species

The matrix can be managed to support broadly distributed populations of many species (deMaynadier and Hunter 1995) (Figure 1.1). Such populations have a lower risk of extinction through demographic stochasticity (Pimm et al. 1988; McCarthy et al. 1994) and environmental variability (Thomas 1990;

FIGURE 1.1. The matrix will be the primary habitat for populations of most temperate forest organisms—or not. The matrix can be managed to provide significant and well distributed populations of many forest species and is essential for maintenance of some species. The conservation of biodiversity in the matrix can have significant positive implications for the maintenance of key ecosystem processes. The conservation of the Australian arboreal marsupial, the mountain brushtail possum (*Trichosurus caninus*), is a classic example. The species is known to consume a wide range of food resources, including hypogeal fungi that form a mycorrhizal association with the root systems of eucalypt trees. Photo by E. Beaton.

Tscharntke 1992) (Figure 1.2). Large populations also have greater levels of genetic variation (e.g., Billington 1991; Madsen et al. 1999b) and are less likely to suffer extinction as a result of genetic stochasticity (Lacy 1987, 1993a; Young et al. 1996) (Figure 1.3). For example, Saccheri et al. (1998) demonstrated that low levels of genetic variation and subsequent inbreeding depression significantly increased the risk of extinction of fragmented populations of the Glanville fritillary butterfly (*Melitaea cinixia*) in Finland.

The maintenance of large, well-distributed populations also reduces the risks that an entire population will be extinguished in a single catastrophic event such as a wildfire (Gilpin 1987; McCarthy and Lindenmayer 1999a). In the forests of southeastern Australia, the maintenance of populations of Leadbeater's possum (*Gymnobelideus leadbeateri*) in many habitat

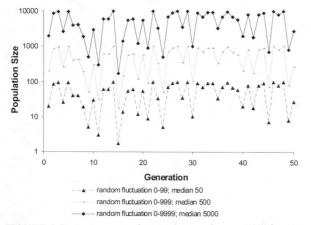

FIGURE 1.2. Larger populations have a lower risk of extinction as a result of environmental variability as indicated in this diagram (redrawn from Thomas 1990). Larger populations distributed across reserves and the matrix have a greater chance of long-term persistence.

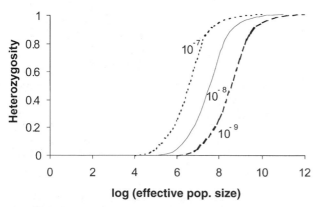

FIGURE 1.3. Relationships between effective population size and genetic variability as reflected by higher levels of heterozygosity in larger populations (redrawn from Frankham 1996). Larger populations distributed across reserves and the matrix should retain higher levels of genetic variability and, in turn, have a greater chance of long-term persistence.

patches is predicted to reduce extinction risks as a result of wildfire (Lindenmayer and Possingham 1995a).

Maintaining populations of species in the matrix can supplement populations in reserves. Species that persist in the matrix will also be those most likely to reside in reserves or remnant patches (Diamond et al. 1987; Laurance 1991a; Ås 1999; Renjifo 2001). The contribution of matrix populations to the persistence of populations within reserves is illustrated by the bald eagle (*Haliaeetus leucocephalus*) in Yellowstone National Park. Although Yellowstone is a large reserve (more than 1 million hectares), the long-term persistence of the species within the park is dependent on dispersal by animals from off-reserve populations (Swenson et al. 1986) (Figure 1.4).

Evidence of rapid species turnover within protected areas (e.g., Margules et al. 1994a) also suggests that individuals dispersing from populations in the

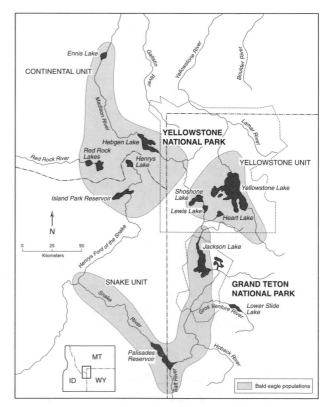

FIGURE 1.4. Populations even in some of the largest national parks often need to be supplemented by populations in matrix lands. This is illustrated by movement patterns among bald eagle populations in the Greater Yellowstone Ecosystem in the central United States (redrawn from Swenson et al. 1986).

matrix can help reverse localized extinctions within reserves (Thomas et al. 1992a; Hanski et al. 1995). Many studies show that the occupancy of reserves and habitat patches by biota is strongly related to their abundance at larger spatial scales (i.e., throughout regions) (e.g., Askins and Philbrick 1987; Askins et al. 1987; Freemark and Collins 1992; McGarigal and McComb 1995; Schmiegelow et al. 1997; Arnold and Weeldenburg 1998; Boulinier et al. 2001).

Regulating the Movement of Organisms

The matrix has a significant effect on connectivity in forest landscapes (Figure 1.5). In most temperate forest landscapes, the matrix will be the most important factor influencing connectivity—the movement of organisms

and genes will be either facilitated or obstructed by the conditions in the matrix (Taylor et al. 1993).

Noss (1991) defined connectivity as "linkages of habitats . . . communities and ecological processes at multiple spatial and temporal scales."

Connectivity in forest landscapes embodies concepts such as

- Persistence of species in cutover areas
- Species recolonization of cutover areas
- Exchange of individuals and genes among subpopulations in a metapopulation
- The role of suboptimal habitat (which may or may not be logged) in maintaining links with optimal habitat for particular species

FIGURE 1.5. Matrix conditions are the primary controllers of connectivity in landscapes, either facilitating or impeding movement of organisms. In these contrasting views (H. J. Andrews Experimental Forest, western Cascade Range, Oregon, United States): (A) Dispersed retention of 15 percent of dominant trees and woody debris on this cutover facilitates movement of many organisms. (B) Clearcutting provides hostile conditions for movement of many organisms. Photos by J. Franklin.

Facilitating connectivity in the matrix may prevent populations of species in reserves from becoming isolated and fragmented (Burkey 1989). It also can allow populations to maintain or increase their demographic and genetic size (Lacy 1993a; Saccheri et al. 1998), thereby enhancing chances of long-term persistence (Scotts 1994). Connectivity is also important because of the role of movement in shaping distribution and abundance patterns (Stenseth and Lidicker 1992)—it underpins processes such as localized extinction and recolonization dynamics (Brown and Kodric-Brown 1977) and influences patterns of gene flow (Leung et al. 1993; Mills and Allendorf 1996).

For plants, connectivity may include not only movements of species and populations, but also the movement of propagules such as spores, pollen, and seeds. In the case of animals, connectivity involves five broad types of movement (modified from Hunter 1994):

1. Day-to-day movements, such as those within home ranges or territories. These can be small for species such as adult frogs, or large in the case of wide-ranging animals like bats (e.g., Lumsden et al. 1994) or large vertebrates like the black bear (*Ursus americanus*; Klenner and Kroeker 1990).
2. Dispersal events between the natal territory and suitable habitat patches (Wolfenbarger 1946). These are typically made by juvenile or sub-adult animals attempting to establish new territories (Stenseth and Lidicker 1992).
3. Annual patterns of long-distance migration, which can span continents and/or hemispheres (Keast 1968; Flather and Sauer 1996).
4. Nomadic movements made in response to temporal and spatial variability of important resources (e.g., food; Price 1999).
5. Large shifts in distribution patterns in response to climate change. These have typically been slow in the past (Keast 1981), but more-rapid and extreme changes are expected in response to global climate change (Peters and Lovejoy 1992; see Chapter 5).

Connectivity is controlled by conditions such as appropriate vegetation cover or key structures (e.g., logs) in the matrix. Connectivity relates, in part, to the extent of matrix hostility, or "permeability," for movement (Wiens 1997a; Hokit et al. 1999). Matrix hostility and an associated lack of connectivity may result in suitable habitat remaining unoccupied, meaning that the spatial distribution of a species may not directly correspond to the spatial distribution of available habitat (Wiens et al. 1997). The connectivity role of the matrix is illustrated by a lack of gap-crossing ability among some forest birds (Dale et al. 1994; Desrochers and Hannon 1997), resulting in habitat fragmentation. The reluctance of some species of forest birds to move through open areas has been documented in many studies (e.g., Martin and Karr 1986; van Dorp and Opdam 1987; Bierregaard et al. 1992). Conversely, fragmentation-tolerant species will typically be those that can readily cross matrix lands and colonize isolated patches (Villard and Taylor 1994; Robinson 1999).

A matrix that provides a high degree of connectivity is critical, because habitat loss and habitat fragmentation are major contributors to biodiversity loss (Wilcove et al. 1986; Groombridge 1992). For example, Angermeier (1995) showed that a lack of connectivity contributed to extinction proneness in fish. Because natural forest landscapes are typically characterized by high levels of connectivity (Noss 1987; Lindenmayer 1998), the connectivity role of the matrix assumes even greater importance. Species that were abundant and well distributed in such well-connected landscapes may not have evolved well-developed dispersal mechanisms. Such taxa with relatively low mobility may be vulnerable to landscape change and fragmentation because their dispersal systems are maladapted to reduced levels of connectivity.

Buffering Sensitive Areas and Reserves

Completely open conditions in the matrix produce significant biotic and abiotic edge effects in adjacent forest patches (e.g., Lovejoy et al. 1986; Murcia 1995) with substantial negative implications for biodiversity conservation (Paton 1994; Richardson et al. 1994). The intensity of the edge interactions between two landscape units (e.g., a patch and the surrounding matrix) is typically directly related to their level of structural contrast. Matrix management strategies that reduce the contrast in structural and biophysical conditions between neighboring areas can, therefore, significantly reduce the intensity and depth of edge effects (Parry 1997; Matlack and Litvaitis 1999) (Figure

FIGURE 1.6. Matrix conditions determine the degree to which reserves and other sensitive areas are buffered. (A) Retention of tree aggregates in this cutover buffer conditions in an adjacent forest patch (Weyerhaeuser Company lands, eastern Vancouver Island, British Columbia, Canada). (B) Sharply defined boundary between reserved federal lands (right) and industrial forest lands (left) (Cascade Range, Washington, United States). Photos by J. Franklin.

1.6). Managing the matrix-to-buffer edges can substantially increase the effective size of small or medium-sized reserves and other protected areas embedded within the matrix (Janzen 1983; Schonewald-Cox 1988; Nelson 1991; see Chapter 6); processes that can negatively influence reserves can be reduced and the area available for species requiring forest-interior habitats expanded.

To illustrate the magnitude of buffering, Harris (1984) believed that an old-growth patch bounded by a recent clearcut would need to be ten times larger than one surrounded by mature forest to achieve the same area of interior forest habitat (Figure 1.7). In the case of fire risks and edge effects, mature forest buffers may reduce the chance of a fire burning into an old-growth

patch (Harris 1984) because the probability of ignition and spread declines with increasing age in some forest types (Agee and Huff 1987). Similarly, an old-growth forest surrounded by a mature stand may support different species or larger populations of a given taxon than the same stand bordered by a recently clearcut forest (Lindenmayer et al. 1999a,b).

Maintaining the Integrity of Aquatic Systems

Aquatic ecosystems support much of the biodiversity in forest landscapes. Aquatic ecosystems include surface water bodies (rivers, streams, ponds, lakes, and swamps) as well as saturated subterranean habitats such as the hyporheic zone (the zone below and

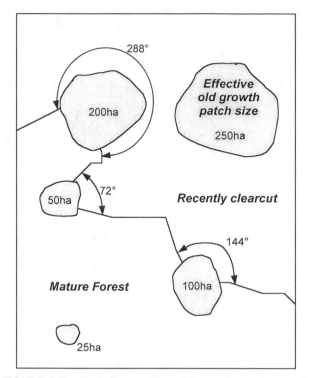

FIGURE 1.7. The buffering effect of the matrix on old-growth forest—the diagram shows the size of an area of old-growth forest needed to maintain interior conditions in a matrix dominated by recently cut forests (250 hectares in size) contrasted with one surrounded by mature forest (25 hectares in size). Redrawn from Harris 1984.

adjacent to the surface stream; Stanford and Ward 1993; Stanford et al. 1994). Aquatic ecosystems and their associated biodiversity have not received as much attention as terrestrial ecosystems in conservation biology, even though they can be heavily, and sometimes permanently, impacted (Michaelis 1984; Forest Ecosystem Management Assessment Team 1993). Maintaining and/or restoring the integrity of aquatic ecosystems must, therefore, receive high priority.

As the dominant patch type in most temperate landscapes, the matrix strongly influences the condition of aquatic ecosystems and water quality (Doeg and Koehn 1990; Naiman 1992) (Figure 1.8). Vegetation conditions in a watershed, especially the type and density of forest cover, directly influence the structure, environment, and diversity of associated aquatic ecosystems (Naiman 1992). Terrestrial vegetation also

FIGURE 1.8. Matrix conditions determine the degree to which the integrity of aquatic ecosystems are maintained. (*A*) Valley of the south fork of the Hoh River, which is buffered by federal park lands (Olympic National Park, Washington, United States). (*B*) Steep mountain slopes that have been roaded and clearcut (eastern Vancouver Island, British Columbia, Canada). Photos by J. Franklin.

regulates the paths and rates of water movement, erosion, and sediment transport through a watershed.

Natural forests typically provide a stable landscape context for the development of aquatic ecosystems and organisms (Likens 1985; Naiman and Bilby 1998). Forest cover mutes environmental extremes, such as in-stream temperature fluctuations; provides energy and nutrient inputs; filters sediments; and provides large woody debris, which is an essential structural element of many aquatic ecosystems (Harmon et al. 1986; Maser et al. 1988). Forest cover can influence storm response such as by reducing peak flood flows. Forests also can extend runoff in watersheds, such as those dominated by spring snowmelt. Erosion is also minimized in natural forest landscapes, resulting in high-quality water with low levels of sediment and dissolved and suspended materials (e.g., Ghassemi et al. 1995).

Harvesting practices, rotation lengths, and the density and quality of road systems are significant variables influencing the integrity of associated aquatic ecosystems (e.g., Vos and Chardon 1998; see Chapter 7). Decisions about what constitutes a significant watercourse, the extent of stream or riparian buffers (e.g., width and lineal extent), and forestry practices allowed within the buffers (e.g., levels of tree harvest) are also critical (Haycock et al. 1997). Clearcutting on short rotations, extensive and poorly constructed and maintained road systems, and the limited use of stream buffers can lead to the degradation of associated aquatic ecosystems (e.g., Silsbee and Larson 1983; Graynoth 1989). Conversely, extensive buffering, restrictions on harvesting on steep slopes and unstable soils, and limited road densities of well-constructed and well-maintained roads are practices that contribute to the diversity and integrity of aquatic ecosystems (Clinnick 1985; O'Shaughnessy and Jayasuriya 1991; Barling and Moore 1994). One difficulty in assessing linkages between terrestrial and aquatic ecosystems is that cause and effect are often highly displaced in time and place with regard to sources, sinks, and movement of sediment and coarse woody debris (e.g., Bormann and Likens 1979; Langford et al. 1982). However, long-term assessments of material flows in watersheds are an essential part of a forest planning process.

Providing for the Production of Commodities and Services

The environment returns an estimated \$US33 trillion in goods and services to human society each year (BirdLife International 2000). In temperate forest regions, the matrix is the primary zone for the production of goods and the provision of services (Franklin 1993a). Management practices and conditions in the matrix will determine the quality, quantity, and sustainability of goods and services obtained from forests (Chapin et al. 1998) (Figure 1.9). Humans derive a variety of goods from forests (Costanza et al. 1997). Production of wood fiber is a major one—wood products contribute \$US400 billion annually to the world market economy (or about 2 percent of total gross domestic product) (World Commission on Forests and Sustainable Development 1999). Services from forests include the regulation of streamflow, soil protection, and nutrient retention and cycling. Forests are also recognized as a major carbon sink—another important ecosystem service (Harmon et al. 1990; Brown et al. 1997; Pinard and Putz, 1997; Wayburn et al. 2000; Harmon 2001).

Many elements of biodiversity need to be conserved within the matrix to sustain the long-term production of wood and other products, as well as ecosystem services (Pimentel et al. 1992, 1997). Losses of elements of forest biodiversity may impair essential ecosystem functions. Examples include organisms that play key roles in the decomposition of organic matter (McGrady-Steed et al. 1997), pollination (e.g., Prance 1991; Robertson et al. 1999), seed dispersal, and the formation of mycorrhizal associations (Maser et al. 1978). Changes in biodiversity could influence the long-term floristic composition and stand architecture of forests (Claridge 1993), which could have negative ramifications for the sustained production of commodities. This is related to the "insurance hypothesis," which suggests that higher levels of biodiversity should lead to the maintenance of more reliable ecosystem functions, particularly when environmental conditions change (Naeem 1998).

Matrix management is also important for conserving ecosystem processes by emphasizing the importance of biodiversity conservation in the matrix as well as conservation of genes, species, and populations for

FIGURE 1.9. The matrix is the source of most commodities, such as wood, and services, such as well-regulated flows of high-quality water. Maintaining long-term productivity of such lands and their ability to maintain natural levels of hydrologic and geomorphic processes is critical (managed forest landscape in Olympic State Experimental Forest, Washington, United States). Photo by J. Franklin.

their own sake (Simberloff 1998). This is why Conner (1988) recommended that organisms be conserved at functionally viable numbers to ensure their ecological "effectiveness" in the maintenance of ecosystem processes.

Many of the components of forest biodiversity that play important roles in ecosystem processes are inconspicuous invertebrates (Recher et al. 1996), microbes (Torsvik et al. 1990), and cryptogams (Ashton 1986; Vellak and Paal 1999). These taxa have received limited attention in conservation programs, and even when they are considered (e.g., Taylor 1991; Brown et al. 1994) they can be difficult to assess and manage (Forest Practices Board 1998). Such species play pivotal roles in such processes as nutrient cycling and pollination (Goldingay et al. 1991). Lichens, for example, are valuable nitrogen-fixing organisms in many

forest ecosystems as are vascular epiphytes in Australian rainforests and wet sclerophyll forests (Lamb 1991). Similarly, fungi that form mycorrhizae promote the regeneration and growth of trees in most forests, as has been demonstrated in the Douglas-fir forests of the northwestern United States (Perry 1994). Retained patches of vegetation within harvested forests can provide a reservoir of mycorrhizal fungi and soil microbes that subsequently inoculate regenerating forests on a cutover site (Perry and Amaranthus 1997).

Vertebrates also may facilitate ecosystem processes that sustain forest productivity. Hummingbirds pollinate many forest plants (Pauw 1998). Pollen from numerous plant species is carried in the fur of some arboreal Australian animals such as the sugar glider (*Petaurus breviceps*); these animals can be significant plant pollinators (Carthew and Goldingay 1997; Goldingay 2000a). Some species of rodents (e.g., voles, shrews, and squirrels) and small and medium-sized forest mammals (e.g., Australian rodents and marsupials) disseminate spores of mycorrhizal-forming hypogeal fungi (Maser et al. 1977; Claridge and Lindenmayer 1993; Mills et al. 1993). In the case of symbiotic small mammal–fungal interrelationships, activities such as poisoning rodents to enhance tree survival can negatively affect forest growth and ecosystem processes (Maser et al. 1978).

Reduced production of goods and services in the matrix due to impaired ecosystem processes has substantial social and economic costs (Costanza et al. 1997; Pimentel et al. 1997). For example, populations of natural parasites and predators are estimated to accomplish the equivalent of $100–200 billion worth of pest control—compared to $20 billion expended on artificial control measures such as spraying (Pimentel et al. 1992). Thus, the loss of natural populations of parasites and predators has massive financial implications for timber and pulp production from native and plantation forests (Pimentel et al. 1992). A useful example of the intersection of pest control and matrix management comes from exotic eucalypt plantations that cover 5 million hectares in Brazil. Strips of native vegetation left within eucalypt plantations support a diverse insect biota, including many natural predators of defoliating lepidopteran caterpillars. These predators may reduce pest populations in the surrounding eucalypt plantations (Zanuncio et al. 1997).

Synergies among Roles of the Matrix

The five roles of the matrix are interrelated. For example, managing the matrix to buffer sensitive areas, such as riparian zones, promotes the conservation of aquatic ecosystems, contributes to improved connectivity for wildlife, and increases the ability of the matrix to support populations of species. This is illustrated in the forests of southeastern Australia, where stream buffers promote the protection of aquatic ecosystems (Clinnick 1985) and its associated biodiversity (Doeg and Koehn 1990) and act as conduits for the movement of terrestrial vertebrates (Hewittson 1997). They also provide habitat for resident animals (Recher et al. 1987; Fisher and Goldney 1997), which may subsequently reinvade post-logging regrowth

stands (Kavanagh and Turner 1994). Some of the species inhabiting riparian buffers may help control populations of forest pests, such as the large, defoliating phasmid "stick" insects.

Limitations of a Reserve-Only Conservation Strategy for Biodiversity Conservation

Despite the critical roles of the matrix for forest biodiversity outlined above, the traditional emphasis of conservation biology and forest management has been the establishment of large ecological reserves. Reserves are unquestionably important where particular ecosystem types, vegetation communities, or forest age classes are being rapidly modified or eliminated (McNeely 1994a), or where only small amounts of the original cover of forest types remain (Pressey et al. 1996; Lindenmayer and Franklin 2000). The establishment of reserves is also critical for species intolerant of even limited levels of human disturbance and for those that can only be conserved in large ecological reserves. For example, even limited levels of human-related forest disturbance can lead to marked changes in stream conditions (McIntosh et al. 2000) with concomitant negative impacts on some aquatic biota, such as fish (Baxter et al. 1999). However, as important as reserves are, we feel that a conservation strategy based primarily or exclusively on reserves will fail because of its inherent limitations (as discussed in Chapter 5). As proposed by Soulé and Sanjayan (1998), 50 percent of tropical taxa would be extinct within the next few decades, even if more than 10 percent of tropical forests were protected in well-designed reserve systems.

We are greatly concerned that approximately half the world's native forests have been cleared in the past forty years—slowing the rate of clearing is critical for forest biodiversity conservation. Hence, the establishment of reserve systems is essential despite their limitations. However, large ecological reserves and matrix-based efforts to conserve forest biodiversity need to be implemented simultaneously. We do not propose to replace large ecological reserves with matrix management, but rather we emphasize the complementarity of the two broad strategies (Figure

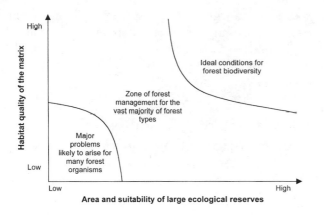

1.10). We stress the need for both reserves and matrix management as part of a comprehensive strategy for biodiversity conservation across multiple spatial and temporal scales.

FIGURE 1.10. Forest biodiversity will require multiscaled strategies that encompass large ecological reserves and management of matrix lands. This is highlighted in this conceptual diagram, which shows the complementarity between the two broad approaches (from D. Perry personal communication).

The Matrix and Major Themes in Landscape Ecology and Conservation Biology

[A] species or system may simply not operate in the way envisioned by the theories applied to it.

DOAK AND MILLS (1994)

Landscapes have been conceptualized using two main models—the *corridor-patch-matrix model* of Forman (1995) and the *landscape continuum model* developed by McIntyre and Hobbs (1999). The two models differ in their relative emphasis. In the corridor-patch-matrix model, landscapes are viewed as varying mosaics of different types of patches and corridors. In the landscape continuum model, landscapes are characterized by having different levels of vegetation cover with a continuum or gradient of possible conditions that range from an intact cover of native vegetation through to relictual levels of cover. The focus of the corridor-patch-matrix model is on the form or structure of landscapes, whereas the landscape continuum model emphasizes the function of a landscape across varying structural gradients of vegetation cover. Simultaneous consideration of both models is useful because it can lead to greater awareness of the range of conditions that occur in real landscapes and, in turn, the diversity of responses to such varying conditions by different biota. Both models have limitations. In particular, landscapes are usually treated (intentionally or otherwise) in very simple terms as having two components—patches (habitat) and remaining land (nonhabitat). Real landscapes are more complex than this. Such complexity matters—particularly when attempting to predict the response of species to landscape modification.

The simplification of landscape conditions pervades many major themes in conservation biology in which the importance of the matrix has received little or no consideration. Ecological theories such as island biogeography, nested subset theory, and metapopulation biology often take a highly reductionist approach and largely ignore the complexity found in real landscapes. This limits the explanatory and predictive value of such theories because

- Habitat fragments and reserves are treated as islands and the matrix is often considered as either totally unsuitable habitat or neutral habitat.
- There is no recognition that the matrix is inherently heterogeneous in both form and function and thus variable, for example, in habitat quality.
- Interrelationships between fragment–matrix dynamics are disregarded or greatly oversimplified.

Conditions in the matrix are pivotal to studies of habitat fragmentation, metapopulation dynamics, extinction proneness, edge effects, and reserve design. The responses of biota in habitat fragments are not isolated from conditions in the surrounding matrix. Matrix conditions may even be more important in determining the survival of some species than factors that are traditionally examined, such as fragment size and patch isolation. Consideration of the matrix can highlight fundamental differences in species responses between fragmented forest landscapes versus fragmented agricultural landscapes, such as the magnitude of edge effects and threshold vegetation cover impacts on species loss.

In order to develop more useful ecological theory and to advance biodiversity conservation efforts, it is important to

shift the emphasis from the fragments to the management of the matrix in which they are embedded. If the biota in the fragmented landscape is to persist then the management of the matrix becomes all important. Ameliorating the matrix may be the most important way to manage fragments. (Crome 1994)

In this chapter, we address two main themes. First, we describe two models that have been developed to classify patterns of landscape cover: Forman's 1995 *corridor-patch-matrix model* and McIntyre and Hobbs' (1999) *landscape continuum model*. We discuss why it is useful to consider both models and also how both are often (mis)interpreted in ways that oversimplify the complexity and range of habitat conditions within real landscapes. Such oversimplification is not confined to these two landscape models—it pervades many themes in conservation biology. In the second part of this chapter, we explore interrelationships between the role and importance of the matrix and major themes in conservation biology. Such conservation-related topics as island biogeography, extinction proneness, habitat fragmentation, metapopulation dynamics, connectivity, reserve selection, and edge effects have developed in recent decades. Their frequent omission of a key element—conditions in the matrix—has often limited the comprehensiveness of some of these concepts and curtailed their predictive value in real landscapes.

Models of Landscape Cover

The matrix is often the most extensive patch type in a landscape (Forman 1995; see Chapter 1). It is considered in both the general conceptual models of landscape cover discussed below—the corridor-patch-matrix model and the landscape continuum model. The aim of both models is to incorporate the range of conditions in a landscape.

The Corridor-Patch-Matrix Model

Forman (1995) developed the corridor-patch-matrix model, in which landscapes are conceived as mosaics of three components: patches, corridors, and the matrix (Figure 2.1). Forman defined these landscape units as follows:

- Patches are . . . relatively homogeneous nonlinear area[s] that differ from their surroundings.

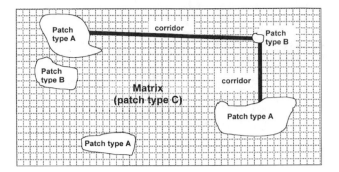

FIGURE 2.1. A landscape perspective based on the corridor-patch-matrix model (sensu Forman 1995). The model has a number of patch types, including the most extensive background matrix (patch type C).

- Corridors are narrow, linear patches of a patch type that differ from those on either side.
- The matrix is the dominant patch type in a landscape. It is characterized by extensive cover, high connectivity, and/or a major control over dynamics.

The matrix is often intersected by corridors or perforated by smaller patches (Forman 1995). In Forman's model, patches and corridors are two types of readily identifiable landscape components distinguished from the background matrix (Forman and Godron 1986; Kotliar and Wiens 1990) (Figure 2.1).

Forman (1995) noted that "every point in a landscape was either within a patch, corridor, or the background matrix," and in allowing for the complexity of many landscapes he states that patches within the matrix can be "as large as a national forest or a single tree" (Dramstad et al. 1996). Similarly, the matrix can be "extensive to limited, continuous to perforated, and variegated to nearly homogenous" (Forman 1995).

The Landscape Continuum Model

In some landscapes, the boundaries between patch types are not obvious, and differentiating them from the background matrix may not be straightforward. The landscape continuum model was developed in response to this problem (McIntyre 1994; McIntyre and Hobbs 1999) (Figure 2.2). The model was originally

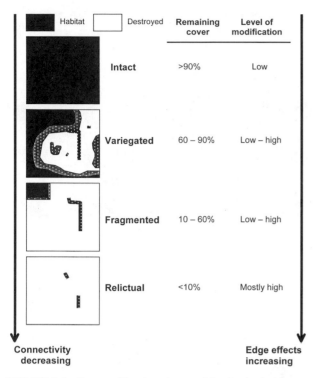

		Remaining cover	Level of modification
■ Habitat	□ Destroyed		
Intact		>90%	Low
Variegated		60 – 90%	Low – high
Fragmented		10 – 60%	Low – high
Relictual		<10%	Mostly high

Connectivity decreasing

Edge effects increasing

FIGURE 2.2. States of landscape condition in the landscape continuum model. Redrawn from McIntyre and Hobbs 1999.

types of landscapes, such as tropical and temperate forests.

The landscape continuum model developed by McIntyre and Hobbs (1999) recognizes four broad cover classes. At opposite ends of the continuum are the intact landscape and the relictual landscape. Intermediate between these conditions are the variegated landscape and the fragmented landscape (Figure 2.2):

> In variegated landscapes, the habitat [not necessarily an unsuitable environment] still forms the matrix, whereas in fragmented landscapes, the matrix comprises "destroyed habitat." (McIntyre and Hobbs 1999)

McIntyre and Hobbs (1999), and other workers before them (e.g., Wiens 1994; Pearson et al. 1996), believed that landscapes were more complex than could be described by the corridor-patch-matrix model. Furthermore, they contended this approach to be anthropocentric because human disturbances give rise to a greater range of landscape conditions. McIntyre and Hobbs (1999) argued their model was needed because the corridor-patch-matrix model did not "capture the richness of possible landscape configurations, since it always assumes that the matrix is 'nonhabitat' . . ." (McIntyre and Hobbs 1999). They equated the corridor-patch-matrix model to only one landscape condition—fragmentation—a condition in which they considered patches and corridors to be habitat "islands" embedded within a hostile background matrix (McIntyre and Hobbs 1999).

McIntyre and Hobbs (1999) believed that by recognizing different landscape conditions, conservation strategies could be better focused. In their view, many forest landscapes will conform to an intact or variegated condition for numerous elements of the biota if the matrix is managed appropriately. Failure to embrace matrix management means that the background matrix can be readily distinguished from the patches and corridors it contains (McIntyre et al. 1996) and the system will degenerate from a variegated to a fragmented or even relictual one (sensu McIntyre and Hobbs 1999). By recognizing that some landscapes were variegated, matrix management strategies would attempt to prevent landscapes deteriorating into a "fragmented" condition that is significantly less suitable for biota (McIntyre et al. 1996). Conversely, by

proposed for semi-cleared grazing and cropping landscapes in parts of rural eastern Australia characterized by small fragments of woodland habitat and relatively isolated native trees scattered throughout grazing lands (McIntyre and Barrett 1992; McIntyre et al. 1996) (Figure 2.3). Here, from a human perspective, "patches" and "corridors" are difficult to identify among the loosely organized and spatially dispersed arrays of trees (or other vegetation cover such as native grassland). Although single paddock trees in isolation may not provide habitat or act as micropatches for many taxa (but see Fischer and Lindenmayer 2002), numerous trees scattered across a landscape will *collectively* provide habitat for some species (e.g., for some woodland birds; Barrett et al. 1994). In this way, the landscape continuum model takes account of small habitat elements that might otherwise be classified as "unsuitable habitat" in the background matrix (Tickle et al. 1998). More recently, McIntyre and Hobbs (1999) extended their model to include other

FIGURE 2.3. Variation in vegetation cover in different landscapes. (*A*) A strongly patchy landscape composed of forest stands of different ages and composition (conifers, hardwoods, and mixtures of the two groups) (Harvard Forest in central Massachusetts, United States) (photo by J. Franklin). (*B*) A variegated landscape (sensu McIntyre and Barrett 1992) of scattered paddock trees in southeastern Australia (photo by D. Lindenmayer). (*C*) A landscape dominated by a matrix of old-growth forest of Douglas-fir and other coniferous species (Mount Hood National Forest, Oregon, United States) (photo by J. Franklin).

distinguishing fragmented landscapes from variegated ones, appropriate matrix management protocols would "re-variegate" fragmented landscapes through steps such as restoring connectivity and halting degrading processes (Hobbs et al. 1993; McIntyre and Hobbs 1999).

We believe that some of the assertions about the advantages of the landscape continuum model fail to recognize aspects of Forman's (1995) original corridor-patch-matrix model. The interpretation of this model by McIntyre and Hobbs (1999) is often a landscape composed of patches and corridors of suitable habitat embedded within a hostile (nonhabitat) ma-

trix. However, Forman (1995) stated that the matrix was often the largest and most dominant landscape component "characterized by extensive cover, high connectivity, and/or a major control over dynamics" (see definition given in Chapter 1). The corridor-patch-matrix model is actually "interpretation-free" in terms of the response of a given species to particular landscape components (Forman 1995). The corridor-patch-matrix model does not rely on a single patch and corridor type or matrix condition of uniform quality and also allows for a continuum of conditions by recognizing that landscapes are a mosaic of many types of patches and corridors that cover different landscapes to varying extents.

Congruence between the Corridor-Patch-Matrix and Landscape Continuum Models

The corridor-patch-matrix model and the landscape continuum model are congruent on many levels. Both clearly recognize a gradient in the complexity of landscape conditions and a continuum of landscape "naturalness." Also, the matrix will usually be the dominant landscape component irrespective of whether the corridor-patch-matrix model or the landscape continuum model is applied.

The primary distinctions between the two models are differences in their relative perspectives on landscapes. The focus of the corridor-patch-matrix model is on the *pattern* or *form* of different landscape units (i.e., the size and shape of different patches and patch types) that compose a landscape mosaic. McIntyre and Hobbs (1999) have recognized that distinct patch boundaries may not always be identifiable in some landscapes (particularly those classified as "variegated"). Therefore, the emphasis of their landscape continuum model is on the specific *function* of a given landscape across a structural gradient of vegetation cover—a focus that requires stipulating the organism or assemblage of interest and specifying the processes influencing species' responses.

Using both models leads to a better appreciation of landscapes and how species respond to them. This is important because landscape changes, such as habitat fragmentation, are both a pattern (e.g., subdivision of habitat cover) and a process (e.g., altered population dynamics) (Wiens 1994; Cale 1999). Looking at attributes of both models encourages landscape ecologists, conservation biologists, and land managers to think not only about the form and the function of a landscape but also about the interrelationships *between* form (or structure) and function. It is also useful because patterns of landscape cover seen from a human perspective may or may not provide a useful framework for interpreting biotic response to landscape conditions (Ingham and Samways 1996). Strips of

vegetation may provide conduits for movements between patches for some animals that avoid the surrounding matrix because of its unsuitability (e.g., kangaroo populations in the wheatbelt of Western Australia; Arnold et al. 1993). In other cases, some taxa will live in, and move readily through, the matrix (e.g., bird taxa in rural northern New South Wales, Australia; Barrett et al. 1994).

A limitation of all landscape models is recognizing that species will have unique responses to the same landscape (Kotliar and Wiens 1990; Villard et al. 1999). Interspecific variation in response will depend on (1) the scale at which species perceive the environment (Andrén 1996), (2) the habitat requirements of a particular species, (3) where suitable habitat occurs in a landscape, and (4) how animals move across (and disperse between) areas of suitable habitat (Tickle et al. 1998). Such an organism-centered view of landscape cover means there is no single most appropriate perspective or scale from which to view landscapes (Haila 1999). Aerial photography interpretation—which both models use to conceptualize landscapes—is only able to represent species-specific responses in a limited sense.

Limitations in the Application of the Landscape Models

Irrespective of the original intentions of the architects of the corridor-patch-matrix model and the landscape continuum model, both models have often been misinterpreted by practicing landscape ecologists and conservation biologists. The models are sometimes simplified to represent landscapes as patches or corridors of habitat, which contrast markedly with remaining areas of nonhabitat. This is a reductionist approach based on a human interpretation of the perceived function of a landscape. Too often, mapping tools (such as geographic information systems, or GIS) have been used to define patches, assuming that species would perceive "patches" in the same way and at the same scale as humans do (Bunnell 1999a). However, the determination of what constitutes a suitable patch and what does not must be based on the habitat requirements, movement patterns, and other attributes of the organism of interest.

The binomial classification of landscapes into habitat and nonhabitat is problematic for other reasons not accounted for by either landscape model (as traditionally interpreted). First, habitat quality for a particular species is often a graded, rather than an either/or, measure. In effect, there can be a gradient of conditions that are suitable to varying degrees (Whittaker et al. 1973; Austin 1999) with some areas of habitat being more important than others (Block and Brennan 1993; Lindenmayer and Cunningham 1996). This is not unlike the source-sink concept of Pulliam et al. (1992) and others (e.g., Howe et al. 1991). Second, patches and corridors (as traditionally defined) are never uniform but rather are always characterized by within-patch heterogeneity (Forman 1995). Such heterogeneity matters (Fox and Fox 2000)—as demonstrated in many studies that have found that within-patch quality can have a significant influence on species occurrence as well as (or instead of) attributes like patch size (e.g., Lindenmayer et al. 1993a, 1999b; Mac Nally et al. 2000; Sieving et al. 2000). In these cases, within-patch processes as well as between-patch processes can be important (Harrison and Taylor 1997).

The simplifications in landscape models are understandable given the complexity of real landscapes and the difficulty ecologists face in tackling such complexity. However, it is important to know the assumptions underlying these models and the limitations in their application.

Landscape Models and the Terminology Used in This Book

Although this book is largely about the function (rather than the pattern or form) of landscapes for forest biodiversity conservation, we have not adopted the terminology used by McIntyre and Hobbs (1999). Instead, we use the terms *patches*, *corridors*, and *matrix* because they will be more familiar to readers responsible for implementing matrix-based management. It is crucial to emphasize that the "matrix" in the corridor-patch-matrix and landscape continuum models can range from a very small entity, such as the area around

a clump of vegetation (e.g., for beetles; Wiens et al. 1997), to a large one, such as an entire national forest (Forman 1995) or an entire geographic region such as the area surrounding Yellowstone National Park (the Greater Yellowstone Ecosystem). This means that the scale of the matrix will vary according to the organism or ecological process under scrutiny and that strategies for biodiversity conservation within the matrix also will be variously scaled from single trees to entire landscapes (see Chapters 3, 6, 7, and 8).

The Matrix and Major Themes in Conservation Biology

Reductionist approaches to the complexity of real landscapes are not confined to models of landscape cover. It is commonplace in many aspects of conservation biology where the importance of the matrix has often been overlooked. Failure to recognize the richness of the matrix concept has the potential to limit the applicability of some theoretical constructs of conservation biology to practical resource management and conservation problems.

Island Biogeography

An observation made early in the study of ecology was that large areas support more species than smaller ones do (e.g., Arrhenius 1921; Preston 1962)—a fact that was later often referred to as the *species-area relationship* (Rosenzweig 1995). Species-area relationships are common for some groups, such as birds (Newton 1998). The *theory of island biogeography* (Macarthur and Wilson 1963, 1967) was developed, in part, to explain the species-area phenomenon, particularly for island biotas. Part of this theory considers aggregate species richness on islands of varying size and isolation from a mainland source of colonists (Shafer 1990). The balance between extinction and recolonization is predicted to produce an equilibrium number of species on a given island, which is a derivative of island size and island isolation. Larger and less-isolated islands are predicted to support more species. Diamond (1975) (among many others) likened oceanic islands to reserves and extended the theory of island biogeography as a basis for generic design principles for protected areas. This became a highly controversial topic in conservation biology for several decades (see reviews by Gilbert 1980; Burgman et al. 1988).

Island biogeography theory does not account for conditions and processes in the matrix, such as disturbance regimes (Baker 1992), the magnitude of edge effects (that are, in turn, related to patch-matrix contrasts), and matrix suitability for habitat and movement—all of which strongly influence the distribution of species in forest landscapes (Gascon and Lovejoy 1998). Uncritical adoption of "general rules" for reserve design derived from island biogeography theory can even have a negative effect on conservation values (Simberloff 1988). The theory is too simplistic to have practical application in most real-world landscape mosaics (Zimmerman and Bierregaard 1986; Doak and Mills 1994; see Chapter 5).

Island Biogeography Theory and General Reserve Design Principles

As reserves and habitat fragments have been likened to oceanic islands, some elements of the theory of island biogeography have been proposed as general principles to guide the design of nature reserves (e.g., Terborgh 1974; Diamond 1975; Diamond and May 1976; Shafer 1990; Noss and Cooperrider 1994). The IUCN's 1980 World Conservation Strategy (IUCN 1980) adopted these principles, and in Australia they were recommended for wildlife management in forests (Davey 1989).

Six general principles (based on the unrealistic assumption that there are identical places from which protected areas can be selected) are derived largely from island biogeography theory:

Principle 1. Large reserves are better than small reserves.

Principle 2. A single large reserve is better than a group of small ones of equivalent total area (the basis for the so-called SLOSS [Single Large Or Several Small] debate; see Gilbert 1980; Burgman et al. 1988).

Principle 3. Reserves close together are better than those far apart.

Principle 4. A compact cluster of reserves is better than a line of reserves.

Principle 5. Circular reserves are better than long, thin ones.

Principle 6. Reserves connected by a corridor are better than reserves not connected by a corridor.

Effects of the matrix on biodiversity are ignored, whether positive or negative. Yet, matrix conditions can significantly influence biodiversity conservation. For example, the outcomes of Principles 2, 3, and 4 (reserve size and spatial location) depend upon interacting factors that include

- *Demographic interactions between populations in reserves and populations from the surrounding matrix lands.* These may include the dispersal capabilities of those taxa targeted for conservation and their ability to colonize reserves from the matrix (Swenson et al. 1986).

- *Disturbance regimes in both the matrix and those parts of a landscape set aside for reservation.* Where disturbance regimes do not impact landscapes or are not fundamental to persistence or regeneration, the outcome of Principle 2 may be quite simple—a single large reserve will often function more effectively than a number of smaller ones of equivalent size. More-complex trade-offs occur in environments where disturbance regimes are important (Baker 1992). For example, a single large reserve may be more susceptible to destruction by a single catastrophic event than a set of smaller, spatially separated ones (Quinn and Hastings 1987; Quinn et al. 1989; Lindenmayer and Possingham 1995b). Organisms in patches unaffected by a disturbance may be able to recolonize patches where localized extinctions have resulted from a catastrophic event. Such a lack of correlation in disturbance regimes between different patches has promoted the persistence of a number of species (Berger 1990; Murphy et al. 1990; Stacey and Taper 1992).

Conditions in the matrix, including the level of structural contrast between reserves and matrix lands, also influence the effect of reserve shape (Principle 5).

If negative edge effects emanating from the matrix are limited, a long, narrow reserve system might better capture sensitive areas, particularly those characterized by a linear spatial configuration like a riparian system (Burgman and Ferguson 1995).

Island Biogeography and Other Issues Associated with Matrix Conditions

One significant issue ignored in island biogeography theory is that the *composition* of a faunal community can be more important than species richness (Noss 1983). The contribution of the matrix to patterns of species diversity and the response of individual species and species assemblages to landscape change are not taken into account. The theory may fail where the matrix provides even temporarily suitable habitat (Zimmerman and Bierregaard 1986; Ricketts 2001). It also has limited predictive ability when the importance of fragment size is outweighed by other factors such as (1) disturbance regimes or habitat conditions within and outside patches (e.g., Fitzgibbon 1997; Fox and Fox 2000), (2) edge effects at the patch-matrix interface (Saurez et al. 1998), and (3) levels of heterogeneity and connectivity across entire landscapes (Brereton 1997; Metzger 1997).

Examples that illustrate the shortcomings of island biogeography theory include the Biological Dynamics of Forest Fragments Project in Brazil, where edge effects and matrix conditions had a substantially greater influence than fragment size for some animals (Brown and Hutchings 1997; see Chapter 14). In that study, there was an *increase* in frog, small mammal, and butterfly species richness following the isolation of rainforest patches—a result opposite to predictions from island biogeography theory (Gascon and Lovejoy 1998). Species richness was elevated by an influx of taxa capable of using the changed matrix. Estades and Temple (1999) similarly recorded increased bird species richness in small rather than large *Nothofagus* spp. remnants embedded within an exotic radiata pine (*Pinus radiata*) plantation in Chile. An increase in species number in fragments may be due to the influx of organisms displaced from the adjacent matrix (Bierregaard and Stouffer 1997; Darveau et al. 1995) or a predominance of edge or generalist taxa from the surrounding matrix (Gascon and Lovejoy 1998; Ås

1999)—taxa that are sometimes called *invaders* (see also Halme and Niemelä 1993; Saurez et al. 1998).

In summary, some of the influences of the matrix that limit island biogeography theory are that (after Davies et al. 2001): (1) the matrix can influence dispersal and recolonization rates (Ricketts 2001; Vandermeer et al. 2001); (2) the matrix can provide suitable habitat for some of the original taxa (Laurance 1991a; Gascon et al. 1999); (3) invading species can use the matrix as habitat and colonize habitat fragments (Aldrich and Hamrick 1998; Saurez et al. 1998); and (4) the matrix can affect the type and magnitude of edge effects (see the section that follows).

Given the problems with the six general principles from island biogeography theory outlined above, Margules et al. (1982) warned against their widespread application. This was because their uncritical adoption without consideration of matrix conditions and other factors can have negative implications for biodiversity conservation.

A problem for reserve design that stems from the uncritical application of island biogeography theory has been excessive emphasis on large reserves (for a detailed discussion, see Chapter 5). While large reserves are unquestionably important, small remnants also can have considerable value for biodiversity (Gascon 1993; Powell and Björk 1995; Turner 1996; Angelstam and Pettersson 1997; Semlitsch and Bodie 1998; McCoy and Mushinsky 1999; Palmer and Woinarski 1999; Schwartz 1999; Abensperg-Traun and Smith 2000; Mac Nally and Horrocks 2000). Indeed, sometimes the protection of small reserves is the only conservation option, particularly in highly productive landscapes subject to significant human alteration.

Nested Subset Theory

Ignoring the matrix and treating reserves and habitat fragments as equivalent to oceanic islands not only limits the validity of island biogeography theory, but also affects the broadly related concept of *nested subset theory* (Patterson and Atmar 1986). This theory attempts to extend the species-area relationship underpinning island biogeography theory by tracking both the numbers of species and their identities on islands. The premise is that species-poor small islands should support assemblages that are subsets of larger species-rich islands (Patterson 1987; Cutler 1991). The concept has been (often uncritically) extended to reserves and habitat fragments, where it is implied that species-poor small fragments should support assemblages that are subsets of larger species-rich fragments—assuming that islands (including habitat fragments) are unaffected by matrix conditions (Doak and Mills 1994). Nested subset theory would not predict, for example, the results of empirical studies such as those from the Biological Dynamics of Forest Fragments Project in Brazil (see Chapter 14), or the outcomes for birds in the Tumut Fragmentation Experiment in southeastern Australia (see Chapter 13). Advocates of nested subset theory often fail to test the null hypothesis that small patches merely sample organisms from a larger regional pool of taxa (but see Haila et al. 1993).

Constructs such as island biogeography theory and nested subset theory that assume that reserve "islands" are located in a sea of "unsuitable [matrix] lands" (Gascon and Lovejoy 1998; Gascon et al. 1999) may be too pessimistic in their predictions of species loss and fail to recognize the importance of matrix-based conservation strategies and/or the recovery of forests following disturbance. This led Wiens (1994) to conclude that

> fragments of habitat are often viewed as islands and are managed as such; however, habitat fragmentation includes a wide range of spatial patterns. . . . Fragments exist in a complex landscape mosaic and dynamics within a fragment are affected by external factors that vary as the mosaic structure changes and/or the recovery of forests following disturbance. The simple analogy of fragments to islands, therefore, is unsatisfactory.

Habitat Loss, Habitat Fragmentation, and Landscape Composition

Natural processes can cause habitat loss and the fragmentation of landscapes and habitats. Geological events (e.g., volcanic eruptions; Croizat 1960;

Franklin et al. 1985), long-term climatic change (Cunningham and Moritz 1998), and wildfires (Williams and Gill 1995; Agee 1999) are some examples. But human landscape modification, including vegetation clearing, is by far the most significant modern factor resulting in habitat loss and habitat fragmentation (Saunders et al. 1987; Groombridge 1992) and is often a significant threat to the persistence of forest fauna worldwide (Wilcox and Murphy 1984; Wilcove et al. 1986).

Proper interpretation of fragmentation effects requires consideration of habitat fragments and conditions in the matrix *and* the use of both by the organisms targeted for investigation (Wiens 1994; Gascon et al. 1999). Fragmentation effects and metapopulation dynamics (see the following section) only become relevant when habitat suitability in the matrix undergoes changes that creates patch-matrix differences that significantly influence the species of interest (Fahrig and Merriam 1994). Crome's (1997) recommendation is most apt here: "[A]s much, if not more emphasis [should be] placed on the matrix as on the fragments in research."

Processes Associated with Habitat Loss and Habitat Fragmentation

The dynamics of populations are influenced by four interrelated processes associated with habitat loss and subsequent habitat fragmentation: habitat loss, subdivision of habitat, patch isolation, and edge effects (Figures 2.4 and 2.5).

- *Habitat loss.* After habitat loss occurs, habitat fragments become "samples" of the original

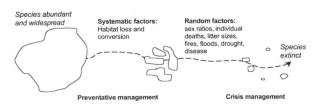

FIGURE 2.4. Interrelationships between the loss of a species and habitat loss (as a systematic threatening process) and subsequently cascading (often random) fragmentation effects that operate in addition to habitat loss and low levels of habitat cover (modified from Clark et al. 1990).

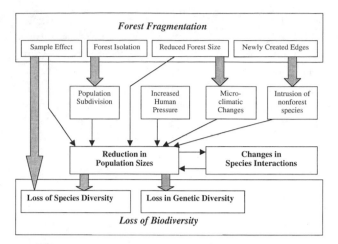

FIGURE 2.5. Graphical summary of the interacting effects of habitat loss and habitat fragmentation (modified from Zuidema et al. 1996).

(pre-fragmentation) vegetation cover (Connor and McCoy 1979; Zuidema et al. 1996). Recent studies indicate the total amount of forest habitat (and hence also the amount of suitable habitat in the matrix) can be a more important factor influencing the distribution of forest organisms than the spatial configuration of habitat (Delin and Andrén 1999; Fahrig 1999; Trzcinski et al. 1999; Villard et al. 1999). For some generalist species, which can use disturbed environments, the amount of suitable habitat may increase with habitat loss as the area between patches expands (Saunders and Ingram 1995; Lindenmayer et al. 2001a).

- *Subdivision of habitat.* Greater levels of habitat loss can lead to increased subdivision of remaining habitat, resulting in smaller habitat patches—the habitat-fragmentation process (Fahrig 1997; Bender et al. 1998). This may reduce populations of some species to nonviable levels (Shaffer 1981; Shaffer and Samson 1985; Armbruster and Lande 1993).
- *Patch isolation.* Increasing habitat loss and habitat fragmentation can lead to larger distances between remnant patches and, therefore, to greater levels of patch isolation (Fritz 1979; Smith 1980; Hanski 1994a). This may impede interfragment movement (Powell and Powell 1987; Bierregaard and Lovejoy 1989; Desrochers and Hannon 1997). The relation-

ship between forest habitat loss and patch isolation is not necessarily linear. A modeling study by Franklin and Forman (1987) forecast that patches of old-growth forest can become fragmented even when approximately 70 percent of the landscape cover remains. Percolation models (e.g., Gardner et al. 1987) predict that habitat fragmentation begins to occur when about 60 percent of original vegetation cover remains. Below some threshold level of habitat loss, distances between habitat patches can increase exponentially (Gustafson and Parker 1992).

- *Edge effects.* The ratio of patch perimeter to interior area increases in fragmented environments. This can result in edge effects within remaining patches. Such edge effects can include weed invasion (Brothers and Spingarn 1992; Burdon and Chilvers 1994), altered microclimatic conditions (e.g., modification of wind speeds, light fluxes, and temperature regimes; Chen et al. 1992; Matlack and Litvaitis 1999), increased nest predation and brood parasitism (Paton 1994; Robinson et al. 1995), and lowered rates of fledging success among birds with territories located at habitat edges (Hutha et al. 1998).

Theoretically, habitat loss can occur without habitat fragmentation, as shown by Forman's (1995) landscape modification models (such as perforation, dissection, and shrinkage) that were tested empirically on the use of an artificial ecosystem by grassland insects (Collinge and Forman 1998). However, in the majority of cases it will be extremely difficult to separate the effects of habitat fragmentation from the effects of habitat loss (Andrén 1997; Harrison and Bruna 2000). The two processes are nearly always confounded. Although theoretical models associated with landscape-levels of vegetation cover and patterns of species loss attempt to tackle this problem (Fahrig 1997, 1999), from a practical perspective it is essential to be aware that several processes accompany landscape modification (Bunnell 1999a; Fahrig 1999).

Determining the relative importance of habitat loss versus habitat fragmentation (i.e., the amount of habitat left versus the spatial configuration of remaining habitat) is an important area of research in forest landscapes. If habitat loss is the primary factor influencing the persistence and occurrence of species (which appears to be true for the majority of taxa; Fahrig 1997), then a thrust of forest management must be to maintain suitable habitat at multiple spatial and temporal scales—which is what, in our view, the overarching objective of matrix management should be (see Chapter 3).

Habitat Loss and Species Loss

Over 80 percent of the world's endangered birds are threatened by habitat loss (Temple 1986). In central Amazonia, for example, habitat loss has caused an immediate and obvious decrease in species richness (Zimmerman and Bierregaard 1986; Klein 1989). Habitat loss is also considered to be the chief factor in invertebrate extinctions (Thomas and Morris 1995). The total amount of habitat lost can be a better predictor of species loss than the spatial arrangement of habitat in many landscapes, especially those with relatively high levels of remaining cover (Fahrig 1998; see the section that follows). The loss of species is believed to track the loss of habitat in these cases.

Cascading Fragmentation Effects below Threshold Habitat Cover Levels

With and King (1999) defined critical thresholds as abrupt, nonlinear changes (such as the rate of loss of species) that occur in some measure across a small amount of habitat loss. Below threshold levels of habitat cover, the loss of species and population declines of individual taxa are more substantial than predicted from habitat reduction alone (Andrén 1994; With and Crist 1995; With 1997; McCarthy and Lindenmayer 1999b). Empirical studies of the capercaillie grouse (*Tetrao urogallus*) identified threshold habitat cover responses, with populations of this species declining faster than by habitat tracking alone when less than 30 percent habitat cover remained (Rolstad and Wegge 1987). Other studies of birds support these findings (Enoksson et al. 1995). However, Mönkkönen and Reunanen (1999) recognized that threshold responses to vegetation cover will vary on a species-by-species basis.

In a review of studies on birds and mammals, Andrén (1994) calculated that threshold levels for remaining habitat were 10–30 percent of original levels of vegetation cover. That is, populations declined

more rapidly than expected by habitat loss alone when less than 10–30 percent of habitat cover remained. Threshold levels will vary among landscape types and species (Andrén 1999), from 60 percent for some taxa to 10 percent for others (Bennett and Ford 1997). They will also vary according to the number and spatial arrangement of patches in fragmented environments (Thomas et al. 1990; Lamberson et al. 1994), the movement capabilities of the species targeted for investigation (Andrén 1994, 1999), and the spatial scale at which organisms use the landscape (Cale and Hobbs 1994; Pearson et al. 1996). It is important to recognize that 10–30 percent threshold levels in habitat availability will be an underestimate for many species (Mönkkönen and Reunanen 1999), and that some species could be lost above particular threshold cover levels, simply as a function of habitat loss per se.

As levels of habitat cover drop below critical threshold levels, a wide range of factors in addition to continued habitat loss can influence both the reduction in species diversity and the decline of populations of individual taxa. These are sometimes called *cascading fragmentation effects* and are summarized in Table 2.1. They include demographic stochasticity, genetic stochasticity, and environmental variability. Additional factors that become important below threshold levels of habitat cover are altered ecosystem processes (e.g., pollination; Cunningham 2000), disrupted species interactions (e.g., aggressive interspecific behavior), and altered intraspecific dynamics (e.g., Allee effects; Allee et al. 1949). For example, Ims et al. (1993) found that with increasing habitat loss and habitat fragmentation, the social system of the capercaillie grouse changed from a lekking system to one of solitary displaying males.

Threshold Habitat Levels, Species Loss, and Matrix Conditions

Conditions in the matrix substantially influence threshold responses, for example, through their impacts on permeability to movement (Taylor et al. 1993), and thus colonization probabilities for habitat fragments (Andrén 1999). A reappraisal of the seminal paper by Andrén (1994) showed that threshold values depended significantly on landscape context—in other words, whether the matrix surrounding habitat fragments was an agricultural one or one composed of re-

TABLE 2.1.

Potential "cascading" impacts of habitat loss and habitat fragmentation that occur below threshold habitat cover levels.

IMPACT AND EXAMPLES	SOURCE
Destabilized population dynamics (e.g., disrupted sex ratios and Allee effects)	McCarthy et al. 1994; Allee 1931
Altered movement patterns (e.g., modified home ranges)	Barbour and Litvaitis 1993
Altered social systems	Ims et al. 1993
Food shortages in fragments	Robinson 1998; Zanette et al. 2000
Altered breeding success (e.g., smaller young)	Hinsley et al. 1999
Increased short-term population densities in habitat fragments	Darveau et al. 1995
Altered vulnerability of populations to catastrophic events	Burgman et al. 1993
Increased human hunting pressure	Redford 1992
Altered gene frequencies in populations	Lacy 1993a; Sarre 1995
Altered interactions between species (e.g., interspecific aggression) or weed/parasite invasions	Catterall et al. 1991; Lavorel et al. 1999
Disrupted ecological processes (e.g., pollination)	Kapos 1989; Klein 1989
Increased potential for some types of disturbance (e.g., windthrow)	Lovejoy et al. 1986
Altered evolutionary processes (e.g., morphological changes in organisms)	Hill et al. 1999; Weishampel et al. 1997

growth forest regenerated after timber harvesting (Mönkkönen and Reunanen 1999).

Although the threshold concept is an appealing one and has some empirical evidence to support it (e.g., Jansson and Angelstam 1999), the concept is not free of problems.

First, threshold theory (like most fragmentation paradigms) treats landscape cover as a categorical variable with only two states—habitat and nonhabitat. This is too simplified. Not all habitat patches are alike, and internal differences between them can have significant effects on species occurrence (Lindenmayer et al. 1999b). Similarly, not all human-perceived nonhabitat is alike—even limited suitability of conditions in some parts of the matrix can influence the ability of some species to persist in a landscape.

Second, threshold levels for habitat loss will vary for each species and will depend on the scale at which it interacts with the patch mosaic in a landscape (With 1999). For some taxa there may not be distinct thresholds at all but rather responses might be linear, curvilinear, or some other form.

Third, there are often interactions between matrix conditions and habitat loss—habitat loss does not occur in isolation from other processes in the land-

scape. This is highlighted in the biotic interactions model developed by Ambuel and Temple (1983) and empirical studies such as those on fragmented populations of small mammals (Dunstan and Fox 1996) and birds (Mac Nally et al. 2000). In these cases, species composition is influenced as much by matrix conditions and habitat suitability within fragments as it is by fragment size per se.

Finally, patterns of habitat loss rarely occur in a random fashion across a landscape that results in a given level of habitat cover of uniform quality. Rather, fragmentation is a nonrandom process driven by human land-use practices—vegetation remnants are not therefore representative of the pre-fragmentation landscape (Saunders et al. 1987). The most productive parts of a landscape are almost always those modified first (Woolley and Kirkpatrick 1999; Smith et al. 2000; Scott et al. 2001a,b). In these cases, the response of some individual taxa (and the number of species persisting) will be strongly affected by the quality of *what* remains as well as *how much* remains. Threshold theory ignores the critical role of some (often small) areas of a landscape for the persistence of biodiversity. For example, the limited area of riparian systems can be essential for a large amount of the biodiversity in some ecosystems (e.g., Mac Nally et al. 2000), and degradation of this habitat can have impacts highly disproportionate to their areal extent (see Chapter 6). Similarly, specialized habitats like cave systems can be small but support an important and often unique component of the endemic biota of a region (Culver et al. 2000; see Chapter 6).

In summary, there are no generic rules for threshold levels of vegetation or habitat cover (e.g., 10, or 30, or 70 percent). Rather, thresholds will depend on the landscape in question (a forest-forest or forest-agriculture system), the species of interest, and the ecological processes in question (the extent of tree health or the hydrological impacts of rain-on-snow events as a function of the extent of landscape-wide clearcutting).

A useful illustration of the effects of different processes in different systems and different species comes from Australian forest and woodland systems. There can be extensive loss of an array of small and intermediate-sized mammals from forest ecosystems whose vegetation appears to be relatively intact. The wide distribution of feral predators (the red fox [*Vulpes vulpes*] and feral cat [*Felis catus*]) in these landscape

(and their use of the road system) could be important factors driving species loss despite limited apparent habitat loss (May 2001). Threshold cover levels and species decline and loss do not appear to be related in this case. Conversely, threshold levels of cover of native woodlands of 30 percent or more have been found to be important for limiting the effects of tree dieback and death (McIntyre et al. 2000).

Compositional Studies, Habitat Fragmentation, and Matrix Conditions

Many compositional and modeling studies clearly demonstrate that matrix conditions have a significant bearing on species occurrence in fragmented ecosystems (Szaro and Jakle 1985; Fahrig and Merriam 1994; Fitzgibbon 1997; Saari et al. 1998; Saab 1999; Renjifo 2001). A species for which the matrix is suitable or even partially suitable will be significantly less affected by fragmentation than one for which the matrix is totally lacking in value (Davies and Margules 1998; Ås 1999; Ricketts 2001) (Table 2.2).

Landscape composition and patch-matrix contrasts also influence the effects of disturbance regimes. For example, the spatial juxtaposition of fire-prone pastures

TABLE 2.2.

Examples of matrix effects on species responses.

ORGANISM, STUDY SITE	DESCRIPTION	REFERENCES
Hazel grouse (*Bonasa bonasia*), Scandinavia	In areas where the matrix was dominated by agricultural land, hazel grouse were absent from suitable habitat patches >100 m from a population source area; hazel grouse occupied patches up to 2 km from a population source if the matrix was an intensively managed plantation forest	Åberg et al. 1995
Small mammals, Brazil	The response of small mammals in rainforest fragments was different if the matrix was cleared pasture compared with regrowth rainforest	Malcolm 1997
Birds, North America	Up to 65% of the variation in the distribution of some taxa could be explained by matrix conditions	Pearson 1993
Invertebrates, English heathlands	Populations changed according to the nature of the surrounding landscape matrix	Webb et al. 1984; Webb and Hopkins 1984
Red-listed fungi, Norway	The occurrence of fungi at the stand level was strongly related to its occurrence in the surrounding forest. Where there was long-term (>140-year) persistence of host logs in a landscape, fungi were likely to occur in a stand.	Sverdrup-Thygeson 1999

can significantly increase wildfire risk in adjacent rain-forest fragments (Kauffman and Uhl 1991). Similarly, relatively nonflammable patches in Yellowstone National Park burned when the surrounding matrix supported high-intensity fires (Romme and Despain 1989).

The preceding examples clearly illustrate that matrix conditions are fundamental to interpreting fragmentation effects. This can help explain why two patches of similar shape and size, but with different landscape contexts, may have very different levels of biodiversity (Lindenmayer et al. 1999b,c; Vandermeer et al. 2001). Even if the matrix is suitable only for foraging (Landers et al. 1979; Loman and von Schantz 1991; Yahner 1983; Rodenhouse and Best 1994; McAlpine et al. 1999), or allows for the extension of home ranges, it may facilitate the occupancy of (other) breeding sites (McCarthy et al. 2000).

History of the Matrix and Fragmentation Effects

Historical factors in the matrix also can be very useful for interpreting fragmentation effects and associated population phenomena (Thomas et al. 1992a; Díaz et al. 2000). For example, forest fragmentation effects can appear and then disappear depending on regeneration dynamics in the surrounding matrix (Fahrig 1992). In other cases, the length of isolation of remnants created by a high contrast with conditions in the surrounding matrix is a significant predictor of species occurrence in some fragments (Suckling 1982; Bennett 1990b). Loyn (1987) showed that one bird species was lost from habitat fragments for every decade of isolation in a modified landscape in eastern Victoria, southeastern Australia. Forest fragmentation, therefore, has both a temporal and a spatial domain (Lord and Norton 1990), and it is dependent on matrix conditions and the length of time the matrix has been characterized by a particular condition.

The Use of the Matrix and Fragmentation Effects

Several studies have shown that a species' ability to exploit the matrix can strongly influence the nature and strength of fragmentation effects (Mills 1995; Sisk et al. 1997). In some cases, new species will occur in the matrix when it is modified, which may affect the dy-namics of populations residing within habitat fragments. For example, clearing of moist forest in Brazil led to the colonization of the matrix by new species (Brown and Hutchings 1997; Tocher et al. 1997). These included both predators and competitors of the original inhabitants of the fragments (Gascon and Lovejoy 1998).

In places where there is high contrast in suitability between a patch and the surrounding matrix, a *fence effect* can occur—levels of patch occupancy and within-patch population density may be considerably higher than expected. Here, animals are reluctant to disperse from a patch of suitable habitat into neighboring areas of unsuitable habitat; that factor can inflate levels of abundance (e.g., "crowding") and/or patch occupancy (Wolff et al. 1997; Bayne and Hobson 1998). Animals restricted to areas of favorable habitat when dispersing (e.g., Garrett and Franklin 1988; Merriam and Lanoue 1990) may be particularly prone to this effect.

Many fragmentation studies (e.g., Sieving et al. 2000) have focused almost exclusively on species use of remnants or corridors and ignored the role of the surrounding landscape matrix (see critiques by Simberloff et al. 1992; Laurance and Bierregaard 1997). This is because the vegetation that characterizes the landscape matrix is often different from that in habitat remnants. However, such matrix areas may support populations of species also found in the remnant patches, and they may make a substantial contribution to population persistence (Gascon et al. 1999). On this basis, Wiens (1989) noted that

> a focus exclusively on fragmentation of habitats misses the point that it is often the structure of an entire landscape mosaic rather than the size or shape of individual patches [that matters]. . . . The likelihood that dispersal can occur between fragments and forestall the extinction of sensitive species on a regional scale is influenced by the configuration of the fragments and the landscape mosaic in which they are embedded.

Sampling of the landscape matrix may reveal that some areas considered to be "remnants" are, in fact, part of a habitat continuum and may not be acting as discrete habitat patches at all. Studies in the Biological Dynamics of Forest Fragments Project showed that frogs could breed and travel across the matrix (cleared pasture) that surrounded the rainforest fragments, indicating that the remnants were not true isolates for these taxa (Tocher et al. 1997; see Chapter 14). The matrix may be particularly important in forest ecosystems where either (1) remnants (e.g., old-growth patches) are surrounded by a matrix that regenerates following harvesting and can eventually provide suitable habitat for a significant proportion of the forest-dependent biota, or (2) enough structural attributes of the original stand are retained in the matrix to allow forest-dependent taxa to persist there.

Hence, the effects of fragmentation in forest environments where protected areas are embedded within a matrix of regenerating or structurally enriched forest may be quite different from those in many agricultural systems where the surrounding matrix is often thought to be unsuitable, or "hostile," for forest biota (McGarigal and McComb 1995; Estades and Temple 1999).

In summary, traditional perspectives on fragmentation effects like those based on ocean-island models and on island biogeography theory are often inappropriate where the matrix is not wholly inhospitable (Pither and Taylor 1998; Ricketts 2001). Understanding fragmentation effects requires an understanding of how biota use all landscape components, including the matrix (Laurance 1991a; Åberg et al. 1995; Flather and Sauer 1996).

Metapopulation Dynamics

A metapopulation is a "set of local populations which interact via individuals moving between local populations" (Hanski and Gilpin 1991). Hanski and Simberloff (1997) contend that the metapopulation paradigm is most useful when successful interpatch dispersal is infrequent and migration distances are limited. Hastings (1993) suggested from simulation modeling that populations could be considered to be "independent" where dispersal rates between them were less than 10 percent.

The concept of metapopulation dynamics has been used widely to describe the spatial arrangement of subpopulations of species in fragmented environments (e.g., Arnold et al. 1993; Hanski and Thomas 1994; McCullough 1997), particularly those taxa susceptible to localized extinction and recolonization (Hanski 1998). Most models of metapopulation dynamics are simplified and do not consider the strongly influential effects of the matrix on patch processes and interpatch dispersal. As noted by Wiens (1997a), most metapopulation models assume that the "the matrix separating subpopulations is homogeneous and featureless." Perhaps this is a relict of island biogeography theory, which some workers consider to be a precursor of metapopulation theory (Wiens 1995). Therefore, predictions from many population models are unlikely to be accurate because of the simplifying assumptions they make about matrix suitability and patch-matrix interrelationships (see Pope et al. 2000). This has led some workers to suggest that simple forms of metapopulation modeling should not be used in population management in real landscapes (e.g., Doak and Mills 1994; Fahrig and Merriam 1994).

By ignoring the matrix in a landscape that is thought to be fragmented, important factors can be overlooked in metapopulation modeling. Consideration of the matrix may reveal that a species is not distributed as a metapopulation at all (Hanski and Simberloff 1997). Including appropriate information on the matrix and how it is used by an organism should improve the understanding of the factors influencing the species distribution in patchy landscapes such as production forests.

Types of Metapopulations

Hanski and Gyllenberg (1993) describe some different forms of metapopulation structure, where a metapopulation structure is defined as "a system of habitat patches which is occupied by a metapopulation and which has a certain distribution of patch sizes and interpatch distances" (Hanski and Gilpin 1991). Two extreme forms of metapopulation structure were recognized by them—the Mainland-Island structure and the Levins structure (Figure 2.6). The Mainland-Island metapopulation structure has two major features: (1) a large "mainland" area in which populations are secure and rarely (if ever) suffer extinction, and (2) an array of small patches in which extinctions can be relatively common events. The mainland provides a source of dispersalists to recolonize the patches and

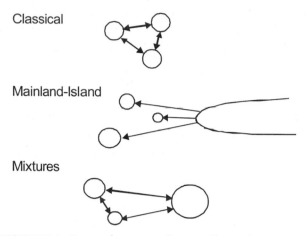

FIGURE 2.6. Types of metapopulations. Classical metapopulations are demonstrated by the *Levins (1970) model* (see text) in which all patches are the same size and equidistant. "Mainland" in the *mainland-island model* refers to a large area of habitat whose populations never suffer extinction. The *mixtures model* falls between the classical and mainland-island models in terms of patch size and distance.

reverse localized extinctions. A number of species of butterflies are thought to exhibit a Mainland-Island metapopulation structure (Harrison et al. 1988; Hanski and Thomas 1994).

The Levins metapopulation structure (from Levins 1970) is characterized by all the patches being the same size. There is no "mainland." Localized extinctions in the patches are reversed by recolonization from other patches (Hanski and Gyllenberg 1993). Harrison (1991) believed that most metapopulation structures will be characterized by patches that vary in size and spatial location—in other words, they will fall between the two extremes illustrated by the Mainland-Island and Levins models.

Metapopulations in the Real World

An increasing number of species have been identified that display various types of metapopulation dynamics (e.g., Hill et al. 1996; Carlson and Edenhamn 2000). Nevertheless, it is not clear whether metapopulations are common in real landscapes (Hastings and Harrison 1994; Harrison and Taylor 1997; but see Niemien and Hanski 1998). Dispersal is essential to the maintenance of metapopulation dynamics (Hanski 1999b), but for some species it may not occur across a hostile

matrix. Therefore, it is not appropriate to assume that all patchily distributed taxa function as metapopulations (Hanski and Simberloff 1997). For example, a study of geckos in Western Australia showed that animals were isolated populations confined to woodland patches with no movement across the matrix (Sarre et al. 1995).

Studies of the matrix also have revealed that some patchily distributed species are not distributed as metapopulations. If a species exhibits metapopulation structure, then patches close to occupied patches should be more likely to be occupied than more distant ones (see Smith 1994; Koenig 1998). There should, in turn, be spatial dependence in its distribution pattern. Statistical tests for seventy-six species of birds living in eucalypt patches of the Tumut Fragmentation Experiment in southeastern Australia (see Chapter 13) found no evidence of spatial dependence, suggesting that metapopulation dynamics were not occurring in any of these taxa (Lindenmayer et al. 2001a). In the same study, extensive surveys of the radiata pine matrix surrounding eucalypt fragments showed that many species used the matrix (see Chapter 13).

The occurrence of species in the matrix can create problems in the application of metapopulation models that ignore matrix effects. For example, Lindenmayer et al. (1999d) found poor predictive ability for Hanski's (1994b) incidence metapopulation model for two species of arboreal marsupials that persisted in the matrix. McCarthy et al. (2000) found that the ability of their target species (the white-throated treecreeper [*Cormobates leucophaeus*]) to forage both in habitat fragments and in the surrounding matrix (termed "landscape complementation"; Dunning et al. 1992) significantly reduced the fit between model-predicted and actual values for patch occupancy. Estades (2001) found that when matrix habitat had value as a food resource, traditional patch size effects tended to dissipate significantly.

The Matrix, Interpatch Dispersal, and Interpatch Movements

Interpatch dispersal can be strongly influenced by the surrounding matrix (Wiens 1997b). Andrén and Delin (1994) found that managed forests had no negative effects on interpatch dispersal by the red squirrel (*Sciurus vulgaris*). Thus, conditions in the matrix can strongly influence the degree of isolation of patches, which may be more or less than typically revealed by simple euclidean distances in landscape models of population dynamics (Ricketts 2001). Most existing computer models for metapopulation dynamics assume that the matrix is totally unsuitable breeding and foraging habitat and does not influence the success of interpatch dispersal and that taxa have random dispersal patterns. Nonrandom dispersal patterns and the effects of matrix conditions on dispersal were found to be significant in the dynamics of butterfly populations (Conradt et al. 2000). Lindenmayer et al. (2000a) found greater congruence between the predictions of a metapopulation model and actual field data for patch occupancy when the effects of matrix conditions on dispersal were included in the model specifications. Most metapopulation models do not consider movement in the context of the mosaic nature of most real landscapes (Cale 1999). Linear distance may actually be a poor measure of connectivity and/or isolation (Wegner and Merriam 1979). Two fragments far apart with high connectivity will be less isolated than closely spaced ones with low interpatch connectivity (Haila 1999).

Forecasts from metapopulation models will also be very poor for animals that cannot use the matrix but can readily disperse through it. Metapopulation models generally do not accommodate changes in home range sizes that can occur with habitat fragmentation (e.g., Barbour and Litvaitis 1993), changed landscape mosaics (Milne et al. 1992), or frequent movements between many patches to gather spatially separated food resources (e.g., Boone and Hunter 1996). Concepts such as the *central place foraging theory* are relevant to how matrix conditions and their relationships with patch configuration influence patchily distributed organisms (Recher et al. 1987; Lindenmayer et al. 1993a). For example, modeling of an Australian kingfisher, the laughing kookaburra (*Dacelo novaeguineae*), produced inaccurate estimates of patch occupancy because birds appear to alter home range movements in response to landscape change, and to move between many different patches to gather

food—in other words, classic metapopulation models did not fit (Lindenmayer et al. 2001b).

Spatial and Temporal Changes in Matrix Suitability

Most metapopulation models assume that the matrix and the patches it contains are homogeneous (Wiens 1997a) and that the habitat suitability of both landscape elements will not change on a spatial or temporal basis. Yet, in wood production forests the reverse will often occur and habitats will become suitable for many taxa with increasing time after harvesting, especially if matrix-based management strategies are adopted. Spatially explicit models like the one developed by Possingham and Davies (1995) can be useful in this regard. Different habitat-quality values can be assigned to any number of patches in this model. These habitat-quality values can, in turn, have varying temporal trajectories depending on, for example, regrowth of cutover stands in the matrix (Lindenmayer and Possingham 1995a).

All models invariably simplify the systems they attempt to portray (Burgman et al. 1993). Metapopulation models are no different in this regard (Hanski 1999a), and models like Hanski's (1994b) incidence function model consistently favor simplicity above model complexity (Ludwig 1999). However, it is essential to consider temporal and spatial variation in matrix conditions when applying metapopulation modeling to dynamic managed landscapes. It is also vital to check model assumptions before models are applied to real landscapes and real conservation problems (Wiens 1994; Pope et al. 2000). In some patchily distributed populations there will be no conformity to a metapopulation structure (1) where the matrix is so hostile that it precludes movement and relict populations are confined only to totally isolated habitat patches, (2) where the matrix is sufficiently suitable to facilitate frequent movements between habitat patches, and (3) where conditions in the matrix are suitable enough to allow organisms to forage or live there.

The incorporation of these sorts of considerations into modeling frames may ultimately lead to new generations of spatially explicit population models that include some of the complexity that characterizes real landscapes and the responses of biota to it (e.g., Gustafson and Gardner 1996; Hokit et al. 1999).

Connectivity and Corridors

Determining what constitutes connectivity for a given species or group of organisms is a critical issue in conservation biology. Much work has focused on corridors, yet connectivity in forests involves much more than just the establishment of wildlife corridors (Wiens 1997a), and the connectivity role of the matrix (see Chapter 1) has often been overlooked. This is particularly true in forest environments—the majority of corridor studies have been in agricultural landscapes (Saunders and Hobbs 1991; Gustafsson and Hansson 1997). Dynamic matrix conditions can have a fundamentally important influence on connectivity in forest ecosystems (Lindenmayer 1998).

Although the establishment of wildlife corridors is common in wood production forests (reviewed by Lindenmayer 1998), assessments of their effectiveness in contributing to connectivity cannot be made without consideration of the matrix (Noss 1987; Simberloff and Cox 1987; Beier and Noss 1998). Moreover, if there is continued habitat loss in the surrounding matrix, the establishment of corridors may make only a limited contribution to biodiversity conservation (Rosenberg et al. 1997; Harrison and Bruna 2000).

If conditions in the matrix are hostile, then large corridors linking large retained patches may be required to retain connectivity for some species. Conversely, given suitable matrix conditions, reliance on corridors to provide connectivity may be minimized (Rosenberg et al. 1997) and corridor widths reduced (Forman 1995; Lindenmayer 1998). Only a limited amount of movement may then be required to "rescue" populations in reserves or habitat remnants and limit losses of genetic variability or limit population decline (Stacey and Taper 1992; Mills and Allendorf 1996). For example, an average of only one migrant per generation (every three to four years) was predicted to stem losses of expected heterozygosity from patchy populations of arboreal marsupials (Lindenmayer and Lacy 1995a).

The matrix will be a filter rather than a complete barrier to movement for many organisms (Gascon and Lovejoy 1998). Forman (1995) termed the control of the matrix over connectivity as "resistance" and hypothesized that barriers to movement across the matrix will be reduced if patch conditions resemble those in the matrix (see Stouffer and Bierregaard 1995 for an example). However, matrix resistance also will depend on the dispersal mechanisms of a given species (whether it is random or influenced by habitat quality), and the scale at which a species moves (Ricketts 2001). If a species is very vagile and populations are not subdivided among patches, then the landscape will be patchy and continuous. Connectivity is therefore a species-specific phenomenon—an outcome of species-specific dispersal behavior and movement patterns (e.g., whether the main form of movement is flying or crawling) and how these interact with patterns of landscape cover. For these reasons, a simple map of vegetation cover may not correspond to a map of connectivity for a given species (Ingham and Samways 1996), and it may not reveal whether a landscape is connected or fragmented (Wiens et al. 1997; With 1999).

Consideration of matrix conditions is particularly important for connectivity in forest landscapes. This is because, unlike agricultural areas, logged and regenerated lands will often immediately or eventually provide suitable habitat for many species (Pattemore and Kikkawa 1975; Loyn 1985a; Smith 1985). Even the structural complexity provided by exotic plantations can provide connectivity for forest species (including some that otherwise could not move through an agricultural landscape) (Renjifo 2001). Hence, changes in matrix conditions that might even be relatively minor may substantially enhance connectivity for many taxa. There are many factors influencing not only the suitability of the matrix as habitat, but also the suitability of the matrix for movement in managed forests. These include the type and extent of silvicultural practices in the matrix (Bennett 1990a; Machtans et al. 1996; Vesely and McComb 1996; see Chapter 8), partial retention of original stands (Taylor 1991; Thiollay 1997), and rotation length (Curtis 1997; see Chapter 7).

Understanding relationships between matrix conditions and the movement behavior of target organisms is fundamental to analyses of connectivity. Even if a habitat patch is close to another occupied patch and well within typical dispersal distances, it may be isolated because of hostile conditions in the matrix (Taylor et al. 1993). In many other cases, dispersal pathways used by organisms will not always conform to designated corridors (Gustafson and Gardner 1996; Lindenmayer 1998; Cale 1999).

Conservation management strategies applied throughout the matrix may provide better connectivity than wildlife corridors for species that disperse randomly (e.g., cabbage butterfly [Fahrig and Paloheimo 1988] and eastern screech-owl [*Otus asio*; Belthoff and Ritchison 1989]). In Australia, matrix suitability can allow migratory and nomadic species such as native pigeons and honeyeaters access to resources that fluctuate in temporal and spatial availability (Date et al. 1996; Price 1999). Connectivity via stepping stones or dispersed islands of potentially suitable habitat may be the best way for these and other mobile species (like butterflies and bats), which can move readily across the matrix between roosting areas and foraging locations (Lumsden et al. 1994; Schultz 1998; Law et al. 1999). Individual structures, such as trees and logs, can also function as stepping stones.

Most existing knowledge about connectivity and the matrix comes from theory (e.g., Murphy and Noon 1992) and simulation modeling (e.g., Burkey 1989; Boone and Hunter 1996; Lindenmayer and Possingham 1996). Since robust empirical data linking matrix conditions and connectivity are presently limited (Nicholls and Margules 1991; Bennett 1998), carefully designed studies of the interrelationships between matrix conditions and connectivity are badly needed (see Chapter 17).

Extinction Proneness

Species vary widely in their response to habitat loss and fragmentation; some decline, while others remain unchanged or even increase (Saunders and Ingram 1995; Dooley and Bowers 1998). In an effort to better predict species responses to landscape change, correlations between extinction proneness and particular traits of species (such as diet, body size, life history strategy, longevity and fecundity, habitat require-

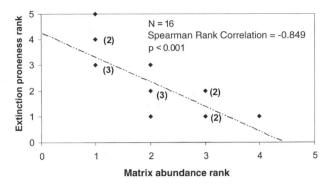

FIGURE 2.7. Correlations between extinction proneness and use of the matrix by sixteen species of nonflying mammals in North Queensland (redrawn from Laurance 1991a). The solid dots show plotted values between matrix abundance and extinction proneness. The values in brackets are the number of species with the same plotted value. The diagonal line is the fitted relationship between the two parameters. Extinction proneness (y-axis) is ranked from high (5) to low (1). Matrix abundance is scaled from 1 to 5, with 1 indicating that a species was never detected in the matrix and a value of 5 indicating that a species was commonly observed in the matrix.

ments, and type of locomotion) have been the focus of many studies (e.g., Terborgh 1974; Karr 1982; Pimm et al. 1988; Robinson and Quinn 1988; Burbidge and Mckenzie 1989; Lawler 1993; Gaston and Blackburn 1995). Extinction-prone species appear to include large wide-ranging taxa (often predators), rare species, or species that are sparsely distributed. One crucial factor in assessing the likelihood of species persistence is occurrence in matrix habitats.

Sixteen nonflying species of Queensland mammals were studied to determine correlations between extinction proneness and ecological traits such as body size, longevity, fecundity, trophic level, diet, and population abundance both in rainforest patches and in the surrounding modified landscape matrix (Laurance 1991a). The best predictor of persistence was the abundance of the species in stands of regrowth forest and pastures in the landscape matrix surrounding the rainforest fragments (Figure 2.7). Species that were rare or absent in the matrix were also those most likely to have been lost or to have declined substantially in the rainforest fragments (e.g., brown antechinus [*Antechinus stuartii*], Atherton antechinus [*Antechinus godmani*], and tiger quoll [*Dasyurus maculatus*]) (Laurance 1991a, 1997a). Several factors may explain these findings. Animals lost from the matrix may be lost from an entire region (and hence also habitat fragments). Conversely, individuals dispersing from matrix habitats may "rescue" declining populations in the fragments (Laurance 1997a).

Persistence in the matrix and resilience to extinction proneness has also been demonstrated in other studies. The Wog-Wog Fragmentation Experiment in southeastern Australia (Margules 1992) showed that the species of beetles that inhabited the radiata pine matrix surrounding fragments of eucalypt forest were most likely to persist in the fragments (Davies et al. 2000). Similar outcomes have been found for insects in Scandinavian forests (Ås 1999), as well as in studies of temperate and tropical bird populations (Blake 1983; Howe 1984; Diamond et al. 1987; Renjifo 2001).

Reserve Selection

Around the world, reserve systems have generally been developed on an ad hoc basis (Terborgh 1992; see Chapter 5). In an attempt to improve systems of protected areas (e.g., to enhance their representativeness), reserve selection methods have been developed, such as gap analysis (e.g., Scott et al. 1993; Stoms et al. 1998) and mathematical algorithms (Kirkpatrick 1983; Williams et al. 1996; Pressey 1997). There are numerous hypothetical examples of these approaches for the selection of reserves (Bedward et al. 1992; Nicholls and Margules 1993; Noss and Cooperrider 1994; see Chapter 5). However, the potential contribution of the matrix to species persistence is often ignored by these methods (Burgman and Lindenmayer 1998). The focus is on species captured within reserves, usually without regard to their status in surrounding matrix lands. However, many of the organisms that persist in the matrix will also be the ones most likely to occur in the reserve system. The inclusion of the matrix will add considerably to the efficiency of reserve design because reserve selection routines can be focused principally on taxa that can be conserved *only* in the protected areas. Diamond (1976) recommended that the selection of reserves should be made on the basis of the species that require them for survival. Therefore, while the stated goal of a network

of reserves has traditionally been to adequately represent the biodiversity of a region, more efficient and refocused reserve design approaches could be those that target species not captured adequately in sensitively managed matrix lands.

Lewis et al. (1991) described some reserve design algorithms that better account for matrix conditions and showed how the buffering potential of adjacent matrix lands could significantly influence protected area outcomes in the Australian state of Tasmania (see also Nix 1997). As noted by Wiens (1989):

> To establish reserves according to ecological insights requires both a consideration of broad-scale landscape configurations and knowledge of the ecological requirements of species that are important in particular situations.

Edge Effects

Distinct edges or boundaries are created between clearcut and unlogged areas (Matlack 1993; Esseen 1994; Parry 1997). Profound modifications of biological and physical conditions can occur at these boundaries; these modifications are typically referred to as *edge effects* (Wilcove 1985; Temple and Cary 1988; Yahner 1988; Reville et al. 1990; Chen 1991; Chen et al. 1992; Kremsater and Bunnell 1999).

The magnitude of many types of edge effects is often related to the level of contrast between the matrix and other landscape units; where the contrast is high, there will be more intense interactions and spatially extensive edge influences (Laurance and Yensen 1991; Mesquita et al. 1999). The extent of the area of the matrix supporting such high-contrast conditions can also influence the magnitude of edge effects. For example, microclimate edge effects may be greater where a large clearcut abuts a retained patch than where the cutover is small (Lindenmayer et al. 1997a). An assessment of conditions in the matrix is therefore fundamental to an understanding of the impacts of various kinds of biotic and abiotic edge effects. It is also vital to the development of buffering protocols in the matrix that aim to better protect sensitive habitats in forest landscapes. For example, the width of buffer strips to protect riparian areas from pesticides and to

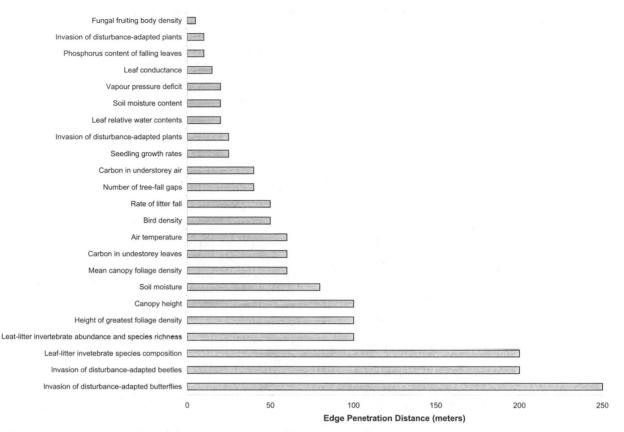

FIGURE 2.8. Variation in edge penetration for a range of measures in the Biological Dynamics of Forest Fragments Project in Brazil. See Chapter 14 for further details (modified and redrawn from Laurance et al. 1997).

reduce in-stream invertebrate mortality (more than 50 meters in Australian eucalypt plantations; Barton and Davies 1993) may be quite different from that required to mitigate changed wind patterns that may otherwise penetrate several hundred meters from cut-over boundaries (Harris 1984; Saunders et al. 1991).

The intensity of edge effects, or the area of a patch subject to significant edge influence or modification, depends on many variables. The magnitude of an edge effect is dependent on the parameter of interest (which has its own unique pattern of response; see Figure 2.8)—whether it is an environmental variable (e.g., air temperature), an ecological process (e.g., rate of organic matter decomposition), or a community interaction (e.g., predation of one species by another). The magnitude of edge effects will also be influenced by the extent of the area of the matrix supporting

high-contrast conditions. For example, Lindenmayer et al. (1997a) found that windthrow significantly increased in retained strips of forest when the adjacent clearcut was large rather than small cutblocks.

Edge effects are not uni-directional processes. In addition to edge effects penetrating from the matrix into adjacent patches, forest patches can exert edge influences on adjacent harvested areas. The effects of retained trees on light and other microclimatic factors are well known in the forestry literature. Similarly, Lindenmayer et al. (2001a) found the occurrence of birds in the matrix was strongly related to their occurrence in adjacent retained patches (see Chapter 13). It is important to be cognizant of the reciprocal nature of edge influences, but in the remainder of this section we concentrate on the impacts of edge processes emanating from the matrix and penetrating into adjacent retained patches.

Edges of forest along boundaries with cutovers may be subject to significant microclimatic changes, such as increased temperatures and decreased humidity that extend for varying distances from an edge. Edge influences may be limited or extend for tens or hundreds of meters into a forest depending upon the environmental variable, physical nature of the edge, and weather conditions at the time of measurement (Moore 1977; Miller et al. 1991; Chen et al. 1992; Parry 1997). In some cases, edge effects can lead to the degradation of habitat in protected areas such as wildlife corridors (Lindenmayer et al. 1997a) and can disrupt connectivity between larger reserved areas (Lovejoy et al. 1986).

Wildlife biologists have known for many decades that certain game species have strong preferences for edge environments where foraging and hiding/thermal cover are close together (Leopold 1933; Patton 1974; Matlack and Litvaitis 1999). Large populations of white-tailed deer (*Odocoileus virginianus*) in the northern United States may exert considerable grazing pressure on food plants that occur at forest edges (Johnson et al. 1995), including endangered taxa (Miller et al. 1992). This can limit stand regeneration (Tilghman 1989) and alter patterns of species diversity in plants (Miller et al. 1992) and birds (McShea and Rappole 2000).

Studies in the Northern Hemisphere have documented negative vertebrate responses to forest edges, including interior forest species (i.e., those requiring habitats away from edge environments) (Gates and Gysel 1978; Andrén and Angelstam 1988; Telleria and Santos 1992; Terborgh 1992; Rudnicky and Hunter 1993; Paton 1994; Rich et al. 1994; Robinson et al. 1995). Many bird species in tropical forests are interior species (Terborgh 1989; Frumhoff 1995) and tend to avoid edges. This is, in part, because the efficiency of foraging patterns is impaired if home ranges or territories are transgressed by nonhabitat or edge conditions (McCollin 1998). Another important factor influencing species responses in edge environments is a lack of food—Burke and Nol (1998) provide an example in which the abundance of arthropods was significantly lower along the edges of small woodlots than in

the interior of larger patches, resulting in a negative effect on bird populations.

Elevated levels of nest predation along edges may not be a general phenomenon characteristic of all habitats and landscapes (Berg et al. 1992; Hanski et al. 1996, reviewed by Lahti 2001). Many studies in which significant nest predation effects have been observed are in agricultural landscapes where there are large contrasts between vegetation remnants and the surrounding environment (e.g., Andrén 1992; Bayne and Hobson 1997, 1998; Hannon and Cotterill 1998). Conversely, nest predation at edges may be limited or absent in landscapes with low levels of contrast (Schmiegelow et al. 1997), such as continuous Australian eucalypt forest dissected by minor bush tracks or juxtaposed native forest and exotic softwood plantations (Lindenmayer et al. 1999e). Sargent et al. (1998) showed that nest predation effects in hardwood fragments in South Carolina (United States) were lower if they were adjacent to pine stands rather than agricultural fields.

Certain types of edge effects can vary in their impacts between different forested ecosystems. Forest edges on the boundary of clearcuts in Sweden are characterized by reduced bird species diversity (Hansson 1983), but no such pattern occurs in clearcut forest edges in the northeastern United States (Rudnicky and Hunter 1993). Similarly, patterns of brood parasitism characteristic of edge environments in many northern hemisphere landscapes appear to be rare in parts of the Southern Hemisphere, such as in Australia (but see Luck et al. 1999). Even within North America, increased nest predation and brood parasitism seen in studies on the eastern side of continent are not common in the west (Kremsater and Bunnell 1999; Marzluff and Restani 1999). Notably, a recent review of the literature (Lahti 2001) has shown that the majority of studies did not find increased nest predation at edges.

Differences in invertebrate community composition have been observed between edge and interior environments (e.g., Keals and Majer 1991; Hill et al. 1992; Scougall et al. 1993). Populations of invertebrates such as amphipods responded negatively to increased levels of exposure, moisture, and temperature regimes at the edges of forest fragments in southeastern New South Wales (Margules et al. 1994b). Other examples of edge responses in insects have been documented by Hill (1995), Baur and Baur (1992), and Bellinger et al. (1989).

Edge environments can exert a strong influence on plant species (Laurance 1991b). Plant reproduction, growth, and mortality may respond to edges, depending upon the species and specifics of the edge. Chen et al. (1992) observed positive responses in tree reproduction and growth of surviving mature trees in old-growth forests bordering on recent clearcuts. Negative effects included accelerated mortality of mature trees due to windthrow.

Edge effects vary according to the type of harvesting prescriptions employed in the cutover area (e.g., clearcutting or shelterwood) and will generally decline (but rarely disappear completely) with the development of regenerating forest (Savill 1983). This is because the magnitude of edge effects is strongly related to the contrast in structure between harvested areas and retained patches.

Understanding various types of biotic and abiotic edge effects and their relationships to matrix conditions is fundamental to the development of appropriate buffering protocols that can make the matrix management more protective of sensitive areas in forest landscapes (see Chapters 3 and 6).

The Matrix and the Importance of Habitat

The study of habitat is essential for the evolving fields of landscape ecology and conservation biology. Because habitat is an organism-specific concept (Whittaker et al. 1973), an understanding of habitat is needed to determine what constitutes a suitable patch or a fragmented landscape for a particular species. Habitat information is also fundamental to matrix management and, in turn, to any comprehensive plan for conserving forest biodiversity. The availability of habitat is one of the primary factors influencing the distribution and abundance of organisms (Caughley 1978; Krebs 1978), and an understanding of distribution patterns through the use of habitat analysis can be crucial for determining (1) if the areas suitable for a particular species are also those targeted for forestry operations, (2) if areas that are not logged provide ef-

fective refugia or reserves, and (3) if logging practices alter the characteristics of a forest in ways that (temporarily or permanently) influence the distribution of a species, and, if so, how might logging practices be modified to mitigate such impacts?

Habitat analysis (sensu Morrison et al. 1992) also can be useful for determining relationships between the species' distribution and abundance and structural and floristic conditions in forests (Urban and Smith 1989; Block and Brennan 1993; Pearce et al. 1994). This information is useful for matrix management because it is central to identifying (1) the structural and floristic attributes that need to be retained to conserve particular species (Adams and Morrison 1993; Lindenmayer 1994a; see Chapter 8), (2) how many of these stand and/or landscape characteristics are needed (McComb and Lindenmayer 1999), and (3) how such attributes should be spatially arranged (Nelson and Morris 1994).

Increasingly, the importance of habitat is being recognized in landscape ecology and conservation biology. The loss of habitat is a key driver in species loss (Fahrig 1999) and, conversely, the maintenance of habitat across many spatial scales is a fundamental plank of any comprehensive approach for the conservation of forest biodiversity. We believe it is the overarching goal of matrix management. General principles to achieve this goal are the topic of the next chapter.

Conclusion

In this chapter, we have explored only a small subset of the significant themes in landscape ecology and conservation biology. The aim was not criticize either discipline but, rather, to highlight the importance of the matrix as a fundamental landscape component and to demonstrate the complexity that characterizes all real-world landscapes. Serious deficiencies arise in many aspects of conservation biology theory when the matrix is overlooked or relegated to the status of nonhabitat. Such oversights are understandable given the complexity of landscapes—and conservation biologists and landscape ecologists have often been forced to select relatively simple theories and approaches as part of preliminary attempts to understand landscapes. Much has already been learned from testing existing theories and approaches. Greater efforts should now be made in conservation biology and landscape ecology to include the roles and contributions of the matrix.

We believe that much of ecology (including conservation biology and landscape ecology) is significantly more difficult than other areas of science such as physics and chemistry. Applied ecology (and, hence, ecologically sustainable forest management) is unlike physics and chemistry in that there are few general principles that can be applied generically. Given such complexity, unique approaches to conserving biodiversity will be necessary that reflect different landscapes, species assemblages, and many other factors (including social ones). Accounting for the inherent complexity that characterizes forest ecosystems does not mean that developing robust plans for conserving biodiversity is impossible. In Chapter 3, we propose a framework and a set of associated generic, multiscaled principles to tackle the problem.

Objectives and Principles for Developing Comprehensive Plans for Forest Biodiversity Conservation

Forest management is not rocket science—it is far more complex

J. W. THOMAS AND F. BUNNELL (2001)

Since species loss is predominantly driven by habitat loss, the overarching goal of matrix management must be to prevent habitat loss. Conservation planning for many species often focuses on developing strategies that operate at only one or two spatial scales, but any comprehensive plan for forest biodiversity conservation requires maintaining habitat across the full range of spatial scales. General principles to meet this objective include

- The maintenance of connectivity
- The maintenance of landscape heterogeneity
- The maintenance of stand structural complexity
- The maintenance of the integrity of aquatic systems by sustaining hydrologic and geomorphological processes

Identifying what constitutes, for example, sufficient connectivity or suitable stand complexity for a given species or set of taxa is not a trivial task. Adoption of multiple strategies at multiple spatial scales is important because it increases the chances that suitable connectivity, heterogeneity, stand complexity, and aquatic ecosystem integrity will be provided for most taxa in at least some parts of a landscape. This is the core of an additional guiding theme for developing comprehensive plans for forest biodiversity conservation: risk-spreading. Risk-spreading involves the implementation of a range of strategies at different spatial scales. It has particular value because if one strategy subsequently proves to be ineffective, others will be in place that might better conserve the entities targeted for management. Risk-spreading is also important because the wide range of different habitat and other requirements of many species requires the implementation of multiple strategies at different

scales—from large ecological reserves to the retention of individual structures within harvested units.

The risk-spreading approach contrasts fundamentally with the norm of strict production forestry. Production forestry tries to reduce variability at the stand and landscape levels. Risk-spreading, conversely, aims to ensure a range of conditions at all spatial scales; stands and landscapes are not homogenized.

Some authors have proposed alternative strategies to the generic principles outlined in this chapter, such as biodiversity surrogate schemes (e.g., indicator species). The validity of such surrogate schemes is a questionable scientific basis for a comprehensive approach to forest biodiversity conservation.

The role of the matrix for biodiversity conservation has been recognized (often unwittingly) by legislation and policy directives in some countries; such directives specify that species should be conserved throughout their known natural ranges—in other words, both in large ecological reserves and in matrix lands (e.g., Commonwealth of Australia 1992; Forest Ecosystem Management Assessment Team 1993; Yaffee 1994; British Columbia Ministry of Forests 1995). Maintaining populations of species in large ecological reserves and in the matrix is only possible by maintaining suitable habitat at multiple spatial scales. This must be the overarching objective of any comprehensive plan for forest biodiversity conservation. This objective is fundamental because habitat loss is the primary factor influencing species loss (Novacek and Cleland 2001), and different species perceive habitat over a range of spatial scales (Chapter 2). We will first discuss the importance of the maintenance of suitable habitat at multiple spatial scales and then describe general principles necessary to achieve this goal.

The Primary Objective: Maintenance of Suitable Habitat for a Range of Spatial Scales

Habitat can be broadly defined as "the range of environments in which a species can occur" (Whittaker et al. 1973). In this book, we consider suitable habitat for a species to be that where reproduction occurs at a rate high enough to maintain long-term positive population growth. Access to habitat influences the distribution and abundance of all organisms (Elton 1927; Morrison et al. 1992), and it significantly influences survival, reproduction, and long-term population persistence (Krebs 1978; Block and Brennan 1993).

Maintenance of habitat at multiple spatial scales is essential because

- *Different species have different spatial and other requirements* (Allen and Starr 1988; Wiens 1989; Allen and Hoekstra 1992; Haila et al. 1993). Suitable habitat may vary from extensive intact stands for area-sensitive organisms, such as some wide-ranging carnivores (Milledge et al. 1991; Bart and Forsman 1992), to the moisture and decay conditions provided by individual logs for invertebrates (Økland 1996; Meggs 1997). For example, empirical studies suggest that while the provision of wildlife corridors and retained trees on logged sites will make a major contribution to the conservation of populations of the mountain brushtail possum (*Trichosurus caninus*) in Australian mountain ash forests (Lindenmayer et al. 1994b), areas containing large continuous stands dominated by old-growth trees will be essential for the conservation of the yellow-bellied glider (*Petaurus australis*) in this same forest type (Lindenmayer et al. 1999a; Incoll et al. 2000).
- *Individual taxa respond to factors at multiple spatial scales*. The distribution and abundance of a given individual species is influenced by fac-

tors at multiple scales (Gutzwiller and Anderson 1987; Schneider 1994; Jaquet 1996). Forman (1964) demonstrated how factors at a hierarchy of spatial scales influenced the distribution of the moss *Tetraphis pellucida* from global climate to the microhabitat of an individual log. Similarly, Diamond (1973) showed how the distribution of birds in New Guinea was influenced by multiscaled processes ranging from broad geographic factors to branch sizes of individual trees. With (1994) demonstrated how nymphal stages of a species of grasshopper moved and interacted with the patch structure of landscapes differently than larger, faster-moving adults. Similarly, the management of the capercaillie grouse in the Bavarian Alps of Germany requires the maintenance of both appropriate stand-level conditions and suitable landscape-level patterns of habitat (Storch 1997).

Multiple management scales are needed because there are multiple ecological scales (Poff 1997; Elkie and Rempel 2001), not only for different ecological processes (Figure 3.1) and different species, but also for the same species (Hokit et al. 1999; Lindenmayer 2000). Thus, there is no single "right" or "sufficient" scale for forest and conservation management. A single conservation strategy adopted at a single spatial scale will only meet a limited number of stand and landscape management goals (Christensen et al. 1996; Tang et al. 1997) and will provide suitable habitat for only a limited number of different taxa.

Another reason a multiscaled approach is important is that different processes at different spatial scales are interdependent. What happens at the stand level cannot be divorced from what takes place at the landscape level and vice versa. A stand of old growth surrounded by other old-growth stands will behave quite differently (and support different species assemblages) than an old-growth stand embedded within an extensive region of continuous clearcutting (Harris 1984) (Figure 1.4 in Chapter 1). Similarly, a landscape is composed of an array of stands (as defined in Chapter 1) and the structural composition of these stands can influence species occurrence at the landscape level. A lack of suitable habitat within many different stands may combine to preclude a species from entire landscapes (Lindenmayer et al. 1999a).

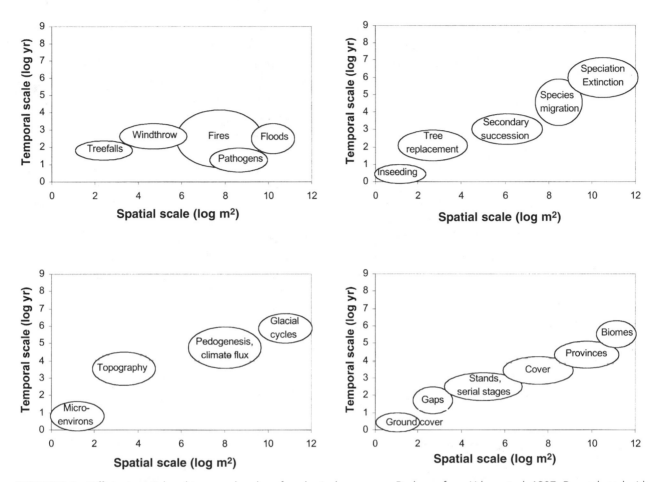

FIGURE 3.1. Different spatial and temporal scales of ecological processes. Redrawn from Urban et al. 1987. Reproduced with permission of the American Institute of Biological Sciences.

Finally, multiscale conservation strategies may produce a heterogeneous landscape (Forman 1995) containing the spatially dispersed array of resources needed by some species (Krusic et al. 1996; Law and Dickman 1998). Baudry (1984) describes such a circumstance in the field/hedgerow landscapes of France.

Principles for Maintaining Suitable Habitat at Multiple Spatial Scales

Angelstam (1996) recognized that habitat loss and habitat fragmentation can occur at several spatial scales in forests. First, at a landscape scale, there can be a direct loss of habitat per se. Second, within "intact" forest cover, formerly continuous areas of distinct forest types or successional stages (e.g., old-growth stands) can become fragmented. Finally, structural and floristic elements can be lost within given forest types (Angelstam 1996). As a result of these multiscaled effects, habitat conservation at multiple scales encompasses implementing management strategies at scales from a few square meters to thousands of hectares—from individual trees to large ecological reserves. General principles to meet this objective include (1) maintenance of connectivity across a landscape, (2) maintenance of landscape heterogeneity, (3) maintenance of structural complexity and plant species diversity within managed stands, and (4) maintenance of the integrity of aquatic ecosystems, including hydrological and geomorphological processes.

Each of these principles is, in itself, multiscaled. For example, the maintenance of connectivity may entail the provision of individual decayed logs for invertebrates (Barclay et al. 1999) through to intact networks of riparian vegetation for some vertebrates (Lindenmayer and Peakall 2000). Similarly, the maintenance of heterogeneity may require the appropriate spatial juxtaposition of structures within a stand for some species (Haila et al. 1993; Lindenmayer et al. 1997b) and the spatial juxtaposition of habitat patches across entire landscapes for others (Price 1999).

Habitat conservation at multiple spatial scales underscores the critical importance of a fifth guiding principle—the need to implement an array of management strategies. This is a risk-spreading approach that is essential in case any single strategy is subsequently found to be ineffective.

The five general principles discussed below and strategies for their practical implementation are explored in Chapters 6, 7, and 8. The integration of strategies as part of a comprehensive plan for forest biodiversity conservation is the primary topic of Chapter 9.

Principle 1. Maintenance of connectivity

Connectivity in the landscape for most species will be determined by conditions in the matrix because connectivity is fundamentally controlled by the degree to which it is hostile or permeable (Wiens 1997a; Hokit et al. 1999; see Chapter 1). Connectivity will influence processes such as population persistence and recovery after disturbance (Lamberson et al. 1994), the exchange of individuals and genes in a population (Leung et al. 1993), and the occupancy of habitat patches (Villard and Taylor 1994).

What constitutes effective connectivity varies among species (Hobbs 1992; Bennett 1998). This is because of interspecific differences in movement patterns and dispersal behavior (Wolfenbarger 1946). How these species-specific traits intersect with patterns of landscape cover will determine the ultimate level of connectivity (Wiens et al. 1997). Connectivity may be achieved by strips of retained habitat, or "wildlife corridors," for some taxa (Bennett 1990a; Hewittson 1997; Beier and Noss 1998), but not others (e.g., Thomas et al. 1990). Research supporting the use of corridors by some species has been documented

(Hobbs 1992; Lindenmayer 1998; Gilbert et al. 1998; Aars and Ims 1999; Laurance and Laurance 1999). For other species, small discrete patches of suitable habitat may act as "stepping stones" and thus provide connectivity (Forman 1995; Dramstad et al. 1996) (Table 3.1).

Although establishing corridors or stepping stones will be valuable for some taxa (Metzger 1997), maintaining connectivity for other species requires retaining appropriate vegetation cover throughout the entire matrix (Murphy and Noon 1992; Franklin 1993a). Example taxa include the northern spotted owl (*Strix occidentalis caurina*), which disperses randomly and does not remain within defined strips of habitat (corridors) (Murphy and Noon 1992; see Chapter 11).

Riparian corridors or stream buffers can make a substantial contribution to the maintenance of connectivity in managed forest landscapes. They provide habitat for large numbers of terrestrial and aquatic fauna and flora (Loyn et al. 1980; Naiman et al. 1993; Spackman and Hughes 1995). Populations of some species can be much more fecund in riparian corridors (Soderquist and Mac Nally 2000) and provide dispersalists to other less-productive parts of the landscape. Riparian corridors may also act as dispersal routes for some terrestrial animals (Lindenmayer and Peakall 2000). Although riparian corridors are useful for some terrestrial taxa inhabiting forests, linkages outside the riparian zone may be required to maintain connectivity for upland species (McGarigal and McComb 1992; Claridge and Lindenmayer 1994; Whittaker and Montevecchi 1999).

Riparian corridors also fulfill a number of roles unrelated to the maintenance of connectivity for terrestrial organisms. They can support substantial levels of forest biomass—up to 100 times more than other parts of forest landscapes (Catling and Burt 1995)—which means they play an important role in carbon storage. Riparian corridors also (1) contribute extensively to the maintenance of intact watersheds (Naiman et al. 1993; Barling and Moore 1994; see Chapter 6), (2) provide protective buffers for aquatic taxa (Silsbee and Larson 1983; Doeg and Koehn 1990), (3) provide habitat for plant taxa restricted to the riparian zone (Michaelis 1984), and (4) provide energy (organic matter) and nutrients to maintain aquatic ecosystem function—another guiding princi-

TABLE 3.1.

Examples of the use of stepping stones to aid connectivity.

ORGANISM, LOCATION	DESCRIPTION	REFERENCES
Butterflies, Europe	Stepping stones contributed to the connectivity in butterfly populations in Europe.	Nève et al. 1996
Fender's blue butterfly (*Icaricia icaroides fenderi*), Oregon, United States	It was speculated that stepping stones rather than corridors would facilitate dispersal in Fender's blue butterfly.	Schultz 1998
Brown Kiwi (*Apteryx australis mantelli*), New Zealand	The brown kiwi used small forest remnants in the matrix as stepping stones.	Potter 1990
Fruit pigeons and bats, Australia	Mobile species like fruit pigeons and bats are able to utilize resources in small habitat patches located many kilometers apart to help them move across the landscape.	Date et al. 1996; Law et al. 1999
Plants, worldwide	Stepping stones may assist connectivity in plant populations as part of range shifts in response to climate change.	Collingham and Huntley 2000

ple for matrix management (see later in this chapter). For example, litter fall from trees in riparian vegetation can regulate energy flows, provide a source of organic carbon, and significantly influence physical and chemical conditions in stream systems (Campbell et al. 1992; Thomas et al. 1992b; Haycock et al. 1997). The retention of riparian corridors can be justified on this basis of these roles alone.

To sum up, the maintenance of connectivity is critical for any comprehensive plan for forest biodiversity conservation and essential for successful matrix-based biodiversity management. Managing the matrix to increase its suitability as habitat and increase its permeability to movement is fundamental to the maintenance of connectivity.

Principle 2. Maintenance of landscape heterogeneity

Another essential principle for conserving biodiversity is maintaining appropriate levels of spatial complexity, or landscape heterogeneity. The diversity, size, and spatial arrangement of habitat patches are important for some taxa (e.g., Hanski 1994a; Halley et al. 1996;

Turner et al. 1997; Saab 1999; Debinski et al. 2001). Strategies to set aside larger conservation areas, such as the Greater Yellowstone Ecosystem, are based on the need for large, heterogeneous areas (Clark and Minta 1994). On a smaller scale, prairie-wetland birds in North America have significantly smaller area requirements in heterogeneous than in homogeneous landscapes (Naugle et al. 1999). Landscape heterogeneity is also important because of its relationships with the impacts of habitat fragmentation. Such effects may be less crucial for taxa that evolved in landscapes where frequent natural disturbances create a naturally heterogeneous environment (Hansson and Angelstam 1990; Hansen et al. 1991; Rudnicky and Hunter 1993; Schieck et al. 1995; Schmiegelow et al. 1997).

Natural forest landscapes are rarely homogeneous (Figure 3.2). They are typically a mosaic of patches representing different forest composition and age classes where different structural conditions occur (Spies and Turner 1999). This is due to (1) geographic variation in environmental conditions such as moisture, temperature, light (Waring and Major 1964; Zobel et al. 1976; Mackey 1993), slope, aspect,

FIGURE 3.2. Unmanaged forests are rarely homogeneous, as illustrated by this area of mountain ash forest in the Australian state of Victoria. Stands of different ages, fire histories, and levels of biological legacies (such as snags) are apparent across the topographic gradient from the ridgeline to the bottom slope. Photo by D. Lindenmayer.

altitude, and soil type (Austin et al. 1990; Ohmann and Spies 1998), and (2) spatial and temporal variation in disturbance regimes (Wardell-Johnson and Horowitz 1996), which in turn are often influenced by environmental conditions (Angelstam 1997; Lindenmayer et al. 1999f).

Environmental gradients and existing forest patch conditions and patterns strongly influence disturbance regimes. These have an additional indirect impact on landscape pattern (Foster et al. 1997, 1998; Mackey et al. 2002).

Recognizing that not all parts of a landscape are created equal in their productivity and biodiversity is implicit in environmental patterning (e.g., biodiversity "hotspots," Hansen and Rotella 1999; see Chapter 6). Understanding these spatial inequities is often vital in planning matrix management for biodiversity conservation. For example, areas with the highest timber productivity are also often those where biodiversity values are highest (Braithwaite et al. 1993).

Great care should be taken in assessing the extent and type of landscape heterogeneity required in the matrix. This topic has frequently been misrepresented and abused by proponents of intensive wood production (Scientific Panel on Ecosystem Based Forest Management 2000). For over a century, production foresters have favored the "fully regulated forest" in which there is a perfectly balanced age-distribution of forest stands (varying in age up to the rotation time). The aim of the fully regulated forest is to provide an even flow of wood products (Oliver et al. 1997). The notion of creating such "equilibrium landscapes" persists today, albeit in modified forms. Desirable landscape heterogeneity has been represented as sixty different age classes of managed pine forest (*Pinus* spp.) in the southeastern United States (e.g., Boyce 1995). Oliver et al. (1997) proposed generic forest policies based on the creation of four or five structural types of managed forest over the entire forest estate. Such regulated or equilibrium forest landscapes can have significant negative impacts on biodiversity in the matrix, particularly when it is assumed that essentially all commercial forest land will eventually be harvested. These are not the types of "heterogeneous landscapes" needed for biodiversity. Many negative consequences of such oversimplified approaches have been documented ranging from the complete elimination of some forest structures (Linder and Östlund 1998) to species loss (Lindenmayer et al. 1999a).

Further details of approaches to maintain landscape heterogeneity in matrix lands are explored in Chapters 6 and 7.

Principle 3. Maintenance of stand structural complexity

Stand structural complexity includes a wide variety of structural features such as

FIGURE 3.3. Structural complexity is a characteristic feature of unmanaged forests worldwide. This stand of mountain ash forest supports a dense understory of 65-year-old *Acacia* spp. trees and tree ferns (at least 200 years old), and an overstory of large 350-year-old snags together with 65-year-old fire-regrowth stems. This stand also provides valuable habitat for the endangered Leadbeater's possum—the snag on the right of the photograph supports a nesting colony of the species. Photo by D. Lindenmayer.

- Trees from multiple age cohorts within a stand
- Large living trees and snags
- Large-diameter logs on the forest floor
- Vertical heterogeneity created by multiple or continuous canopy layers
- Canopy gaps and anti-gaps (i.e., areas with very dense canopy coverage under which understory development can be limited; Halpern and Franklin 1990; Franklin et al. 2002)
- Thickets of understory vegetation

FIGURE 3.4. Structural complexity can be lost from intensively managed stands as has occurred in this stand near Copenhagen in Denmark. This area of forest has been harvested many times in the past centuries and is characterized by an almost complete lack of coarse woody debris and understory plants as well as by trees of similar diameter and uniform spacing. There are no decadent trees in the stand. Photo by D. Lindenmayer.

Structural complexity embodies not only particular types of stand attributes, but also the way they are spatially arranged within stands. For example, the juxtaposition of overstory and understory trees from multiple-age cohorts within a stand contributes to vertical heterogeneity in forests (Figure 3.3). Structural complexity per se is a common feature of all natural temperate forests throughout the world (Franklin et al. 1981; Berg et al. 1994; Noel et al. 1998), and high levels of spatial heterogeneity are characteristic of essentially all old-growth forests (Franklin et al. 2002), although each type differs in specific details.

Logging may lead to marked medium- to long-term changes in stand structure and plant species composition (Halpern and Spies 1995; Lindenmayer and Franklin 1997a) (Figure 3.4) that can negatively impact taxa dependent not only on particular structural attributes but also on presently abundant generalist species (Niemelä et al. 1993). For example, it can impair the suitability of foraging habitat for vertebrates, such as birds and bats (Brown et al. 1997; Woinarski et al. 1997; Brokaw and Lent 1999).

Active management to maintain structural complexity is vital to prevent the decline and eventual loss of key structural attributes. This problem is not new. Almost 120 years ago, Gayer (1886) expressed

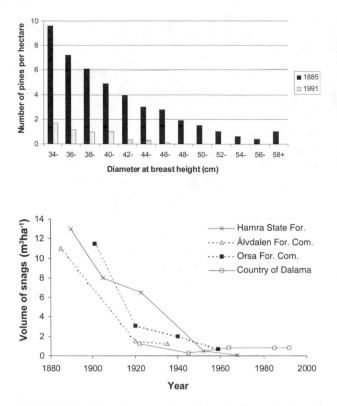

FIGURE 3.5. Altered structural attributes in four districts of Swedish boreal forest subject to a century of intensive harvesting. Data are shown for the number of pine trees in different diameter classes (A) and the volume of snags (B). Redrawn from Linder and Östlund 1998. Reprinted with the permission of Elsevier Science.

FIGURE 3.6. Structures typically selected for retention at the time of timber harvest typically include structures that cannot be reproduced during subsequent rotations, such as the large old trees, snags, and logs shown here (Willamette National Forest, western Oregon, United States). Photo by J. Franklin.

concerns about the simplification of German forests. In Sweden, a century of intensive management in a 123,000-hectare area of boreal forest transformed stand structure from one dominated by widely spaced, large-diameter trees to young, densely stocked forests. The number and volume of large trees and snags were reduced by 90 percent and the extent of old stands by 99 percent (Linder and Östlund 1998) (Figure 3.5). Recently it was recognized that large, dead trees are particularly valuable for biodiversity in Scandinavian forests (Samuelsson et al. 1994). Berg et al. (1994) calculated that almost 50 percent of the threatened (or red-listed) species in Sweden were dependent on snags or logs. Similarly, mature deciduous trees are also now a rare element of managed stands in Scandinavia (Esseen et al. 1997), but they are a key component of forest com-

position for a wide range of animal and plant groups (Enoksson et al. 1995).

The negative effects of logging on biodiversity may be partly mitigated by the retention of structural elements at the time of harvest (Hansen et al. 1991; Franklin et al. 1997; Hazell and Gustafsson 1999) (Figure 3.6). Bunnell (1999b) listed a wide range of species benefits from the maintenance of biological legacies in logged areas. Enhancement of stand structural complexity is considered essential to reverse the decline of many red-listed species in Scandinavia (Berg et al. 1994; Linder and Östlund 1998).

Maintenance of stand structural complexity can be valuable in four ways:

1. It may allow organisms to persist in logged areas from which they would otherwise be eliminated—a "lifeboating" function (Franklin et al. 1997). Many species will remain in logged areas if some of the original structures are retained or microclimatic conditions are maintained within tolerance levels. Examples include (a) species that display long-term site affinity (Van Horne 1983), such as parrots (Webster 1988) and some types of arboreal marsupials (Tyndale-Biscoe and Smith 1969); (b) plants that persist on large trees, such as epiphytes, and including lichens and mosses (Hazell and Gustaffson 1999); and (c) populations of small mammals that use windrows of logs within otherwise unsuitable areas of exotic softwood (Friend 1982). Very long-term persistence or continuity of particular structural features through many successive generations on the same site may allow insects and threatened species of fungi to persist within logged Norwegian forests (Sverdrup-Thygeson and Lindenmayer 1999, 2002). Such ecological "continuity" is regarded as a measure of forest sustainability by forest managers in that country. Retaining freshly cut logs within harvested forests facilitates the persistence of diverse groups of fungi that might otherwise be lost from production landscapes (Niemelä et al. 1995). Dead trees left in these environments are also used by saproxylic beetles (Niemelä et al. 1993, 1995; Kaila et al. 1997). Importantly, retaining selected structures as part of a harvesting operation can provide the resources required by non-autotrophic organisms, such as mycorrhizal fungi, that inoculate the remainder of the cutover area and facilitate stand regeneration (Perry 1994; Simard et al. 1997).

2. It may allow logged and regenerated stands to more quickly return to suitable habitat for species that have been displaced—a "structural enrichment" function (Franklin et al. 1997). This can limit the time logged areas remain unsuitable habitat (Lindenmayer and Franklin 1997a). Several studies have shown that retained trees can promote the recolonization of logged and regenerated forests by birds (e.g., Recher et al. 1980; Smith

1985; Kavanagh and Turner 1994; Hansen et al. 1995a).

3. It may enhance dispersal of some animals through a cutover area—a "connectivity" function (Franklin et al. 1997). For example, retention of logs provides travel routes for microtine rodents and allows animals to disperse into, and through, disturbed areas (Maser et al. 1977). This has been termed *softening the matrix* (Franklin 1993a). It may be particularly useful for taxa that employ random dispersal strategies and do not use wildlife corridors (Thomas et al. 1990; Murphy and Noon 1992).

4. It may be essential to provide the within-stand variation in habitat conditions required by some taxa—a "habitat heterogeneity" function. Structural complexity (e.g., trees of multiple ages or multiple layers of understory and overstory vegetation) can provide optimum habitat for a range of forest taxa, including some habitat specialists (Franklin 1993b). It also may provide more niches within a stand with corresponding benefits for species richness (Lindenmayer et al. 1991a; Niemelä et al. 1996).

Practical matrix management approaches to maintain stand structural complexity are discussed in detail in Chapter 8.

Principle 4. Maintenance of the integrity of aquatic ecosystems

A central goal of matrix management is preserving aquatic ecosystem integrity and the hydrologic and geomorphological processes upon which much biodiversity depends. Given its fundamental importance to human societies, the maintenance of a well-regulated, high-quality supply of water is (or should be) one of the chief objectives in the management of forest lands (O'Shaughnessy and Jayasuriya 1991; Fenger 1996).

The degree to which the integrity of aquatic ecosystems and associated processes is maintained is largely determined by conditions in the matrix. Forests have powerful influences on hydrological processes such as the interception of rainfall and snow, and the condensation, evapotranspiration, and infiltration of moisture (Swanson et al. 1988;

FIGURE 3.7. Forest cover is critical to the physical and biological integrity of small streams because it regulates light and temperature regimes and provides inputs of organic matter and large woody debris (Gifford Pinchot National Forest, western Washington, United States). Photo by J. Franklin.

Naiman 1992; Wardell-Johnson and Roberts 1993). Forest conditions also strongly affect nutrient retention and soil stability, especially on slopes. Forests adjacent to aquatic ecosystems stabilize banks and provide sediment filters (Barling and Moore 1994). They also have a direct influence by controlling light and temperature regimes and providing inputs of organic matter and nutrients in the form of litter. Forests also provide large woody debris, which is a significant structural element of riparian, riverine, and many wetland and pond ecosystems (Harmon et al. 1986; Koehn 1993) (Figure 3.7) that affects stream hydrodynamics (Gippel et al. 1996) and habi-

tat suitability for aquatic biota (Maser et al. 1977; Abbe and Montgomery 1996).

Inappropriate rates or patterns of forest harvest and poorly constructed road systems can have a negative impact on hydrological and geomorphological processes and biodiversity in aquatic ecosystems (Doeg and Koehn 1990; Naiman 1992; Trayler and Davis 1998). Harvesting schedules resulting in extensive areas of recently clearcut forest can increase flood flows (Jones and Grant 1996) with massive effects on aquatic ecosystems (e.g., Silsbee and Larson 1983; Graynoth 1989).

Roads, relative to the area they occupy, exert disproportionate, persistent, and intense impacts on aquatic ecosystems (Forman 1998). Several studies have shown that road density has a negative effect on aquatic fauna (Vos and Chardon 1998). Many hydrological changes arising from road networks are permanent because subsurface flows and patterns are interrupted and altered flows rerouted into extensive constructed surface channels (e.g., road ditches and culverts) (Jones and Grant 1996; Walker 1999).

Riparian buffers and other strategies to protect the integrity of aquatic ecosystems have been in place in some jurisdictions for several decades (e.g., wood production forests in southeastern Australia; Recher et al. 1980; Forestry Commission of Tasmania 1993). In other jurisdictions, the relationships between forest management practices and aquatic ecosystems are often unacknowledged by the wood products industry. This is changing, however, with increased attention to the effects of forest practices on fisheries and water quality in many regions, such as the Pacific Northwest in the United States (see Forest Ecosystem Management Assessment Team 1993).

In summary, it is clear that in watersheds that are predominantly matrix lands—meaning most temperate zone watersheds—issues such as the type and rate of forest harvesting, quality and density of road networks, and levels of buffering will determine the degree to which aquatic biodiversity and water quantity and quality will be maintained (Doeg and Koehn 1990; Naiman 1992; Barling and Moore 1994; Naiman and Bilby 1998).

We consider some practical matrix-based management strategies for maintaining intact aquatic ecosystems in Chapter 6.

Principle 5. Risk-spreading and the importance of different conservation strategies at different spatial scales

Embracing the four general principles outlined above requires adopting multiple approaches at multiple scales. This has many advantages. First, the adoption of multiple approaches is more likely to provide conditions needed by different species in at least some parts of a landscape. Management for diversity calls for diversity of management (Evans and Hibberd 1990). This is critical because, as outlined earlier, suitable connectivity, stand complexity, landscape heterogeneity, and aquatic ecosystem integrity will be defined on a species-specific basis and can vary markedly between species. Since defining these variables for a large set of species is essentially impossible, creating a range of conditions is a practical response to this problem.

A multifaceted approach to management has another advantage. If any one strategy is found to be ineffective (e.g., the establishment of wildlife corridors), others (such as tree retention on logged areas) will be in place that might better protect sensitive elements of forest biodiversity. This is a form of *risk-spreading* in forest management (Lindenmayer and Franklin 1997a); it reduces overreliance on a single strategy that may subsequently be found to be of limited value in meeting specific conservation objectives. Risk-spreading is particularly appropriate for biodiversity conservation because it is often extremely difficult to accurately forecast the response of species to landscape modification (see Mac Nally et al. 2000).

The risk-spreading approach highlights not only the need for different conservation strategies at different spatial scales, but also a need for variation in the protocols governing the way any given strategy is actually implemented at a particular spatial scale. Another advantage of multiple management strategies is that a given approach may generate positive benefits for another strategy implemented at a different spatial scale. For example, increased levels of stand retention on logged sites can reduce rates of windthrow and vegetation loss in adjacent wildlife corridors, riparian areas, wetland buffers, and small reserves within the matrix.

The risk-spreading approach contrasts fundamentally with the norm of strict production forestry. Production forestry tries to reduce variability at the stand and landscape levels. Risk-spreading, conversely, aims to ensure a range of conditions at all spatial scales; stands and landscapes are not homogenized.

The Indicator Species Concept

The guiding principles in this chapter provide a general framework for enhanced biodiversity conservation. They are the broad principles to help develop comprehensive plans for conserving forest biodiversity and, in turn, guide more focused strategies for matrix management at the landscape level (Chapters 6 and 7) and the stand level (see Chapter 8). Some workers have suggested that indicator species will be a useful approach to conserve biodiversity in matrix forests. However, we regard the use of indicator species as a simplistic substitute for the adoption of the guiding principles outlined above.

Landres et al. (1988) defined an indicator species as

an organism whose characteristics (e.g., presence or absence, population density, dispersion, reproductive success) are used as an index of attributes too difficult, inconvenient, or expensive to measure for other species or environmental conditions of interest.

The term *indicator species* has been used to mean different things (Spellerberg 1994; Hilty and Merenlender 2000). Some examples of types of indicator species include

- A species whose presence indicates the presence of a set of other species and whose absence indicates the lack of that entire set of species
- A keystone species (sensu Terborgh 1986) that is a species whose addition to, or loss from, an ecosystem leads to large changes in abundance or occurrence of at least one other species (e.g., Mills et al. 1993)
- A species whose presence indicates human-created abiotic conditions such as air or water pollution (Spellerberg 1994)

- A dominant species, meaning one that provides much of the biomass or who numerically dominates an area
- A species that indicates particular environmental conditions, such as certain soil or rock types (Klinka et al. 1989)
- A species believed to be sensitive to, and therefore an indicator of, environmental changes such as global warming (Parsons 1991) or modified fire regimes (Wolseley and Aguirre-Hudson 1991)
- A management indicator species, which is a species believed to reflect the impacts of a disturbance regime or the efficacy of efforts to mitigate disturbance impacts (Milledge et al. 1991)

There are many problems with the indicator species concept. These have been reviewed by several authors (Landres et al. 1988; Simberloff 1998; Lindenmayer et al. 2000b), and an extensive appraisal is beyond the scope of this book. Briefly, however, there are three major problems with the concept of indicator species.

The first problem is that different species can respond differently to disturbance—even taxonomically or ecologically similar species like guild members (Simberloff and Dayan 1991; Morrison et al 1992; Thiollay 1992). Each species has different habitat requirements and responds differently to forestry practices (see Berg et al. 1994) and to the effects of landscape change and habitat fragmentation (Robinson et al. 1992; Villard et al. 1999; Debinski and Holt 2000). Therefore, the response of a given species to human activities may not be indicative of that of other taxa (Kotliar and Wiens 1990). Even the same species can respond in different ways in differently modified landscapes (Dooley and Bowers 1998; Lindenmayer et al. 2001c). Hence, the possibility that there are valid indicator species seems remote. For example, Caro (2001) showed that protected areas set aside for large mammals in East Africa did not lead to the conservation of small mammals. Hence, large mammals were not useful indicators or umbrella species for small mammals.

The second problem is that so-called indicator species might lack sensitivity to change. Indicators with a high threshold response may result in some en-

vironmental problems being well advanced (and difficult to reverse) before they are detected (Lindenmayer et al. 2000b). Similarly, although Milledge et al. (1991) believed that the maintenance of suitable habitat for a management indicator species would conserve other taxa with similar requirements, Landres et al. (1988) gave an opposing view. They noted that any species specifically targeted for conservation by particular management actions can no longer be an independent yardstick of those actions for other species.

The third and last problem is that we may not sufficiently understand the causal relationships between a nominated indicator species and the process for which it is supposed to be indicative. Although many workers have contended that particular taxa are indicator species (e.g., Davey 1989; Johnson 1994; Hill 1995), the specific entities for which they are supposed to be indicative are often not explicitly stated. For example, is the presence of an indicator species indicative of the occurrence of a wide range of other species, the abundance of some selected other taxa, or the absence of a given threatening process (Lindenmayer and Cunningham 1997)? Even where the indicator species concept has been best developed—in the case of pollution indicator species (Spellerberg 1994)—the behavior of some indicator species may prove to be contrary to what was first expected. Significantly, the causal relationships between disturbance (or other forms of perturbation) and species response have not been established for any taxon recommended as an indicator species (Lindenmayer et al. 2000b).

Indicator species, and related concepts such as umbrella and focal species, are used as surrogates for biodiversity (e.g., Lambeck 1997, 1999). It is sobering to consider the results of a study of biodiversity surrogates by Andelman and Fagan (2000). They examined the efficacy of an array of biodiversity surrogates including indicator species, flagship species, and umbrella species and found that none captured more species or better protected habitat areas than a given organism selected at random from the large databases they had assembled to conduct their tests.

Although it is clear that there are major technical problems with the indicator species concept, it may nevertheless have some value as a social mechanism to promote biodiversity conservation. For example, its use may stimulate increased conservation management or applied management activity. However, difficulties will arise if attempts are made to actively implement biodiversity surrogate schemes such as the indicator species and focal species approaches on the ground. The uncritical application of such approaches could have considerable negative impacts for biodiversity conservation (Lindenmayer et al. 2000b) by giving managers a false impression that they have adequately conserved biodiversity when they have not. There is strong evidence to suggest that the conservation of one or several indicator species will not necessarily lead to the conservation of all elements of biodiversity (Prendergast and Eversham 1997; Pärt and Söderström 1999; Andelman and Fagan 2000; Carroll et al. 2001). In addition, the indicator species concept could focus particular management actions on individual forest components. Modern forest management entails managing forests for all their components, not just for selected ones (Franklin 1993a). Ecological systems are multiscaled (e.g., aquatic ecosystems; Poff 1997), ecological processes are multiscaled (see Figure 3.1), and factors at different scales influence the distribution of a particular species (and also different species). The indicator species concept oversimplifies the need for sophisticated and carefully considered approaches for matrix management applied at a range of spatial and temporal scales.

The inherent problems with the indicator species approach and similar biodiversity surrogate schemes do not mean that matrix management should overlook the need to implement strategies targeted at conserving particular species and then subsequently monitor the effectiveness of such strategies (see the case studies on spotted owls in Chapter 11 and Leadbeater's possum in Chapter 13). But it is crucial to recognize that such actions directed at particular species may not automatically conserve other elements of biodiversity.

Finally, indicator species are only one form of indicator that might be applied in managing forests. Other indicators such as water quality or levels of soil quality are not linked to the indicator species concept (Noss 1999) but can be targets for monitoring and management under agreements such as the Montreal Process for ecological sustainability (e.g., Santiago Declaration 1995; Commonwealth of Australia 1998).

Other Principles

Other general principles for developing comprehensive plans for forest biodiversity conservation agreements may emerge from future research. One particularly valuable principle not addressed in this chapter is the use of knowledge of disturbance regimes in natural forests to guide matrix management. The general philosophy of this principle is that strategies for biodiversity conservation are most likely to be successful in cases where human disturbance regimes (such as logging) are similar in their effects to natural disturbance (Hunter 1994). This is a massive topic in its own right with far-reaching implications at the stand and landscape levels. It is addressed in Chapter 4.

Using Information about Natural Forests, Landscapes, and Disturbance Regimes

Disturbances are a major driver of vegetation change and not necessarily rare events that are "outside the system." . . . Using a framework of disturbance ecology, it is possible to evaluate how well disturbances that derive from direct and indirect human sources match disturbance regimes of the past which have shaped many ecosystems and to which many species are adapted.

—SPIES AND TURNER (1999)

Natural disturbance regimes and their interactions with climate and terrain determine the size, shape, location, and types of patches that provide heterogeneity in unmanaged forest landscapes. These disturbance regimes create a rich array of biological legacies, such as logs, intact thickets of understory vegetation, and large living and dead trees, which provide within-stand structural complexity and habitat for many organisms.

Natural disturbance regimes provide insights into silvicultural approaches that can reduce impacts on biodiversity. From detailed observations of natural forests and natural disturbance regimes, inferences can be drawn that are useful in developing forest management approaches that achieve the general principles outlined in Chapter 3. Two guiding principles are particularly relevant in this regard—maintenance of landscape heterogeneity and maintenance of stand structural complexity.

Human disturbance regimes will never be exact duplicates of natural disturbances, and this should not be the goal in incorporating knowledge from natural disturbance events. Rather, the objective is to use such information to develop silvicultural systems that better achieve the guiding principles outlined in Chapter 3, thereby enhancing the conservation of biodiversity in matrix lands while achieving other human objectives, including commodity production.

Effective conservation of forest biodiversity requires the maintenance of suitable habitat at multiple spatial scales (Chapter 3). Knowledge and inferences from natural forests and disturbance regimes can be used in designing on-the-ground management approaches that better conserve biodiversity. We agree with the view that impacts of human disturbances on biodiversity are generally less when these disturbances resemble natural ecological disturbances (Hunter 1994), the premise being that organisms are best adapted to the disturbance regimes under which they have evolved (Bergeron et al. 1999; Hobson and Schieck 1999). Conversely, organisms may be less adapted to novel ecosystem disturbances, including those involving different disturbance agents, different frequencies or intensities of disturbance, or new combinations of disturbances (Paine et al. 1998).

In this chapter, we briefly discuss natural disturbance regimes in forests, but a thorough treatment is beyond the scope of this book. We then outline the value of using knowledge and information from natural disturbances to guide development of human disturbance regimes (i.e., management) so that better outcomes for biodiversity conservation can be achieved. We examine both the landscape and the stand levels. The chapter ends with a comparison of human and natural disturbances for biodiversity conservation and exploration of the consequences of differences. We also briefly review the concept of historic or natural ranges of variability (NRV) with regard to its value in guiding management regimes.

We emphasize our view that natural ecosystems, disturbance regimes, and ranges of variability are primarily of value as information sources and guides in planning for management of natural resources. Using such natural models as templates is rarely appropriate in managing matrix lands. There are many reasons for this, including the goal-directed nature of matrix management and the highly altered conditions that currently exist in most forested landscapes as a result of human influences at local to global scales. Matrix management objectives rarely include the re-creation of a past condition; we question whether this is an appropriate or feasible goal, even on reserves. Our view is that we should learn from natural models and apply that information in designing management regimes to achieve multiple goals, including the conservation of biodiversity.

Natural Disturbances in Forests

Natural disturbances are characteristic of all ecosystems (Pickett and Thompson 1978; Agee 1993). We define natural disturbances as discrete events that are not primarily of human origin and which alter ecosystem structure and resource availability (see White and Pickett 1985). The composition and structure of forests at the tree, stand, landscape, and ecosystem scales are shaped by disturbance events such as wildfires (Luke and McArthur 1978; Wadleigh and Jenkins 1996; Gill et al. 1999), windstorms (Foster and Boose 1992; Peterson and Pickett 1995; Foster et al. 1997; Schnitzler and Borlea 1998), volcanic eruptions (Franklin 1990; Franklin and MacMahon 2000), floods (Calhoun 1999), landslides (Ogden et al. 1996; Veblen et al. 1996), disease (Holling 1992b; Cogbill 1996), and drought (Gordon et al. 1988).

Natural Disturbance and Landscape Heterogeneity

Nonuniform patterns of organism distribution and abundance result (in part) from spatial variation in environmental regimes, such as climate, terrain, soils, and nutrients (e.g., Croizat 1960; Woodward and Williams 1987; Nix and Switzer 1991; Prentice et al. 1992; Mackey 1994; Hansen and Rotella 1999). Such spatial variation leads to natural forests being heterogeneous at the landscape scale (see Chapter 3). Disturbance regimes overlay and interact with the patterns created by environmental regimes to further shape heterogeneous landscapes and influence the

FIGURE 4.1. Variability in fire effects related to topography in lodgepole pine (*Pinus contorta*) forests burned in the 1988 wildfires in Yellowstone National Park, Wyoming, United States. Photo by J. Franklin.

distribution of species (Wardell-Johnson and Horowitz 1996; Spies and Turner 1999; Beaty and Taylor 2001). Factors such as topographic variability result in undamaged patches of forest within the broad boundaries of a disturbance event, such as a fire or windstorm (Syrjänen et al. 1994; Eberhart and Woodard 1987; Delong and Kessler 2000) (Figure 4.1).

Natural Disturbance and Biological Legacies

Natural disturbance regimes not only create considerable heterogeneity at the landscape level but also create heterogeneity within stands (Angelstam 1996; Noel et al. 1998). This is because most natural disturbances leave traces and features of the original stand in the form of *biological legacies*. Biological legacies are organisms, organically derived structures, and organically produced patterns that persist from the predisturbance ecosystem (Franklin et al. 2000b), and they include logs, intact thickets of understory vegetation, large living trees, and snags (Franklin et al. 1985; Hansen et al. 1991; Cascade Center for Ecosystem Management 1995) (see Table 4.1, Figure 4.2). Even intense catastrophic disturbances, such as the Mount St. Helens eruption in Washington state (northwestern United States), can leave enormous numbers and varieties of biological legacies (Franklin and MacMahon 2000).

FIGURE 4.2. Natural disturbances in forests typically leave behind extensive biological legacies (including structures from the original stand) that are incorporated into the young recovering forest. The types and quantities vary with the type of disturbance. Contrasting types and levels of biological legacies between disturbances are apparent in this photographic series: (*A*) Standing dead trees and logs in areas subject to wildfires (Yosemite National Park, California, United States). (*B*) Organic matter on forest sites disturbed by intense windstorms (Bull Run River drainage, Mount Hood National Forest, Oregon, United States). (*C*) Legacies of snags, logs, and sapling trees in the scorch zone of the Mount St. Helens 1980 eruption (Washington, United States). (*D*) Above-ground biological legacies associated with clearcutting (Willamette National Forest, Oregon, United States). Photos by J. Franklin.

TABLE 4.1.

Broad categories of biological legacies (modified from Franklin et al. 2000b).

TYPE OF BIOLOGICAL LEGACY	EXAMPLES
Organisms	Whole organisms
	Perenating parts
	Propagules (seeds, spores, eggs)
Organic matter	Dissolved and particulate organic matter
	Feces
Structures	Snags
	Logs and coarse woody debris
	Large soil aggregates
	Termite mounds
	Bodies of dead animals
Patterns (Plant or Animal Created)	Root mounds
	Burrows
	Root channels
	Understory community patterns (gaps and anti-gaps)
	Wallows/yards
	Soil chemical, microbiological, and physical patterns

Biological legacies have a wide range of functions. They can

- *Survive, persist, and regenerate after disturbance and be incorporated as part of the recovering stand.* Multi-aged stands of forest (i.e., stands with trees from different age cohorts) result from biological legacies (live trees) surviving a disturbance and being incorporated into a post-disturbance stand (e.g., Lindenmayer et al. 1991d, 1999f; Franklin and Fites-Kaufmann 1996).
- *Assist other species in persisting in a disturbed area through a variety of mechanisms* (often termed a *life-boating function*). The Mount St. Helens volcanic eruption provided excellent examples of this phenomenon (Franklin et al. 1985; Franklin and MacMahon 2000).

- *Provide habitat for species that eventually recolonize a disturbed site.* For instance, legacy trees facilitate the return of vertebrates to logged and regenerated stands (e.g., Carey 1995; Gibbons and Lindenmayer 1997). This phenomenon has been referred to as *structural enrichment* of the post-disturbance stand.
- *Influence patterns of recolonization in the disturbed area.* Biological legacies within a disturbed area can provide foci that facilitate population recovery. That is, stand recovery can occur not only via colonization from neighboring disturbed areas, but also from organisms and structures persisting within a disturbed area. These multi-foci or "nucleated" recovery processes (sensu Turner et al. 1998) have been observed by Franklin and MacMahon (2000) within the eruption-affected area at Mount St. Helens and following the conflagrations that burned 45 percent of Yellowstone National Park in 1988 (Spies and Turner 1999).
- *Provide a source of energy and nutrients for other organisms.* This function is particularly important as it relates to maintaining a flow of energy into the soil to maintain the rich array of soil organisms, including mycorrhizal-forming fungi (Amaranthus and Perry 1994; Hooper et al. 2000).
- *Modify or stabilize environmental conditions in the recovering stand.* Perry (1994) demonstrated how resprouting trees in disturbed forests in the Pacific Northwest of the United States helped stabilize soil conditions and soil microbes.

The concept of biological legacies is highly relevant in assessing forestry impacts on biodiversity. Naïve comparisons of the biota of logged and unlogged stands have been made that ignore the influence of legacies from natural forests in cutover areas on recolonization of species (Macfarlane 1988). If legacies are not considered, an incorrect conclusion may be drawn that there are no logging impacts. Quantities and types of legacies are depleted by logging and other silvicultural practices over several cutting cycles, which will ultimately have detrimental impacts on species dependent on these stand attributes (Lindenmayer and Franklin 1997a). Hence, differences in the types and numbers of biological legacies left by human and natural disturbances and the impacts of such differences on biodiversity need to be quantified before inferences are drawn about effects of logging (see below).

The severity of natural disturbance regimes influences the numbers, types, and spatial patterns of biological legacies that remain in a disturbed area (Franklin et al. 2000b). This is because disturbances can be regarded as "editors"—selectively removing or modifying stand components to varying degrees (Franklin et al. 2000b). Types, numbers, and spatial arrangement of biological legacies also influence the trajectory of succession and recovery processes following disturbance (Turner et al. 1998). A combination of chance and the environmental factors affecting disturbance regimes (see above) will influence the mix of colonizing species, propagules, and other legacies at a site following disturbance. Subsequent changes in stand conditions reflect a sorting of species according to life histories, interspecific interactions, and other factors, such as the presence of herbivores (Egler 1954). Recovery following disturbance almost never starts at "zero." For example, the persistence of large living trees in disturbed stands leads to a multicohort forest (Lindenmayer et al. 1999f). Similarly, the floristic composition of a post-disturbance stand will be strongly influenced by predisturbance vegetation and its persistence in the form of individuals, seeds, and other propagules (Franklin et al. 2000b).

Variability in Disturbance Regimes

Natural disturbances vary substantially in their timing (e.g., the time of the year or time of day when they occur), frequency (or return interval), intensity, size, heterogeneity (i.e., variation in intensity and impact within the limits of the total area affected), and duration (e.g., fires and windstorms can be relatively short-lived whereas drought and insect attack can be prolonged).

These factors act together rather than in isolation. For example, fire frequency and fire intensity usually co-vary—few forest landscapes naturally experience frequent high-intensity wildfires, but many are characterized by recurrent low-intensity disturbances. Factors such as climatic conditions and topography further influence how variables interact within particular landscapes and stands (Lindenmayer et al. 1999f).

Variation in disturbance regimes leads to marked differences in landscape and stand conditions by significantly influencing the number, type, and spatial distribution of habitat patches (e.g., age cohorts of stands) as well as stand-level biological legacies (see Hansen et al. 1991; Lindenmayer and Franklin 1997a). Such temporal and spatial variability in natural disturbance regimes helps explain regional differences in species assemblages. For example, areas of western Canada subject to high natural fire frequencies support more taxa typical of early successional forests than regions with longer disturbance return intervals (Bunnell 1995).

Disturbance regimes are inherently variable, so no two disturbances will be identical. This variability also produces multiple (and often simultaneously acting) disturbance pathways (Noble and Slatyer 1980; Turner et al. 1998; Spies and Turner 1999). This is illustrated in many mesic forest types, such as those found in northwestern North America (Halpern 1988; Morrison and Swanson 1990). Although some disturbance pathways have higher probabilities of occurring, the variability in natural recovery patterns suggests that variation in human (logging) disturbance regimes in forests is also appropriate (discussed later in this chapter).

Old-growth Australian mountain ash forests reveal the structural complexity that is related to inherent variability in natural disturbance regimes. Few stands are pure old growth; the majority of these stands incorporate trees of multiple ages and also show evidence of understory rejuvenation (Lindenmayer et al. 2000c). Old-growth Douglas-fir stands in the Cascade Range of the Pacific Northwest often have similar multi-aged structures (Franklin and Hemstrom 1981). Late-successional stands of pine and mixed conifer in the Sierra Nevada in the western United States incorporate the full range of developmental stages within individual stands as a mosaic of structural units (Franklin and Fites-Kaufman 1996; Franklin et al. 2002).

To summarize, the interplay of different disturbance factors over different temporal and spatial scales creates complex stand conditions that develop over prolonged periods (e.g., over multiple natural life spans of a given tree species). Therefore, it may take a very long time (more than 1,000 years) to restore aspects of structural complexity lost as a result of traditional management regimes that simplify forests, such as large-scale clearcutting (Resource Assessment Commission 1992; Stohlgren 1992).

Natural Disturbance as a Guide for Enhanced Biodiversity Conservation

Many scientists and resource managers believe that conservation of biodiversity is best achieved by forest management regimes that are as consistent as possible with natural ecological processes, including disturbance regimes (e.g., Hunter 1994). The assumption is that logging should have minimal effects on biodiversity when operations are within the bounds of natural disturbance regimes (Attiwill 1994; Bunnell 1995).

Three properties of natural disturbance regimes that can guide harvesting are frequency, spatial pattern, and levels of legacies (Hunter 1993). Hence, determining congruence between human and natural disturbance regimes requires knowledge of

- Size, pattern, and composition of patches in different forest types and successional stages of those forest types (Mladenoff et al. 1993; Angelstam and Pettersson 1997).
- Impacts of natural disturbances on ecosystem processes and properties, such as soil nutrient regimes.
- Types, quantities, and spatial arrangements of biological legacies that are left behind by natural disturbances at both the stand and the landscape scales (Franklin et al. 2000b). This includes the size distribution, composition, and geographical distribution of undisturbed (refugial) patches that are left within the boundaries of disturbance events (e.g., Delong and Kessler 2000).

Forest harvesting contrasts with typical natural forest disturbances in several important characteristics:

- Harvesting usually occurs with higher frequency and greater regularity than natural disturbances (McCarthy and Burgman 1995), at least in forest types that are subject to episodic, stand-replacing disturbances. As a consequence of frequent harvesting, age class distributions are skewed toward younger stands.
- Harvesting alters plant species composition in terms of both tree and non-tree components (Halpern and Spies 1995). Some species (especially commercial tree species and aggressive understory shrubs and herbs) are favored while other species decline. For example, red cedar (*Toona australis*) has been virtually eliminated from natural forests in northern New South Wales and southern Queensland (Australia) by logging (Boland et al. 1984).
- Harvesting simplifies stand structures, sometimes drastically, such as when clearcutting is utilized (Lindenmayer and Franklin 1997a) (Table 4.2).
- Harvesting alters the spatial distribution of structural attributes of stands such as snags and understory (Lindenmayer et al. 1991e; Mladenoff et al. 1993).

Comparisons with natural disturbance regimes highlight the limitations of intensive forest management regimes (Lindenmayer et al. 1991e; Franklin et al. 2000b). The magnitude of differences between natural disturbances and intensive timber management regimes is clearly the basis for many problems in biodiversity conservation in managed forest landscapes (see Box 4.1). For example, Thompson and Angelstam (1999) described problems for woodland caribou (*Rangifer tarandus*) in Canada arising from creation of patches smaller than those created by natural disturbances. In another case, regeneration failures in logged, high-elevation forests in northeastern Victoria (Australia) occur where high-intensity slash fires are used (purportedly to promote stand regeneration) instead of the characteristically less-intense natural fires.

Attempts to create greater congruence between natural and human disturbance have usually focused on mesic forest types where harvesting methods, such as clearcutting, are traditionally applied (Rülcker et al. 1994; Lindenmayer and Franklin 1997a; Bergeron et al. 1999). However, the importance of matching disturbance regimes applies equally to forests where other harvesting systems (such as selection cutting) are deployed. These silvicultural approaches can also have long-term impacts when applied repeatedly over a longer period (Meredith 1984; Thiollay 1992; Bader et al. 1995; Gibbons and Lindenmayer 1997). For ex-

TABLE 4.2.

Differences in the effect of clearcutting and natural wildfires on stand structure in Australian mountain ash forests.

VARIABLE	FOREST RESPONSE AFTER NATURAL WILDFIRES	FOREST RESPONSE AFTER CLEARCUTTING
Forest floor architecture	Large diameter logs often occur	Average number, size, and volume of logs reduced
Spacing of hollow trees in the forest	Regular or random	Clustered
Standing life of hollow trees	Up to, or more than, fifty years	Trees removed during logging or destroyed by regeneration fire
Range of forms of living trees and snags	Often two or more morphological forms present	Trees removed during logging
Survival of hollow trees	Variable—it depends on stand age and fire intensity	Stems removed or severely burnt
Age class structure	Multi-aged stands may occur	Even-aged stands
Plant species composition	Variable depending on fire intensity and frequency	Shrubs and ground plants typical of wet environments lost. Tree ferns, fire-resistant understory thickets, and rainforest trees depleted.

BOX 4.1.

Lack of Congruence between Natural and Human Stand-Level Disturbance in Australian Mountain Ash Forests

Australian mountain ash forests provide a useful example of the need to create greater congruence between natural and human disturbance regimes. The most common form of natural disturbance is wildfire, and the most common form of human disturbance is clearcutting followed by a very high-intensity slash fire to burn logging debris. The two forms of disturbance have substantially different impacts at the stand and landscape levels (Table 4.2). These differences can have severe implications for biodiversity conservation (Lindenmayer and Franklin 1997a).

Wildfires in mountain ash vary in intensity. In some stands, high-intensity fires kill many trees and only very few living trees remain. In others, where less-intense fires occur, a multi-aged stand develops that contains a mixture of fire-damaged living and dead trees. Large living trees and snags are key habitat components, because they contain cavities that are used as den, nest, and shelter sites by many vertebrate and invertebrate taxa (Gibbons and Lindenmayer 2001).

Clearcutting in mountain ash forests produces simple even-aged stands on the cut area (Lutze et al. 1999). Trees with cavities are scarce and numbers are typically much lower than in stands subject to natural disturbance regimes (Lindenmayer et al. 1991b). This has negative impacts for more than 100 species of vertebrates dependent on these structures. Long-lived understory plants like tree ferns (*Dicksonia antarctica* and *Cyathea australis*) are reduced by 95 percent in cutover areas (Ough and Ross 1992). These plants can tolerate natural wildfires and can also tolerate natural physical disturbances such as mechanical damage resulting from windstorms and snowfalls. However, intensive mechanical disturbance from logging machinery coupled with high-intensity regeneration fires is essentially a novel form of perturbation in mountain ash forests—the reason it has such detrimental impacts on the understory (Ough and Ross 1992).

Clearcutting is well outside the bounds of natural disturbance regimes in mountain ash forests, particularly with respect to its effects on biological legacies such as cavity trees, snags, and understory vegetation (Lindenmayer and McCarthy 2002a). By modifying silvicultural prescriptions, conditions in logged areas can be made more similar to those characteristic of naturally disturbed mountain ash forests (see recommendations in Gibbons and Lindenmayer 1997, Ough and Murphy 1998, Ball et al. 1999). This, in turn, should make a positive contribution to biodiversity conservation.

ample, the demise of red spruce (*Picea rubens*) forests in the northeastern United States and their replacement with short-lived tree species have been attributed to recurrent partial cutting (Seymour and Hunter 1999).

In summary, in order to use natural forests as models for managing matrix forests it is necessary to document spatial and temporal patterns created by natural disturbance as well as biophysical and environmental processes that underpin and influence natural disturbance regimes (Spies and Turner 1999).

Natural Disturbance as a Guide for Forest Management at the Landscape Level

Patch types, sizes, and shapes and the internal complexity of patches (i.e., biological legacies) created by disturbance regimes in natural landscapes can be used to provide guides for planning patch patterns in matrix landscapes (Mladenoff et al. 1993; Wegner 1994) (Table 4.3), such as the size, location, spatial arrangement, and rotation period of harvest units (Franklin 1993b). Selecting forest management regimes that maintain or restore landscape heterogeneity (the second guiding principle in Chapter 3) is particularly important where forest biodiversity conservation is a concern (Hunter 1990; Haila et al. 1993; Welsh and Healy 1993; McNeely 1994b). Historical fire regimes

and resulting landscape patterns have been used to develop novel approaches for landscape management that have significant benefits for biodiversity conservation (e.g., Cissel et al. 1999). Harvest units were actually enlarged (to 500 hectares) to more closely approximate the size of patches created in pre-European disturbance regimes in forest landscapes in Wisconsin (United States) (Parker 1997). This was done concurrently with increased retention of structures and small patches within harvest units.

Some management practices in the forests of British Columbia attempt to mimic landscape patterns created by natural disturbances (British Columbia Ministry of Forests 1995). For example, forest patches are retained within clearcut harvest units to mimic the unburned patches left by wildfires (see Chapter 8 for further details). Retained patches in regenerating harvest units are valuable in promoting recolonization of wildlife and other organisms, even though they are not identical to remnant patches in burns (Delong and Kessler 2000).

Identifying disturbance refugia—areas that were rarely subjected to natural disturbances—and protecting or carefully managing them is another way to achieve greater congruence between natural and human disturbance regimes at the landscape level. The ASIO model developed for Swedish forests (Rülcker et al. 1994) attempts to do this for parts of the landscape that formerly would have been fire refugia. Additional approaches are explored in Chapters 6 and 7.

TABLE 4.3.

Landscape-level differences in patches between clearcut logging and wildfire.

ATTRIBUTE	FORM OF DISTURBANCE	
	Wildfire	*Clearcut Logging*
Patch numbers	Variable, but can be small depending on spatial contagion	Deterministic and set by prescription for number of harvest units
Patch size	Highly variable, but can be very large	Deterministic and set by prescription for cutover size
Patch location	Variable depending on climate, terrain, and other factors	Set by prescription and accessibility
Patch pattern	Often displays contagion	Usually dispersed
Patch boundary	Often diffuse	Sharp

Establishing Baseline Landscape Patterns for Comparison with Managed Forests

Contrasting patterns of landscape heterogeneity and composition in managed and unharvested landscapes have been the focus of many studies (e.g., Ambrose and Bratton 1990; Ripple et al. 1991, 2000; Williams and Marcot 1991; Wegner 1994). Models can also be useful tools in such comparisons. A model was used in order to estimate the proportion of old-growth forest in the northern Rocky Mountains ecosystems prior to European settlement (Lesica 1996). Computer simulations were used to demonstrate that declines in late-successional forest could be reversed simultaneously with reduced risks of high-intensity wildfires in the mixed conifer forests of the Sierra Nevada in California through active management (Johnson et al. 1998).

Large differences in the mean age of the forest and patch characteristics were found when wildfires were compared with traditional forms of dispersed logging using models (McCarthy and Burgman 1995). Differences between fire and harvesting are present even if the area disturbed and the mean disturbance frequency are the same (Figure 4.3a). More and larger areas of old forest occur in burned than in logged landscapes, a conclusion identical to that of Bergeron et al. (1999) in their study in Canadian forest (Figure 4.3b).

Variability in natural disturbance regimes is as important as the mean return interval in examining congruence between human and natural disturbance regimes (McCarthy and Burgman 1995; Bergeron et al. 1999). Seymour and Hunter (1999) devised an interesting method for calculating rotation times based on the variability in natural disturbances rather than on the mean return interval (see Chapter 6). This is the basis for two important inferences, assuming that congruence between natural and forest landscapes is a goal: (1) regeneration of newly cut forests should create new stands at rates similar to those produced by natural disturbances, and (2) the rotation period in parts of a landscape subject to traditional clearcutting needs to be at least twice as long as the mean return rate for natural disturbances in forests. This is necessary to achieve the right-skewed distribution of age classes generated by natural disturbances and to provide for some areas of very old forest that would be

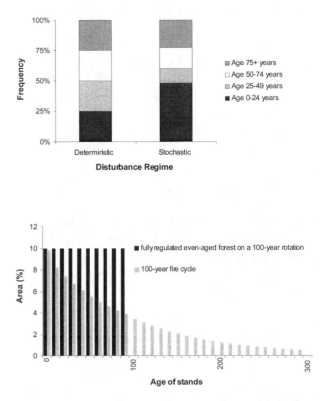

FIGURE 4.3. (*A*) Differences between deterministic (logging) and stochastic (fire) disturbance on the area of forest in different age classes in Australian forests (from McCarthy and Burgman 1995). (*B*) An example for a Canadian forest (Bergeron et al. 1999).

otherwise absent (Van Wagner 1978; Spies and Turner 1999).

Other forms of vegetation analysis have been used to reconstruct presettlement conditions and disturbance regimes in natural forest landscapes. Premanagement landscapes in Sweden have been reconstructed using analyses of pollen, charcoal, and plant fragments (e.g., Björse and Bradshaw 1998; Linder and Östlund 1998). Species-rich deciduous forests were found to have been dramatically reduced by human activity; such forests currently compose less than 1 percent of forested land but support more than 55 percent of all the threatened species in Swedish forests (Berg et al. 1994). A logical inference is that forest planning to improve biodiversity conservation should include the expansion of deciduous forest (Björse and Bradshaw 1998).

Comparisons of Scandinavian forests with ecologically similar but unmanaged Russian forests have been important in documenting natural disturbance regimes, natural landscape patterns, and natural levels of structural complexity in forests (e.g., Siitonen and Martikainen 1994; Angelstam et al. 1995; Hanski and Hammond 1995; Martikainen et al. 1996; Uuttera et al. 1996). Extensive late-successional forests still exist in Russia, in contrast to Sweden and Finland, where young managed stands dominate as the result of intensive and recurrent forestry operations. Pine forests in Russia currently contain thirty-three times more snags, forty-six times more logs, and eight times more large trees than equivalent forests in Sweden (Angelstam 1996). Differences in disturbance regimes between the two areas is a basis for understanding differences in faunal assemblages. Similar cross-nation comparisons were conducted for boreal forests in eastern Canada and Fennoscandia (Imbeau et al. 2001). An important outcome of that work was that the cavity-nesting birds that have severely declined in the latter area have been identified as being among those species likely to be under threat with the expansion of industrial forestry in eastern Canada (Imbeau et al. 2001).

Natural Disturbance as a Guide for Forest Management at the Stand Level

As traditionally practiced, existing silvicultural systems—from clearcutting to selective harvesting (including group selection and gap-phase removal)—do not incorporate landscape- and stand-level complexity characteristic of natural disturbance regimes (Franklin et al. 1997, 2002). There is, of course, no reason that management regimes focused exclusively on wood production should follow a natural rather than an agricultural model. The degree of differences (see Box 4.1) can be inferred by comparing the types, quantities, and spatial arrangements of biological legacies (Franklin 1993b; Haila et al. 1993; Franklin et al. 1997; Fries et al. 1997; Lindenmayer and Franklin 1997a). Such legacies, which are essential habitat components for wildlife, may be totally eliminated from harvested areas (Lindenmayer and Franklin 1997a). Hence, a key question is, How different are the types and numbers of biological legacies remaining after multiple logging rotations in comparison with multiple natural disturbance events?

Young stands regenerated following natural disturbances typically have high levels of structural complexity resulting from biological legacies from the previous stand (Figure 4.4). Clearcuts are unlike most natural disturbances in this regard (Lindenmayer et al. 1991e; Esseen et al. 1997; but cf. O'Neill and Attiwill 1997; Tuckey 2001) (Table 4.4). Wildfires typically consume less than 10 percent of the wood, leaving huge quantities of dead and down timber (Foster 1983; Payette et al. 1989), whereas up to 95 percent of wood volume may be removed in clearcutting (Angel-

---

BOX 4.2.

Caveats for Identifying Landscape Patterns from Natural Fire Regimes

Efforts to identify natural patterns in forest landscapes are complicated by inherent variability in fire frequencies and fire sizes. Fire intervals and fire sizes are variable over time (Chou et al. 1993; Gill and McCarthy 1998), making it difficult to determine baseline patterns against which to compare (and modify) patterns created by logging regimes. Also, historical landscape patterns may be influenced by disturbances by native peoples (King 1963; Flannery 1994; Bowman 1998), making it difficult to establish "natural" patterns for benchmarking (Hunter 1996). A number of studies have shown that forests thought to have been pristine are now known to have been influenced by human activity for a prolonged period (Lamb 1966). Another problem for benchmarking is that landscapes and ecosystems change in relation to changes in long-term climate—what a system was like 500 years ago will be different from 5,000 years ago (Lawton 1997). Long et al. (1998) showed that fire frequency has varied continuously in the Oregon Coast Range (northwestern United States) on millennial time scales. Similarly, Bergeron et al. (1998) noted that large changes in the fire frequency in the boreal forests of Canada during the Holocene meant there was not a (single) characteristic fire regime for this system.

Finally, it is essential to recognize that few disturbance regimes in modern forest landscapes are "natural"; fires, for example, are routinely extinguished to limit threats to human life and property. Indeed, fire has been virtually eliminated as a natural disturbance regime in Swedish forests (Fries et al. 1997).

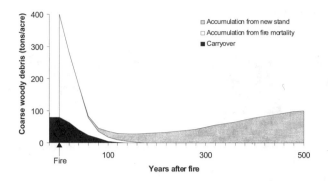

FIGURE 4.4. Temporal patterns of accumulation of coarse woody debris following fire in the Douglas-fir forests of the Pacific Northwest region of the United States. Redrawn from Maser et al. 1978.

stam 1996). Perhaps the only significant similarity between clearcutting and wildfire is in creating a light environment (i.e., near full sunlight) desired for the regeneration and rapid early growth of shade-intolerant plants (Franklin et al. 2002).

There are also temporal contrasts between traditional silvicultural regimes and natural disturbances. For example, legacies, such as overstory trees and understory plants, are known to survive repeated natural disturbance events based on carbon dating and dendrochronology (Banks 1993; Mueck et al. 1996). Large logs can persist through many natural disturbances. In Scandinavia, such persistent logs sustain populations of many red-listed species (Bader et al. 1995). The inference is that forest management regimes should be designed to maintain such biological legacies through multiple rotations. Infor-

mation on the longevity of individual biological legacies (e.g., how long snags remaining standing) can be the basis for planning recruitment schedules necessary to maintain desired levels of structures of particular types, ages, and conditions as part of modified silvicultural systems (Gibbons and Lindenmayer 1996; Ball et al. 1999). Similarly, data on habitat relationships of target organisms assist in defining types of stand conditions that are needed to conserve biota that are sensitive to traditional forest practices.

New silvicultural systems are often a necessity when maintenance of biodiversity and ecosystem processes is a goal, because of the vast differences between natural and human disturbance regimes at the stand and landscape levels. Black spruce (*Picea mariana*) forests in Canada provide a useful example (Bergeron et al. 1999). Fire, competition, and gap dynamics are the primary natural processes influencing stand dynamics in black spruce forests. A sequence of cutting of different cohorts could emulate some of the complex and heterogeneous stand conditions typical of multicohort black spruce forests (Figure 4.5). The aim is to maintain at least some aspects of the natural composition and structure of natural forests and to integrate them into an array of silvicultural systems employed within the one stand. Similar modifications of clearcutting practices to enhance structural complexity have been described for three major forest types in Sweden (Angelstam and Pettersson 1997; Fries et al. 1997) (Table 4.5).

Silvicultural approaches that can be used to create greater congruence between natural and human

TABLE 4.4.

Differences in the quantities of biological legacies following different types of forest disturbances (modified from Franklin et al. 2000b).

LEGACY TYPE	FORM OF DISTURBANCE		
	Fire	*Wind*	*Clearcut*
Snags	Abundant	Few	None
Logs	Common	Abundant	Few or none
Soil disturbance	Low	Patchy	High
Understory plants	Common	Abundant	Few
Soil nutrients	Pulse in nitrogen and phosphorus release	No change	Pulse in nitrogen and phosphorus release

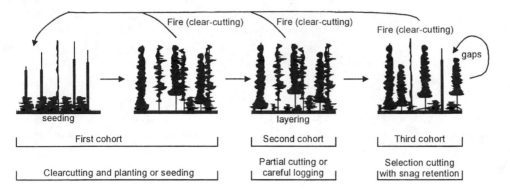

FIGURE 4.5.
A sequence of cutting of different cohorts to emulate the complex and heterogeneous stand conditions typical of multicohort black spruce forests (from Bergeron et al. 1999).

TABLE 4.5.

Suggested modifications of traditional clearcutting forest management based on natural features found in three disturbance regimes of boreal forests (modified from Fries et al. 1997; Angelstam and Pettersson 1997).

SCOTS PINE FORESTS ON DRY OR MESIC SITES NATURALLY REGENERATED BY FIRE	DECIDUOUS OR NORWAY SPRUCE–DOMINATED FORESTS ON MESIC SITES NATURALLY REGENERATED BY FIRE	UNEVEN-AGED NORWAY SPRUCE FORESTS NATURALLY REGENERATED BY GAP-PHASE DYNAMICS
Use of the clearcutting system with green tree retention and more varied rotation period (e.g., 50–200 years)	Use of the clearcutting system with green tree retention and more varied rotation (e.g., 100–200 years)	Use of selection systems or shelterwood systems that assure continuity in tree cover
Use prescribed burning or slight or moderate mechanical disturbance of ground vegetation (scarification)	Use prescribed burning or slight to moderate mechanical disturbance of ground vegetation (scarification)	Leave (parts of) selected stands uncut
Leave trees as relicts; omit, vary, or modify the traditional low thinning to generate self-thinning; girdle, push over, or fell selected trees	Leaves trees as relicts; omit, vary, or modify the traditional low thinning to generate self-thinning; girdle, push over, or fell selected trees	Leave trees as relicts and accept that some of them die by uprooting
Use natural regeneration by means of seed trees of shelterwood system or by seeding	Use of natural regeneration of seeding (after scarification) to complement planting	Leave wind-thrown trees on the site
Leave single or groups of large pines at final felling	Promote a portion of deciduous trees at thinning; no thinning and acceptance of self-thinning	Use natural regeneration or, if necessary, planting
	Leave single or groups of large pines at felling	Use prescribed burning on a portion of this site type (preferable combined with green tree retention)
	Design certain areas for deciduous successions; use modified clearcutting system; promote deciduous trees at cleaning and thinning; no thinning and acceptance of self-thinning	Retain living and dead trees; vary the intensity of tree removal to promote self-thinning; girdle, push over, or fell selected trees
	Leave forest patches of 0.1 to several hectares on clearcut areas	Favor other species than spruce and, above all, deciduous trees
		No ditching

BOX 4.3.

The ASIO System in Sweden

The Swedes have proposed a model called ASIO—the acronym refers to the frequency of fire in a particular part of the forest and it stands for Almost never, Seldom, Infrequently, and Often. *The model uses natural disturbance in boreal forests as a template to guide silvicultural practices. The primary form of natural disturbance in Swedish boreal forests was formerly fire, although it has been virtually eliminated by efficient fire-suppression methods.*

In the past, the prevalence and impacts of fire varied across Swedish forest landscapes—it was common in dry areas but relatively rare in wet areas (Rülcker et al. 1994). Differences in fire regimes have, in turn, strongly influenced patterns of species distribution and abundance across Swedish forest landscapes.

Swedish boreal forests have been classified according to the ASIO system (Angelstam 1997). As shown below, forest management systems are based on the ASIO model to guide variable silvicultural systems so they better emulate natural disturbance regimes (Rülcker et al. 1994) (see Figure 4.8).

- "Almost never" forest includes wetland forest, ravines, small islands in lakes, and northeast-facing slopes. These forests are habitat for a range of species that require stable microhabitats. The recommended management regime for these areas is no forestry activity.
- "Seldom" forest land rarely burned—on average every 200 years. This type of forest occurs not only on watercourses, but also around flat, moist

areas. Selective and shelterwood harvesting is recommended for this category of forest.

- Forest land that burned "infrequently" (every 100 years on average) includes all mesic areas (apart from those assigned to the "almost never" and "seldom" categories). It comprises the majority of boreal forest in Sweden, and the fire frequency was the same as the rotation time now employed. The silvicultural system here is controlled burning of the cutover with the retention of seed trees on site.
- The frequency of fires in "often" forests was approximately fifty years. All dry forest land (e.g., pine forest on sedimentary soils located on flat terrain) is assigned to this class. These areas support trees that survive recurrent low-intensity fires—so controlled burning of the forest is used. Because the fire regimes resulted in multi-aged stands, the final felling operation includes the retention of **seed trees**.

The ASIO strategy is one of the few examples where a practical strategy is used in variable silvicultural systems to create greater congruence with natural disturbance regimes (Rülcker et al. 1994). The application of the ASIO strategy is useful because revised approaches to forest management in Sweden have attempted to maintain high levels of wood production and simultaneously embrace sound environmental practices to conserve biodiversity (Angelstam and Pettersson 1997).

---

disturbances are discussed in detail in Chapter 7 (landscape patterns) and Chapter 8 (stand management).

Natural disturbance regimes typically exhibit high levels of variability within a given forest landscape that should be considered when using them as models for silvicultural systems. Knowing about multiple disturbance pathways and what is, in essence, a continuum of variability is useful in forest management planning. Application of multiple silvicultural prescriptions to different parts of the landscape (Rülcker et al. 1994) (Box 4.3; Figures 4.6 and 4.7) as well as within a single stand (Bergeron et al. 1999) (see Figure 4.5) may be needed to achieve desired levels of structural complexity. The benefits of multiple

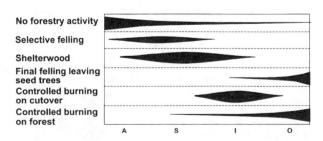

FIGURE 4.6. Variability in fire regimes in Swedish boreal forests and associated variations in harvesting regimes (from Rülcker et al. 1994). The acronym ASIO (Almost never, Seldom, Infrequent, Often) on the x-axis highlights the frequency of fires. The black polygons symbolize the extent of natural disturbance (fire on the x-axis) in relation to human disturbance (on the y-axis). Reprinted with the permission of Skog Forsk, Sweden.

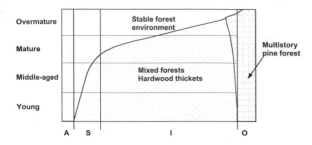

FIGURE 4.7. Application of the ASIO system (see Box 4.3) in relation to different forest types and disturbance regimes in Swedish forests (from Rülcker et al. 1994). Reprinted with the permission of Skog Forsk, Sweden.

silvicultural systems are exemplified in Douglas-fir forest in the Coast Ranges of Oregon (United States), where several different cutting regimes are needed to conserve the full complement of bird species (Chambers et al. 1999).

We can infer that using an array of silvicultural systems would be beneficial to biodiversity from the variability inherent in natural disturbance regimes. This natural variability generates heterogeneity in landscapes and stands in types, quantities, and distribution of biological legacies (Boxes 4.1 and 4.3). For example, since approximately 30 percent of mountain ash stands are multi-aged, biologists have recommended that as much as 30 percent of managed Victorian mountain ash stands should be maintained in a structurally diverse condition by using methods other than clearcutting (McCarthy and Lindenmayer 1998). Even if high levels of stand retention are implemented, the uniform adoption of the same silvicultural system can erase between-stand heterogeneity. Substantial within-forest-type heterogeneity occurs in Western Australian forests due to variations in hydrological regimes and related variability in fire regimes (Wardell-Johnson and Horowitz 1996). Different silvicultural systems may be needed to maintain structural and floristic properties of these forests as well as the associated patterns of localized plant endemism.

Finally, it is important to recognize that the reconstructed disturbance regimes are really approximations. Consequently, basing all management within a landscape on this hypothesized regime is risky (see Hunter 1996). Any untested silvicultural prescription—and even those that have been utilized for many

decades or even centuries—may lead to an "ecological surprise," such as reductions in long-term site productivity and loss of desired biotic components. For example, when staggered-setting clearcutting was adopted in the federal forests of the Pacific Northwest, its role in fragmenting forest habitat was not even considered (Franklin and Forman 1987). For all the reasons given above, variation in silvicultural systems provides for a form of risk-spreading (see Chapter 3) from the viewpoints of long-term wood production as well as biodiversity conservation.

Lessons for Forest Ecosystem Restoration from Intense Stand-Replacing Disturbances

Forest ecosystems subject to intense floods, wildfires, windstorms, and volcanic eruptions provide many valuable lessons regarding aspects of forest management (Franklin and MacMahon 2000; Franklin et al. 2002). Such stand-replacing events can even aid landscape restoration by creating some of the structural complexity and landscape heterogeneity that have been lost through traditional forestry practices. Severe flooding can reshape or "reset" riparian areas through massive mobilization and redistribution of sediment and coarse woody debris (Bayley 1995), thereby revitalizing modified aquatic ecosystems (Gregory 1997). Similarly, major wildfires generate immense legacies of snags and logs and promote the development of cavities in trees (Inions et al. 1989), structural attributes that are depleted by some forestry practices.

Understanding the effects of intense disturbances on such variables as types and numbers of biological legacies provides information to guide management of forest ecosystems recovering after such events, such as salvage and restoration activities. Such information also can help account for historical impacts of past human and natural disturbance. For example, the effects of the 1938 hurricane in the forests of northeastern North America were reinterpreted following an experimental study of uprooted forest; observed impacts on these forests were actually the result of the hurricane and extensive post-disturbance salvage logging, which removed most of the rich array of biological legacies left behind by the storm (Foster et al. 1997). These timber salvage operations, which were the largest in the history of the United States, had enormous effects on hydrology and many other ecosystem processes. In Australian mountain ash forests, many impacts on stand structure attributed to wildfire are actually the results of widespread and intensive post-fire salvage harvesting (Smith and Woodgate 1985; Lindenmayer 1996).

Forest management agencies can capitalize on benefits that can flow from catastrophic natural disturbances (Gregory 1997), although these benefits can be lost as a result of inappropriate post-disturbance practices, such as excessive timber salvage after partial-stand-replacing wildfire (Hutto 1995; Hobson and Schieck 1999) and windstorms. It may be appropriate to limit salvage and reforestation activities on some areas subject to stand-replacing disturbances. The government of Victoria has adopted policies that prohibit salvage logging in the Yarra Ranges National Park following major fires (Land Conservation Council 1994). In the United States, federal forests in the Pacific Northwest are subject to significant constraints on salvage activity following major forest disturbances to ensure that adequate structural legacies are retained (Forest Ecosystem Management Assessment Team 1993). Indeed, naturally developed early-successional forest habitats, with their rich array of snags and logs and nonarborescent vegetation, are probably the scarcest habitat in the current regional landscape. In southeastern Asia, salvage logging of burned rainforests can lead to significant forest deterioration and loss with major negative impacts on the regenerative potential of stands and a wide range of other undesirable effects (van Nieuwstadt et al. 2001).

We conclude this section by noting that rapid and uniform reforestation of large areas subjected to a stand-replacing disturbance may have detrimental consequences for biodiversity, just as in the case of aggressive timber salvage operations. Disturbed forested areas that take many decades to develop closed forest canopies may make valuable contributions to biodiversity, particularly if they contain structural legacies, such as snags, logs, and scattered live trees. Natural elements of biodiversity that require early-successional conditions are more likely to be accommodated in such naturally disturbed areas than they are in clearcuts. Humans are typically much more efficient than nature at promptly, densely, and uniformly reforesting large disturbed areas and can greatly truncate the open pre–tree-canopy-closure successional stage. More consideration and study of the role of large, naturally disturbed areas in maintaining regional biodiversity, such as of song bird populations in western North America, is needed.

The Lack of Congruence between Natural and Human Disturbances

Human disturbance will never exactly mimic natural disturbance regimes, and we view such a goal as inappropriate. Wildfires do not remove the large amounts of biomass that logging operations do (Angelstam 1996). Nevertheless, lessons learned from observing natural disturbances can provide guides for matrix management at both the landscape and the stand levels (Table 4.6). Peterken (1999) has outlined how greater congruence between human and natural disturbance regimes can promote biodiversity conservation, even in heavily modified environments like plantations (see Box 10.1 in Chapter 10).

Ultimately, the objective is to use information from natural disturbances to develop human disturbance regimes that achieve defined goals in balancing biodiversity and commodity production. Creating identical replicas of natural disturbance regimes—even in large biological reserves—is impossible and is certainly not the aim of modified silvicultural systems. As an example, natural fires that cover tens or even hundreds of thousands of hectares (Dyrness et al. 1986) are socially and politically unacceptable as models for harvest patch sizes (Hunter 1993; Haila et al. 1993; see Chapter 7).

Although there is already substantial evidence that modified silvicultural practices can sustain biodiversity better than traditional practices can (e.g., Franklin et al. 1997; Ough and Murphy 1998; Hickey et al. 1999), empirical data that quantify relationships between different biotic elements and alternative silvicultural practices are still very limited (Fries et al. 1997). Uncertainties are evident even among foresters, as illustrated by the Society of American Foresters'

TABLE 4.6.

General principles for creating greater congruence between human and natural disturbance in matrix forests.

VARIABLE	PLANNING APPROACH	REFERENCE
Regeneration scheduling	Create new stands at the same rate as those produced by natural disturbance	Seymour and Hunter 1999
Forest type cover	Similar to natural (historical levels)	Björse and Bradshaw 1998
Rotation time	Double the mean interval between major disturbance events	Seymour and Hunter 1999
Harvest unit size	Similar range of sizes to those created by natural disturbance	Parker 1997
Harvest unit location	Parts of landscapes more subject to natural disturbance	Hansen and Rotella 1999
Harvesting intensity at the stand level (silvicultural system)	Leave similar types, amounts, and patterns of biological legacies to those arising from natural disturbance	Franklin et al. 2000b
Variation in harvesting intensity at the landscape level (silvicultural system)	Silvicultural systems vary across the landscape depending on spatial variation in natural disturbance	Rülcker et al. 1994

(1984) observation that it is unlikely that forest managers can create functional old growth by silviculturally manipulating younger forests. The inappropriateness and impossibility of matching natural and human disturbance regimes in the matrix are reasons why large ecological reserves are needed. The value and limitations of large ecological reserves are the topic of the next chapter.

Historic or Natural Range of Variability

Our final topic in this chapter is the historic or natural range of variability (HRV or NRV) and its use in designing and implementing plans for ecosystem management, including maintenance of biodiversity. We use the term *HRV*, although many other terms have been applied, including *range of natural variability* (Aplet and Keeton 1999). HRV is widely used in North America to describe "the bounded behavior of ecosystems prior to the dramatic changes in state factors prior to settlement" (see Aplet and Keeton 1999) or, more broadly, human manipulation. For example, HRV can be applied to such widely dis-

parate parameters as variability in the percentage of old-growth forest within a region during the last millennium or in monthly mean temperatures in stream ecosystems over decades of record prior to initiation of extensive logging.

The objective in utilizing HRV is to identify the range of specific conditions that have been experienced by biota prior to significant modern human activity and that are presumed, therefore, to be within their tolerance range. The assumption is that "restoring and maintaining landscape conditions within distributions that organisms have adapted to over evolutionary time is most likely to produce sustainable ecosystems" (Manley et al. 1995). Activities of indigenous peoples are usually accepted as "natural" and are incorporated within the HRV.

HRV can be (and has been) calculated and applied at multiple scales, including the watershed, landscape, and regional levels. In fact, the "scale dependence" of HRV is one of the issues that complicates its application and has been a source of criticism, along with difficulties in acquiring valid information about HRV for many ecosystem parameters, and the effects of climatic variability during the historic period. Generally, the smaller the area considered in calculating HRV, the greater the range of variability (Aplet and Keeton 1999). The proportion of old-growth forest in the Pacific Northwest provides an example; at the scale of a small watershed, the proportion of old growth in a typical small watershed (e.g., 100 to 1,000 hectares) varied from close to zero to nearly 100 percent during the last 1,000 years, but only 30 to 75 percent at the level of the entire Douglas-fir region.

An important consideration in use of HRV is the difficulty in obtaining and interpreting relevant data (Aplet and Keeton 1999). There can be significant differences of opinion with regard to interpretation of evidence and, consequently, conclusions about what truly represents the HRV. The historic pattern of disturbances in the Black Hills of South Dakota—a region characterized by forests of ponderosa pine (*Pinus ponderosa*)—provides an excellent example (Shinneman and Baker 1997). The traditional interpretation of the presettlement landscape was that of frequent, low-intensity wildfires that maintained open, park-like pine stands. A careful reassessment found evidence that there were also large, intense (stand-replacing) fires in the presettlement landscape, which periodically generated large patches (exceeding 5,000 hectares) of relatively dense, closed-canopy forests, including old growth (Shinneman and Baker 1997). Hence, there was a mosaic of disturbance regimes and forest conditions present in the Black Hills, and management plans need to reflect this complexity.

HRV values have been developed for a wide variety of parameters or ecosystem elements and subsequently applied in management (Manley et al. 1995). For example, HRV has been utilized in development of the Northwest Forest Plan (USDA Forest Service and USDI Bureau of Land Management 1994a,b), such as in setting objectives for desired levels of late-successional forest within the range of the northern spotted owl. Similarly, HRV concepts were applied in developing the current strategy for managing national forest lands within the Sierra Nevada (California and Nevada) (USDA Forest Service 2001a,b). HRV has also been utilized on much smaller spatial scales to develop ecologically based management regimes, such as a disturbance-based management regime proposed for federal forest lands in the Oregon Cascade Range (United States) (Cissel et al. 1998) (see Chapter 7).

We conclude this section by noting that HRV—like other natural models—should be viewed as tools and guides in developing and evaluating management regimes that are intended to sustain biodiversity. Neither HRV nor any of the concepts discussed earlier in this chapter should be taken as strict models or bounds that are to be precisely followed. Rather, information on natural disturbance regimes and their effects and HRV are indicative of conditions under which biota have persisted and are, therefore, useful guides to approaches and conditions that may sustain biodiversity and ecosystem processes in the future.

PART II Biodiversity Conservation across Multiple Spatial Scales

In Part I, we argued that the maintenance of suitable habitat across multiple spatial scales is fundamental for comprehensive approaches to the conservation of forest biodiversity (Chapter 3). In the second part of this book, we describe three major elements of a comprehensive approach—setting aside large ecological reserves (Chapter 5), adopting landscape-level matrix management strategies (Chapters 6 and 7), and adopting stand-level matrix management strategies (Chapter 8). All three approaches are needed because each one will be inadequate by itself for some elements of biodiversity. If matrix management is practiced only at the stand level, key landscape-level values may be lost. Conversely, many processes, such as the movement of ungulates and the migration of fish, cannot be dealt with at the stand level. Therefore, simultaneous consideration of the strategies in Chapters 5, 6, 7, and 8 is needed to accommodate the extraordinary biodiversity that characterizes forest ecosystems. As an example, if stand-level issues, such as a limited supply of snags, are disregarded in many cutblocks, the consequences can be cumulative and result in large areas of unsuitable habitat, which will eliminate opportunities for species to recolonize regenerating forest. In Chapter 9, we briefly revisit ideas associated with the integration of strategies at scales ranging from individual stands to large ecological reserves as part of comprehensive approaches to biodiversity conservation.

Chapters 6, 7, and 8 relate primarily to matrix management in natural or near-natural forests. However, the landscape-level and stand-level approaches to matrix management that we have outlined are also relevant to intensively managed plantation landscapes, where even moderate changes in management regimes can contribute positively to biodiversity conservation (Chapter 10).

Importance and Limitations of Large Ecological Reserves

It is certain that the . . . park and reserve system will be totally inadequate for conservation of the products of four billion years of evolution. . . . A new ecocentric paradigm, that integrates biodiversity conservation with the totality of human activities, is long overdue.

—N<small>IX</small> (1997)

During recent decades, conservation biologists have focused primarily on the design, selection, and establishment of large ecological reserves. In the many cases where the objectives are broader than the conservation of an individual taxon or assemblage, four general principles—the CARR principles—provide a framework for the design and selection of reserve systems. Reserves should be Comprehensive, Adequate, Representative, and Replicated. Mathematical algorithms and gap analysis can be used to guide the selection of reserves to make them more representative and comprehensive and therefore help them better represent the biodiversity of an area or region. However, these methods often fail to address the adequacy of reserves and ignore the dynamics of ecosystems, including those associated with large-scale environmental change. Other approaches, such as risk assessment methods, are needed to take these factors into account.

Despite the merit of the CARR principles and associated analytical methods, it is extremely difficult to create reserve systems that are truly comprehensive, representative, and adequate for all elements of biodiversity. This reality is the major limitation of a reserve-only focus. If reserves lack representation of many ecosystems, and opportunities for their expansion are limited, then matrix management is essential for the conservation of many species.

The fact that existing reserve systems have limitations does not justify substituting matrix management for the retention of existing reserves or the establishment of additional reserves. What it does mean is that comprehensive plans for forest biodiversity conservation must increasingly incorporate matrix-based approaches that complement reserve-based approaches.

Although much of the focus of this book is on the matrix (see Chapter 1), the conservation of forest biodiversity requires both large ecological reserves and matrix management. Credible (and comprehensive) conservation programs must include ecological reserves wherever possible (McNeely 1994a; Christensen et al. 1996) to ensure the maintenance of habitat across multiple spatial scales (see Chapter 3).

In this chapter, we examine issues associated with large ecological reserves. First, we discuss why large ecological reserves are a fundamental element of any comprehensive plan for forest biodiversity conservation. Second, we outline approaches to the design and selection of reserves. Third, we consider why reserves sometimes need to be managed. Finally, we discuss the limitations of a reserve-only approach to the conservation of forest biodiversity. The limitations of reserves are treated in some detail because they highlight the need for matrix-based conservation strategies. For the purposes of this book, we distinguish between large ecological reserves (the focus of this chapter) and smaller areas of protected habitat that are embedded within the matrix (see Chapter 6).

The Importance of Large Ecological Reserves

Large ecological reserves are extensive tracts managed primarily to perpetuate natural ecosystems and related processes, including biota. They are an essential part of all comprehensive biodiversity conservation plans and are needed in all ecosystems and vegetation types. They are important because

- Large ecological reserves support some of the best examples of ecosystems, landscapes, stands, habitat, and biota and their interrelationships. They also provide important opportunities for the maintenance of natural evolutionary processes.
- Many taxa find optimum conditions within large ecological reserves, which become strongholds for these species.
- Some species are intolerant of human intrusions, making it imperative to retain some areas that are exempt from human activity.
- Large ecological reserves are needed to provide control areas against which the impacts of human activities in managed forests can be compared.
- Effects of human disturbance on biodiversity are poorly known, and some impacts may be irreversible. Others such as synergistic and cumulative effects can be extremely difficult to quantify or predict. Ultimately, this makes large ecological reserves valuable as "safety nets" relatively free from human disturbance.

Large Ecological Reserves and Restricted Communities

The priority on establishment of large ecological reserves in conservation programs is often warranted because near-natural examples of many ecosystems and vegetation types are rapidly disappearing. They must be identified and protected now or they will be lost forever. Hence, an emphasis on ecological reserves is necessary where particular ecosystems, vegetation communities, or forest age classes are being extensively and rapidly modified (McNeely 1994a) and only limited amounts of the original cover remain (Pressey et al. 1996). Examples come from all over the world and include old-growth forests in Scandinavia (Jarvinen et al. 1977; Uuttera et al. 1996) and the northwestern United States (Thomas et al. 1990; Kohm and Franklin 1997), deciduous forests in Cen-

tral Europe (Hannah et al. 1995), dry forests of western Central America, and brigalow (*Acacia harpophylla*)–dominated vegetation communities in eastern Australia (Nix 1994; Fensham 1996). Indeed, we conclude that there are threats wherever a forest type capable of commercial exploitation exists.

Large Ecological Reserves and Human Disturbance

The establishment of large ecological reserves is essential for ecological processes and taxa negatively impacted by even low levels of human disturbance. Some taxa may be conserved only within large reserves. Putz et al. (2000) recognized this need for tropical forest ecosystems and stressed that some areas should never be logged. In another example, populations of the arboreal marsupial yellow-bellied glider (*Petaurus australis*) within the montane ash forests of Victoria appear to be virtually confined to extensive unfragmented stands of old-growth forest (Incoll 1995; Lindenmayer et al. 1999a). Extensive old-growth stands now occur only in closed water catchment reserves where no logging has occurred. Similar types of old-growth species occurrence relationships have been reported for other taxa elsewhere around the world (e.g., Jarvinen et al. 1977; Benkman 1993; Niemelä 1997).

Large ecological reserves can be useful to protect some species from human activities, such as hunting. For example, the ibex (*Capra ibex ibex*) was exterminated from France prior to the development of a system of national parks and survived only in a hunting reserve in Italy (Grodinski and Stüwe 1987; Skonhoft et al., in press).

Large ecological reserves are valuable because the impacts of human activities, such as forest harvesting on biodiversity, are poorly known and for many species will probably never be known. Some effects will almost certainly be irreversible. As noted in Chapter 3, for most species we have little or no understanding of what constitutes suitable connectivity, heterogeneity, stand structural complexity, and aquatic integrity. In other cases, the ability to detect responses to human disturbance may vary depending on the assemblages targeted for study—some may show a rapid response (Niemelä et al. 1993) whereas for other taxa, extinctions may occur a prolonged period after disturbance has occurred (even if such perturbations are halted; Lamberson et al. 1994). This can lead to what Tilman et al. (1994) termed an "extinction debt." This makes it important to maintain large ecological reserves in case errors are made in matrix management (i.e., the risk-spreading approach discussed in Chapter 3).

Large Ecological Reserves as Reference Points to Quantify Human Impacts

Large ecological reserves are fundamental to quantifying the impacts of natural and human disturbances (Norton 1999). They are needed as reference points to contrast with disturbed areas as part of scientific experiments and observational studies (Christensen et al. 1996; Norton 1999; Lindenmayer et al. 2000b). The numerous unexpected responses of plant communities and biota following the 1988 fires in Yellowstone National Park and the 1980 eruption at Mount St. Helens are two of many examples that highlight the contribution of large ecological reserves to increased scientific understanding. The value of large ecological reserves as reference points is further demonstrated in places such as Scandinavia where the extent of landscape change and stand simplification has been so substantial that it is difficult to envisage original conditions. Angelstam (1996) noted that

> In some heavily modified landscapes, such as those in parts of Scandinavia, particular species have become so rare that it is impossible to complete meaningful studies of their habitat requirements. It has become necessary to examine intact landscapes elsewhere that still support these species to determine the properties and structures of suitable habitat.

In this case, forest managers and ecologists from Scandinavia examined Russian forests to gain new insights into ways to better conserve forest biodiversity in their own countries (Angelstam et al. 1995; see Chapter 4).

Large ecological reserves are also useful for comparisons of landscape heterogeneity between logged or unlogged landscapes (Ambrose and Bratton 1990;

Mladenoff et al. 1993) and the corresponding responses of biota (McGarigal and McComb 1995; Lindenmayer et al. 1999a). Such work may show it is necessary to increase the size of reserves to accommodate particular types of landscape patterns that cannot be replicated within production landscapes (Baker 1992; Mladenoff et al. 1994).

Large Ecological Reserves and Cumulative Effects

Large ecological reserves are valuable because they are relatively safe from cumulative impacts of human activities (sensu Cocklin et al. 1992a,b). These occur when repeated perturbations accumulate spatially or temporally in a landscape and have negative impacts, particularly when they exceed threshold levels (Risser 1988; McComb et al. 1991). Examples include altered hydrological regimes that arise when the aggregate area of harvest units in a landscape exceeds a given size (Aust and Lea 1992). In the case of the response of the biota, species still present in forest landscapes after an initial harvesting rotation may disappear during subsequent rotations if refugia are destroyed or if habitats needed for recolonization of logged and regenerated areas are eliminated (Crome 1985; Lindenmayer 1995).

Cumulative effects can result in novel types of disturbance with impacts well outside the bounds of natural disturbance regimes with corresponding negative outcomes for biodiversity conservation and ecosystem processes at the stand and landscape levels (Paine et al. 1998). Unfortunately, the ability of matrix management strategies to mitigate or reverse cumulative human disturbance impacts is unknown. This makes it crucial to have some areas set aside as large ecological reserves.

Another reason for setting aside large ecological reserves is that organisms in managed landscapes must sometimes contend with the combined effects of two or more types of perturbation—a deterministic one (logging) as well as a random or stochastic one (e.g., wildfire) (Lindenmayer 1995). There are many documented cases where the impacts of two types of disturbance are independently unimportant but together create significant problems for biodiversity conservation (Taylor 1979; Caughley and Gunn 1995; Paine et al. 1998). The combined effects of logging, fire, and

---

BOX 5.1.

Large Ecological Reserves, Wilderness, and Biodiversity Conservation

The word wilderness *derives from the Old English "wild deer-ness" relating to uninhabited and uncultivated tracts of land occupied only by wild animals (Shea et al. 1997). Wilderness is a human construct that relates to the remoteness of areas from human influence and human infrastructure development (Mackey et al. 1998). The concept is usually linked to the spiritual, aesthetic enjoyment, and recreation needs of western peoples in natural landscapes (Mackey et al. 1999). The views of indigenous peoples (such as those that lived throughout the continents of North America and Australia), who occupied the land before European settlement and who did not have concepts of wilderness, have generally not been taken into consideration (Hammond 1991).*

The role of wilderness in conserving biodiversity is complex. Some authors contend that large wild areas are essential. For example, Noss and Cooperrider (1994) state that "unless it contains many millions of acres, no reserve can maintain its biodiversity for long." (See also initiatives like the Wildlands Project; Foreman et al. 1992). There is certainly evidence to demonstrate that in some jurisdictions, for example, the only places where the stream channel and deep-pool architecture of riparian systems remains unaltered by human activities occur in roadless designated wilderness areas (McIntosh et al. 2000). In turn, the conservation of particular groups (such as fish) can be related to such a lack of human disturbance, highlighting the value of wilderness areas (Baxter et al. 1999).

However, strong relationships between wilderness values and conservation values are not always apparent. Lindenmayer et al. (2001b) found no correlation between the occurrence of any species of arboreal marsupials in central

Victorian forests and measures of the intensity of human development taken from the National Wilderness Inventory (Lesslie and Maslen 1995) such as the distance of field survey sites from roads. Better predictors of species occurrence were found to be such factors as the extent of matrix management practices (e.g., levels of tree retention on harvest units). Similarly, at a continental level there was no relationship between the number of threatened Australian mammal species and wilderness quality (as measured by the National Wilderness Inventory), although significant trends were recorded for vascular plants (Mackey et al. 1999).

Some areas will be important for biodiversity conservation precisely because the remainder of the landscape that contains them has been subject to intensive human use. Schwartz and van Mantgem (1997) demonstrate this for the intensively modified landscapes of Illinois in the Midwest region of the United States, where small reserves (less than 10–20 hectares) have valuable roles to play in conserving many plant and animal taxa. These ecosystems have been so extensively transformed by humans, they will never be wilderness. Yet the remnants support numerous species, many of which can be conserved by appropriate management.

We support the concept of wilderness—wilderness quite clearly has spiritual, aesthetic, and recreation values. Yet wilderness and biodiversity conservation are not always mutually inclusive (Brown and Hickey 1990). There is abundant evidence to show that while large wild areas have considerable conservation value, small reserves in "non-wilderness" areas also have much to contribute to the conservation of biodiversity.

---

grazing of domestic herbivores in forest ecosystems is a useful example (Kirkpatrick 1994). Fire, grazing, or logging acting individually may not have significant detrimental impacts in the forests of southeastern Queensland. However, when combined, the recruitment of new trees to replace harvested stems is impaired—trees are either burned by subsequent fires or eaten by domestic livestock (Smith et al. 1992). The problem of cumulative effects of logging and fire is also increasingly recognized in tropical forests (Holdsworth and Uhl 1997). These effects are inextricably linked because logging opens the canopy of tropical forests, creates additional coarse and fine fuels, dries the understory, and promotes the develop-

ment of fire-prone grasses. Fire frequency can increase dramatically and preclude forest recovery. In addition, fires are not constrained to logged ignition points; Putz et al. (2000) noted that in 1999, the area of logged forest was 10 percent of the total area actually burned by fire in Bolivia. The negative cumulative impacts of fire and subsequent salvage logging in other rainforests such as those in southeastern Asia are also increasingly being recognized (van Nieuwstadt et al. 2001).

Multiple (and sometimes compounding) forms of disturbance mean that while some species presently persist in managed landscapes, their future occurrence is not guaranteed. The concepts of "species richness relaxation" and "extinction debts" are well known in the theoretical literature (Macarthur and Wilson 1967; McCarthy et al. 1997; Brooks et al. 1999) and also from empirical data (e.g., Loyn 1987; Robinson 1999). Since the current existence of a species may not be a good indicator of its future persistence (Lindenmayer 1995), we need to avoid overconfidence in our ability to ensure the conservation of some taxa in the long term (Niemelä et al. 1993; Noss and Cooperrider 1994).

Large Ecological Reserves, Wilderness, and Biodiversity Conservation

The concepts of large ecological reserves, wilderness, and biodiversity conservation have been intertwined by many authors, but as we show in Box 5.1, they are not necessarily synonymous. For example, Dobson et al. (2001) recommend a mix of reserve-based and matrix-based conservation strategies from their studies of plants, birds, and herpetofauna in several states in the United States, noting:

> Our results indicate that, although protecting wilderness is valuable and relatively easy, conserving the most biodiversity will require greater focus on those areas that are also of highest value to humans.

Therefore, while large ecological reserves and wilderness areas are valuable, small protected areas also can be important—a topic discussed at the end of this chapter and in Chapter 6.

The Design of Reserve Systems

A thorough treatment of the design of reserve systems is well beyond the scope of this book. However, since large ecological reserves are a fundamental component of a comprehensive plan for forest biodiversity conservation, we provide below a summary of some of the key themes in reserve design. Detailed treatments of reserve design are found in many books and papers, such as Noss and Cooperider (1994), Margules et al. (1995), Pigram and Sundell (1997), and Margules and Pressey (2000).

Principles for Reserve Design

The "general reserve design principles" derived from island biogeography theory provided much of the early focus on reserves in conservation biology (see Chapter 2). Problems with these design principles are considered in Chapter 2 and in Box 5.2. The remainder of this section will deal with better-developed approaches for reserve design.

The design of a perfect reserve system, if conservation science were the only factor influencing land

---

BOX 5.2.

Island Biogeography and Reserve Design

Almost all conservation biology texts summarize the six "general principles of reserve design" proposed by Diamond (1975) based on the theory of island biogeography (Macarthur and Wilson 1963, 1967). This is not surprising given the considerable effort dedicated to discussing and testing predictions from island biogeography over the past three decades (reviewed by Shafer 1990). However, Zimmerman and Bierregaard (1986) and Doak and Mills (1994) showed why island biogeography theory often has little practical value in reserve design. This is, in part, because reserves do not operate as isolates like oceanic islands (Saunders et al. 1991). Rather, attributes of, and processes in, the matrix can be highly influential (Gascon et al. 1999; see Chapter 2).

Much of island biogeography theory focuses on the numbers of species. However, in many reserve design cases, species diversity per se is not the important issue—the number of species may remain constant or even increase after fragmentation, but large changes in species composition can take place (Bennett 1990b; Hutchings 1991 in Gascon and Lovejoy 1998). Different species can exhibit conflicting (even diametrically opposite) responses to the same ecological process, and these cannot be taken into account with simple (and often misleading) measures such as species diversity. Additional concerns for reserve design should be the identity of taxa that comprise species assemblages in reserve systems (Murphy 1989), the viability of populations of threatened species in reserves (Grumbine 1990; Lamberson et al. 1994), and the need to improve representativeness or comprehensiveness for biodiversity surrogates captured within a set of protected areas (Scott et al. 1993; Mackey et al. 1988).

We believe that rather than apply generic and somewhat simplistic general design principles, it is far better to select and design reserves with specific objectives in mind.

allocation, would depend solely on ecological considerations such as

- Whether the aim is to conserve a particular species, conserve the maximum number of species possible, or protect complete or representative sets of taxa associated with a given vegetation community or forest type
- The viability of populations of a given organism in any reserve or set of reserves
- The threats to the integrity of reserves resulting from activities in the matrix

In most cases, the objectives of a reserve system are broader than the conservation of an individual taxon or assemblage, such as the representation of the full array of biodiversity of a region (Austin and Margules 1986). If this is the goal, three guiding principles—the CAR principles—underpin the design of a reserve system. This acronym stands for Comprehensive, Adequate, and Representative (Dickson et al. 1997). *Comprehensiveness* refers to the need for a reserve system to capture the complete array of biodiversity, ranging from species (and their associated genetic variation) to communities and ecosystems. *Adequacy* relates to the need for a reserve system to support populations of species that are viable in the long term. *Representativeness* means that a reserve system should sample the full range of species, forest types, communities, and ecosystems from throughout their geographic ranges (Pressey and Tully 1994; Anonymous 1995; Burgman and Lindenmayer 1998).

A fourth principle could be added to the CAR principles—*Replication*. This refers to the need for a reserve system to contain multiple protected areas of a given vegetation type, forest community, or species. This limits the risk that all reserved examples of an vegetation type, population, or community could be affected by a single catastrophic event, such as a wildfire (Lindenmayer and Possingham 1994). For this reason, replication is often vital (Pickett and Thompson 1978). The need for replication is influenced by

- The size of a single reserve, especially whether or not it exceeds the maximum size of a single disturbance event
- Whether a biodiversity surrogate (such as a forest type or an age cohort of that forest type) is sufficiently well represented in a given

reserve such that there is a high probability that some unaffected area of the surrogate will remain after a disturbance and, consequently, propagules or offspring can recolonize disturbed areas
- Whether there is high potential for more than one reserve to be damaged by the same catastrophic event

Replication may not always be possible because the targeted organism or entity may be so limited that all individuals are captured in one reserve.

Replication of reserve systems may have other values. Replication may better protect taxa that might otherwise be lost from a single reserve because of species turnover (or localized extinction) or distributional changes enforced by processes such as global warming (see below). Such temporal considerations in reserve design are valuable because although two candidate areas may appear to be similar at the time reserve selection tools are applied, they may subsequently follow different successional pathways and eventually have quite different conservation values. Hence, both merit consideration for establishment.

Reserve Design and Surrogate Schemes

The design of a CARR reserve system will often be based on surrogates for overall levels of biodiversity. These surrogates are attributes thought to represent the distribution and abundance of species and assemblages (Hunter 1994). They can include forest types or other plant communities (Brown and Hickey 1990; Specht et al. 1995), particular climatic parameters (e.g., rainfall), ecoregions (Sims et al. 1995), or climate domains (Mackey et al. 1988; Richards et al. 1990).

Surrogates are essential in reserve design because it is impossible to document comprehensively all biodiversity (Burgman and Lindenmayer 1998). A region or landscape can be classified and mapped according to biodiversity surrogates, providing a spatial dataset against which reserve selection procedures can be applied. Box 5.3 provides an example.

Reserve Selection and Historical Levels of Vegetation Cover

To fully meet the goals of a CARR reserve system, comparisons should be made between current levels of biodiversity or biodiversity surrogates and those at some time in the past. For example, if there are reservation targets of 10 percent for each biome (Scientific Advisory Group 1995), that should mean 10 percent of the original extent of those biomes. In the case of Australian forest reserve planning, this has entailed comparisons with the assumed pre-1750 distribution of particular forest types (Dickson et al. 1997; Austin et al. 2000) in an attempt to ensure that 10–15 percent of the original cover of each type is represented in the reserve system (Anonymous 1995). This period immediately preceded the arrival of Europeans in Australia, which was the beginning of widespread vegetation clearing and landscape change (Walker et al. 1993). A similar approach was used in New Zealand (Awimbo et al. 1996).

Reserve Selection

Although representation of the full array of biodiversity of a region is a traditional goal of a reserve system, existing protected area networks typically contain a biased sample of biodiversity (e.g., Pressey 1995; Norton 1999). Societies tend to reserve land for which there is no other current economic use—the "worthless land hypothesis" (Hall 1988) (see also the section below on the limitations of large ecological reserves). An excellent example is in Chile, where forests at high latitudes and altitudes are well represented while those at lower latitudes and altitudes are not (Armesto et al. 1998).

Methods such as reserve selection algorithms (Kirkpatrick 1983; Margules et al. 1995; Pressey 1997) and gap analysis (Scott et al. 1993; Noss and Cooperrider 1994) can be used to resolve problems associated with a lack of representativeness (e.g., Williams et al. 1996; Pressey 1997). Gap analysis integrates information on biodiversity, land tenure, and management regimes to identify biodiversity surrogates poorly represented in the existing reserve system. Burke (2000) presents a diverse set of recent examples of the application of gap analysis. The two approaches can also be used in combination (Pressey and Cowling 2001). Reserve selection procedures are typically implemented in a stepwise fashion (after Margules and Pressey 2000):

1. Compile data on biodiversity in the target region.
2. Identify conservation goals for the target region.
3. Review existing reserve systems in the target region.
4. Identify additional reserves (using reserve selection or gap analysis techniques).
5. Implement conservation actions.
6. Maintain (manage) the specified values of the selected reserves.

As part of the identification process, three principles govern most reserve selection protocols. These are *complementarity*, *flexibility*, and *irreplaceability* (Pressey et al. 1993).

Complementarity is the degree to which a given area contributes previously underrepresented features (e.g., species, land units, ecosystem types, or environmental climatic domains) to a reserve system. The principle involves identifying a set of areas in which all species or land units are represented (or represented a number of times). Heuristic algorithms or mathematical programming can be used to implement complementarity (Kirkpatrick 1983; Margules et al. 1988). These operate by identifying new reserves that are to be added to a set of protected areas until all species or units are represented.

It may also be possible to develop a representative network of reserves from a range of different combinations of areas. This is the principle of *flexibility*, which allows different areas to be substituted in a reserve system if they contribute the same previously underrepresented taxa or land units (Pressey et al. 1993).

The concept of *irreplaceability* has two meanings in reserve selection: (1) the degree to which an area is essential to the goal of a completely representative reserve system, and (2) the contribution a given area makes to representativeness (Pressey et al. 1993). Irreplaceability both addresses the potential conservation value of a given area and provides a means to explore different planning options.

The principles of complementarity, flexibility, and irreplaceability can be applied to any type of biodiversity unit—species, species assemblages, ecosystem types, land units, environmental classes, and the like (Margules et al. 1995). The result of a reserve selection analysis is a map of selected biodiversity surrogates, subsets of taxa, assemblages, ecosystems, environments, or combinations of these.

One example of how reserve design algorithms can help improve the representativeness of reserve systems is provided in Box 5.3. However, there are a number of problems associated with using reserve selection algorithms to improve representativeness.

First, extensive baseline data are often required to provide detailed coverage of surrogates (e.g., forest types or vegetation communities) from which to instigate reserve selection protocols. These data are unavailable for many areas, particularly developing nations (Norton 1999). Pressey and Cowling (2001) argue that this should not preclude the application of reserve design algorithms, although users need to be aware of data limitations.

Another problem is that the units used as a basis for reserve selection (e.g., forest types, plant communities or ecological vegetation units, climatic domains) may not be good surrogates for other forms of biodiversity (Burgman and Lindenmayer 1998). Would a set of areas set aside for a suite of forest birds also adequately sample the reptile taxa? Good correspondence was found by Kirkpatrick and Brown (1994) between areas selected on the basis of environmental domains and target plant communities in Tasmania, although many rare species missed selection. Dominant forest type was found to be a reasonable surrogate for bryophytes in New South Wales (Pharo and Beattie 2001). But in many cases, the ability of one measure to act as a surrogate for another remains unknown (Pressey 1994b; Wardell-Johnson and Horowitz 1996; Gustafsson et al. 1999; Lindenmayer et al. 2000b). Forest type has been used as a surrogate for reserve selection in Australia. However, some widespread dominant tree species occur across a wide range of environments and species assemblages vary between these different environments (Hunter 1991). Some rare plants are associated with particular soil or rock types rather than with overstory vegetation. A good example of this is the array of rare plant species endemic to serpentine soils (Lyons et al. 1974; see Chapter 6). Correspondence between faunal distribution and environmental classifications (such as forest types or land systems) is also unclear. In addition, some taxa such as invertebrates and small mammals are closely associated with understory plants (Gullan and Robinson 1980; Woinarski and Cullen 1984) rather than with dominant tree species.

Finally, reserve design algorithms often attempt to identify a minimum set of areas to achieve representativeness (see Box 5.3). This minimalist approach to reserve design (Crome 1994) can mean that protected areas are vulnerable to species loss (Virolainen et al.

BOX 5.3.

An Application of a Reserve Selection Algorithm

One application of a reserve selection procedure comes from the forests of southern New South Wales in southeastern Australia (Nicholls and Margules 1993; Margules et al. 1995). The area examined was approximately 3,800 square kilometers, and the study aimed to identify the minimum area required to sample 10 percent of the twenty-six environmental domains that occur in the area as well as 10 percent of the area occupied by each of thirty-one communities of forest trees. A grid of 1-square-kilometer pixels was overlaid on the study area and a series of steps was used to achieve the goal of sampling 10 percent of the area of the various environmental domains and forest communities. These were (after Margules et al. 1995):

Step 1. Identify a subset of grid cells that sampled 10 percent of the area of each of the thirty-six environmental domains.

Step 2. Determine the extent to which the thirty-six different forest communities had also been sampled by the procedure in Step 1.

Step 3. Add more grid cells until 10 percent of the area of each forest community had been sampled.

This work showed that thirty-four grid cells achieved the 10 percent sampling target after the completion of Step 1 and that the addition of a further three grid cells was required to sample the tree communities to the same level. The thirty-seven grid cells were equivalent to 8.7 percent of the entire study area. The flexibility of the approach used by Nicholls and Margules (1993) and Margules et al. (1995) was highlighted by the fact that some of the grid cells chosen could be substituted for others in the region and still achieve the same level of representation with a similar number of grid cells.

Margules et al. (1995) then compared the spatial distribution of the thirty-seven grid cells that had been selected with current land tenure. This showed that a large number of the grid cells (and thus particular tree species and environmental conditions) were not represented in the national park system as it existed at that time (Margules et al. 1995).

1999; Rodrigues et al. 2000)—the problem of reserve adequacy discussed in the nest section.

Reserve Adequacy

Using biodiversity surrogates as part of reserve selection procedures can facilitate the process of reaching a nominated target for reserve allocation (e.g., 10 percent of a given forest type). But reserve selection algorithms often fail to consider some of the issues associated with the adequacy criteria of the CARR principles. They do not, for example, reveal how much land needs to be reserved in total. They also typically ignore the potential for environmental changes (such as those forecast to occur with global warming—see below) and ecosystem and population dynamics (Cabeza and Moilanen in press). For example, reserve selection algorithms do not provide an indication of the viability of populations of species that initially occur in a reserve system (Witting and Loeschcke 1995; Burgman and Lindenmayer 1998). Extinction risk assessment tools such as population viability analysis are useful for exploring issues associated with the viability of populations in large ecological reserves (Armbruster and Lande 1993). Other approaches can be valuable, such as the optimization methods recently developed by Moilanen and Cabeza (in press), which optimize reserve selection by specifically incorporating the likely persistence of target species in a reserve network. Long-term empirical studies of the dynamics of populations of different sizes (Berger 1990) and the presence of species within reserves of different sizes (Newmark 1985, 1987; Gurd and Nudds 1999) also may provide an indication of the chances of persistence (and hence the viability) of populations residing within reserves.

The adequacy of a reserve system is also underpinned by such considerations as the size, number, and shape of individual protected areas, connectivity between protected areas (Burkey 1989), natural disturbance regimes within reserves (Pickett and Thompson 1978), and analyses of the impacts of threatening processes in the matrix and how they may threaten the integrity of reserves (Nelson 1991). How these factors actually influence adequacy will depend on the particular objectives of a reserve system and the biodiversity surrogates targeted for conservation. Generally, many adequacy issues can be addressed by

making reserves as large as possible (Soulé and Simberloff 1986; Hunter 1994). This is because

- Large reserves will generally (although not always) support more species than smaller ones. This view is supported by the species-area relationship elucidated by Preston (1962) and more recently reviewed by Connor and McCoy (1979) and Rosenzweig (1995).
- Large reserves will generally support larger (and thus less extinction-prone) populations of particular taxa (Armbruster and Lande 1993). However, population size will also be a function of other factors such as environmental productivity, habitat suitability, and the spatial arrangement of suitable habitat (Braithwaite 1984). Large populations are also likely to have greater genetic variability than small ones (Billington 1991) and are less likely to suffer extinction as a result of genetic stochasticity (Lacy 1987; Saccheri et al. 1998).
- Large reserves will have a better chance of incorporating natural disturbance regimes (Pickett and Thompson 1978; Baker 1992). This, in turn, has implications for the amount of replication that is needed across a system of reserves (see the CARR principles outlined above).
- Large reserves will contain a greater area of interior habitat buffered from negative edge effects (Janzen 1983) associated with the boundaries of reserves (e.g., weed invasions; Richardson et al. 1994).

In some cases, large reserves may *not* be superior to small ones. Cinnamon fungus (*Phytophthora cinnamomi*), which causes severe disease in plant communities, is now widespread through the extensive wilderness areas of southwestern Tasmania and is dispersed by hikers. In contrast, the risk of spreading the fungus can be more readily controlled within smaller protected areas that are targeted for careful proactive management (such as midspatial-scale reserves within wood production forests) (M. Brown personal communication).

Reserve adequacy involves much more than the total area or size of protected areas. Processes taking place in the matrix impinge on reserve integrity (Nelson 1991). For example, up to two-thirds of World Heritage sites suffer from threats originating outside their boundaries (World Resources Institute 1992). Disturbance regimes do not recognize reserve-matrix boundaries. For example, pollutants and pests flow from the matrix into reserves via river systems and wetlands, and surrounding incompatible land uses affect species inside reserves (Calhoun 1999). Determining the extent to which threatening processes originating in the surrounding matrix can be controlled or mitigated can be a good way to assess whether or not a reserve is adequate (Janzen 1983).

Connectivity between protected areas will also strongly influence reserve adequacy (Norton 1999). Matrix-based strategies, such as the provision of corridors and stepping stones and the retention of original cover throughout harvested areas, may contribute to the dispersal of organisms between reserves.

FIGURE 5.1. Active management is sometimes required to maintain or restore important ecological values or conserve particular taxa. Prescribed fire in Wilson's Promontory National Park in the Australian state of Victoria is used to promote habitat suitability for the rare bird the ground parrot (*Pezoporus wallicus*). Photo by D. Lindenmayer.

Management within Reserves

Simply setting aside a set of reserves will not necessarily ensure the conservation of biodiversity. Management in reserves is often needed to maintain habitat suitability for some species (see Box 5.4; Figure 5.1). Some reserves may need to be actively managed to maintain ecological processes or limit the impacts of a threatening process (Barker et al. 1996). For example, the active exclusion of fire from reserves can be imperative. The endemic Tasmanian conifer, King Billy pine (*Arthrotaxis selaginoides*), is highly sensitive to fire, and living trees can exceed 1,000 years of age. The species is fully protected but nevertheless remains susceptible to fire in reserved areas (Brown 1988) (Figure 5.2).

Active incorporation of fire may be essential in other reserves. For example, the complete suppression of fires within reserves may lead to large changes in vegetation structure and plant species composition with negative impacts on biodiversity (Zackrisson 1977; Shinneman and Baker 1997; Miller and Urban 2000). In some cases, there may have to be restoration of past burning regimes (Burrows et al. 1989; Fulé and Covington 1999) (Figure 5.1). For example, Parks Canada has reintroduced fires in the form of prescribed burning to some national parks as has the National Park Service in many parks in the United States.

FIGURE 5.2. Reserving species in strictly protected areas may not mean they are conserved. A classic example is the highly fire-sensitive tree species King Billy pine, in Tasmania, Australia. The photo shows extensive fire-damaged stands of the species. A fire occurred in 1914, and there has been minimal regeneration of the pine since then. Photo by J. Hickey.

Active management in the Kruger National Park in South Africa involved the installation of permanent water bodies within the park to prevent animal die-off during drought years. As dry-season habitats for large ungulates occur in highland areas outside the boundaries of the national park, the animals needed access to nearby water.

Management Within and Outside Reserves

The highly endangered western swamp tortoise (Pseudomydura umbrina) near Perth in Western Australia is a useful example of both the need to manage reserves and the extrinsic processes in the matrix outside reserves.

The western swamp tortoise is regarded as Australia's most endangered reptile (Cogger 1995), with a distribution covering only about 100–150 square kilometers. Much of this has been extensively modified for other uses such as agriculture and urban development (Kuchling et al. 1992). Two swamps, one covering 65 hectares and the other 155 hectares, were set aside as reserves for the western swamp tortoise. The species disappeared from the larger reserve, and by the late 1980s only a few dozen animals remained in the smaller swamp (Kuchling et al. 1992). In addition to the restricted range of the species, four factors were thought to have contributed to its decline (after Kuchling et al. 1992):

1. Predation by the introduced red fox (*Vulpes vulpes*).
2. Specialized requirements that made the species dependent on habitat that was itself rare.
3. Reliance on reserves that supported marginally suitable habitat susceptible to periodic desiccation.
4. Low rates of fecundity.

This example highlights the fact that simply setting aside a reserve and failing to address the need for management action both within the reserve and in the neighboring matrix will not always guarantee the long-term survival of a species (Burgman and Lindenmayer 1998).

Animal populations within reserves sometimes have to be controlled (e.g., Wilson et al. 1992) (Box 5.4). Examples include feral predators (Kinnear et al. 1988), overabundant native animals, such as kangaroos in Australian reserves (Shepherd and Caughley 1987), and deer in small North America forest reserves (Johnson et al. 1995).

Although active human management of reserves is often needed, it has to be balanced against the importance of maintaining all or portions of reserves as free as possible from human activities. This is because the full impacts of human actions are not known and some may have unexpected negative effects on biodiversity. For example, when restoring past fire regimes, only part of a reserve or a subset of protected areas might be targeted for deliberate burning.

Reserve Selection and Design in the Real World

Large ecological reserves are fundamental to a comprehensive conservation plan, but as demonstrated above, the selection and design of reserves is a demanding task. There are many factors to consider—chief among them are the objective(s) for setting aside a reserve. Is the reserve intended to capture a representative sample of all biota or a subset of biota, to conserve a particular threatened species, or to protect a scenic feature? Issues such as representativeness, adequacy, replication, reserve size, and reserve shape follow logically from the objectives of setting aside a reserve. Other considerations include the type and size of natural disturbance regimes, conditions in the surrounding matrix, and the need for management within reserves.

The CARR principles can provide a useful theoretical framework for reserve design that addresses some (but not all) of these issues. When selecting a reserve, planning tools such as reserve selection algorithms are useful, but they are not panaceas because

- They are data intensive and require information that is almost never available in the majority of jurisdictions.
- Data limitations require the use of biodiversity surrogates, which is problematic.
- They typically do not address reserve adequacy.
- They typically do not address connectivity between reserves.
- The individual circumstances of a particular reserve selection case will dictate the most appropriate algorithm to use (Pressey et al. 1997).

Only some of these problems are relevant to gap analysis, because the start of the procedure is the existing reserve system. However, where the aim is to minimize the amount of land captured in protected areas the outcome of gap analysis will still have significant limitations (Pressey 1994a). Nevertheless, the use of gap analysis and reserve selection algorithms is far better than an ad hoc approach to reserve design (Margules and Pressey 2000; Pressey and Cowling 2001).

Despite the potential value of reserve selection methods, their use has often been hypothetical (Ehrlich 1997; Kremen et al. 1999; Prendergast et al. 1999). One exception is the work by Brown and Hickey (1990), who used forest type, geology, elevation, and broad climatic regions to identify gaps in the coverage of forest reserves of Tasmania and then to identify areas called "Recommended Areas for Protection" (M. Brown personal communication). The CARR criteria for reserve design also have been applied (in part) to the forest estate in Australia during the Regional Forest Agreement process (e.g., Pressey et al. 1996).

One major problem in reserve selection is that no two parts of a landscape or region are the same (they differ in climate, landscape context, and a host of other factors)—the environmental continuum concept of Austin (1999) and others. Competing human demands and the extent of past impacts will almost always be different among parcels of land. Hence, in the real world there is never an array of identical candidate reserves from which to select. Prendergast et al. (1999) acknowledged this for the United Kingdom and noted the impossibility of picking reserves from a set of candidate areas. Even more important is that political, economic, and social factors almost always take precedence over ecological goals and requirements when land is considered for reserve allocation (Hunter 1994; Pressey et al. 1996; Lawton 1997; Margules and Pressey 2000). The tension between economic, social, and ecological objectives makes reserve selection highly idiosyncratic. What is appropriate in one jurisdiction will be entirely unsuitable in others. For example, in the United Kingdom, reserve design is influenced by the fact that there are many different types of reserves that are acquired by many different institutions, including government and voluntary bodies (Prendergast et al. 1999). In contrast, for many less-developed nations, simply setting aside reserves without considering the interests, behavior, and attitudes of the local human population typically leads to conflicts and ultimately degradation within a reserve

(e.g., Western and Gichohi 1993; Smith et al. 1997). The reality of successful reserve design in many parts of the world is neatly revealed in a recent paper on the Masoala National Park in Madagascar, in which the design was based on a unique blend of ecological and socioeconomic criteria including strong local community support (Kremen et al. 1999). Tellingly, the final section of that paper also acknowledged the support of the U.S. ambassador and the president of Madagascar. Such idiosyncrasies are perhaps best summed up by Prendergast et al. (1999), who noted that "no single procedure for identifying areas of conservation interest is likely to be universally appropriate."

The limitations of a reserve-only approach and the importance of matrix management for the conservation of forest biodiversity are highlighted by the largely theoretical approaches to reserve selection and the complexities of reserve design in the real world.

Limitations of a Reserve-Only Conservation Strategy for Biodiversity Conservation

Reserves are an essential part of any comprehensive regional strategy for forest biodiversity conservation. Conversely, a plan that ignores the matrix and is based exclusively on reserves will fail because of the inherent limitations of a reserve-only approach. These include

- Limited area available to allocate to reserves
- Limited size of most reserves
- Lack of representativeness of reserves
- Social and economic impediments to expansion and management of reserve systems
- Difficulties in capturing highly mobile taxa, such as migratory or nomadic species
- Difficulties in capturing taxa with fine-scale or patchy distribution patterns
- Instability of abiotic and biotic conditions within reserves
- Potential impacts from intensification of matrix exploitation once reserve systems are established, including reduced matrix contribution to conservation and potentially degraded productivity on matrix lands

Limited Area Available for Reserves

The limited area available for forest reserves is a significant limitation on a reserve-only focus for biodiversity conservation. The limitation relates to both the absolute amount of suitable area available and the socially acceptable area. Recent statistics indicate that approximately 38×10^6 square kilometers, or about 8 percent of the world's forest, is protected in IUCN category 1b reserves (i.e., strictly protected areas) (Commonwealth of Australia 1999) (Figure 5.3).

Reserves are not likely to exceed 10–15 percent of the land base in the vast majority of forest regions around the world, even in wealthy countries such as the United States, Canada, and Australia. Usually, it will be significantly less. For example, the total area of strictly protected forests in twenty-six countries in Europe is about 3 million hectares, or roughly 1.7 percent of the total forest area (Parviainen et al. 2000). The figures at national levels in Europe are similarly small. In Finland in 1994, only 0.7 percent of the southern and mid-boreal coniferous forest was in dedicated nature reserves (Virkkala et al. 1994). Currently, reserves in the entire forest estate in Finland total approximately 3.5 percent, similar to levels in Sweden (National Board of Forestry 1996a; Hazell and Gustafsson 1999).

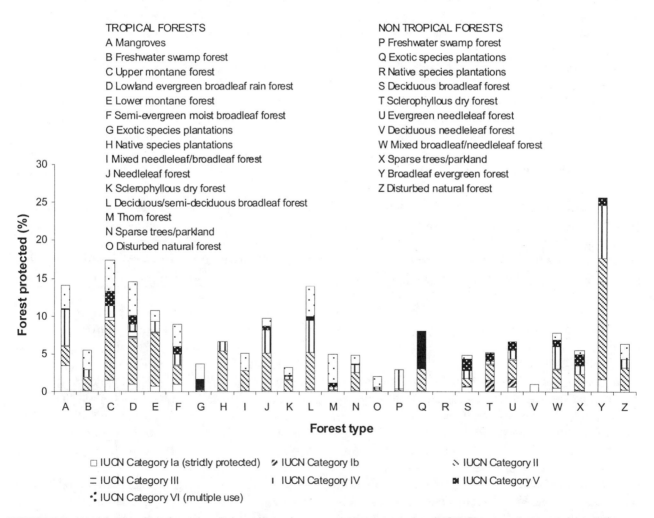

TROPICAL FORESTS
A Mangroves
B Freshwater swamp forest
C Upper montane forest
D Lowland evergreen broadleaf rain forest
E Lower montane forest
F Semi-evergreen moist broadleaf forest
G Exotic species plantations
H Native species plantations
I Mixed needleleaf/broadleaf forest
J Needleleaf forest
K Sclerophyllous dry forest
L Deciduous/semi-deciduous broadleaf forest
M Thorn forest
N Sparse trees/parkland
O Disturbed natural forest

NON TROPICAL FORESTS
P Freshwater swamp forest
Q Exotic species plantations
R Native species plantations
S Deciduous broadleaf forest
T Sclerophyllous dry forest
U Evergreen needleleaf forest
V Deciduous needleleaf forest
W Mixed broadleaf/needleleaf forest
X Sparse trees/parkland
Y Broadleaf evergreen forest
Z Disturbed natural forest

☐ IUCN Category Ia (strictly protected)   ⤳ IUCN Category Ib   ⤫ IUCN Category II
═ IUCN Category III   ı IUCN Category IV   ⊠ IUCN Category V
⦂ IUCN Category VI (multiple use)

FIGURE 5.3. Worldwide, most forest types have limited representation in reserves. IUCN Category 1a areas are strictly protected reserves, and the data clearly show that global levels of protection of most broad forest types are minimal (less than 5 percent). Data from Commonwealth of Australia 1999 and World Conservation Monitoring Centre.

Even in the rare cases where reserve targets of 40 percent are achieved for a particular region or forest type, the majority of forest land and its associated biodiversity will still be in the matrix. Where reservation levels are low, as in Scandinavia, matrix management is even more crucial for the conservation of biodiversity.

The total area of reserved land is relatively meaningless unless representativeness and comprehensiveness are considered (Armesto et al. 1998). The IUCN has set reservation targets of 10 percent for each biome on a global basis (Scientific Advisory Group 1995). Reservation levels at or below such percentages are likely to be typical, except for some high-profile forest types with limited distribution, such as the giant sequoia (*Sequoiadendron giganteum*) forests of the Sierra Nevada in California and the King Billy pine forests of Tasmania (Brown 1988). Approximately 30 percent of the forest land in the Douglas-fir (*Pseudotsuga menziesii*) region of the United States has been reserved, including most remaining late-successional forest.

The kind of targets adopted for reserved areas, such as 10 percent of the land base, are arbitrary and were not developed using any scientific criteria (Scientific Advisory Group 1995; Soulé and Sanjayan 1998). In fact, such levels are generally viewed as totally inadequate (Rodrigues and Gaston 2001), and perhaps as much as 50 percent of a region and even 100 percent of an ecosystem may need to be reserved in some cases (Noss and Cooperrider 1994). No single minimum reserve target will be appropriate because, for example, areas, ecosystems, or nations with higher levels of endemism or species richness may require much larger proportions of the land base to be protected (Rodrigues and Gaston 2001).

Increased reserve levels (such as 20–30 percent) will still be insufficient for some individual species. For example, approximately 20 percent of the montane ash forests of the Central Highlands of Victoria has been set aside as reserves (Commonwealth of Australia and Department of Natural Resources and Environment 1997). Despite this level of protection, there is a threat that recurrent disturbances such as wildfire will extinguish the endangered Leadbeater's possum from large ecological reserves (Lindenmayer and Possingham 1995a). Wide-ranging carnivores and ungulates are particular challenges in terms of the necessary size of reserves (Noss and Cooperridder 1994).

In summary, due to limitations in the area available for the reserve system, the majority of forest biodiversity will either exist in the matrix outside protected areas—or may cease to exist (Franklin 1993a; Lindenmayer and Franklin 2000).

The Limited Size of Any Given Reserve

The limited area that can be committed to reserves also limits reserve size in many cases. While small and medium-sized reserves still have significant conservation value (Zuidema et al. 1996), they may not support viable populations of some species in the long term (Wilcove 1989; Armbruster and Lande 1993). Even small species, such as some invertebrates, may be viable only in large numbers (Thomas 1990; Tscharntke 1992) or may need large areas for long-term persistence (Økland 1996; Didham 1997). This problem may not only be a function of the size of a given reserve size per se (Newmark 1985; Clark and Zaunbrecher 1987), but also the predominance of low-productivity environments in reserves (see below).

Small reserves are likely to be too small to fully incorporate natural disturbance regimes, such as wildfires. Many cannot incorporate disturbance initiation areas as well as the surrounding places where disturbances would naturally spread (Baker 1992). Therefore, natural landscape patterns and levels of biological legacies (see Chapter 4) will not be captured within such reserves.

Long-term persistence of some species in small reserves depends on supplementation from populations in adjacent matrix lands. Metapopulations of butterflies in the United Kingdom provide an example (Thomas et al. 1992b). This also applied to the northern spotted owl in northwestern United States prior to adoption of current forest plans (Yaffee 1994; see Chapter 11). In such cases, protecting isolated reserves is not sufficient to maintain resident populations (Baillie et al. 2000).

Small reserves are a problem for mobile species that require resources which vary in their temporal and spatial availability (reviewed by Law and Dickman 1998) and occur both within and outside reserves (Redford and da Fonseca 1986; Woinarski et al. 1992;

Mac Nally and Horrocks 2000). Examples are provided by a number of Australian forest birds (Woinarski and Tidemann 1991), such as pigeons and honeyeaters (Keast 1968; Date et al. 1996), as well as fruit bats (Palmer and Woinarski 1999). The swift parrot (*Lathamus discolor*) in the Australian state of Tasmania is another classic example; only 2 percent of the nesting habitat of the species occurs in dedicated conservation reserves with the rest on private and publicly owned production forests (Brereton 1997). Such problems are related to difficulties in reserving highly mobile taxa and not a lack of representativeness of the reserve system. The long-term conservation of the swift parrot depends almost entirely on management actions outside reserves (Brown and Hickey 1990; Swift Parrot Recovery Team 2000).

Various strategies have been proposed to deal with current limitations on potential reserves sizes, including the restoration of large subregions and the use of buffers around core areas (Noss and Harris 1986), such as the biosphere reserve approach (Noss and Cooperrider 1994). However, the potential for such approaches appears limited, particularly by social and economic factors.

Our view is that the area limitations for reserves require integrated conservation strategies across reserves and matrix lands, including production environments, to assist movement between temporally suitable habitats and protect roosting and foraging sites for highly mobile taxa.

Lack of Representativeness

Existing reserve systems are almost always an unrepresentative sample of the range of environments, biomes, biogeographic provinces, plant communities, forest types, and rare taxa that occur in a region (e.g., Crumpacker et al. 1988; Johnson 1992; Khan et al. 1997) (Table 5.1; Figures 5.4 and 5.5). In the vast majority of countries, most of the productive and often biologically rich forest lands have been long since occupied and converted to agricultural and urban environments as well as to managed forests. Consequently, land captured within reserves is typically characterized by one or more of the following: steep terrain, high elevation, high latitude, or low productivity.

Reserve systems in North America, Europe, Australia, New Zealand, and Asia are typically in areas of low productivity (Braithwaite et al. 1993; Hunter and Yonzon 1993; Pressey 1995; Norton 1999) (Figure 5.4). In northwestern North America, highly productive sites were occupied by private landowners more than a century ago, and so they are poorly represented in forest reserves. Consequently, protected areas are

TABLE 5.1.

Lack of representativeness of biogeographic provinces in reserves in Australia (after Johnson 1992; Burgman and Lindenmayer 1998).

BIOGEOGRAPHIC PROVINCE	AREA (KM^2)	% PROTECTED
Queensland coastal	300,000	27.2
Tasmanian	68,000	20.5
Southern mulga/saltbush	837,000	7.0
Southern sclerophyll	246,700	6.9
Western sclerophyll	410,800	6.5
Eastern sclerophyll	643,000	5.6
Northern savannah	580,900	4.5
Northern coastal	350,400	3.7
Western mulga	778,100	2.9
Eastern grassland/savannah	527,800	1.9
Brigalow	231,600	1.7
Northern grasslands	96,700	0.7

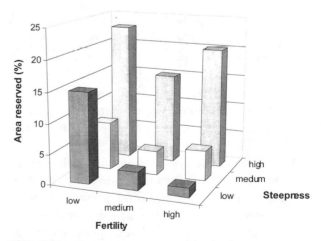

FIGURE 5.4. Reserve systems worldwide do not contain a representative samples of species, communities, or ecosystems. This is clearly evident from the bias of forest reserves to steep, infertile land in northern New South Wales (Australia). Redrawn from Pressey 1995. Reproduced from Web search (www.control.com.au/search).

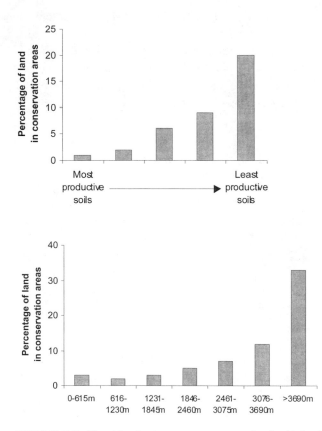

FIGURE 5.5. The bias in the reserve system in the United States is typical of reserve systems around the world in being dominated by low-productivity and high-elevation areas. Redrawn from Scott et al. 2001a,b.

dominated by steep, high-elevation lands with low vertebrate diversity (Harris 1984; Scott 1999) (Figure 5.5). Similar circumstances exist in Scandinavia (Virkkala et al. 1994). In Chile, more than 90 percent of the protected land is concentrated at latitudes above 43 degrees and areas of highest biodiversity have virtually no representation in the reserve system (Armesto et al. 1998).

The bias toward low-productivity environments has significant implications for biodiversity conservation. First, there is theoretical and empirical evidence that suggests that species diversity of plants and animals is often (although not always) positively associated with productivity (Harris 1984; Srivastava and Lawton 1998). Second, populations of particular species may be limited in low-productivity areas, such as those on poor soils (Braithwaite 1984), in steep terrain (Lindenmayer et al. 1991a), or at high elevations.

For example, bioenergetics research illustrates the high costs of animal movement on steep terrain (Taylor et al. 1972) and several studies have demonstrated than populations of forest-dependent animals are lower in such areas (e.g., Lindenmayer et al. 1997c, 1999b).

Lack of representativeness has occurred because reserves have typically been set aside for reasons other than nature conservation (Pressey 1994a; Khan et al. 1997) such as (1) low value for commodity production or human settlement (Chindarsi 1997), (2) high value for recreation (Pouliquen-Young 1997; Sax 1980), or (3) high scenic and aesthetic values (Recher 1996).

As a result, matrix (unreserved) lands can sometimes have higher current value for particular elements of biodiversity than formally protected large ecological reserves (Armesto et al. 1998).

Social and Economic Impediments to Expanding Reserve Systems

Although many reserves are unrepresentative and too small to support viable populations of some species (East 1981; Grumbine 1990), there are limited opportunities to improve their size and representativeness. Many poorly represented forest types and other plant communities are often in private ownership and are valuable for other land uses—making them difficult and/or expensive to add to the reserve system (Scott et al. 2001b). For example, in the Great Lakes region of Wisconsin, in the United States, simply absorbing remaining old-growth forest stands within a large ecological reserve was not a feasible management option (Mladenoff et al. 1994). In the monsoon rainforests in northern Australia, it was politically untenable to reserve and manage all representative samples (Price et al. 1995). A combination of strategies was recommended that included both monsoon rainforest reserves and matrix-based management in savannah landscapes surrounding the rainforest patches.

Difficulties in Capturing Taxa with Fine-Scale or Patchy Distribution Patterns

Even extensive reserve systems may fail to adequately capture species-rich assemblages with fine-scale or restricted distributions. This is, in part, related to the *sampling effect* (Connor and McCoy 1979); that is, reserves "sample" a subset of a landscape or region. Hence, reserves may fail to capture some species or a key environment for those species. Localized biodiversity hotspots that support concentrations of species or are critical for species persistence (e.g., overwintering or calving grounds; Hansen and Rotella 1999) may be left out. Conservation of velvet worms (Phylum Onychophora) in southern Australia illustrates this problem. Onychophorans are a species-rich group with substantial localized variability in genetic diversity (Rowell et al. 1995; Tait et al. 1995). Numerous "cryptic species" occur (P. Sannucks personal communication). Many species have patchy and often highly limited distributions. Some species of onychophorans have poor dispersal ability and depend on rotting log habitats in wet forests, making them susceptible to such disturbances as prescribed burning and woodchipping (New 1995). Setting aside reserves for these and many other species-rich assemblages with restricted distributions is problematic as there are invariably taxa confined to matrix lands. Matrix-based conservation strategies are essential in these cases, such as small reserves within the matrix (see Chapter 6). For onychophorans, stand-level strategies, such as retaining logs within clumps of intact vegetation cover and using low-intensity prescribed fires for site preparation after logging (Forest Practices Board 1998), can have considerable conservation value (see Chapter 8). Examples for other species exist as well. For instance, in East Africa, protected areas set aside for large mammals do not adequately conserve small mammals, making matrix lands valuable for the protection of these species (Caro 2001).

Instability of Conditions

The persistence of populations within reserves can be dependent on the maintenance of populations in the surrounding matrix. Natural ecosystems are dynamic with concomitant changes in species composition (e.g., Woinarski et al. 1992; Margules et al. 1994a) and changes in species distribution patterns. Therefore, species will sometimes disappear from a reserve (Witting and Loeschcke 1995; Rodrigues et al. 2000).

Most reserves are subject to recurrent disturbances such as wildfires—hence, the proposition that reserves should exceed the maximum size of disturbed areas (Pickett and Thompson 1978; Hobbs and Huenneke

1992). Larger areas have a greater probability of supporting areas that escape the effects of a single catastrophic fire event (Seagle and Shugart 1985). Relatively large-scale catastrophic disturbances in small reserves may limit their ability to support viable populations of some taxa (McCarthy and Lindenmayer 1999a; Norton 1999). In such circumstances, the maintenance of suitable habitat in the matrix is important to maximize the prospects of population persistence.

Margules et al. (1994a) studied plant communities on natural pavements of limestone in northern England and discovered that the set of areas needed to conserve all taxa in 1985 no longer supported them in 1994, just nine years later (Table 5.2). If the respective limestone pavements were treated as reserves and no other populations of those organisms existed, the network of reserves identified in 1985 would not have conserved many species of plants (Burgman and Lindenmayer 1998).

Studies of birds (Rodrigues et al. 2000) and plants (Virolainen et al. 1999) have produced similar results to those of Margules et al. (1994a). Rodrigues et al. (2000) suggested selecting the best areas for a particular species to overcome these problems and focusing on rare species. However, selecting the best areas for

---

TABLE 5.2.

Plant species turnover on limestone pavements in Yorkshire (United Kingdom) over a nine-year period (modified from Margules et al. 1994a).

CHANGE IN THE NUMBER OF SPECIES PER PAVEMENT	NUMBER OF PAVEMENTS
No Change	7
Net Loss of Species	
1–5	21
6–10	10
11–15	10
>15	5
Net Gain of Species	
1–5	10
6–10	9
11–15	2
>15	0

reserves is usually not possible and, moreover, this approach still ignores the landscape context of reserves and, hence, the contribution of the matrix to biodiversity conservation.

Anticipated changes in species ranges associated with global climate change (Peters and Lovejoy 1992; Parmesan 1996) also have major consequences for species currently occurring in reserves—they may disappear from reserves in the relatively near future (Peters and Darling 1985). Climate strongly affects broad-scale distribution patterns of plants and animals (Nix 1978; Woodward 1987) through its fundamental influence on environmental resources such as the availability of water and heat regimes (Mackey 1994). Given this, rapid changes in climate associated with global warming will likely result in large shifts in the distributions of many species (Parmesan 1996; McCarty 2001); a temperature increase of 3°C may produce a habitat shift of 250 kilometers in latitude or 500 meters in elevation (Macarthur 1972).

Some of the potential changes stimulated by global climate change include

- Contractions or expansions in the range of a species
- Altered breeding times (e.g., flowering patterns, bird migration behavior, frog spawning; reviewed by McCarty 2001)
- Altered growth patterns (e.g., growing seasons for plants) and changed palatability of

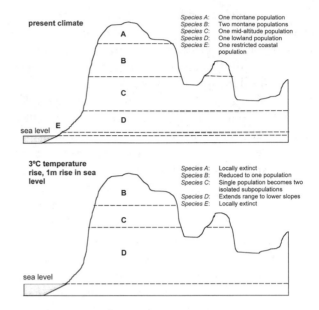

FIGURE 5.6. Conditions in reserves do not remain constant but can change in response to altered environmental conditions such as climate change. Species presently captured within a reserve may not persist there if climatic conditions change. This is illustrated by hypothetical changes in species distribution patterns in response to global warming. Redrawn from Mansergh and Bennett 1989.

plant-based food for herbivorous animals (Kanowski 2001)
- Fragmentation of a continuous range into a patchy and disjunct distribution
- Complete extirpation of a species (Mansergh and Bennett 1989) (Figure 5.6)

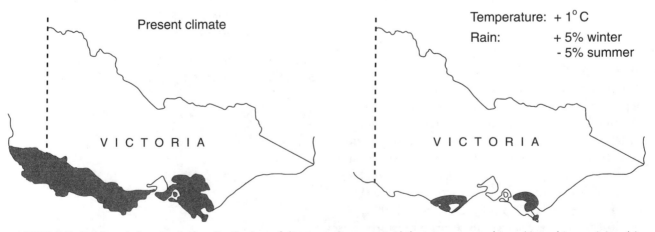

FIGURE 5.7. Predicted changes in the distribution of the Australian mammal the swamp antechinus (*Antechinus minimus*) in response to one possible climate change scenario under global warming (redrawn from Brereton et al. 1995). These results highlight the importance of maintaining populations of species both within and outside reserves in case major changes in climate take place as has been predicted as a result of global warming. Courtesy of Victorian Department of Natural Resources and Environment.

Altered distribution patterns have significant implications for the conservation of biodiversity (Peters and Darling 1985). Most reserves will be too small, lack sufficient variation in terrain and elevation, or both, to accommodate widespread within-reserve range shifts (Lindenmayer et al. 1991c; Brereton et al. 1995). If conditions in the surrounding matrix are unsuitable or preclude species from dispersing between reserves, then taxa will be unable to migrate to areas supporting suitable bioclimatic conditions and suffer severe declines or extinction (Westman 1990) (Figure 5.7). This prospect underscores the importance of the overarching principle of Chapter 3—maintaining critical habitat across multiple spatial scales.

Intensification of Exploitation in the Matrix

An expanded system of protected areas may shift additional harvesting pressures onto remaining matrix forests (Davie 1997). If levels of timber harvest are not modified to reflect reductions in available land, the benefits of expanded reserve systems may be largely offset by increased production pressures (intensive harvesting) in the matrix. There are signs of this problem in Australian forests with a policy of harvesting "intensification" in matrix lands following the establishment of new reserves (Bauhaus 1999) (see Box 5.5). Intensive, recurrent mechanical thinning operations are now being added to clearcutting in some Australian forests (Churton et al. 1996). Intensification has the potential to threaten both biodiversity and ecosystem productivity in matrix lands. For the Regional Forest Agreements in Australia, it may have been better to improve matrix management practices in the larger area of wood production forest than to expand the reserve system (see Box 5.5).

Downgrading the Conservation Value of Matrix Lands

Discounting the gains made by appropriate matrix management may mean the downgrading of the conservation value of matrix lands in some quarters. Establishing large ecological reserves can distract attention from comprehensive multiscaled plans needed to ensure the long-term conservation of species, communities, or ecosystems (Schwartz 1999). Focusing solely on large ecological reserves can be counterproductive when small habitat areas within the matrix also may make a valuable contribution to conservation (Franklin et al. 1997; Hale and Lamb 1997; Mac Nally and Horrocks 2000). In South American forests, Armesto et al. (1998) argued that priority should be given to managing forest remnants in intensively managed landscapes—remnants that currently have high levels of unprotected biodiversity. Many other

---

BOX 5.5.

Intensification of Management in the Matrix—the Land Allocation Problem in Tasmania

The situation in the Australian state of Tasmania highlights the problems of focusing solely on reserve systems and neglecting the importance of the matrix for biodiversity conservation. Here, the conservation movement has focused extensively on increasing the size of the reserve system—particularly those forests with high "wilderness value." Under the recently signed Regional Forest Agreement for Tasmania, which locks in decision making for the forest estate, the size of the reserve system increased by approximately 400,000 hectares (now encompassing 40 percent of the forest in the state). However, the trade-off was that up to 85,000 hectares of native eucalypt forest will be cleared for exotic plantations of radiata pine and fast-growing eucalypts from mainland Australia. These will be managed on a 15- to 30-year rotation (Forestry Tasmania 1999). The percentage of the state's forests managed for plantations will increase from 2.8 percent to 5.1 percent in the next ten years—up to 20 percent of a given biogeographic province can be converted to plantations under the regional forest agreement. Although the reserve system is now more representative than it was previously, the areas targeted for plantation establishment are the highly productive parts of the northwest and northeast of Tasmania—areas already heavily fragmented by agriculture and existing plantations.

The focus on expanding the reserve system in Tasmania while neglecting the importance of the matrix will have a negative effect on biodiversity values. This is of considerable concern given that there are 114 forest-associated threatened species in Tasmania. A far better outcome for biodiversity conservation arising from the Regional Forest Agreement process would have been to improve management in the matrix and forgo increases in the reserve system.

examples highlight the importance of small areas of habitat in matrix lands (Kirkpatrick and Gilfedder 1995; Prober and Thiele 1995; Angelstam and Petersson 1997; Fries et al. 1998; Semlitsch and Bodie 1998; Law et al. 1999; Lindenmayer et al. 1999b, McCoy and Mushinsky 1999; Abensperg-Traun and Smith 2000). The habitat remnants in all these examples are not candidates for inclusion in large ecological reserves. They are too small, spatially dispersed, and embedded in production landscapes. Tenure constraints also limit their potential to become reserves.

If the conservation values of matrix lands are not considered, the wrong message can be sent to owners of unreserved land. Barrett et al. (1994) and McIntyre and Barrett (1992) recognized this in conserving bird populations in northern New South Wales, Australia. They recommended influencing management on woodland remnants and other small habitat patches within the agricultural matrix rather than securing a few large ecological reserves. Focusing only on large ecological reserves would have concentrated efforts on approximately five species that depend on large and relatively undisturbed areas of forest. In contrast, conserving small protected areas in the matrix helped protect a larger number of species and strongly signaled to private landowners that biodiversity conservation could be enhanced without evicting them from their land and returning it to a natural state (Barrett et al. 1994).

Many species can and will persist in forests managed for timber and pulpwood production with improved matrix management (e.g., Hunter 1990; Lunney 1991; Gustafsson and Ahlen 1996; Recher 1996). A comprehensive review of timber harvesting in tropical zones highlighted the conservation values of harvested tropical forests (Putz et al. 2000). Other workers have called for reassessment of these managed areas for biodiversity (Cannon et al. 1998). New criteria for ecological sustainability and biodiversity conservation in temperate and tropical production forests acknowledge the importance of matrix lands (e.g.,

Convention for Sustainable Development Intergovernmental Panel on Forests, Helsinki Process, Montreal Process; see Arborvitae 1995). Conservation values of the matrix are further highlighted by codes of forest practice throughout the world that promote the conservation of biodiversity and wood production as joint objectives (e.g., British Columbia Ministry of Forests 1995; Department of Natural Resources and Environment 1996; Fenger 1996; Asia-Pacific Forestry Commission 1997).

Matrix Management Versus Reserves

We emphasize again that, despite the limitations of reserves, we are not arguing to substitute matrix management for reserves. Rather, we stress the potential contributions of the matrix for biodiversity conservation and hence the need for both reserves and matrix management as part of any comprehensive conservation strategy to maintain habitat across multiple spatial scales. This is consistent with principles outlined in Chapter 3, where we highlight the importance of simultaneously implementing a range of conservation strategies to spread risk in conservation management and meet the demands of a more diverse array of forest-dependent taxa.

When matrix management is properly implemented, the total land base needed for strictly protected ecological reserves may be less because

- There will be fewer species that can be conserved only in reserves.
- The overall amount of land where biodiversity conservation is considered will be increased.
- Biodiversity conservation occurs across a broader and more representative set of environments.
- Threatening processes emanating from the matrix will be reduced.

CHAPTER 6

Landscape-Level Considerations within the Matrix: Protected Habitat at the Patch Level

. . . the patchy structure of landscapes is important to ecological functioning at a variety of levels of biological organization, and is itself worthy of conservation and management attention.

—BAKER (1992)

A comprehensive approach to the conservation of forest biodiversity incorporates a continuum of conservation approaches from the establishment of large ecological reserves through an array of conservation measures within the matrix, including the maintenance of individual forest structures within the managed stand. In this chapter, we provide a checklist of elements that should be considered as part of a multiscaled approach to forest biodiversity conservation, elements that are consistent with the general principles outlined in Chapter 3. The checklist contains four broad categories of approaches to landscape-level matrix management: (1) protection of aquatic ecosystems and networks, specialized habitats, biological hotspots, and remnants of late-successional or old-growth forest found within the matrix; (2) establishment of landscape-level goals for retention, maintenance, or restoration of particular habitats or structures as well as limits or thresholds for specific problematic conditions; (3) design and subsequent management of transportation systems (generally a road network) to take account of impacts on species, critical habitats, and ecological processes; and (4) selection of the spatial and temporal pattern for harvest or other management units. The first of these—the importance of midspatial-scale protected areas within the matrix—is the focus of this chapter. The remainder are addressed in Chapter 7.

Midspatial-scale protected areas within the matrix are often critical habitats for biodiversity and fundamental to the maintenance of ecosystem processes. They can include (1) aquatic ecosystems, such as streams, lakes, and ponds; (2) wildlife corridors; (3) specialized habitats, such as cliffs, caves, thermal habitats, meadows, and vernal pools; (4) biological hotspots or places of intense biological activity, such

as calving sites, overwintering grounds, and spawning habitats; and (5) remnants of late-successional and disturbance refugia. Other places in the matrix may need to be protected or specially managed for cultural and social reasons, although these too can sometimes have considerable value for many elements of biodiversity.

The value of midspatial-scale protected areas is often highly disproportionate to the area they occupy, but they can be severely compromised by inappropriate practices within the matrix. These areas typically require special management, including the establishment of networks of buffers (such as riparian corridors) and protection from routine logging and other silvicultural activities. Zoning systems can sometimes be useful to signify the need for particular care to maintain the integrity of midspatial-scale protected areas in the matrix.

The importance of both large ecological reserves and the appropriate management of matrix lands for forest biodiversity was covered in Chapter 5. Since most existing or potential reserve systems are going to be limited, however, the long-term fate of many forest-dependent organisms will depend on activities and conditions in the unreserved portions of forested landscapes, which we refer to as the matrix. The challenge, then, is to develop and implement management strategies that sustain rather than reduce biodiversity in the matrix (deMaynadier and Hunter 1995).

We begin this chapter with a checklist of elements that should be considered in forest management that are consistent with the general principles outlined in Chapter 3 (Table 6.1). The remainder of Chapter 6 is an exploration of patch-level habitats that require landscape-level consideration within the matrix, including aquatic ecosystems, wildlife corridors, and specialized habitats. Additional landscape-level considerations within the matrix, such as road systems and the arrangement of harvest units in space and time, are identified and discussed in Chapter 7. Finally, approaches to management of harvested forest stands within the matrix that contribute to maintenance of biodiversity, such as modified silvicultural regimes and long rotations, are considered in Chapter 8.

A Checklist for Forest Biodiversity Conservation

A comprehensive approach to the conservation of biodiversity incorporates a continuum of conservation approaches from the establishment of large ecological reserves through an array of conservation measures within the matrix, including the maintenance of individual forest structures at the stand and substand scale (Table 6.1). Elements in the checklist are approximately hierarchical, progressing from very large spatial-scale strategies (large ecological reserves) to midspatial-scale or landscape-level strategies (e.g., protected areas within the matrix) to stand-level silvicultural approaches (e.g., structural retention). The checklist is intuitive, but its elements have rarely been presented in a comprehensive and hierarchical list (Figure 6.1).

The general principles (see Chapter 3) of maintaining connectivity, landscape heterogeneity, stand complexity, and the integrity of aquatic systems apply to all matrix lands. However, any one of the elements of matrix management identified in Table 6.1 may contribute to these principles (Table 6.2).

How the components listed in Table 6.1 are actually addressed in a real landscape—individually and collectively—depends upon many considerations. These include the nature of the landscape—physical and biological conditions, human developments (such as roads), objectives of the landowner(s), regulatory and social directives, and species targeted for conservation. Since each landscape is unique in the mix of such considerations, we cannot offer generic landscape- and stand-level prescrip-

---

TABLE 6.1.

Checklist of factors for matrix management.

LARGE ECOLOGICAL RESERVES (CHAPTER 5)

CONSERVATION STRATEGIES WITHIN THE MATRIX: LANDSCAPE-LEVEL CONSIDERATIONS (CHAPTERS 6 AND 7)*

*Protected habitat within the landscape matrix: protected areas at intermediate-spatial scales**
*Aquatic ecosystems and riparian buffers**
   Springs, seeps, vernal features, lakes, ponds, wetlands, streams and rivers, and associated buffers
*Wildlife corridors**
Culturally sensitive areas within the matrix
*Special habitats**
   Cliffs, caves, rockslides, thermal features, meadows, vernal features
*Remnant patches of late-successional forest**
*Biological hotspots**
   Calving areas, source areas for coarse woody debris, populations of rare species
*Fire, wind, and other disturbance refugia**

Other landscape-level considerations
Transportation systems (e.g., roading networks)
Landscape-level goals for specific structural features (e.g., large snags) or vegetative conditions
Spatial and temporal patterns of timber harvesting
   Dispersed versus aggregated
   Size of harvest units
   Rotation lengths
Restoration and re-creation of late-successional forests or other habitat features

CONSERVATION STRATEGIES WITHIN THE MATRIX: STAND-LEVEL CONSIDERATIONS (CHAPTER 8)

Habitat within management units or stands
Retention of structures and organisms at time of regeneration harvest
Creation of structural complexity through stand management activities
Lengthened rotation times

Strategies addressed in this chapter are indicated by an asterisk (*).

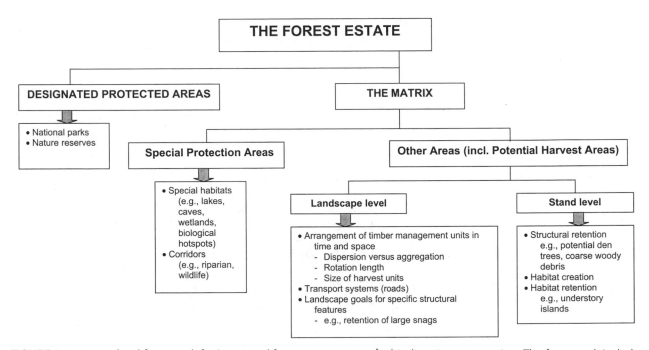

FIGURE 6.1. Hierarchical framework for integrated forest management for biodiversity conservation. The framework includes large ecological reserves and landscape-level and stand-level matrix-based management.

tions. We do feel that each element of the checklist needs to be appropriately considered in developing a conservation strategy for a forest landscape. The management regimes that emerge from such a planning effort will vary widely in different natural and social settings, as illustrated by the case studies in Chapters 11 to 15.

Approaches to Landscape Management

Considerations in landscape-level matrix management can be assigned to four broad categories:

1. Protection of aquatic ecosystems and networks, specialized habitats, biological hotspots, and remnants of natural forest found within the matrix
2. Establishment of landscape-level goals for retention, maintenance, or restoration of particular habitats or structures as well as limits or thresholds for specific problematic conditions
3. Design and subsequent management of transportation systems (generally a road network) to take account of impacts on species, critical habitats, and ecological processes

4. Selection of the spatial and temporal patterns for harvest or other management activities

Spatially explicit databases on conditions in the existing landscape are an essential starting point for landscape-level matrix management. This may take the form of GIS data layers or maps. One important layer is the forest types (or other mappable vegetation units) that occur and their condition (e.g., age class, harvest history, etc.). The data layers or maps should include the location of roads and stream systems as well as rare and significant vegetation types or places (e.g., biological hotspots; see below). Spatially explicit information of this type provides the basis for analyzing and displaying management alternatives. In some jurisdictions, maps of pre-European settlement patterns of vegetation cover can be useful for comparison with existing cover (see the discussion in Chapter 4 on contrasting natural patterns and those created by human disturbance). However, such maps are difficult to produce in places with a prolonged history of human use, such as in many parts of Europe. The fundamental importance of spatially explicit data layers or maps to landscape analysis and management planning is obvious and nontrivial; however, it is an immense topic in itself, which we will not deal with further here.

TABLE 6.2.

Matrix management and the achievement of the general principles outlined in Chapter 3.

PRINCIPLE	STRATEGY
Principle 1. Maintenance of connectivity	• Riparian and other corridors • Protection of sensitive habitats with the matrix • Vegetation retention on logged areas throughout the landscape • Careful planning of roading infrastructure • Landscape reconstruction
Principle 2. Maintenance of landscape heterogeneity	• Riparian and other corridors • Protection of sensitive habitats within the matrix • Midspatial-scale protected areas • Spatial planning of cutover sites • Increased rotation lengths • Landscape reconstruction • Careful planning of roading infrastructure • Use of natural disturbance regimes as templates (see Chapter 4)
Principle 3. Maintenance of stand complexity	• Retention of structures and organisms during regeneration harvest • Habitat creation (e.g., promotion of cavity-tree formation) • Stand management practices • Increased rotation lengths • Use of natural disturbance regimes as templates
Principle 4. Maintenance of intact aquatic ecosystems	• Riparian corridors • Protection of sensitive aquatic habitats with the matrix • Careful planning and maintenance of roading infrastructure
Principle 5. Risk-spreading	• Adoption of array of strategies critical to principles listed above • Ensuring that strategies are varied between different stands and landscapes ("don't do the same thing everywhere")

The rest of this chapter focuses on identification and protection of midspatial-scale areas within the matrix—category 1 of the earlier list. Remaining landscape-level matrix issues are addressed in Chapter 7.

The Roles of Midspatial-Scale Protected Habitats within the Matrix

Identifying and protecting sensitive and ecologically important habitats within the matrix is essential in developing comprehensive strategies for biodiversity conservation. Some of these habitats are widely distributed, such as stream and river networks and associated riparian zones, and lakes, ponds, and wetlands and associated littoral zones. Others, such as caves, rock outcrops, and areas of specialized rock types, may

be rare but important for species found nowhere else (Culver et al. 2000). These types of habitat may not be adequately represented in a system of large ecological reserves; in fact, with some very rare habitats the goal may be to protect all extant examples. Also, large ecological reserves often cannot mitigate loss of such habitats elsewhere in a region or landscape, such as in the case of spawning habitat for anadromous fish. Finally, remnant patches of natural vegetation, such as stands of late-successional forest, may need to be identified and protected.

Protected areas within the matrix can be considered to be midspatial-scale reserves managed primarily for conservation goals. These "meso-reserves" contribute to biodiversity conservation by

• Providing critical habitat for some biota, for example, organisms that are totally dependent upon unique features such as caves and thermal features.

- Protecting aquatic and semi-aquatic ecosystems and, in some cases, providing essential inputs for those ecosystems, such as coarse woody debris.
- Providing refugia for forest organisms that subsequently provide propagules and offspring for recolonizing surrounding forest areas as they recover from timber harvest (Kavanagh and Webb 1998).
- Maintaining landscape heterogeneity (Principle 2 in Chapter 3). For example, the protection of even small grassland patches within intensively cropped landscapes enhanced invertebrate conservation by providing landscape heterogeneity (Halley et al. 1996).
- Facilitating connectivity (Principle 1 in Chapter 3). For example, meso-reserves can be stepping stones that facilitate movement of at least some animals across managed landscapes (Nève et al. 1996).
- Providing nuclei or nodes for the restoration and expansion of key habitats, such as increased amounts of late-successional forest (McCarthy and Lindenmayer 1999a).
- Significantly increasing protection for habitats, vegetation types, and organisms poorly represented or absent in large ecological reserves.

Protecting and Sustaining Aquatic Ecosystems

Aquatic features of landscapes—streams, rivers, wetlands, lakes, and ponds—are critically important to biodiversity conservation and ecosystem function, although they are often not specifically considered as part of traditional reserve design (Pringle 2001). A very large proportion of the biodiversity found in forested landscapes is associated with aquatic ecosystems—all aquatic and many of the terrestrial organisms (Loyn et al. 1980; Naiman et al. 1993; Naiman and Bilby 1998; Brinson and Verhoeven 1999; Calhoun 1999; Pharo and Beattie 2001). Considerations in conservation planning must extend substantially beyond the most obvious aquatic

ecosystem components, which are the surface waters occupying a stream channel or depression. For example, the hyporheic zones (sensu Stanford and Ward 1993) of streams and rivers and the benthic environments of lakes are integral functional elements of aquatic ecosystems and harbor rich and distinctive biotas.

The habitat and functional relationships between spatially adjacent terrestrial and aquatic habitats have rarely received sufficient consideration in forest management and conservation planning. Adjacent terrestrial habitats (e.g., riparian and littoral zones) should be viewed as integral components of aquatic ecosystems because of the extensive functional relationships between adjacent terrestrial and aquatic communities. For example, the half life of large woody debris in the Queets River (Olympic National Park in Washington state, United States) was found to be approximately twenty years, which means that virtually all wood disappears within fifty years; the inference is that harvesting all large coniferous trees within riparian zones of large streams could have adverse impacts on the structure of aquatic ecosystems within three to five decades (Hyatt and Naiman 2001). Many organisms utilize both aquatic and terrestrial habitats during their life cycles. For example, in the Sierra Nevada range in California (United States), nearly 100 percent of the aquatic invertebrates are closely associated with adjacent terrestrial (e.g., riparian) areas (Sierra Nevada Ecosystem Project 1996).

Aquatic influences on adjacent terrestrial habitats need to be considered along with terrestrial influences on aquatic processes and biota. For example, riparian forests are important habitats because of their unique microclimate and vegetative structure and composition. Recognition of this has increased interest in the conservation of riparian habitats for their own sakes rather than simply as protective buffers for adjacent streams. In another example from the Sierra Nevada, approximately 17 percent of the plants and 21 percent of the vertebrate species are closely associated with riparian or wet areas (Sierra Nevada Ecosystem Project 1996). In southeastern Alaska, 74 percent of all plant species in the 145-square-kilometer Kadashan River drainage were found in the riparian zone. In the state of Washington, 60 percent of 480 species of wildlife occur in wooded riparian zones (Raedeke 1988). These levels of riparian contribution to biodiversity are probably common in forest landscapes.

Aquatic ecosystems have been immensely impacted—physically, chemically, and biologically—by a wide variety of human activities. Some activities have been intentional and direct, such as the creation of dams and other structures and diversion of water, which alters flow regimes and impedes movement of organisms and materials. Many other activities have been incidental, such as pollution and additional sedimentation (Metzeling et al. 1995).

Aquatic and riparian habitats were found to be the most altered and impaired habitats in the Sierra Nevada (Sierra Nevada Ecosystem Project 1996), and such circumstances are probably widespread. This resulted from many factors, including dams and water diversions, mining, timber harvesting, and repeated release of fish into areas from which they were previously absent. Massive declines in amphibian abundance, distribution, and diversity and probably similar (but unknown) impacts on aquatic invertebrate diversity have been a consequence of fish introductions. The deterioration of forested wetlands in Finland is another example of the problems created by intensive timber harvesting. Protective measures are now being developed for endangered plant communities there that often support high levels of plant biodiversity (Aapala et al. 1996).

A fundamental premise of this book is that aquatic features will be identified and appropriate measures adopted for their conservation. A detailed consideration of the conservation values and appropriate management of aquatic ecosystems is far too large and specialized a topic to address here. Our focus is on the management of the terrestrial environment (the forest) as it interacts with aquatic ecosystems. It is important to ensure that conditions (e.g., habitat and water quality) essential to aquatic biodiversity are maintained along with those conditions critical for terrestrial organisms and riparian health.

Stream and River Networks

Riverine networks are large-scale networks of streams and rivers that occur in all forested regions. They typically consist of complex (often dendritic), strongly longitudinal, hierarchically structured, large-scale networks that we refer to as *streamscapes* and *riverscapes*. Streams and rivers are broadly (but not sharply or consistently) differentiated on the basis of "volumes of water moving within a visible channel, plus subsurface water moving in the same direction and the associated floodplain and riparian vegetation" (Naiman and Bilby 1998).

Vegetation associated with streams and rivers is variously described as a riparian or streamside influence zone. Traditionally, riparian zones were defined by the presence of water-loving plants intolerant of high moisture stress and tolerant of seasonal flooding. They also have distinctive disturbance regimes (Johnson et al. 2000). However, the area that is now considered relevant to aquatic ecosystems is usually much broader than the zone characterized by riparian plants (Swanson et al. 1982; Gregory et al. 1991; Naiman and Bilby 1998). The term *streamside influence zone* recognizes the broader region that has significant interactions with streams and rivers, including shading influences and primary source areas of allochthonous inputs. Our riparian focus is on the larger interpretation. The term *streamside corridor* incorporates the aquatic ecosystem—including the hyporheic zone—and terrestrial near-stream areas that have significant functional relationships with the aquatic ecosystem including a more narrowly defined riparian zone.

There are important and mutual relationships between aquatic and terrestrial ecosystems (Swanson et al. 1982; Gregory et al. 1991; Naiman and Bilby 1998; Naiman et al. 2000). Streams and rivers have major impacts on streamside terrestrial assemblages in terms of microclimate (Chen 1991), movement corridors for wildlife, habitat for terrestrial organisms (such as breeding areas for amphibians), sources of energy and nutrients (e.g., from returning anadromous fish), and agents of erosion, deposition, and flooding. However, as noted earlier, riparian zones are increasingly recognized as distinctive and important habitats for biodiversity (Doeg and Koehn 1990; Kondolff et al. 1996;

Naiman et al. 2000). Consequently, riparian zones are increasingly proposed for protection on the basis of their intrinsic value rather than simply as a buffer zone for aquatic ecosystems (e.g., Recher et al 1980; Sierra Nevada Ecosystem Project 1996).

Multiple Roles of Forests

Forests and other terrestrial vegetation within a streamside corridor have extensive and substantial influences on aquatic ecosystems. The nature of these relationships changes (often systematically) along the gradient of stream size (Vannote et al. 1980; Swanson et al. 1982; Gregory et al. 1991) (Figure 6.2).

The main roles of riparian forests in supporting biodiversity and maintaining the function of riverine ecosystems include

- Environmental controls, such as on light and temperature regimes (Budd et al. 1987)
- Inputs of organic matter and nutrients (Graynoth 1989; Barling and Moore 1994)
- Physical protection, such as stabilization of stream banks and filtering sediments
- Provision of coarse woody debris—a key structural component of streams and rivers for habitat and its effects on hydrological processes (Silsbee and Larson 1983; Harmon et al. 1986)

Terrestrial vegetation exerts a major control on environmental conditions of small streams. Closed-canopy forest can completely shade streams up to the third or fourth order in size (sensu Strahler 1957), resulting in light levels that sustain only low levels of aquatic primary productivity, such as by algae and mosses (Figure 6.2). Shading also prevents solar heating of water

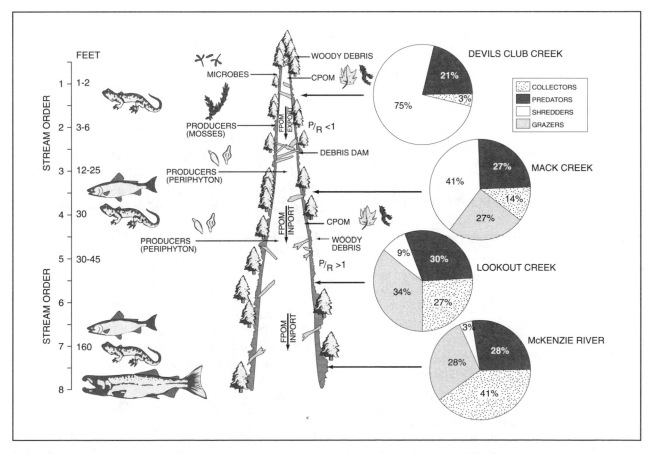

FIGURE 6.2. Compositional, structural, and functional attributes of aquatic ecosystems, including their interrelationship with the riparian zone, tend to vary longitudinally with stream order as represented here with dominant predators, producer groups, P/R (production/respiration ratio), importance of wood, and proportion of invertebrate functional groups. Redrawn from Sedell et al. 1988.

(Barton et al. 1985). One of the early rationales for forest buffers on streams in northwestern North America was to prevent water temperatures from reaching lethal levels for salmonid fish (Beschta et al. 1987; Hicks et al. 1991).

Since primary productivity in shaded streams is limited, terrestrial plants and animals provide the energy and nutrient base in the form of allochthonous materials—leaves, branches, feces, and other detritus, root exudates, and so forth (Campbell et al. 1992). Streamside corridors that incorporate a variety of life forms (herbs, shrubs, and trees) are particularly valuable in providing these materials that vary in both quality and timing and are the basis for the food chain and a primary source of nutrients (Gregory et al. 1991). Products from the small stream systems are, of course, exported to downstream aquatic ecosystems.

Streamside vegetation also can contribute to the stabilization of stream channels. For example, root systems bind stream banks, and stems and organic debris may reduce the potential for erosion. Uprooting of tree stems also can contribute to erosion when streamside trees are windthrown or undermined by water.

Large woody debris derived from a streamside corridor is an important input to stream and river ecosystems (Swanson et al. 1982; Maser et al. 1988; Naiman and Bilby 1998; Naiman et al. 2000) (Figure 6.3). The numerous roles of large woody debris in small streams include

- Creation of stream complexity and habitat diversity, such as in the form of debris jams, sediment accumulations, and plunge pools. Large woody debris is the primary determinant of channel form and habitat in small streams (Bilby and Bisson 1998; Naiman and Bilby 1998).
- Increasing the retention capacity of the stream for allochthonous inputs and sediments.
- Providing a long-term source of energy and nutrients in the form of organic matter as well as sites for nitrogen fixation.
- Reducing channel erosion by dissipating energy, such as by creating plunge pools.
- Providing large woody debris for downstream areas.

FIGURE 6.3. Large woody debris fulfills many critical roles in streams, including the retention of sediment and organic matter (Stanislaus National Forest, California, United States). Photo by J. Franklin.

- Contributing mechanically to disturbance of the aquatic system (Swanson et al. 1998; Johnson et al. 2000).

A major objective of riparian buffers in many forest landscapes is to provide a continuing source of large woody debris.

Large woody debris also plays valuable (although different) roles in larger river systems (Harmon et al. 1986; Maser et al. 1988; Naiman et al. 2000). Unlike small streams, large streams and rivers have substantial capacity to float and redistribute large woody debris. Large woody debris in rivers generates habitat diversity by

- Modifying channel flow, resulting in the creation of deep pools (Gippel et al. 1996). These "dead water zones" can be valuable habitats for fish (Fetherston et al. 1995).
- Influencing sediment deposition to generate large midchannel bars or channel margin bars.
- Facilitating vegetation development on bars by providing protective structures.
- Creating productive, protected, off-channel habitat by establishing log jams on main channel segments.
- Regulating flows into secondary channels.

River systems lacking large woody debris tend to develop simplified profiles dominated by riffles and a reduced number of pools (Sedell et al. 1988; Bilby and Bisson 1998). The habitat diversity generated by large

FIGURE 6.4. Large woody debris is an important structural component of large river ecosystems, contributing to the creation of habitat complexity. Large woody debris is delivered to a river bed by bank cutting and subsequent fluvial processes (South Fork Hoh River, Olympic National Park, Washington, United States). Photo by J. Franklin.

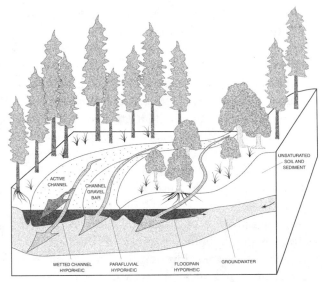

FIGURE 6.5. Location of hyporheic zones beneath and adjacent to a river. Redrawn from Naiman et al. 2000. Reproduced with permission of the American Institute of Biological Sciences.

woody debris greatly enhances both biodiversity and the productivity of aquatic ecosystems.

Streamside and floodplain forests are the primary source areas for large woody debris in stream and river ecosystems. Dominant input mechanisms are bank cutting and windthrow (Figure 6.4). Fluvial transport of wood from smaller, high-gradient streams, via debris torrents and mass wasting, is another major pathway for delivery of large quantities of large woody debris to larger streams and rivers (Keller and Swanson 1979; Bilby and Bisson 1998). An important point to consider in managing river ecosystems for biodiversity is that, over long time periods, the entire floodplain functions as the source of large woody debris and sediments (see below).

The Hyporheic Zone

The hyporheic zone is the "volume of saturated sediment beneath and beside streams and rivers where ground water and surface water mix" (Edwards 1998) (Figure 6.5). It has been recognized relatively recently and has great importance for biodiversity conservation and maintaining water quality and productivity in riverine ecosystems (Stanford et al. 1994). The water in porous, hydraulically conductive sediment is intimately involved with the water in the open channel. It supports a diversity of invertebrate species, referred to as *hypogean fauna* (Edwards 1998), which occupy in-

terstitial spaces and are actively involved in breaking down organic matter and releasing nutrients.

The subterranean and invisible hyporheic zone is not isolated from the impacts of surface activities. It can be affected by the vegetation and disturbance processes in the floodplain as well as activities that occur on the surface of the floodplain, such as logging (Edwards 1998). For example, floodplain and gravel-bar vegetation can influence the nutrient content of subsurface waters, such as nitrogen enrichment due to fixation associated with red alder (*Alnus rubra*). Flood events can result in significant modification of hyporheic zones (Wondzell and Swanson 1999). Timber harvest can compact the substrate, reducing the volume and size of interstitial spaces, thereby reducing the rate of water infiltration and flow and the quality and amount of habitat available to the hypogean community. For example, clearcutting in karri (*Eucalyptus diversicolor*) forests in Western Australia led to significant alterations in the interstitial community below the streambed (Trayler and Davis 1998).

Streamside Corridors

Streamside corridors are the preferred conservation strategy for streams and rivers in forested landscapes.

These are protected zones, or "buffers," of terrestrial vegetation within which little or no exploitative activity, such as logging, is allowed (Clinnick 1985; Davis and Nelson 1994; Haycock et al. 1997). The objectives of streamside corridors typically include protecting the aquatic ecosystem from the direct impacts of logging and other activities, but often these objectives are broadened to include maintenance of functional relationships between terrestrial and aquatic ecosystems. Streamside buffers are known to be effective in limiting sediment and nutrient run-off into aquatic ecosystems (e.g., Daniels and Gilliam 1996).

Riparian processes considered in the development of the streamside buffering system of the Northwest Forest Plan in the northwestern United States (Forest Ecosystem Management Assessment Team 1993) include root strength for soil and bank stabilization, delivery of large woody debris to streams and riparian habitat, inputs of leaf and other particulate organic material (allochthonous inputs), shading of stream and maintenance of riparian zone microclimate, maintenance of water quality, including sediment buffering, and provision of wildlife habitat.

Adoption of streamside buffers has been accepted—in principle—as an important tool in the management of forest landscapes, despite substantial earlier resistance by many foresters and forest-land owners. However, there are still wide differences of opinion about how extensive stream and river buffers need to be to achieve conservation objectives. Specifically, questions are raised regarding the size and longitudinal extent of buffer widths necessary to achieve desired objectives (e.g., how wide should buffers be, and are they needed on all channels?) and the degree to which management activities, such as timber harvesting, can be allowed in streamside buffers. Issues associated with buffer widths include (1) the necessity of having a buffer that can sustain natural disturbance regimes and still continue to function in its protective roles, and (2) the desirability of buffering the riparian habitat itself.

The science of streamside corridors is still evolving. In northwestern North America, buffering of streams began with a focus on larger, fish-bearing streams and rivers and on relatively narrow protective zones immediately adjacent to the channel. Primary objectives typically included controlling stream temperatures by shading and protection of stream banks from disturbance. Approaches and objectives are changing rapidly, largely in response to emerging knowledge about geomorphological and hydrological processes (especially at larger spatial scales), the nature and locale of aquatic biodiversity and ecosystem processes, and, especially, the numerous functional relationships of terrestrial and aquatic ecosystems. Consequently, proposed and adopted stream corridors and specific riparian zone management prescriptions are migrating upward in river drainages into small headwater streams (and even intermittent channels) as well as laterally so as to incorporate more (or all) of the floodplain.

The increasing extent of stream corridors reflects several emerging concerns, including recognition of the importance of providing source areas for large woody debris and sediments (including attention to the large spatial scales and long temporal scales that need to be considered in delivery and movement of these materials), protecting the riparian zone itself, protecting secondary channels, and protecting hyporheic habitats and organisms.

Width of Streamside Corridors

Selection of the appropriate widths for streamside corridors must be based on the objectives of the buffer and the spatial pattern of relevant influences. For example, if a primary goal of a streamside buffer is the provision of large woody debris to the adjacent channel, the buffer needs to be of sufficient width to incorporate an appropriate source area based on relationships of the type illustrated in Figure 6.6. If an objective is to conserve a particular species, then its movement patterns relative to the riparian zone need to be established. In the case of the giant barred frog (*Mitophyes iteratus*) in the production forests of New South Wales, radio-tracking data showed that 30-meter-wide unlogged stream buffers would capture the typical extent of its movement (Lemckert and Brassil 2000). To maintain the entire breeding bird community, Pearson and Manuwal (2001) recommended a minimum buffer of 45 meters each side of second- and third-order streams in northwestern United States conifer forests; narrow buffers failed to maintain prelogging bird communities. Riparian buffers extending at least 45 meters on both sides of streams were viewed

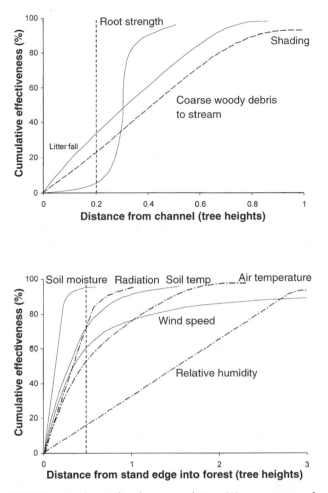

FIGURE 6.6. Generalized curves relating (A) percentage of riparian ecological functions and processes occurring within varying distances from the edge of a forest stand and (B) percentage of microclimatic influences occurring within varying distances from the edge of a forest stand. Redrawn from Forest Ecosystem Management Assessment Team 1993.

as necessary to maintain natural microclimatic conditions along streams based on a study of microclimatic gradients associated with forested streams in moderate to steep topography in western Washington state (United States) (Brosofske et al. 1997).

For influences such as shading and allochthonous inputs, narrower streamside corridors may suffice than are required in the case of large woody debris. Tree heights and rates of lateral channel change are critical with regard to both of these issues. Other influences or goals, such as buffering of the riparian or floodplain habitat itself, may require wider streamside corridors.

Proposed and adopted widths for streamside corridors vary widely from near-stream buffers of 10 meters or less to buffers that cover 100 meters or the entire floodplain (Forest Ecosystem Management Assessment Team 1993). These differences are typically related to differences in management objectives, such as between public and private land, and the trade-offs between conservation goals and commodity production. However, narrow buffers (e.g., 10 meters or less) generally are not believed adequate to address the biological and physical interactions known to exist between a riparian forest and a stream or river. Some authors recommend linking prescriptions for minimum corridor widths with stream order. However, appropriate buffer widths depend not only on stream size but also on

- Forest type (e.g., the potential tree height; see below).
- Lateral extent of aquatic systems and habitat types to be conserved.
- Topography.
- Soil conditions.
- Types of species and ecological processes targeted for protection (Spackman and Hughes 1995; Lemckert and Brassil 2000). For example, in parts of the wood production estate in Australia, 20-meter buffers may limit extensive sediment flow into aquatic systems (Clinnick 1985) but much wider ones (exceeding 100 meters) are needed to limit increases in stream salinity due to rising groundwater and the mobilization of salts in the soil profile (Borg et al. 1997; Trayler and Davis 1998).
- The silvicultural system and harvesting machinery employed in the surrounding forest (Chen 1991; Barling and Moore 1994; Gregory 1997). For example, specially designed harvesting machinery can limit damage to wet soils and aquatic systems (Stokes and Schilling 1997), which, in turn, can influence the width of buffers required.

These considerations must all be placed in the context of the conditions within the remainder of the landscape and watershed (Gregory 1997).

Many modern proposals for streamside corridors are scaled to the heights of trees growing in the riparian zone, such as some fraction or multiple of the height of a mature tree. For example, streamside corridors adopted for federal lands in the northwestern United States (identified as "riparian reserves") are based upon the "height of a site-potential tree," which is defined as the "average maximum height of the tallest dominant trees (200 years or more) for a given site class" (Forest Ecosystem Management Assessment Team 1993). Ultimately, the adopted widths of riparian reserves were multiples of site-potential tree height (USDA Forest Service and USDI Bureau of Land Management 1994a,b): two site-potential tree heights on both sides of fish-bearing streams (four tree heights in total); one potential tree height on perennial, nonfish-bearing streams; and one potential tree height on seasonally flowing or intermittent streams. The rationale for this approach was the relationship between distance and functional influence (Figure 6.6) plus consideration of other riparian values and the need to "buffer the buffer" on large channels (Forest Ecosystem Management Assessment Team 1993).

Variable-width streamside corridors (e.g., in British Columbia; see Fenger 1996) are sometimes referred to as *smart buffers*. As an example, streamside corridors might be much wider along unconstrained reaches where substantial lateral movement of the stream channel is possible, and narrower along areas where the stream is constrained, such as in bedrock-controlled reaches. In western Canada, the Clayoquot Sound Scientific Panel (Scientific Panel for Sustainable Forest Practices in Clayoquot Sound 1995) proposed a sophisticated approach to "hydro-riparian reserves" for streams based upon whether the channel was alluvial or nonalluvial, entrenched or not entrenched, and upon stream channel gradient and width. The widths of reserved areas varied substantially with geomorphological conditions as well as stream width; occurrence of fish was not a consideration.

The riparian reserves of the Northwest Forest Plan—sometimes viewed as a model of riparian buffering—were originally intended to be an interim streamside protection system. The design team of aquatic biologists, hydrologists, and geomorphologists expected the interim reserves to be modified (i.e., varied) following detailed watershed analyses (Forest Ecosystem Management Assessment Team 1993; USDA Forest Service and USDI Bureau of Land

Management 1994a,b). For example, buffers could be removed from some stream reaches and protection of unstable landforms and soils extended beyond near-stream areas. In practice, significant modifications of the riparian reserves have never occurred, in part because of the perceived importance of the riparian reserves for terrestrial biodiversity. As noted below, this universal application of streamside corridors to the landscape can create problems, and alternative approaches that are considered to be consistent with the goals of the Northwest Forest Plan have been proposed (Cissel et al. 1999).

Selection of wider streamside corridors obviously relates to objectives that extend beyond simply protecting aquatic ecosystems from direct impacts of harvesting. Protecting large portions of floodplains on larger rivers reflects at least three concerns: (1) the potential importance of the entire floodplain as a source area for large woody debris and sediments for the river system and its estuary due to the long-term meandering of the river channel, (2) the extent and importance of hyporheic habitats in large river valleys, and (3) the importance of the diverse aquatic habitats (e.g., off- or side-channels, seeps, and wetlands, and low-gradient tributaries) found on large floodplains.

Longitudinal Extent of Buffered Streamside Corridors

The longitudinal extent of stream buffering is another significant issue in designing a system of streamside corridors. That is, what lengths or reaches of a riverine system should have buffers? In North America, the focus has traditionally been on major fish-bearing streams and rivers and on the main channel. However, it is clear that within many river systems, small streams actually support most of the biodiversity, represent the bulk of the aquatic habitat, generate most of the aquatic productivity, and are the source areas for large woody debris, sediments, and water (e.g., Forest Ecosystem Management Assessment Team 1993; Naiman and Bilby 1998). Consequently, low-order headwater streams and even stream channels that have intermittent flows are receiving increasing attention in design of stream-buffering strategies. This is a challenge in wood-production landscapes because of the lineal extent and density of low-order

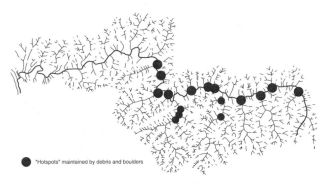

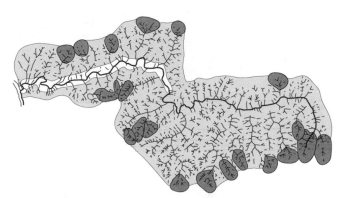

FIGURE 6.7. (*A*) Fish-spawning habitat (primarily associated with debris jams) within the Knowles Creek drainage in the Coast Ranges of Oregon, United States. (*B*) Headwater drainages that need to be managed so as to ensure periodic landslides and torrents contain large quantities of large woody debris for renewing the debris jams. Note: most of this habitat is located where high-gradient tributaries have a high angle of intersection with the larger stream reach.

stream channels and consequent impacts on harvest levels and logging practices that would be associated with any buffering of this part of the streamscape. In effect, if rainfall is adequate to support production forestry, the drainage density and extent of the landscape that will be in buffer zones is likely to be substantial.

Federal policies in the northwestern United States again exemplify an approach that is ecologically conservative. All channels (including intermittent ones) have one site-potential tree height buffer on each side or, where trees are short, a streamside corridor width of 100 meters. The extension of buffering to low-order (and, often, high-gradient) streams and intermittent channels is based on several ecological objectives, two of which we will elaborate on.

First, the aquatic habitat and biota associated with these streams and the associated riparian habitats represent a large proportion of the habitats present within a drainage area and are the most widely distributed geographically. The importance of the riparian zone to terrestrial as well as aquatic biodiversity has become a critical issue in regional conservation planning (Scotts 1991; Forest Ecosystem Management Assessment Team 1993; Sierra Nevada Ecosystem Project 1996).

Second, there is a greatly increased appreciation of the importance of low-order streams in mountainous topography as source areas for sediments and, especially, large woody debris that ultimately influences processes and habitats within larger stream reaches. Studies in the Coast Ranges of Oregon, in the United States, provide examples of the importance of low-order stream systems and headwalls as source areas for large woody debris and sediment that maintain critical habitat and processes within larger stream and river reaches (Figure 6.7). Periodic mass failures of these steep landforms and high-gradient streams generate debris torrents that are deposited in larger stream reaches, typically at points where the tributary channels intercept larger channels at high angles. These resulting debris jams create primary spawning habitat for anadromous fish. Without the periodic input of large woody debris, these productive reaches would decline or even disappear. Management of source areas aims to ensure that when geomorphic failures occur they are "loaded" with large woody debris.

Limitations of Streamside Corridors and Other Considerations in Riparian Conservation

Universal adoption of streamside buffers throughout a drainage system can have the undesirable effect of fragmenting upland habitats and creating a landscape that has a high level of structural contrast between riparian and upland habitats. The fragmentation problem is particularly notable in areas of moderate to high stream density (Figure 6.8). Management of the small, narrow, and isolated upland forest fragments that are left behind may be impractical for both physical reasons (e.g., access and maneuvering room for harvesting) and economic reasons. Where

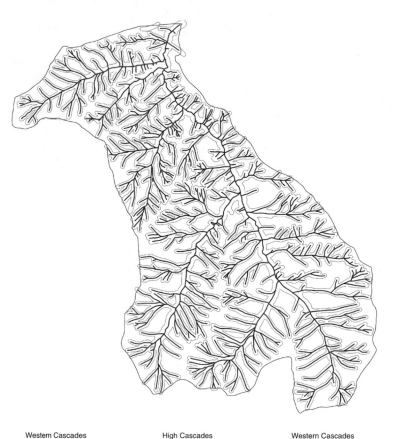

FIGURE 6.8. The Northwest Forest Plan streamside corridors (riparian reserves) in an exemplary watershed (Augusta Creek) in the Cascade Range of Oregon. Although intended as an interim buffering system that would be reconfigured following watershed analysis, it has become the de facto permanent system. All channels, including intermittent stream channels, are assigned buffers about 110 meters in width (55 meters on both sides of the channel). As illustrated here, upland habitats may be highly fragmented by a universal buffering scheme of this type, significantly constraining options for their management and creating the potential for a landscape with a high level of contrast between riparian and upland sites. Redrawn from Forest Ecosystem Management Assessment Team 1993.

Western Cascades
Resistant Rock

Coffin Mountain
Williamette NF
5.63 mi./sq. mile
4000 feet (msl)

High Cascades
Resistant, other

Wolf Peak
Mt.Hood NF
2.66 mi./sq. mile
4000 feet (msl)

Western Cascades
Weak Rock

Abbott Butte
Umpqua NF
6.87 mi./sq. mile
3900 feet (msl)

Western Cascades
Weak Rock

Buckeye Lake
Umpqua NF
6.55 mi./sq. mile
3600 feet (msl)

Western Cascades
Weak Rock

Reynold's Ridge
Umpqua NF
8.27 mi./sq. mile
2440 feet (msl)

Western Cascades
Weak Rock

Sinker Mountain
Williamette NF
7.75 mi./sq. mile
2900 feet (msl)

Western Cascades
Weak Rock

Smith Creek Butte
Gifford Pinchot NF
11.57 mi./sq. mile
2200 feet (msl)

Olympics
Intermediate Sediments

Deadman's Hill
Olympic NF
9.82 mi./sq. mile
1000 feet (msl)

Coast Range
(Oregon and Washington)
Resistant, other

Daniel's Creek
BLM-Medford
12.8 mi./sq. mile
1000 feet (msl)

Unconsolidated deposits

Trout Lake
Gifford Pinchot NF
5.42 mi./sq. mile
2500 feet (msl)

management of upland forests occurs, the result can be a high-contrast landscape in which densely vegetated riparian corridors are embedded in a managed forest that is much younger and simpler in structure (Carey et al. 1999b; Cissel et al. 1999) (see Figure 6.8). Windthrow within riparian buffers is also likely to be a more serious problem under such scenarios.

An additional consideration is that most streamscapes in forested regions did not have streamside zones that were uniformly and continuously dominated by closed-canopy forests under natural conditions. The historic range of conditions of streamside forest doubtless included periodic creation of reaches with open canopy conditions by disturbances. Some aspects of aquatic ecosystems (e.g., primary productivity) and some aquatic organisms (e.g., many invertebrate, fish, and amphibian species) benefit from the increased insolation associated with openings. Hence, riparian management (including buffering strategies) needs to consider natural patterns of vegetative cover rather than focusing exclusively on maintaining closed-canopy, late-successional forests throughout stream corridors.

The extent of streamside buffers should also reflect management goals in a practical way—including not only biodiversity but also socioeconomic considerations (see Brinson and Verhoeven 1999, in their example of the Luquillo Experimental Forest in Puerto Rico). For example, in some dissected landscapes, wide riparian buffers could eliminate any timber harvesting (Bren 1997). In these cases, altered buffer widths with ecologically conservative harvest practices outside the riparian zone can afford some protection of aquatic systems while providing for some wood production.

Other Considerations

Retaining biodiversity and ecological functions within stream and river systems involves more than the management of streamside corridors. Attention to road systems—particularly where they intersect with the stream network—is also imperative. Even with a very conservative buffering plan (i.e., wide buffers through the longitudinal extent of the drainage), the aquatic ecosystem may be subject to road-related catastrophes that travel down the stream channel in the form of massive landslides and debris torrents. Even in the ab-

sence of catastrophes, roads in riparian zones have chronic impacts (e.g., sediment and dust production) and may modify channels and limit various riparian functions. Therefore, any aquatic conservation strategy in roaded landscapes must be two-pronged—addressing the protection of the stream influence zone while also minimizing negative impacts of the road network (see Chapter 7 for more details).

Comprehensive strategies for the conservation of aquatic ecosystems typically incorporate additional components such as

- Identification and protection of *key watersheds*, which are designated areas providing high-quality aquatic habitat (Johnson et al. 1991; Moyle and Sato 1991; Reeves and Sedell 1992; Sierra Nevada Ecosystem Project 1996). These are watersheds that typically represent "the best of what is left and have the highest potential for restoration" (Forest Ecosystem Management Assessment Team 1993). Key watersheds also may be chosen because they contain evolutionarily significant fish stocks.
- *Watershed analysis*, which is "a systematic process for characterizing watershed and ecological processes to meet specific management and social objectives . . . [and] a key . . . [to] melding social expectations with the biophysical capabilities of a specific landscape" (Forest Ecosystem Management Assessment Team 1993). Watershed analysis is traditionally applied at the scale of 500 to 5,000 square kilometers.
- *Watershed restoration*, which must be considered because conditions of streams and aquatic ecosystems in many forested landscapes have been degraded as a result of clearcutting of streamside forests, removal of coarse woody debris from stream channels, and construction of roads that have increased the frequency and intensity of higher magnitude fluvial events. For example, streamside-influence zones of entire forested drainages in the northwestern United States have been converted from forests of large, decay-resistant conifers to woodlands of much smaller and decay-prone red alder (Figure 6.9a). Hence, active restoration programs are often necessary to reestablish the structures and

BOX 6.1.

Coarse Woody Debris and the Health of Australian Aquatic Ecosystems

A significant problem in some Australian ecosystems has been the loss of coarse woody debris from aquatic systems. These structures regulate patterns of stream flow, provide habitat for biota, and stabilize stream-bank conditions. Systematic de-snagging of rivers and streams has had negative impacts on a wide range of aquatic biota and ecological processes (Lovett and Price 1999). The quantities of large woody debris vary substantially throughout river systems depending on catchment position, adjacent riparian vegetation, channel size, bank conditions, stream power, and log decomposition rates. Processes of re-snagging aquatic systems have attempted to account for the pattern and variability in snag type and abundance throughout river systems as part of landscape-level restoration efforts of these key ecosystems (Rutherford et al. 2000). The ultimate aim of this work is to develop guidelines to help restore and then maintain large woody debris throughout aquatic systems. One requirement must be to develop restoration programs so they can evolve without continued human intervention. In the case of in-stream snags and woody debris, this entails maintaining vegetation in appropriate successional stages in the riparian and adjacent upslope zones.

FIGURE 6.9. Timber harvesting has eliminated large coniferous trees, which are the primary source of persistent large woody debris for streams and rivers, from riparian zones in the northwestern United States. (*A*) Hardwood forests (such as the stand of red alder depicted here) that have replaced the coniferous stands are inferior sources of large woody debris (north fork of the Stillaguamish River, Washington, United States) (photo courtesy D. R. Berg). (*B*) River habitat restoration in such rivers often includes direct reintroduction of large woody debris into channels, such as this log jam being constructed on the north fork of Stillaguamish River in northern Washington, United States. Photo by P. Stevenson.

processes characteristic of high-quality aquatic habitat. These may include such diverse activities as modifying or eliminating road effects; restoring large woody debris and its source areas to streams and small rivers (Figure 6.9b; Box 6.1); preventing trees from being felled in the riparian zone (e.g., Macfarlane and Seebeck 1991); restoring vegetation in riparian areas by thinning or underplanting

to accelerate growth of desired trees and understory plants to replace shrubs with desired overstory trees (Gregory 1997; Brinson and Verhoeven 1999); or reconstructing "natural" stream-channel conditions. In many rivers, sediment regimes have been altered dramatically by human land use (e.g., Metzeling et al. 1995), leading to the simplification of streambed conditions. In Australia, rock deflectors, or groins, have been constructed out from the riverbank of some streams to narrow the river's channel and create deep pools that provide habitat for large crustaceans and fish (e.g., the endangered trout cod, *Maccullochella macquariensis*).

Lakes, Ponds, and Wetlands

Lakes, ponds, and wetlands are aquatic features of many forested landscapes that also require attention as enclaves of biodiversity and sites for key ecosystem functions. Included here are a broad variety of aquatic and semi-aquatic ecosystems, particularly in the case of wetlands. Wetlands include swamps, marshes, and bogs and are broadly identified as (after Cowardin et al. 1979)

> lands transitional between terrestrial and aquatic ecosystems where the water table is usually at or near the surface or the land is covered by shallow water. . . . [A wetland] must have one of more of the following three attributes: (1) the land supports predominately hydrophytes (at least periodically), (2) the substrate is predominantly undrained hydric soil, and (3) the substrate is nonsoil [organic] and is saturated with water or covered by shallow water at some time during the growing season of each year.

We will consider here only the management of the adjacent terrestrial habitats since we assume that appropriate actions will be taken to conserve biodiversity within the lakes, ponds, and wetland ecosystems themselves.

Terrestrial communities adjacent to lakes, ponds, and wetlands (i.e., in a broadly defined littoral zone) have numerous interactions with these aquatic ecosystems, many comparable to those between riparian forest and riverine ecosystems. For example, littoral forests provide inputs of organic matter and nutrients, habitat for adult life stages of aquatic invertebrates, inputs of large woody debris, influences on meteorological conditions such as shading and sheltering from winds, filtering of sediments and other materials from upland areas, and shoreline stabilization.

Reciprocally, aquatic ecosystems provide resources such as food and water for many animals and breeding habitat for amphibians. They also influence environmental conditions in the littoral forest.

Conservation management strategies for lakes, ponds, and wetlands should have several objectives. First, they should be designed to protect the aquatic ecosystem and their biota from the direct impacts of terrestrial activities, primarily logging and roads. These would include altered environmental (e.g., meteorological) conditions and overland movement of sediments, nutrients, and pollutants to shorelines. Second, conservation strategies need to provide for

continued inputs of organic materials and large woody debris that have naturally flowed from the terrestrial to the aquatic ecosystem and, equally, try to avoid significant changes in the quality and quantity of these direct inputs. Third, conservation strategies need to maintain the habitat conditions essential to life cycles of both aquatic and terrestrial organisms that utilize littoral forests during all or part of their life cycles.

Numerous organisms, including many water birds, depend heavily upon the structurally complex ecotonal area between upland forest and aquatic ecosystem. Littoral forests should be protected for their unique intrinsic values as well as for the role they play in healthy functioning of the adjacent aquatic ecosystems as with the riparian forests.

Protected forest buffers are the primary strategy in many forest plans for conserving biodiversity in all aquatic features. The Northwest Forest Plan (USDA Forest Service and USDI Bureau of Land Management 1994a,b) provides for buffers (riparian reserves) that are two site-potential tree heights in width around lakes and natural ponds, beginning at the extent of riparian vegetation or seasonally saturated soil, and buffer widths of one site-potential tree height around all constructed ponds and reservoirs and all wetlands greater than 0.4 hectare in size. In the Rio Condor Project in Tierra del Fuego (Arroyo et al. 1996), a minimum buffer of 10 meters is recom-

FIGURE 6.10. Wetlands require special consideration in conservation planning. For example, many Key Woodland Habitats in Sweden (which are extremely valuable for biodiversity; see Box 6.2) are adjacent to lake systems such as this one in the south-central part of the country. Photo by D. Lindenmayer.

mended around all peat bogs, which are a valuable wetland feature of those landscapes (see Chapter 15). In Clayoquot Sound, in British Columbia, Canada, a protective zone is required around lakes, which extends either 50 meters or over the entire area under hydroriparian influence, whichever is greater (Scientific Panel for Sustainable Forest Practices in Clayoquot Sound 1995).

Road networks are the other major consideration in wetland management in matrix lands. As with streams and rivers, roads can significantly modify hydrological regimes and geomorphological processes (see Chapter 7). One result of poorly designed and maintained roads is the increased delivery of sediments and organic materials into lakes and ponds even when roads are not directly affecting the aquatic ecosystem, such as a roadbed that is on a shoreline or that uses a fill to cross a wetland. Road networks need to be located, constructed, and maintained to minimize such effects.

Other potential impacts of human activities include the effects of harvesting activities in a drainage basin on hydrological and geomorphological regimes and, hence, sediment, water, and nutrient budgets and temperature regimes of the streams, rivers, lakes, ponds, and wetlands. These indirect impacts can be profound (see Chapter 7). Another category of direct impacts is that of human use of the area for a variety of activities, including recreation and grazing for domestic livestock. Such effects on riparian and littoral zones need to be considered as a part of any overall conservation plan for streamside or lakeside forest buffers.

Wildlife Corridors

Wildlife corridors are another category of midspatial-scale protected areas that can contribute to the maintenance of biodiversity in the matrix. There is a substantial ongoing debate about the merits of wildlife corridors to facilitate connectivity for biodiversity (Noss 1987; Simberloff and Cox 1987; Simberloff et al. 1992; Beier and Noss 1998), and confusion as to the ecological roles anticipated for wildlife corridors. In some cases, the most effective wildlife corridors

will be those that (1) support breeding populations of animals and facilitate dispersal of their offspring, and (2) facilitate the general movement of biota (Lindenmayer 1998; Sieving et al. 2000).

Many authors view wildlife corridors as linear refugia within production forests where populations can persist and their offspring can eventually recolonize adjacent cutovers (e.g., Lindenmayer et al. 1993a; Kavanagh and Webb 1998). While there is abundant evidence in the literature of the value of linear strips of forest as habitat (reviewed by Bennett 1998), significantly fewer studies have demonstrated the use of wildlife corridors for movement (Bunnell 1999b). Many species do not appear to require or respond to corridors when dispersing (Rosenberg et al. 1997; Danielson and Hubbard 2000). For example, the wolf (*Canis lupus*) has recolonized parts of Montana (United States) where there are no formal wildlife corridors (Forbes and Boyd 1997).

There are also issues associated with the cost-benefit ratio of wildlife corridors as compared with other conservation strategies (Lindenmayer 1998). Consequently, *wildlife corridors should not automatically be assumed to be an essential component of all conservation strategies;* their value depends upon many issues, including the taxa in question and the overall intensity of management within a landscape. As noted in Chapter 3, *the best general strategy to facilitate connectivity for some biota may be to improve structural conditions throughout the matrix.*

Some species *do* appear to benefit from wildlife corridors linking areas of suitable habitat. Examples include species that avoid dispersing through open areas (Martin and Karr 1986; Desrochers and Hannon 1997) and species for which habitat suitability is a primary factor influencing dispersal (e.g., Baur and Baur 1992; Nelson 1993). Wildlife corridors also may enhance dispersal success by reducing mortality during such movements (Beier 1993).

Although riparian buffers may act as wildlife corridors for some species, these may not be suitable for organisms requiring midslope and ridge-top habitats (McGarigal and McComb 1992; Claridge and Lindenmayer 1994; Vesely and McComb 1996; Catterall et al. 2001). In other cases, corridors outside of streamside areas will be appropriate for taxa, such as some frogs, that are dependent upon aquatic habitat

FIGURE 6.11. A wildlife corridor in logged forest that provides a connection from a stream to a ridgeline in Victoria. These cross-topography corridors were found to support significantly greater populations of arboreal marsupials than corridors restricted to only one part of the landscape (i.e., a gully only or a midslope only). Photo by R. Meggs.

during part of the year but overwinter elsewhere in the landscape. Arboreal marsupials were found to significantly be more likely to occur in wildlife corridors that connected two or more landform positions in a landscape (e.g., gully and midslope positions) than where corridors were confined to a single topographic position (Lindenmayer et al. 1993a). This may relate to a need for access to food resources that occur in different parts of a forest landscape (Claridge and Lindenmayer 1994). The establishment of wildlife corridors across a topographic gradient also can link areas in different subbasins of a watershed—a potentially valuable strategy for maintaining connectivity for some species (Lindenmayer 1998) (Figure 6.11).

Corridor Width

Much wildlife corridor research has focused on identifying minimum corridor widths (e.g., Harrison 1992). This is because of the positive correlation between corridor width and the abundance and/or species richness for birds, mammals, and invertebrates (e.g., Stauffer and Best 1980; Dickson and Huntley 1987;

Cale 1990; Keals and Majer 1991; Keller et al. 1993; Vermeulen and Opsteeg 1994). Corridor widths can also influence the dispersal behavior of some species (Baur and Baur 1992; Arnold et al. 1993), which can result in changes in home range size, shape, and use (La Polla and Barrett 1993; Lynch et al. 1995).

However, corridor width is only one of several factors influencing wildlife corridor use (Table 6.3). For a set width, wildlife corridor effectiveness will co-vary with other attributes, such as length, habitat continuity, habitat quality, and topographic position in the landscape. It will also vary for different species (Harrison 1992; Lindenmayer 1994a; Mech and Hallett 2001) and may vary among forest types, even for the same species (Lindenmayer et al. 1994b). For these reasons it is not possible to provide generic guidelines for minimum corridor widths. Nevertheless, wide corridors are generally more effective than narrow corridors (Lindenmayer 1998; Brinson and Verhoeven 1999) because

- Wider wildlife corridors better approximate interior forest conditions and minimize edge effects (Moore 1977; Steinblums et al. 1984; Laurance 1990).

TABLE 6.3.

Factors influencing wildlife corridor use (based on Lindenmayer 1994a, 1998).

FACTOR	EXAMPLE
Target species characteristics	Foraging strategy; colonial versus solitary social system (Recher et al. 1987)
Biotic interactions	Aggressive interspecific behavior (Catterall et al. 1991)
Abiotic edge effects	Microclimatic conditions (e.g., light regimes) reduce habitat suitability (Hill 1995)
Dispersal behavior	Random dispersal versus movement along habitat gradients (Murphy and Noon 1992)
Habitat suitability within corridor	Structural features influence movement (Lindenmayer et al. 1994b; Bowne et al. 1999)
Corridor characteristics	Corridor width and length (Andreassen et al. 1996)
Topographic location and variation	Confined to a gully or capturing multiple topographic positions (Claridge and Lindenmayer 1994)
Vegetation gaps	Roads and tracks through corridors pose barriers to movement (Lindenmayer et al. 1994c)
Size of areas connected	Small connected patches provide few dispersalists to move through corridor (Wilson and Lindenmayer 1996)
Number of other corridors	Influences chances of corridors being contacted during movement (Forman 1995)
Matrix condition	Clearcut versus selectively logged adjacent forest

- Wider wildlife corridors may maintain plant species composition over long time periods thereby increasing their long-term conservation value as compared with narrow wildlife corridors (Harris and Scheck 1991).
- Wider wildlife corridors may capture a greater array of habitat types (Lindenmayer 1994a), since these are often associated with different topographic positions in the landscape (e.g., McGarigal and McComb 1992). Consequently, they are more likely to provide for the habitat requirements of specialist species (Darveau et al. 1995; Forman 1995), although there are presently few data to support this hypothesis.
- Wider wildlife corridors have a higher probability of supporting populations of resident animals than narrow corridors do (Scotts 1991; Bennett et al. 1994), particularly of wide-ranging species (Shepherd et al. 1992). Species with large home ranges often do not survive with narrow species corridors (e.g., Recher et al. 1987; Reiner and Griggs 1989 in Forman 1995).

Although wide wildlife corridors appear to have several advantages, narrow corridors may have conservation value and should not be categorically rejected simply because they are narrow. For example, narrow corridors still promote the movement of some taxa and provide habitat for others as exemplified by the roles of hedgerows in agricultural lands in the United Kingdom (Eldridge 1971; Arnold 1983; Osborne 1984). Narrow corridors also may be useful as nuclei in programs to restore and expand corridor systems (Crome et al. 1994).

Corridor Densities

The desirable density and type of wildlife corridors for a forest landscape depends upon many factors. These include objectives of the wildlife corridor system, the silvicultural system utilized in the surrounding landscape, rotation length, spatial arrangement of wildlife corridors, and biology of species targeted for conservation (e.g., their dispersal behavior). The required density will be lower if timber harvesting is based on a selective approach, incorporates high levels of structural retention, or retains habitat islands that can function like stepping stones in contrast to clearcutting.

Taxa with poor dispersal ability and landscapes in which logged areas are highly unsuitable as habitat require higher densities of wildlife corridors to provide a greater array of potential movement routes and, possibly, enhanced dispersal success (Lefkovitch and Fahrig 1985; Merriam et al. 1991). Predictions from simulation models are that some species moving through a landscape are more likely to detect and use wildlife corridors when linear landscape elements are common (Stamps et al. 1987a,b). In The Netherlands, population densities of some bats are positively related to increased densities of wildlife corridors (Verboom and Huitema 1997).

Objectives of Establishing Wildlife Corridors

Networks of wildlife corridors need to be developed around specific objectives and the array of factors influencing wildlife corridor use (Table 6.3). Key questions about their design and establishment include the following:

- Which species move between habitat patches without corridors and which species are dependent on corridors and to what degree (Beier and Noss 1998)?
- How is corridor use influenced by the suitability of the production forest landscape in which they are embedded (Rosenberg et al. 1997)?
- Which species are supposed to benefit from the corridors?
- Is a corridor to function solely as a conduit for movement or is it also to provide suitable habitat?
- What types of areas are being connected by the corridor and how suitable are they for the species of interest?
- What is the condition of the surrounding landscape in which the corridors are embedded?

In assessing corridor needs within a forested landscape, these questions should be addressed along with analysis of the costs and benefits of alternative methods for facilitating connectivity (see below). Unfortunately most studies of wildlife corridors have been conducted in agricultural landscapes where corridors create a stark and often permanent contrast with the surrounding fields. Extrapolation from agricultural to managed forest landscapes is problematic because conditions surrounding wildlife corridors offer lower contrasts and can be dynamic as the result of forest regeneration and development (Lindenmayer 1998).

Alternatives to Wildlife Corridors

A principal reason for establishing wildlife corridors is the promotion of connectivity for biodiversity. However, connectivity and corridors are not synonymous. Corridors may be particularly useful where limited vegetation retention on cutovers (e.g., clearcuts) provides a high level of contrast between the retained strip and adjacent harvest units (Lindenmayer 1998). But corridors are only one approach to facilitating connectivity and are not effective for some species (see Chapter 3).

The northern spotted owl exemplifies a vertebrate for which use of corridors for facilitating movement of dispersing birds among reserves was explicitly rejected because scientific data indicated the species dispersed randomly (Thomas et al. 1990; Murphy and Noon 1992). The conservation strategies for the northern and California spotted owls are, in fact, based upon the maintenance of connectivity by maintaining specific forest structural conditions throughout the matrix (Verner et al. 1992; Forest Ecosystem Management Assessment Team 1993; see Chapter 11).

Another example of an alternative approach to facilitating connectivity is found within the Tumut Fragmentation Experiment in southern Australia. Patches of remnant native vegetation embedded within an exotic radiata pine plantation have acted as stepping stones to facilitate the movement of some bird and arboreal marsupial species (see Chapter 13). Structural retention in harvested units is a primary strategy being used to promote connectivity in *Nothofagus* forests in Tierra del Fuego (see Chapter 15).

Although corridors have limitations, and other strategies can better provide the connectivity required by some biota, existing wildlife corridors should be retained because of the connectivity that they do provide (Beier and Noss 1998). However, corridors and other strategies, such as setting aside stepping stones, may have limited value if habitat loss and degradation continue in the surrounding matrix (Rosenberg et al. 1997).

Other Important and Unique Areas

Specialized Habitats

Most forest landscapes incorporate a number of unique physical habitats that are critical for the persistence of highly specialized or unique species and communities of organisms. They also may provide roosting, nesting, or hibernating places for more widespread species. Special habitats include cliffs, caves, rock slides, rare rock types, meadows, thermal features, and vernal pools. Conflicts between the protection of special habitats and commercial activities are often (but not always) negligible because such sites have limited value for resource exploitation and typically occupy only small portions of a landscape (Thomas 1979; Brown 1985). For example, species-rich cave environments are extremely limited in their spatial extent and tend to be concentrated in particular areas (Culver et al. 2000). Management plans need to identify (and map) specialized habitats and develop protective measures to ensure that timber harvesting, road building, and other activities do not inadvertently impact upon them and their associated flora and fauna (Gerrand 1997; Culver et al. 2000).

Our category, "specialized habitats" does overlap with that of "biological hotspots." Specialized habitats are primarily geophysical features of a landscape, whereas biological hotspots are places within a forest landscape or streamscape that have extraordinary value for biodiversity conservation. Both types of habitats can be broadly considered to be biological hotspots in that they contribute considerably to the biological richness and productivity of a landscape. The conservation value of both types of areas is often highly disproportional to the area that they occupy (Zimmerman and Bierregaard 1986).

Cliffs

Cliffs are steep, vertical, or overhanging rock faces (Thomas 1979; Larson et al. 2000) (Figure 6.12). They support a variety of niches such as cracks, ledges, and caves. Cliffs provide physical protection from climatic conditions and their special attributes are often related to concentrations of biodiversity within a relatively small and stable portion of the landscape (Larson et al. 2000). The distinctive biota and its strong interrelationships with the physical environment of cliffs have led to their recognition as distinctive and important ecosystems (Larson et al. 2000). Rock type, cliff height, and degree of isolation typically influence the effectiveness of cliff habitats for biota. Attributes that affect the presence of thermals or atmospheric updrafts are important for raptors (Craighead and Craighead 1969; Olsen 1995), and distance to permanent water is essential for some other vertebrates (Thomas 1979; Churchill 1998).

Cliffs are utilized as habitat by many vertebrate species. For example, raptorial birds such as condors and falcons are strongly dependent upon cliffs as habitat elements (Thomas 1979; Olsen 1995). Denton (1976) reported that the prairie falcon (*Falco mexicanus*) used high cliffs because of predictable thermals and security from ground-dwelling predators. Other examples of species groups that utilize cliffs are swifts and bats (Thomas 1979; Churchill 1998). In Australia, the highly specialized rock-wallaby group is strongly associated with cliff habitats.

Distinctive floras, including ferns and lichens, are also associated with cliffs and rock outcrops (Duncan and Isaac 1986; Anderson et al. 1999). For example, cliffs of calcareous rocks in Sweden support specific flora and are often registered as Key Woodland Habitats (National Board of Forestry 1995) (see Box 6.2).

Talus and Scree

Talus and scree are accumulations of broken rocks found at the base of cliffs or other steep slopes (Thomas 1979) (Figures 6.12 and 6.13). Physical variables include rock type, rock size, pore size, depth, width, and length, all of which influence the potential of talus and scree as habitat. For example, in northwestern North America, a large, deep, stable talus of igneous intrusive rock is considered to be a more

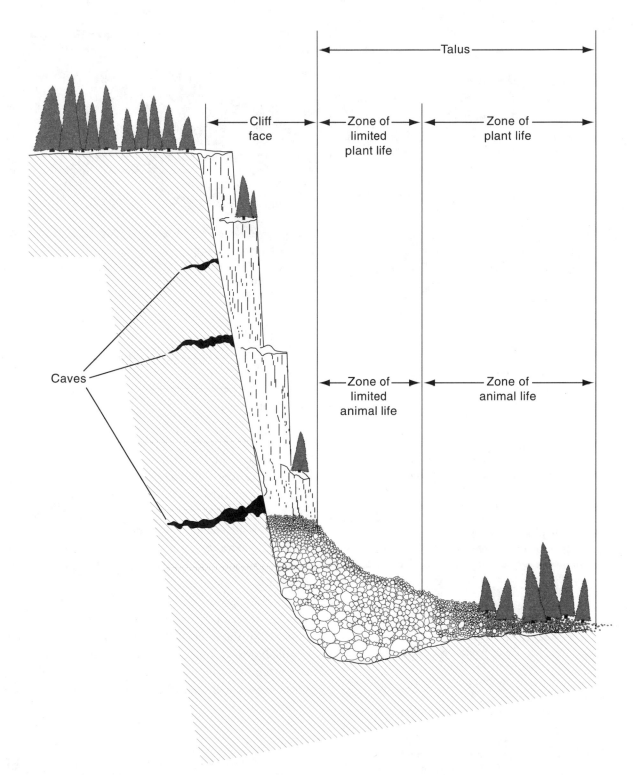

FIGURE 6.12. Common topographic relationship of cliffs, talus, and scree, and caves. Redrawn from Brown 1985.

FIGURE 6.13. Rock slides, outcrops, and cliffs provide critical habitat for organisms in forested landscapes (Columbia River Gorge National Scenic Area, Oregon, United States). Photo by J. Franklin.

valuable wildlife habitat than a small talus or one composed of sedimentary rock (Thomas 1979). The depth of these rock formations can influence animal movements and help them to find suitable environmental conditions; deep talus has also been found to be more stable (Krear 1965). Areal extent determines whether there is sufficient habitat for viable populations or breeding pairs of organisms.

Some species are virtually restricted to talus, scree, and rock outcrops. Weathered rocky sandstone outcrops are daytime shelter sites for the endangered broad-headed snake (*Hoplocephalus bungaroides*) near Sydney (Australia) (Shine and Fitzgerald 1989). In North America, the American pika (*Ochotona princips*) is strongly dependent upon talus (Smith 1980). In other cases, concentrations of widespread plants and animals occur on talus, scree, and rock outcrops. For example, in the interior Columbia Basin region of the Pacific Northwest, in the United States, large numbers of vertebrates such as snakes, lizards, amphibians, birds, and small mammals occur in talus habitats (Thomas 1979).

Caves

Caves are natural underground chambers that are open to the surface. The Federal Cave Resources Protection Act of 1988 (United States) defined caves as

> any naturally occurring void, cavity, recess, or system of interconnected passages which occur

beneath the surface of the earth or within a cliff or ledge . . . and which is large enough to permit an individual to enter, whether or not the entrance is naturally formed or man-made.

There are many different types of caves, and they vary depending upon parent material, mode of origin, depth, and overall size. Caves can be very important for the biodiversity of the landscape in which they are located. They provide a dark, stable internal environment sheltered from extreme weather conditions and offer protection from predators. Pools, streams, and even rivers also may occur in caves.

Caves provide roosting, hibernating, and hiding habitat for many animals such as bats, which utilize cave systems, sometimes in immense numbers (Dwyer 1983). For example, in northwestern North America, caves provide breeding, nursery, and hibernation sites for the western big-eared bat (*Plecotus townsendii*) (Thomas 1979). Enormous congregations of bats, in turn, attract many predators, creating diverse vertebrate assemblages and food webs.

Caves are habitats for some highly specialized organisms found nowhere else. Culver et al. (2000) calculated there were 927 species in the forty-eight contiguous states of the United States that were limited entirely to caves and associated subterranean habitats. The distributions of many of these species is highly restricted; almost half (44 percent) of the cave-dependent aquatic species and subspecies have ranges limited to a single county (Culver et al. 2000).

Deep caves are inhabited by a distinct flora and fauna adapted to total darkness (Brown 1985). In very low light environments, various kinds of algae may be present. Under conditions of total darkness, the only producers found will be bacteria and similar organisms that use chemicals rather than light as an energy source (chemosynthetic autotrophs). Cave taxa are adapted to extremely stable environments and include an array of species of worms, insects, small crustaceans, snails, fish, and amphibians (Culver et al. 2000).

Human activities at the surface can affect caves in many ways (Brown 1985). Even limited activity within a cave can impact negatively on the biota—Thomas (1995) documented the loss of body condition in bats disturbed during hibernation by humans. The destruction of habitats at cave entrances can impact species, such as crickets, that live inside caves but forage in adjacent areas. Cave ecosystems depend upon inputs from above the ground, and modifications in those inputs can have profound consequences. For example, small changes in the amount of organic materials entering a cave can lead to drastic changes in the resident species (Wilson 1978), and alterations in wind currents and light around the entrance can significantly affect the cave environment. One observer (Nieland in Brown [1985]) noted

> a marked decrease in nutrient input into shallow lava tube caves . . . following timber harvest over them [and speculated] that tree roots that penetrate through cracks and hang from the ceilings of these caves provide a nutrient source for invertebrate species. The harvest of trees . . . has an obvious effect on the species living in the cave.

Serpentines and Other Geological Formations with Unusual Properties

Highly distinctive ecosystems and biotas are often associated with geological formations having unusual chemical properties, such as ultramafic or highly calcareous rock types (e.g., Proctor and Woodell 1975; Gibson et al. 1992; Anderson et al. 1999). This is a consequence of the chemically and physically distinctive substrates associated with such formations (Klinka et al. 1989; Harrison et al. 2000). Because such rock and associated soil types often occur in small outcrops and narrow seams, management plans should take special note of their occurrence in the matrix.

Serpentines provide a widespread example of such a habitat (Lyons et al. 1974; Anderson et al. 1999; Knowles and Witkowski 2000). Serpentine areas are habitats with soils derived from ultramafic rocks either as peridotite and dunite (igneous forms) or serpentinite (the metamorphic derivative) (Kruckeberg 1967). Such soils are typically low in total and adsorbed calcium and high in magnesium, chromium, and nickel (Walker 1954; Whittaker 1954a). As a consequence, serpentine supports compositionally and structurally distinctive ecosystems, including plants tolerant of low calcium levels that are rare or absent from adjacent habitats (e.g., Whittaker 1954b;

FIGURE 6.14. Serpentines are often occupied by structurally and compositionally distinct ecosystems that contrast with the surrounding forested areas and are, consequently, important contributors to landscape-level biodiversity. Avenues of grass trees (*Xanthorrhea* spp.) often occur on belts of serpentine rock within woodlands of southeastern Australia. Photo by D. Lindenmayer.

Kruckeberg 1967; Barbour and Major 1977; Franklin and Dyrness 1988) (Figure 6.14).

Cedar glades and barrens provide other examples of habitats distinctive in structure and composition associated with limestone, sandstone, and granite outcrops in the eastern deciduous forests of North America (Anderson et al. 1999; Delcourt and Delcourt 2000). The cedar glades, named for the conspicuous presence of eastern redcedar (*Juniperus virginiana*), occur on thin-bedded, dolomitic limestone. These grassy islands within a heavily forested region support a distinctive flora that includes winter annuals and prairie grasses and forbs.

Deep deposits of sand (whether in the form of dunes or not) are another example of a geological formation that may support conditions which contrast sharply within the surrounding vegetation, thereby providing unique habitats and biota (e.g., Barbour and Major 1977; Christensen 2000). Deposits of sand can make it possible for the development of forest cover within steppe regions that are generally considered too dry for forest. An example is forests of ponderosa pine (*Pinus ponderosa*) within shrub steppe zones in the interior Columbia Basin region of the Pacific Northwest (Franklin and Dyrness 1988).

The direct impacts of forest management on unusual geological formations are often minimal because of the relatively low productivity of many of these types of habitats. However, there can be significant direct impacts from grazing, harvesting, or casual collection of plant products, and, especially mining (since many formations such as serpentines are associated with highly mineralized rock formations). Indirect effects, such as those associated with road networks or fire control programs, also need to be identified and considered in management planning.

Thermal Habitats

Thermal habitats, such as hot springs, support distinctive floras and faunas that make a unique and irreplaceable contribution to the biodiversity of a land-

FIGURE 6.15. Thermal habitats can have extremely important values for biodiversity—sometimes in unexpected ways. The thermal areas in Yellowstone National Park are often attractive as warmer resting places for large charismatic mammals such as bison (*Bison bison*) toward the end of a difficult winter. Photo by D. Lindenmayer.

scape (Figure 6.15). Overall plans for conservation of biodiversity within a landscape should ensure their identification and the adoption of appropriate protective measures.

Meadows within Forest Landscapes

Communities dominated by herbaceous species, such as meadows and grasslands, or those dominated by shrubs enrich the habitat and biodiversity of otherwise forest-dominated landscapes (Peterken and Francis 1999) (Figure 6.16). The contrasting vegetation conditions (floristic composition and community structure) typically support distinctive floras and faunas, dramatically expanding species diversity within a landscape. In addition, herbaceous and shrubby communities are often critical for some forest-dwelling taxa. Identifying these herbaceous and shrub-dominated communities and adopting appropriate protective measures is a valuable part of a comprehensive approach to biodiversity conservation within a forest landscape.

Appropriate protective measures vary with the nature and contribution of the nonforested communities, the landscape context, and potential impacts of current and projected land use. They may include designing road networks to eliminate or minimize impacts on nonforested communities, designating protective buffers that maintain existing edge or ecotonal environments between the nonforested and forested patch, and regulating or eliminating grazing by domestic livestock.

FIGURE 6.16. Meadows interspersed through forested landscapes provide important resources for forest fauna and support distinctive biotic assemblages. Photo by J. Franklin.

Active management, such as removal of trees, is sometimes necessary to maintain desired conditions in meadows, grasslands, and heaths. In the Wyre Forest National Nature Reserve in the midlands of England, areas of heath are deliberately kept open and the invasion of woodland and forest trees prevented so as to maintain habitat for reptiles such as the adder (*Vipera berus*). Similarly, in the forests of Swiss Jura, a reduction in the abundance and cover of trees and shrubs is considered critical for the survival of populations of the related asp viper (*Vipera aspis*) (Jäggi and Baur 1999).

As with aquatic ecosystems, the ecotone (or transitional zone between the nonforested communities and the forest matrix within which they are embedded) may have intrinsic value for biodiversity as well as for its protective buffering effect on the nonforested feature. For example, raptors foraging in meadows and grasslands may require adjacent nesting and perching habitat (Olsen 1995). Brown (1985) described how the endangered orange-bellied parrot (*Neophema chrysogaster*) in Tasmania nested in cavity trees in forested gully lines but foraged in adjacent buttongrass moorlands. Similarly, appropriate protective cover (thermal and hiding habitat) may be necessary for ungulates that use the nonforested communities. Buffers or other management measures for protecting such ecotonal environments are essential.

Vernal Pools

Vernal pools are seasonally inundated areas that occur within shallow depressions in some landscapes and often are host to distinctive floras and faunas (Holland and Jain 1977; Richardson 2000). They can vary in size from a few tens of square meters to many hectares. Because they are seasonal features that are often small, they can be overlooked in landscape analyses and forest management. Seasonally ponded areas also may occur in forested landscapes as a result of porous dams created by lava flows or landslide deposits that fill during rainy seasons and drain during dry periods (e.g., Franklin and Dyrness 1988). These areas should be identified and appropriate protective measures adopted.

In some forest landscapes, standing water may occur primarily in the form of seasonally ponded wetlands or as accumulated rainwater within the depressions created by the uprooting of trees rather than as true vernal pools.

Biological Hotspots

Biological hotspots are areas within a landscape that have extraordinary importance for organisms or ecosystem processes but that are not typically recognized as distinct patch types (compared with the specialized habitats described in the previous section). Biological hotspots include sites with considerable value for some species or groups of organisms, particularly with regard to essential biological activities, such as reproduction or overwintering, and source areas for resources, such as coarse woody debris for riverine ecosystems.

These biologically important places need to be identified and managed for the special contribution they make to landscape-level biodiversity.

Areas of Intense Biological Activity

Biological hotspots include locations used for reproduction, such as calving sites, high-quality spawning habitat, foraging sites with rare but essential food resources, and overwintering habitat. Biological hotspots can act as important population source areas from which offspring and/or other age cohorts can disperse and help to maintain populations over much larger areas of the landscape (e.g., Pulliam et al. 1992).

- *Calving sites for ungulates.* Many ungulates, such as the North American elk (*Cervus elaphus*), seek out particular habitat to give birth (typically protected places with mild environments). Management plans should assure that appropriate structural conditions, such as those that provide visual screening and protection, are maintained. Seasonal restrictions on human use of such sites also may be appropriate.
- *High-quality spawning habitat.* High-quality spawning habitat for fish is another example of localized areas with extraordinary significance for species reproduction and survival. Stream and river reaches vary widely in their potential as spawning habitat, and very high-quality reaches may be the sites of essentially all successful reproduction. Consequently, such areas need to be identified and provided with appropriate protection from the effects of such activities as timber harvesting and

road construction. The geomorphology of river systems is a factor significantly influencing the relative importance of various reaches and features for aquatic diversity and productivity. However, reproductive sites in some riverine ecosystems are often also sites influenced by biota such as beaver (*Castor canadensis*), and by inputs of coarse woody debris (see below).

- *Foraging sites with rare but essential food resources.* Sometimes essential components of the diet of a species can occur in a very limited portion of a landscape—and the places where it occurs can act as a biological hotspot. A classic example is the eucalypt trees that are sap sites for the yellow-bellied glider in eastern Australia. Just a handful of trees (less than 1 percent) in an entire stand have the necessary rates of sap flow to support a colony of gliders (Goldingay 2000b) (Figure 6.17). These trees may be used repeatedly over a decade or more. Special prescriptions have been instituted in the Australian states of New South Wales and Queensland to prevent sap-feeding sites from being cut down during forestry operations.
- *Overwintering habitats, especially for large populations.* Overwintering sites are another example of localized areas essential to maintenance of biota. Some of these may be the specialized habitats mentioned earlier, such as caves, or they may be individual structures, such as logs and snags, which are important to many vertebrates. For example, local populations of the broad-headed snake in southern Australia hibernate communally in a small number of cavity trees. Specific locations within a landscape, region, or continent also may have considerable value as overwintering sites for particular species or groups of organisms. The winter habitat for species that migrate for long distances, such as many waterfowl and the monarch butterfly (*Danaus plexippus*), provide extreme examples of this type. Monarch butterflies congregate at high densities in relatively small overwintering areas (Malcolm and Zalucki 1993), some of which are protected, although they remain under continual pressure for logging (Thompson and Angelstam 1999). A number of species of large herbivores con-

FIGURE 6.17. Locations as small as individual trees can act as biological hotspots, such as this sap tree tapped by a colony of the yellow-bellied glider in an Australian forest. Very few trees generate sufficient sap flows to be suitable feed trees, but these stems are extremely important to the survival of the species. This is because the yellow-bellied glider derives a significant proportion of its dietary carbohydrate from plant sap. Photo by D. Lindenmayer.

gregate in localized overwintering areas that are visited repeatedly. Another example is the winter movement of the California spotted owl to the foothill woodlands of the Sierra Nevada in California.

Biological hotspots can be places that support rare species, species of special interest, or high levels of species richness (see Box 6.3). In parts of tropical South America, high concentrations of endemic birds often occur at ecotones and locations where there is persistent fog (Laurance et al. 1997). In the Greater

Yellowstone Ecosystem, 25 percent of all the bird taxa occur in localized patches of structurally complex forests at low elevations and on good soils (Hansen and Rotella 1999); these are source areas for maintaining population of many taxa.

Biological hotspots can have particular size and geographical attributes that make them critical for the persistence of a rare species in a landscape. Lindenmayer and Possingham (1996) simulated the consequences of the loss of old-growth remnants from a system of old-growth remnants scattered through a wood production–forest area in the Australian state of Victoria. Loss of a few strategically located old-growth patches strongly affected the risk of extinction of arboreal marsupials, such as the endangered Leadbeater's possum, across the entire patch system.

Source Areas for Aquatic Ecosystems

The structure of aquatic ecosystems within forest landscapes is strongly influenced by the availability of essential sediments and coarse woody debris, particularly large logs and entire trees, which need to be delivered at appropriate intervals (Beschta and Platts 1986; Maser et al. 1988). The primary source areas for these materials are headwater reaches and major floodplains of a stream network; these are, therefore, significant features of riverscapes. Some of the best spawning habitat for fish is associated with debris jams and other distinctive topographic steps in stream profiles (Abbe and Montgomery 1996; Maser et al. 1988). In some estuaries in coastal British Columbia and southeastern Alaska, individual downed trees and log jams, derived from upstream floodplain forests, provide habitat needed by juvenile chinook salmon (*Oncorhyncus tshawytscha*) and coho salmon (*Oncorhyncus kisutch*) before migrating out to sea (Maser et al. 1988).

Remnant Patches of Late-Successional or Old-Growth Forest

Protecting areas of late-successional or old-growth forest left within the matrix is an important element of most regional forest conservation strategies. As noted earlier, a significant and integrated system of large-scale forest reserves is a component of many emerging regional conservation plans (e.g., Forest Ecosystem Management Assessment Team 1993; Arroyo et al. 1996; Macfarlane et al. 1998). Such reserves provide the large blocks of habitat needed for many organisms and ecological processes. However, these large reserves have limitations (see Chapter 5). For example, more widely distributed refugia are needed for many forest organisms with limited dispersal capabilities. Some of these needs can be met by altered management regimes within harvested units (e.g., structural retention at harvest; see Chapter 8) but the protection of even relatively small blocks of intact late-successional forest also can contribute substantially to biodiversity conservation.

The contribution that small, protected forest areas can make to biodiversity conservation has been highlighted in several studies (Schwartz 1999). Turner and Corlett (1996) demonstrated the value of protected areas of tropical forest smaller than 100 hectares for plant conservation. Zuidema et al. (1996) reviewed nearly sixty studies of forest fragmentation and found that midspatial-scale reserves often had considerable conservation value (e.g., Heywood and Stuart 1992); based on this review they challenged the notion that forest biodiversity can only be conserved in very large fragments (e.g., 10,000 to 100,000 hectares) (cf. Meffe and Carroll 1995; see also Chapter 5).

In many intensively managed forests, old-growth stands are small remnants within landscapes in which they were previously dominant. Yet, these remnants can support some old-growth taxa that are rare or absent elsewhere in the landscape, provide core areas or nodes for restoration of larger late-successional blocks, and provide propagules and individuals for surrounding managed stands. Many studies have highlighted the importance of small, well-distributed patches of late-successional or old-growth stands for biodiversity in matrix lands, whether they are remnants or areas selected for protection prior to initiation of forest harvesting activities. In northwestern North America, almost 1,100 species, ranging from fungi and cryptogams to vertebrates, were identified as being closely associated with late successional forests within the range of the northern spotted owl (Forest Ecosystem Management Assessment Team 1993).

Although some species depend upon large areas of late-successional forest, many smaller taxa, such as some arthropods, fungi, and cryptogams, will persist in small remnants. Zielinski and Gellman (1999) showed that in California, small remnant areas of old growth redwood (*Sequoia sempervirens*) (exceeding 500 years old) surrounded by younger regrowth stands were valuable habitats for forest bats. This was despite the fact that such remnants were comparatively rare in the landscape and the trees they contained supported fewer smaller cavities than stems in continuous old-growth stands (Zielinski and Gellman 1999). There are many red-listed species associated with small late-successional remnants in intensely managed Scandinavian forests (Esseen et al. 1996). Indeed, the Key Woodlands Habitat initiative in Sweden (see Box 6.2) focuses much of its effort on areas within heavily used matrix lands that still retain structural and floristic elements typical of late-successional forests (e.g., large logs and large snags; Gustafsson et al. 1999).

The importance of well-distributed old-growth forest stands as refugia for many organisms was recognized in the Northwest Forest Plan for federal lands within the range of the northern spotted owl (USDA Forest Service and USDI Bureau of Land Management 1994a,b). The plan provides for the retention of all remaining late-successional forest in midscale watersheds (500 to 5,000 square kilometers) that currently contain 15 percent or less of such forests—even though the plan also provided for an extensive, well-distributed system of late-successional reserves and retention of aggregates within harvested areas.

Finally, late-successional forest remnants can be restoration nuclei for efforts to restore landscapes. McCarthy and Lindenmayer (1999a) suggest that existing protected old-growth mountain ash forests and adjacent secondary stands would be good foci for efforts to expand the area of late-successional forests for the purposes of sustaining arboreal marsupial populations. Thomas et al. (1990) recommended retaining 32 hectares of suitable owl habitat within 0.5 kilometer of known northern spotted owl nest trees, with up to seven sites per township (surveyed land segments totaling about 2,300 hectares) in forests outside of habitat conservation areas (proposed owl reserves). They expected these areas to be utilized occasionally by dispersing owls. However, more important, these old-growth relicts could function as "nuclei of older forest, surviving from the current stand, that will become core areas for breeding pairs . . . as the surrounding forest matrix grows up around them" (Thomas et al. 1990), a phenomenon that has been repeatedly observed for the northern spotted owl.

Disturbance Refugia

The presence of long-term landscape-level refugia from fire, wind, or other catastrophic disturbances overlaps substantially with our concept of late-successional forest remnants, except that these are areas that have been free of both natural and human disturbance for long periods of time. Some parts of a landscape are less likely to suffer certain kinds of natural disturbances or the intensity and frequency of such disturbances is lower (see Chapter 4). Disturbance refugia are areas that have typically experienced markedly different disturbance regimes than the bulk of the landscape. Boose et al. (2001) identified areas in the New England region of North America that had escaped hurricanes during recent centuries. Parts of Pacific Island (Samoan) landscapes (e.g., volcanic cones and deep valleys) were found to be refugia from cyclone damage (Pierson et al. 1996); these refugia contributed to the persistence of frugivorous flying foxes (*Pteropus* spp.).

Fire refugia have been identified in many regions where fire is the dominant disturbance agent. For example, refugia have been identified for Australian forests where low-intensity fire has created complex, multi-aged stands (Mackey et al. 2002); these stands occur in cooler, more shaded parts of the landscape that provide habitat for many vertebrates, including several that are rare and declining (Lindenmayer et al. 1999f).

Cultural Sites

Cultural sites are localities that have special significance to either indigenous or modern cultures. Specific values associated with these sites can vary widely. For indigenous peoples, particular locations may have general religious significance, such as associations

FIGURE 6.18. Partially carved canoe on South Queen Charlotte Island in coastal British Columbia; potential canoe trees as well as living trees utilized as sources of bark and building planks are cultural elements of forests along the north Pacific Coast of North America. In lands managed by Weyerhaeuser Company on the Queen Charlotte Islands, culturally modified trees are used as anchor points around which patches of retained forest are maintained with harvest units where modified harvesting practices are employed (see Chapter 8). Photo by J. Franklin.

with rituals and spirit quests, as sources for medicinal herbs or construction materials (Figure 6.18), or as burial sites. Caves and rock overhangs with rock paintings (Flood 1980) and fire-scarred trees where bark was stripped to construct canoes (Douglas 1997) are some of the many features of indigenous activities in forest landscapes in southeastern Australia (Byre 1991). Cultural sites for modern societies may be related to their historic, aesthetic, or recreational values, religious significance, or value as collection sites for forest products, such as berries and mushrooms.

Although cultural sites are not typically viewed as an issue for biodiversity conservation, they should be part of landscape-level management planning. They also can make surprising contributions to biodiversity conservation. For example, cemeteries and other religious enclaves have functioned as refugia for species and even significant subsets of biotic communities (Prober and Thiele 1995). Similarly, railroad and road rights-of-way sometimes function as refugia for communities or organisms (Bennett 1991) (Figure 6.19). Daubenmire (1970) made extensive use of both cemeteries and railroad rights-of-way in the heavily agricultural Palouse grasslands of eastern Washington

FIGURE 6.19. Cultural sites left by western societies can sometimes make surprising contributions to biodiversity conservation. This old and now unused wooden trestle bridge provides roosting sites for a large number of forest bats. Note that this early form of transportation network extended above the aquatic ecosystem and not directly through it as occurs with modern road networks (see the section on transportation systems in Chapter 7). Photo by D. Lindenmayer.

state, in the United States, to assess the composition and structure of natural, lightly disturbed steppe communities.

Individual structures and habitats created by western societies also may be significant for the conservation of biodiversity in managed landscapes (Saunders and Ingram 1995). For example, mine shafts and abandoned wooden bridges and buildings are roosting and hibernating habitat for several bat species in northwestern North America and elsewhere around the world (Churchill 1998). Consequently, surveys of such features are required prior to human activity under provisions of the Northwest Forest Plan (USDA Forest Service and USDI Bureau of Land Management 1994a,b). This plan requires occupied habitat is to be protected from "destruction, vandalism, disturbance from road construction or blasting, or any other activity that could change cave or mine temperatures or drainage patterns."

Meadows and heaths created and/or maintained by humans may also contribute to regional biodiversity goals. In Sweden, old, fenced pastures can be valuable to maintain because of the species-rich grassy swards they support (National Board of Forestry 1996b). Fire-maintained grasslands created by aboriginal peo-

ple in southern Queensland (Australia) are habitat for an array of flora (Bowman 1999) and some fauna in Tasmania (Brown 1985), and the maintenance of burning regimes could be valuable for conserving desired elements of biodiversity. The nonforested "balds" of the southern Appalachian Mountains in the eastern United States are a similar example (Billings and Mark 1957; Anderson et al. 1999). There are also larger anthropogenic features and entire landscapes such as the moorlands of Scotland and exotic *Eucalyptus* groves of California that have become significant reservoirs of biodiversity despite their "unnatural" origin.

The potential for conflict between cultural objectives, human activities, and conservation management of landscapes as a whole is another reason that the cultural features need to be identified and their relationships with conservation goals subsequently determined. Some jurisdictions have developed databases of cultural sites for indigenous and western societies (McConnell 1993), and their protection has been mandatory under codes of forest practice for some time (e.g., Tasmanian Forestry Commission 1987). In the Australian state of Tasmania, a zoning system has been developed to assist the protection of archeological sites (Gerrand 1997). The system is based on environmental factors as these dictate where most aboriginal sites of cultural significance are likely to occur (McConnell 1995).

Protecting and Managing Sensitive Areas, Special Habitats, and Disturbance Refugia in the Matrix

Forest management needs to take account of biological hotspots, specialized habitats, and individual structures to ensure they are not degraded or lost. Several approaches can be employed to protect sensitive areas or special habitats in the matrix. One is direct and permanent protection. For example, in Tasmania some rare and threatened invertebrates such as the Stimson stag beetle (*Hoplogonus stimsoni*) have highly restricted distributions virtually confined to wood production forests (Meggs 1997). Habitat for this and ecologically related species is provided by wildlife priority areas and habitat strips within the matrix (Forest Practices Board 1998).

Another approach is to zone the forest, with different levels of management emphasis in the different zones. Zoning is used widely in many forest management plans around the world. The zoning system within matrix mountain ash forests gives temporary protected status to habitat hotspots for Leadbeater's possum, a status that could change if the forest becomes unsuitable for the species (Box 6.3). On Vancouver Island in British Columbia, Weyerhaeuser Company uses three distinct zones—the timber zone, the habitat zone, and the old-growth zone—in which different levels of modified silvicultural practices are employed (MacMillan Blodel Limited 1999).

Another type of de facto zoning is the Key Woodland Habitats initiative in Swedish matrix forests (see Box 6.2), which are areas protected under a voluntary code (Gustafsson et al. 1999). For some kinds of sensitive areas and special habitats, harvesting may occur, but the cutting methods need to be carefully applied and/or broadly congruent with natural disturbance regimes. Thompson and Angelstam (1999) recommended the application of targeted selective harvesting techniques in herbivore overwintering grounds to ensure both cover and suitable food resources are maintained in perpetuity. In Tasmania, a geomorphology manual and matching code of practices has been developed to guide harvesting practices around rock types such as karst and, in turn, protect the biodiversity associated with such systems (Forestry Commission of Tasmania 1990, 1993; Gerrand 1997).

Forest planning and harvest practices need to account for natural disturbance refugia to ensure they that are protected or to adopt harvesting regimes that are congruent with natural disturbance regimes (see Chapter 4). For example, the ASIO system developed by Rülcker et al. (1994) for matching cutting regimes to natural disturbance regimes in Swedish forests (see Box 4.3 in Chapter 4) recognized that parts of the landscape were fire refugia and recommended they not be subject to harvesting. In northern Europe, old-growth spruce forest in fire refugia are significant for conservation and management (Esseen et al. 1992). Hansen and Rotella (1999) described a study of wind refugia on the mountainous islands of southern Alaska. Late-successional stands

---

> **BOX 6.3.**
>
> ### Using a Forest Zoning System to Protect Leadbeater's Possum Habitat
>
> *Biological hotspots in mountain ash forests on mainland Australia include areas that support the rare and declining species Leadbeater's possum. The habitat of the species has been well documented and is characterized by a combination of large-diameter, highly decayed snags and living cavity-trees surrounded by a dense understory of regrowth Acacia spp. plants (Lindenmayer et al. 1991a). Almost 80 percent of the species' distribution occurs in areas broadly designated for wood production. A management zoning system within matrix lands is based on structural features, particularly the abundance of snags. The system is used to assign areas of mountain ash forest either for production or for protection from harvesting. This approach is an attempt to limit the number of hotspots for Leadbeater's possum destroyed by logging operations (Macfarlane et al. 1998). Thus, a zoning system is one approach to protecting special habitats within the matrix. Areas assigned to logging exclusion areas under this zoning approach are not permanently reserved. Rather, they can be rezoned for logging if the processes of stand succession (such as snag-fall) result in areas becoming unsuitable habitat (Lindenmayer and Cunningham 1996).*

were found to be significant for biodiversity and were patchily distributed on valley floors, north-facing slopes, and topographic "wind shadows." Harvesting can more closely resemble natural disturbance regimes (see Chapter 4) and take account of wind refugia by (1) focusing on younger, midseral stands located in areas of higher wind exposure, and (2) applying small-gap cutting on long rotations in wind-protected areas supporting late-successional stands (Hansen and Rotella 1999).

Buffering Sensitive Areas and Special Habitats in the Matrix

Simply identifying and then reserving sensitive areas and key habitats in the matrix may not maintain their integrity. For example, Sjörberg and Ericson (1992) noted that a red-listed species of lichen was lost from a protected stand when the surrounding forest was cut. Edge and other effects may threaten protected areas within the matrix and a buffering strategy may be needed for impact mitigation. What acts as a suitable buffer depends on several factors—it could be different if the aim to is limit the impacts of a disturbance regime versus maximize species diversity within a protected area (Baker 1992). Kelly and Rotenberry (1993) formalized a general approach for enhanced buffer zone design through an interrelated set of key questions (modified from Kelly and Rotenberry 1993):

- What external forces or processes are likely to have an impact on the protected entity in question (species, community, or resources)?
- To what extent are the external forces likely to penetrate the boundary of the protected area or sensitive site and result in negative impacts?
- Can these forces be ranked in terms of their impacts to enable a priority list of buffering requirements to be developed?
- Are the potentially negative forces amenable to hypothesis testing?
- How can data be gathered to test these hypotheses?
- How can the external forces be mitigated?

Burgman and Ferguson (1995) considered several of these issues in assessing threats to rainforest fragments within production forests in the Australian state of Victoria. They presented six recommendations to improve the integrity of rainforest:

1. Improve planning and mapping of forest landscapes.
2. Adopt a system of buffers.
3. Provide special protection and exclusion zones around rainforest areas.
4. Modify timber harvesting practices in eucalypt stands adjacent to rainforest.
5. Alter road construction activities to avoid sensitive areas.
6. Conduct new research to address knowledge gaps and assess hypotheses regarding the sensitivity of rainforest communities to human activities (Burgman and Ferguson 1995).

Other approaches can be used to address some of the questions listed above by Kelly and Rotenberry (1993). Noss and Harris (1986) outlined a strategy called *multiple use modules* (MUMs) of concentric management zones in the matrix to buffer a core reserved area. Mladenoff et al. (1994) described 100-meter restoration zones around remnant old-growth patches in second-growth matrix lands designed to buffer and reduce edge effects. Unlogged or selectively harvested stands can have a positive buffering effect for adjacent sensitive areas (Recher et al. 1987; Macfarlane and Seebeck 1991). Similarly, planning of the spatial alignment of harvest units can mitigate impacts of abiotic effects such as wind damage (Lindenmayer et al. 1997a). Aquatic zones can be buffered by staggering logging operations to ensure that both sides of a riparian buffer are not logged at the same time (Recher et al. 1987). All of these approaches have adjacency rules that specify waiting periods before an area neighboring a regenerating harvest unit can be cut.

Conclusion

There are no generic recipes to guide the development of systems of midspatial-scale protected areas within the matrix. The type, size, number, and spatial arrangement of protected areas depend on management objectives (e.g., intensive high-yield wood production versus an emphasis on biodiversity conservation), the taxa targeted for conservation management, and a range of other factors, such as landscape-specific, topographic, and hydrological attributes. This is borne out by the idiosyncrasies of our case studies (Chapters 11 to 15). An important caveat regarding midspatial-scale protected areas within the matrix is that they will *not* conserve all species—some taxa will require larger intact areas, such as large ecological reserves (see Chapter 5).

There are other issues associated with matrix management at the landscape scale that will be addressed in Chapter 7, including the design and maintenance of road systems and the design and distribution of management units in time and space.

Landscape-Level Considerations: Goals for Structures and Habitats, Transport Systems, and Distribution of Harvest Units in Space and Time

If we only consider careful management at the stand level, we will sacrifice the integrity of the forest landscape. —HAMMOND (1991)

In short, we must maintain natural, functioning landscapes. —HUNTER (2002)

We continue the theme of landscape-level matrix management in this chapter, moving from critical areas within the matrix for biodiversity conservation (Chapter 6) to examine (1) landscape-level goals for the maintenance of specific structural or vegetative conditions, (2) the design, construction, and maintenance of transport systems, especially roads, and (3) the spatial and temporal arrangement of harvest units in the landscape.

Negative impacts on biodiversity and ecosystem processes can occur when the extent of recently cutover areas exceeds some threshold or, conversely, the area of some specified forest conditions, such as late-successional forests, falls below critical levels. An important objective of forest management then must be to identify such thresholds and plan management activities accordingly.

The impacts of transport systems—primarily road networks—on biodiversity have only recently received serious attention. Transportation networks can have major impacts on many aspects of landscape functioning, including biotic conservation and aquatic ecosystem processes. Many of these impacts can be reduced by appropriate road location, design and maintenance, reconstruction or elimination of old roads, and adoption of alternative methods of logging, such as by suspended cable systems and helicopters.

Planning the spatial and temporal arrangement of harvest or active management programs on the landscape is a particularly complex aspect of landscape-scale matrix management but one that is often unrecognized and poorly understood. Two extremes along a continuum of approaches are management patterns that either disperse or concentrate activities in time and space. Selection of the most ap-

propriate approach requires an analysis of the interaction of target taxa and habitat conditions with social, economic, and political constraints. Substantive considerations in decisions on spatial and temporal placement of harvesting activities in the matrix include consideration of the size of harvest units, levels of structural complexity retained within those units, and the time interval (rotation period) between regeneration harvests.

The theme of landscape-level matrix management begun in Chapter 6 is extended in this chapter to three additional and disparate issues:

- Setting landscape-level goals for specific habitats (individual structures or areas with specific vegetative or structural conditions) or landscape-level limitations or thresholds for specific conditions
- Analyzing effects of road networks (or other transportation systems) on biodiversity and incorporating that information into management decisions
- Analyzing effects of alternative approaches to scheduling management activities in space and time (e.g., dispersed versus concentrated harvest units) on biodiversity and incorporating that information into management decisions

These three issues differ in nature and complexity. We include them in this chapter primarily because they are issues that have to be addressed at the landscape level. The two largest issues—transportation networks and distribution of management activities in space and time—are related in that decisions on placement of management activities will influence decisions about transportation networks.

At present, knowledge is limited to guide analyses and decision making on these three landscape-level issues in matrix management as compared with many other topics in conservation biology. The important and complex questions surrounding effects of transportation networks and of spatial and temporal scheduling of major management manipulations on biodiversity have received relatively little creative attention, particularly among academic conservation biologists (but see Forman 1998). There are good reasons for this lack of understanding: landscape issues typically occur over larger space and time scales than are readily conceptualized (and observed) by most people.

Landscape-Level Goals for Structures and Habitat Conditions

In a managed landscape, it is sometimes appropriate to establish goals for specific structures and vegetative conditions (e.g., forest successional stages) at some larger spatial scale (e.g., an area of hundreds to thousands of hectares) in order to achieve conservation objectives. First, the distribution of some wide-ranging species may depend on access to suitable numbers of structural attributes across many stands. As an example, black bears often den in large trees (Tietje and Ruff 1980) and can be scattered over a wide area (Klenner and Kroeker 1990). Maintaining certain numbers of these structures at the landscape level is necessary to meet the requirements of such species. Second, setting structural goals at a larger spatial scale may be necessary because it is difficult or impossible to achieve either the desired levels or spatial distributions of structures—such as large-diameter snags—on a stand-by-stand basis (Box 7.1). In such a case, structural goals have to be set at a larger spatial scale, where they can be achieved. For example, overall goals for the density and spatial distribution of large-

---

BOX 7.1.

Landscape Prescriptions for Cavity Trees in British Columbia

Cavity trees are a critical element of wildlife habitat in the production forests of many jurisdictions around the world. British Columbia, Canada, is no exception. A landscape approach has been taken to address issues related to the maintenance of cavity trees in that province's forests. British Columbia has been divided into biogeographic units and cavity-tree retention strategies and then adopted in a proportion of each unit. On-ground prescriptions for any given cutover within the units are based on how the remainder of the unit had been managed in the past (without the implementation of adequate tree retention prescriptions [Anonymous 1995]). If most of the unit was unlogged, then few trees are retained on a given cutover. More trees are retained on a cutover if the other parts of the unit have a history of intensive harvesting.

---

diameter snags might be set at the scale of small drainages 100 to 1,000 hectares in size.

Maximum or minimum amounts of specific vegetative conditions may be necessary to achieve a variety of objectives. For example, where areas of recent clearcuts and young forest are expected to exacerbate peak flows associated with rain-on-snow flood events, setting limits on the allowable proportion of a landscape in that condition may be necessary to reduce their potential influence on peak flows. Alternatively, it may be necessary to maintain some percentage of a landscape in a particular habitat or condition to meet one or more requirements of a target organism, such as northern spotted owls (see below).

Specified Minimum Levels of Structures

Specified minimum numbers of stands or patches providing desired habitat conditions—such as forests with a defined level of structural complexity—may be required within a landscape unit. Such patch types may be needed as core breeding habitat or to facilitate the safe movement of organisms within a landscape. An excellent example of this kind of landscape-level, condition-based, goal setting was the *50-11-40 rule* for the Pacific Northwest region of the United States created by the Interagency Scientific Committee to Address the Conservation of the Northern Spotted Owl (Thomas et al. 1990; see Chapter 11). This rule stipulated that on federal lands located between specified Habitat Conservation Areas (owl reserves), 50 percent of the matrix was to be maintained in stands of timber with a mean diameter at breast height of 11 inches (about 27 centimeters) or greater and at least 40 percent canopy cover. This percentage applied to quarter townships (surveyed land segments totaling about 2,300 hectares). Thus, the matrix (non-HCA area) of every 2,300-hectare block had to meet the 50-11-40 standard. In another owl-related example, the habitat conservation plan developed by the Department of Natural Resources in the state of Washington (United States) had specified minimum area-based objectives for presence of NRF (nesting, roosting, and foraging) habitat and dispersal habitat within landscape units. Both NRF and dispersal habitat have defined levels of structural complexity (Washington State Department of Natural Resources 1997).

There are prescriptions for retaining trees with cavities within harvest units in every forest management jurisdiction in Australia (e.g., Lamb et al. 1998), although the number of trees retained is insufficient to adequately conserve all cavity-dependent taxa (Gibbons and Lindenmayer 2002).

The strategy for landscape-level cavity-tree prescriptions is useful to stimulate forest planners and managers in thinking about vegetation structure at scales beyond the stand level. However, like any management strategy, it is not perfect—issues such as connectivity and the movement of many cavity-dependent organisms remain unresolved.

When developing landscape-level management guidelines for structural features, it is useful first to analyze the density and distribution of such features in naturally disturbed landscapes (e.g., Ohmann et al. 1994. This information can then be compared with patterns found in current or proposed human-disturbed landscapes (see Chapter 4) and goals adjusted to ensure as much congruence as possible between natural and human-disturbed landscapes.

Landscape-Level Thresholds Associated with Cumulative Watershed Impacts

Concerns about cumulative impacts of management activities on hydrologic and geomorphic processes—as well as many other ecosystem components and processes—are often addressed by setting area limits on percentages of particular vegetative conditions within a drainage basin. For example, along the Pacific Coast of North America, rain-on-snow events are the primary cause of winter floods. Forest conditions have a significant influence on peak flood flows (Jones and Grant 1996; Jones 2000a). Specifically, areas of recently cutover forest exacerbate the intensity of the floods whereas late-successional forests tend to minimize their intensity. Consequently, limits, or "caps," may be placed on the percentage of recently cutover forestlands that is acceptable within a drainage basin.

The Mount Baker–Snoqualmie National Forest in western Washington state provides an example where such a threshold was established. In this forest, an acceptable upper limit of 20 percent of recently cutover land was identified for medium-sized river basins. Hydrologic recovery of cutover lands was assumed after twenty years; hence, an average of 1 percent of the landscape could be clearcut within a single year. (Actually, subsequent research has shown that it takes much longer that twenty years for vegetation to develop to the point where its influence on rain-on-snow hydrology is comparable to the late-successional stands that were harvested. Effects of logging roads on hydrologic regimes probably never disappear [Jones and Grant 1996].)

Obviously, landscape-level goals or thresholds are going to vary with the conservation objectives. For example, the appropriate distribution of forest age classes is likely to be very different for maintaining ungulate populations than for minimizing impacts of harvesting on peak flood flows.

Area-based goals or acceptable thresholds have some common features in addition to their application at larger spatial levels. First, such goals are often viewed as constraints on timber harvest or other manipulative activities. Second, where managers have defined goals for particular habitats, the expectation (and sometimes the legal requirement) is that forest managers will maintain targeted levels if they are currently met or manage to create or restore them to targeted levels where the goal is currently not met.

Transportation Network and Logging Systems

Transportation systems are among the most obvious human alterations of landscapes and have a broad array of functional roles for human society. These include networks of highways and secondary roads, railroads, pipelines, irrigation canals, and power lines. All of these can have direct and indirect impacts on biota and ecosystem processes, intended or otherwise. In this section, because of their extent and influence in the forest matrix, we focus on transportation systems—and roads in particular—that were developed to facilitate forest management activities, and on the responsibility that resource managers have for their design, modification, and maintenance.

Forest management requires transportation systems to facilitate protection, management, and utilization of forest resources. Forest road networks often reach relatively high densities for a non-urbanized landscape and require large initial investments for construction and high continuing costs for maintenance. The United States Forest Service maintains and administers over 622,000 kilometers of roads on the national forests with an annual maintenance cost of about $900 per kilometer; the current backlog of deferred maintenance on this road system is $US5.5 billion (USDA Forest Service 2000). This road distance is double the length of the national highway system in the United States and fifteen times the circumference of the earth (Worldwatch Institute 1998).

Road networks can have immense direct and indirect impacts on terrestrial and aquatic ecosystems and their associated biota because they are so extensive and offer such contrast with vegetated areas (Figure 7.1). Road impacts are particularly profound when they intersect the riverine network—the other extensive network found in forested landscapes (Figure 7.2; see Chapter 6). Hydrological and geomorphological disturbance regimes associated with streams and rivers

FIGURE 7.1. Roads have profound effects on biodiversity. They can be barriers to movement, but corridors for others (including exotic species). They are also areas of high mortality for some organisms. Roads can modify physical conditions in adjacent forests (secondary road through ponderosa pine forest on the Navajo Indian Reservation, Arizona, United States). Photo by J. Franklin.

may be highly modified by their interactions with road networks with significant consequence for native biota.

Historically, the transportation and stream networks were synonymous in some jurisdictions (e.g., northwestern North America; Sedell et al. 1988). Early long-distance transport of logs was often achieved primarily by floating logs down rivers and, occasionally, rafting them over larger bodies of water. In some locations, these "log drives" continued through most of the twentieth century. Log movement was sometimes accomplished through the use of

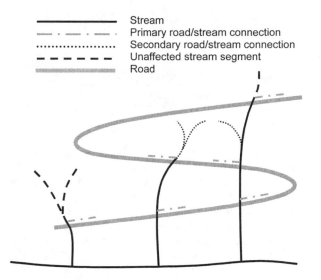

Stream
Primary road/stream connection
Secondary road/stream connection
Unaffected stream segment
Road

FIGURE 7.2. Impacts of road networks on riverine networks can be profound because of the extensive intersection of the two networks in a typical mountain landscape that can lead to altered hydrological flows, including extension of the channel system. Diagram based upon illustrations from Swanson et al. 2000.

"splash dams," which were temporary dams used to accumulate water until it was released as a flood flow that could facilitate log movement downstream. The use of splash dams—and channel simplification carried out to facilitate log drives—significantly impacted downstream riparian and aquatic ecosystems and organisms. Flumes, which are open channels constructed of wood and filled with water, were also used to move logs and timber well into the twentieth century.

Railroads were another common means of transporting logs during the late nineteenth century and the first half of the twentieth century in many parts of the world (e.g., Evans 1993). Although railroads continue to be used for forest transport in some areas, they were gradually replaced during the twentieth century because of the lower capital investment required for tractor- and truck-based logging and hauling. Railroad networks generally had lower impacts on aquatic ecosystems than did road systems because of lower network densities and the use of trestles to bridge streams, rivers, and depressions rather than culverts and extensive landfills. Abandoned timber trestles (the wood used was rarely treated with preser-

vatives) even made positive contributions of coarse woody debris to the aquatic ecosystems. Wildfires were often a problem with logging railroads, however.

Many factors influence the design of a transportation network in managed landscapes, including the management strategy and objectives, economics, and the physical characteristics of an area. Forest managers have often assumed that an extensive permanent road system is essential because they planned to manage all of the forest estate all of the time. This paradigm of "access and management everywhere, forever" is at least partially a reflection of earlier management emphases (e.g., fire protection) and past technologies (e.g., no aerial access). It has only recently has been seriously challenged (Gucinski et al. 2000). In contrast, new approaches may allow portions of road networks to be temporary and subject to closure or even removal following timber harvest.

Logging methods have been a dominant influence on decisions about the characteristics of the road network (e.g., location and total road density) that is needed in a managed landscape. Harvest systems that utilize ground-skidding equipment usually require the most extensive road systems. Cable-logging systems, especially those that use long-span skylines, significantly reduce the road densities necessary in an area. Aerial logging using helicopters can effectively eliminate the need for most roads but is, of course, costly and dependent upon fossil fuels.

It is only recently that biodiversity conservation has been a consideration in the design, construction, and maintenance of transportation systems. During the past several decades, concerns have grown about roads and their impacts (Noss and Cooperrider 1994; McGarigal et al. 2001). Negative relationships between road density and aquatic fauna have been observed in many studies (e.g., Vos and Chardon 1998; Baxter et al. 1999) and include effects on geomorphological (e.g., mass soil movements) and hydrological-processes and, ultimately, on fish and other aquatic organisms (e.g., Forman 2000; Jones et al. 2000; Trombulak and Frissell 2000). Improvements in road location (such as identifying and avoiding unstable landforms), road construction methods, and road maintenance have received increased attention for several decades in many regions. However, roads are viewed as having unavoidable effects on streams even if they are well located, designed, constructed, and maintained (O'Shaughnessy and Jayasuriya 1991; Gucinski et al. 2000; USDA Forest Service 2000). Elimination of existing roads and bans on construction of new roads are also components of current management programs in some forest jurisdictions (e.g., Gucinski et al. 2000; USDA Forest Service 2000).

Transportation networks other than roads—such as railroads, irrigation systems, pipeline and power-line rights-of-way, and hiking trails—are also important conservation influences in some landscapes. Effects of these systems need to be considered in the development of any landscape-level strategy for the conservation of biodiversity. However, with the exception of trails, these networks are not as common as road networks or an integral part of the managed forest infrastructure, so we will not discuss them further here.

Impacts of Roads on Terrestrial Ecosystems and Biota

Forman (2000) estimated that 20 percent of the land area in the United States is influenced directly by the system of public roads based on the conservative assumption that major road effects extend for about 100 meters on either road side. Road networks can have a wide range of negative ecological impacts (Spellerberg 1998; Forman and Deblinger 2000; Gucinski et al. 2000; Trombulak and Frissell 2000; USDA Forest Service 2000) (Figure 7.3). Not all effects of roads are of equal intensity along their length—some areas, such as stream crossings, are particularly heavily impacted.

Major negative impacts of roads on terrestrial biodiversity and ecosystem processes include the following:

- *Roads can be dispersal barriers*. Roads can bisect and fragment habitats and block migration and other travel routes for organisms. They have been found to impede the movement of a broad range of organisms such as invertebrates (Mader 1984; Baur and Baur 1990; Haskell 2000), small mammals (Burnett 1992; Goosem 2001), and large mammals (Cle-

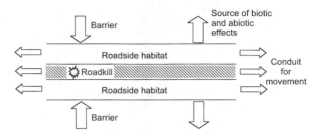

FIGURE 7.3. Roads function as barriers to movement of organisms, habitat modifiers, pathways for exotic movement, and zones of high mortality in essentially all forms, whether they are wide, high-speed roads or narrow, unpaved ones. Redrawn from Bennett 1991. Reprinted from *Nature Conservation 2: The Role of Corridors*. Edited by D. A. Saunders and R. J. Hobbs. Chipping Norton, N.S.W., Australia: Surrrey Beatty and Sons.

venger and Waltho 2000; Forman and Deblinger 2000). Roads can result in isolated and fragmented populations and altered patterns of genetic variability (Reh and Seitz 1990). Even rudimentary roads (tracks and log-skidding paths) can have significant impacts. For example, overgrown logging tracks impeded movement of the small marsupial the brown antechinus (*Antechinus stuartii*) in the forests of New South Wales in Australia (Barnett et al. 1978). Roads through wildlife corridors may be barriers to the animal movement that the corridors are intended to facilitate (Lindenmayer et al. 1994c; Bright 1998).

- *Road traffic is a major cause of animal mortality*. In addition to plant and animal mortality at the time of road construction, collisions with vehicles are a continuing and often substantial source of mortality for many animals (Bennett 1991; Trombulak and Frissell 2000). The problem is particularly acute where road systems intersect regular travel or migration routes for animal populations, which is often the case with amphibians moving to and from breeding habitats. Reptiles often seek roads for thermal heating (Vestjens 1973). It has been estimated that approximately 5.5 million frogs and reptiles are killed annually on paved Australian roads (Ehmann and Cogger 1985). There are many examples of individual species for which impacts of road mortality have been documented, such as the koala (*Phascolarctos cinereus*) in Australia (Lee and

Martin 1988) and the hedgehog (*Erinaceus europaeus*) in The Netherlands (Huijser and Bergers 2000). Huggard (1993) found that elk close to the Trans-Canada Highway were 2.5 years younger than those more than one kilometer away—an effect due to higher rates of animal mortality close to roads.

- *Roads alter ecosystem composition and structure.* The construction of roads results in the direct and sometimes permanent modification of native plant and animal communities in the road right-of-way (Malcolm and Ray 2000). Many existing organisms are killed and the physiological and reproductive status of surviving organisms is altered. New communities develop along rights-of-way that incorporate opportunistic native species and exotic taxa that were not previously present. The abundance and diversity of some groups, such as small mammals, may increase because of the influx of generalist and nonforest species (Adams and Geis 1983). This also may make roadside areas prime foraging habitat for raptors and other predators.

- *Roads fragment and modify the terrestrial physical environment and create edge effects.* The presence of roads results in substantial modification of the physical environment within the road right-of-way as well as in adjacent ecosystems as a consequence of edge effects (Trombulak and Frissell 2000). The alterations include (1) the roadway itself, which is a densely compacted surface, (2) modified environmental conditions (e.g., light, wind, moisture, and temperature regimes), (3) modified chemical environments from the addition of salts, heavy metals, organic compounds, ozone, and other toxic materials, and (4) dust.

Impacts of roads on the environment and behavior of organisms in adjacent areas may create significant edge effects. In one study, the amount of edge habitat created by roads was more than twice that produced by clearcutting (Reed et al. 1996). In another investigation of nine attributes (wetlands, streams, road salt, exotic plants, deer, moose, amphibians, forest birds, and grassland birds), edge effects extended more than 100 meters from the road right-of-way for all measures while some impacts penetrated up to a kilometer (Forman and Deblinger 2000). The road effect zone averaged approximately 600 meters in width, with convoluted boundaries. Increased nest predation can occur along roads, even along narrow bush tracks (Burkey 1993). Windthrow is also often associated with roads (Franklin and Forman 1987).

The impacts of roads on aquatic ecosystems, including modification of hydrology, geomorphological processes, and aquatic organisms, are so profound that they are dealt with separately in the next section.

- *Roads modify the conditions of organisms and animal behavior.* Effects of roads on resident organisms include altered reproductive success and physiological condition (for both plants and animals), altered movement patterns and escape responses, and home range shifts (Trombulak and Frissell 2000). Haskell (2000) showed that the depth of the litter layer of forests adjacent to roads was reduced and this, in turn, was believed to impact a wide range of invertebrate and vertebrate taxa. Road density has been shown to influence the presence and abundance of wildlife (Lyon 1983; Wisdom et al. 1986; Mader et al. 1990; Van der Zee et al. 1992; Thurber et al. 1994), although some of this is probably related to human use (see below). Many studies have shown that bears avoid roads (Mace et al. 1996; Kremsater and Bunnell 1999). Traffic noise may affect some groups of animals such as birds.

- *Roads promote the dispersal of weeds, pathogens, and animal pests.* Roads promote the spread of exotic species by altering physical conditions, removing or stressing native species, and facilitating movement of vectors, such as human beings and domestic animals (Benninger-Traux et al. 1992; Trombulak and Frissell 2000). Animals that utilize roads can disperse weeds (Bennett 1991; Goosem 1997). One investigator recovered viable seeds of more than 220 weed species from a motor-vehicle-washing facility in the Canberra region of southeastern Australia (Wace 1977). Similar results have been reported in Europe (Schmidt 1989). Vehicular traffic and road drainage are implicated in the dispersal of a virulent exotic root rot, *Phytopthora cinnamomi*, which is decimating populations of Port Orford cedar (*Chamaecyparis lawsoniana*) in the Siskiyou Mountains of the northwestern United States (Zobel et al. 1985). Roads may provide conduits for the movement of introduced animal pests (Seabrook and Dettmann 1996), such as the red fox (*Vulpes vulpes*) and feral cat (*Felis cattus*), in the forests of southeastern Australia (May and Norton 1996). This may allow feral animals to gain access to areas from which they were previously absent and increase predation on native animals (Robertshaw and Harden 1989; May 2001).

- *Roads promote alteration and use of habitat by humans.* Road networks facilitate the presence in, and utilization of, the landscape by human beings. The World Commission on Forests and Sustainable Development (1999) estimates that between 400 and 2,000 hectares of forest are lost for each kilometer of new road constructed in Brazilian Amazonia. Other negative consequences can include (1) increased harassment of animals and damage to plants, (2) human predation in the form of hunting and trapping, both legally and illegally (Wisdom et al. 1996; Bennett 2000), and (3) settlement of habitat by human beings (Trombulak and Frissell 2000). Commercial harvest or firewood cutting results in loss of large trees, snags, and logs in areas adjacent to roads, which has adverse effects on cavity-dependent birds and mammals.

- *Predation and harassment of animals can be significant.* More than 70 percent of the ninety-one vertebrate species reviewed in the Interior Columbia River Basin, in the northwestern United States, are negatively affected by one or more road-related effects (Wisdom et al. 1996). Roads facilitate poaching of many large North American animals, including North American elk (*Cervus elaphus*), pronghorn (*Antilocapra americana*), mountain goat (*Oreamnos americanus*), big-horned sheep (*Ovis canadensis*), gray wolf (*Canis lupus*), and grizzly bear (*Ursus horribilis*) (Mech 1970; Stelfox 1971; Yoakum 1978; Dood et al. 1985; Knight et al. 1988; McClellan and Shackleton 1988; Cole et al. 1997). In Africa, roads constructed to aid timber extraction are used by humans to hunt for bushmeat, with huge numbers of animals being taken every year (Bennett 2000). Elevated hunting pressure from roads can also influence the population dynamics of native animals (Thiel 1985; Redford 1992). Many carnivores are sensitive to the presence of human beings, including the grizzly bear (Mace et al. 1996). Gray-wolf packs have a significantly reduced chance of persisting when road densities exceed 0.6 kilometer per square kilometer (Thiel 1985; Jensen et al. 1986; Mech et al. 1988). These kinds of negative interactions with humans increase the mortality of bears and wolves and can cause high-quality habitat near roads to become population sinks (Mech 1973).

- *Roads may promote disturbance events.* Roads may generate new patterns of disturbances in landscapes (Williams and Marcot 1991; Miller et al. 1996). Roads can result in increases in human ignition of fires either accidentally or intentionally (Franklin and Forman 1987). Roads can increase the effectiveness of fire suppression programs, which can have negative conservation consequences in some landscapes (Norse et al. 1986) and positive effects in others. As already noted, roads can significantly increase the potential for windthrow in residual forests. Finally roads have very large impacts on disturbance regimes in aquatic ecosystems (see below).

Impacts of Roads on Aquatic Ecosystems and Biota

The potential impacts of road networks on aquatic ecosystems and biota merit special attention. These impacts range widely from direct effects on biota and habitat to indirect impacts via dramatically modified disturbance regimes. These effects also may extend for very long distances from road-stream intersections both downstream (e.g., in generating sediment) and upstream (e.g., by blocking the movement of organisms).

BOX 7.2.

**Roads, Culverts, and Connectivity
in Aquatic Ecosystems**

The potentially negative effects of roads on aquatic systems are well documented. Road crossings of streams in particular can be problematic. Culverts under roads are usually designed to cope with extremes of water flows in a cost-effective manner. However, they have been found to be barriers to fish movements in some aquatic ecosystems (Walker 1999) with corresponding negative impacts on connectivity and population viability. A New Zealand research team has developed a design to limit the barrier effects of road culverts (Boubeé et al. 1999). Similarly, an Australian study demonstrated that water velocity and distance from the mouth of a culvert to the stream surface significantly influenced the movement of aquatic organisms (Walker 1999). Guidelines for culvert designs have been formulated in the Australian state of Tasmania in an attempt to limit these problems (Forest Practices Board 1999a).

- *Roads destabilize landforms and increase sediment production.* Water quality (e.g., levels of suspended solids and silt concentrations) is reduced in response to frequency of road use (O'Shaughnessy and Jayasuriya 1991). This can have negative effects on aquatic flora and fauna (Silsbee and Larson 1983; Brown and McMahon 1988; Graynoth 1989; Rieman et al. 1997; Baxter et al. 1999). For example, thirteen species of Victorian fish lay eggs on stream beds and are susceptible to smothering by increased sediment deposition (Metzeling et al. 1995).
- *Roads can permanently alter hydrological regimes and accentuate flood flows.* Roads can have profound and permanent impacts on the hydrological regimes of landscapes (USDA Forest Service 2000, 2001a). Road networks intersect, collect, and redirect both surface and subsurface flows of water (Figure 7.4) thereby modifying both hydrological regimes (e.g., timing and level of peak flood flows) and geomorphological processes (e.g., destabilizing landforms by saturating substrates).
- *Roads block movements of aquatic organisms.* Roads often disrupt aquatic habitat connectivity by creating physical barriers to movement at stream crossings. Upstream movements of

fish, amphibians, and some aquatic invertebrates are often prevented by stream culvert installations that create steps (e.g., falls) at the outfalls or create other impassible physical barriers. Accelerated rates of water flow generated by culverts laid with excessive slopes can also impair connectivity for fish and other aquatic organisms (see Box 7.2).

Modifying Road Networks to Reduce Impacts on Biodiversity

There are many strategies for modifying road networks to reduce their impact on biodiversity (Grayson et al. 1992; Reed et al. 1996). These include

- Reducing road densities by either building fewer roads or eliminating existing roads, or both
- Reducing the width of roads and also limiting the extent of associated cleared roadside vegetation (Thiollay 1999)
- Improved location and design of roads, particularly the avoidance of sensitive areas
- Improved construction and maintenance
- Construction of safe passages for organisms

Reduced Road Densities

Reducing overall road densities is a good way to reduce their environmental impacts (Schonewald-Cox and Buechner 1990) (Figure 7.4). For example, limiting the density of roads is considered to be an important strategy for the conservation of bull trout (*Salvelinus confluentus*) in densely forested catchments in Montana (United States) (Baxter et al. 1999). Given the large impacts that roads have on both terrestrial and aquatic biota, excluding roads entirely from currently unroaded areas (or by planning lower densities of permanent roads) should be carefully considered when biotic resources have high conservation value (USDA Forest Service 2000).

Selecting appropriate logging systems is an important variable affecting road densities where timber harvesting is a major landscape-level activity. Cutting methods based on ground skidding of logs requires relatively high road densities; ground skidding is, of

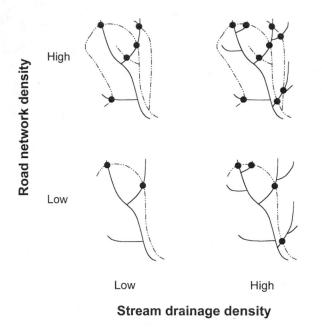

FIGURE 7.4. Schematic illustrating effects of different road and stream densities on levels of interaction between road and riverine networks. Based on illustrations in Swanson et al. 2000.

course, inappropriate on steep and unstable slopes. Short-span cable logging systems reduce soil disturbance in logged areas but generally require road densities nearly as high as ground-skidding logging systems.

Long-span cable and balloon and helicopter logging are systems that can dramatically reduce overall road densities in the landscape (Pengelly 1972; Aust and Lea 1992; Putz et al. 2000). In a recent example, yellow cedar (*Chamaecypars nootkatensis*) and Douglas-fir are being harvested by helicopter in parts of landscapes managed by the Weyerhaeuser Company on Vancouver Island in British Columbia (B. Beese personal communication). These methods also can eliminate the need for roads in highly sensitive areas, such as on steep mountain slopes where roads inevitably affect hydrological and geomorphological processes. Both approaches require higher levels of investment in equipment and skilled workers; however, increased costs can be balanced against the negative environmental impacts and costs of constructing and maintaining the roads that would otherwise be required. In the case of some tropical countries, the use of meth-

ods like helicopter logging needs to be balanced against the conservation benefits of having some areas that would otherwise be inaccessible because of the difficulty of building conventional road systems to reach them (Putz et al. 2000).

Decisions about scheduling activities in time and space are also extremely important variables in determining the density of permanent roads required. Longer rotations and the concentration of management activities in time and space (rather than dispersing them) provide more opportunities for reducing the permanent road network (i.e., substituting permanent roads with temporary ones). Temporary roads can be used during harvest and subsequently decommissioned following initial silvicultural treatments (e.g., regeneration and precommercial thinning). Issues associated with rotation period and the spatial location of harvest units are discussed more thoroughly in a following section.

Decommissioning or removing roads is a strategy for reducing road densities and reducing impacts in landscapes where road systems are already well or excessively developed (Rab 1998). The objective is to mitigate damage, restore hydrological function, and restore the road right-of-way to as near the natural condition as is physically and financially possible. Activities typically undertaken in eliminating existing roads include (after USDA Forest Service 2000)

- Blocking the entrance(s) to vehicular travel
- Creating water bars across the road surface
- Removing culverts
- Reestablishing natural drainage channels
- Removing unstable road fills
- Pulling back the shoulders of the road
- Restoring natural contours

A general goal should be to maintain flows within natural drainage basins and avoid diverting water from one basin to another, which can overload the receiving basin and increase the tendency to develop gullies or landslides (Montgomery 1994; Wemple et al. 1996).

Depending upon the level of restoration undertaken, the cost can be as little as $50 or as much as $30,000 per kilometer. Attempting to fully restore road rights-of-way (e.g., to restore natural contours and reestablish drainage channels) requires levels of effort and expense comparable to those expended in creating a road, but it can be very effective in reducing erosion (Madej 2001).

Improved Location, Design, Construction, and Maintenance of Roads

Improved location and design of roads can significantly reduce the impact of road networks on biodiversity by

- Minimizing their effects on hydrological and geomorphological regimes and thereby on aquatic ecosystems and organisms
- Minimizing the contribution of roads to habitat fragmentation and inhibitions to movement of terrestrial organisms, such as by blocking traditional travel routes (Lindenmayer 1998)
- Avoiding direct and indirect impacts to areas that are disproportionately important for biodiversity, such as meadows, wetlands, and riparian zones (Burgman and Ferguson 1995)

Locating and designing roads so as to minimize impacts on stream networks is particularly important because of their potential for extraordinary negative impacts. A fundamental principle is to avoid locating and constructing roads on geological formations, landforms, and soils that are known to be highly unstable. The negative impacts of roads constructed in such areas is highly disproportionate to their relative extent because of their tendency to cause both chronic and catastrophic disturbances of aquatic features, such as streams.

Midslope roads—roads located on mountain slopes between the ridgeline and valley bottom—typically have significant impacts, even on stable topography, because they intersect subsurface flows. This can bring such flows back to the surface, combining them with other surface flows and directing them into an expanded channel network (i.e., the road ditches). Pathways and rates of flows are affected—often dramatically and permanently—by roads (La Marche and Lettenmaier 2001; Tague and Baud 2001). Therefore, any management approaches (such as use of long-span cable or aerial logging systems) that will significantly reduce or eliminate midslope roads need to be considered seriously. The role

of slope position as well as other variables on the amount and type of erosion and deposition was quantified by Wemple et al. (2001).

Stream crossings are critical points in road location and design. The number, location, and construction of crossings is an important issue (Walker 1999) (Box 7.2). One general principle is that the fewer the stream crossings, the better. Another principle is to utilize structures, such as bridges or constructed fords, that have less potential for blockages at high-water flows, which can cause debris torrents as well as divert water into new channels. Culverts are much more likely to become blocked with organic debris and sediments, which is a process that can lead to temporary ponding and catastrophic failure of road fills and, in turn, destructive debris torrents. Both location and design (e.g., culvert dimensions) are important considerations and must address such diverse issues as sufficient capacity to handle infrequent events (e.g., 100-year floods) and a reduced need for routine servicing.

Road impacts can be significantly reduced by the selection of appropriate designs and construction techniques. Traditional approaches to road construction in steep forested landscapes often involved the use of excavated material to create the outer portions of the roadbed and extensive sidecasting of rocks, soil, and other debris. Roads of this type are subject to gradual or catastrophic failure of the fill portion of the roadbed, particularly when organic materials are included or areas of sidecast (which are sometimes extensive) are unstable. Use of "full bench" designs, whereby the entire roadbed is on an excavated bench, can dramatically reduce road impacts. Excavated materials are removed to locations where it can be utilized as road fill or otherwise safely deposited ("backhauling").

Maintaining road systems regularly can prevent them from deteriorating into sources of chronic problems as well as catastrophic problems. The importance of road maintenance was noted at the beginning of this section; many—probably most—forest road networks have large maintenance backlogs, including reconstruction of chronic problem areas. Stream crossings and other road elements close to surface waters often need particular attention.

Creating Safer Passages for Biota

Structures can be created to provide safe passages across roads for biota (Bennett 1991; Keller and Pfister 1997). Examples are overpasses and underpasses at wildlife crossings on busy roads and railroads (Evink et al. 1996; Canters 1997). Few studies of the effectiveness of such structures have been made. However, human activity around underpasses differentially influenced use by carnivores and herbivores in a Canadian study, with carnivores exhibiting higher sensitivity to humans (Clevenger and Waltho 2000). Hence, human activity needs to be considered along with topography and habitat quality in locating such structures. Structures that deter animals from crossing roads, such as reflectors on marker posts, also can reduce animal injuries and death (Jones 2000b).

Regulations can also lessen impacts of roads on wildlife, such as by limiting legal speed limits and increasing driver awareness of the potential for killing wildlife (Bennett 1991). Slowing the speed of vehicles by 20 kilometers per hour significantly reduced the numbers of road kills of native carnivorous marsupials in the Australian state of Tasmania (Jones 2000b).

Some effects of roads, such as the dispersal of weeds and pathogens, are partially related to where a road originates and terminates. Some of these problems can be curtailed by regulating vehicles using roads. For example, limiting access and cleaning vehicles is one approach used to limit the spread of cinnamon fungus in the forests of southwestern Australia. Similarly, Lindenmayer and McCarthy (2002b) proposed vehicle cleaning as a way to reduce the spread of some types of weeds in plantation landscapes in southeastern Australia.

The Arrangement of Management Activities in Space and Time

The arrangement of management activities in space and time is an extremely important landscape-level consideration in managing biodiversity and maintaining ecosystem processes within the matrix. In this section, we focus on issues related to the spatial and tem-

poral distribution of timber harvesting, particularly even-aged and related approaches that result in significant alteration of forest landscapes. These same general issues are relevant to other large-scale manipulative management programs. An example would be design and implementation of landscape-scale fire management programs, which typically involve spatially variable fuel treatments, fire breaks, and prescribed burning regimes (see, for example, Agee et al. 2000). This is a challenging topic involving many interrelated issues and questions. Should activities be dispersed or concentrated in time and space? Should management units be small or large, uniform or variable in size? How will alternative strategies affect connectivity in the landscape? How do various strategies compare with the natural disturbance regimes that formerly existed in a landscape?

Traditionally, conservation biologists have given little attention to matrix management in general, or to the issue of how biodiversity is affected by alternative approaches to imposing management regimes in time and space. Foresters typically approached the topic from very practical perspectives, such as those related to development of the transportation network, stand priorities for harvest, and economics. Their conceptual goal was the creation of a classic, fully regulated forest that—in the simplified case of a single commercial tree species, uniform and constant site conditions, and a single silvicultural prescription—can be reduced to (after Davis et al. 2001):

Total Area ÷ Rotation Age = Area in Each Age Class

The focus within a regulated forest landscape was on the perpetual even flow of benefits (timber) from a property and not on the conservation of ecological values (Davis et al. 2001). Foresters usually assumed they were working *everywhere* in a landscape essentially *all* of the time. Decisions on patch sizes were usually based on logging methods and harvest economics.

Effects of different spatial and temporal patterns of management on biodiversity and ecosystem processes are now a significant concern of managers and biologists, and conceptual and mathematical models are being developed. For example, dispersed patch clearcutting—a favored approach among many government agencies—results in rapid fragmentation of forest habitats and the creation of landscapes vulnerable to many types of disturbance (Franklin and Forman 1987; Li et al. 1993). The effects of different management regimes on biodiversity at the landscape level have been explored in mathematical models. Management regimes have been modeled on natural disturbance regimes, incorporating trade-offs among rotation length, harvest intensity, and patch size and location (Cissel et al. 1998, 1999). Landscape attributes of harvested and natural landscapes also have been compared (e.g., Mladenoff et al. 1993).

There are many interrelated elements—different spatial-temporal approaches provide different options with regard to the transportation network. For example, concentrating activities in time and space may provide opportunities for a reduced permanent road network. As discussed in following sections, three interrelated issues must be considered in designing management regimes: (1) patch size (e.g., the size of harvest units), (2) patch content (e.g., internal complexity of harvest units), and (3) rotation period.

Significant trade-offs are possible among these variables. For example, concerns over maximum patch size decline as (1) levels of within-patch structural heterogeneity are increased, (2) rotation periods are lengthened, or (3) both (see Chapter 9). Conversely, intensive management that homogenizes patches and shortens rotations will tend to increase ecological (and social) concerns about patch size (i.e., the size of harvest units). A major choice, whether it is explicitly recognized or, as in many cases, simply ignored by managers and stakeholders as a given, is whether to disperse or concentrate management activities in time and space. The answer, which is usually in favor of dispersion, is also an implicit choice between chronic and episodic human disturbance regimes.

We begin this section by considering two contrasting models of disturbance regimes—chronic and episodic disturbance. We then examine traditional management models and their consequences, and consider important variables in designing management regimes in time and space. We conclude by identifying some conservation-based approaches that are being proposed or adopted.

Natural Models of Disturbance

Natural disturbance regimes are useful models to explore when considering spatial and temporal patterns for management regimes in a region. They can provide insight into issues related to patch size, internal heterogeneity of patches, types and extent of edges, period between disturbances (rotation period), and landscape context (Chapter 4). Two contrasting models of natural forest disturbance provide end points along a continuum of disturbance frequency and intensity: (1) intense episodic disturbance regimes, and (2) chronic disturbance regimes.

Episodic disturbance regimes involve large- (stand) scale disturbances that result in a high rate of mortality of trees in an existing stand and occur at relatively long time intervals. These are sometimes described as "catastrophic," but this term is somewhat subjective and is probably inappropriate as a general description of a natural phenomenon. Another term that is used is *stand-replacing disturbances.* Episodic disturbances are typical of many forests characterized by important commercial tree species, such as Douglas-fir (Figure 7.5), red pine, and jack pine in North America, and mountain ash in Australia. Attributes of episodic disturbances typically include

- Large (sometimes very large) patch sizes
- High levels of within-patch heterogeneity reflecting spatial variation in the intensity of disturbance

FIGURE 7.5. Even-aged Douglas-fir stand regenerated following a large catastrophic fire (Yacholt Burn of 1902), which killed a stand dominated by 400-year-old trees. Photo by J. Franklin.

FIGURE 7.6. Forests subjected to chronic light-to-moderate fire regimes typically display a heterogeneous structure that includes gaps created by spatially aggregated mortality (transect through ponderosa pine forest in Bluejay Springs Research Natural Area, Oregon). Drawing courtesy of Robert Van Pelt.

- High levels of biological legacies throughout a disturbed area, but of varying quantities and types consistent with variation in disturbance intensity
- Edges that are complex in shape and often feathered or irregular and diffuse
- Variable and often long return intervals between disturbances

Forests and tree species associated with episodic disturbance regimes have been described as the model for even-aged management approaches despite profound differences in biological legacies and other ecosystem attributes between managed even-aged forests and even-aged stands resulting from natural disturbances (see Chapter 4).

Chronic disturbance regimes involve events that occur at relatively frequent intervals and at low to moderate severity. As a result, overall levels of tree mortality are typically low and the integrity of the forest stand is maintained. Chronic disturbances may affect relatively large areas but do not typically generate new forest patches (i.e., the replacement of the dominant cohort) at a large scale. Rather, some of the mortality generated in such disturbances is spatially aggregated thereby producing small openings or gaps within the stand. Forest stands that have high spatial heterogeneity (i.e., large numbers of gaps) are typically produced by such disturbance regimes (Figure 7.6). Examples include the pine and mixed conifer forests of western North America, the

longleaf pine forests of the southeastern United States, the banksia woodlands of western Australia, and many of the woodland types of eastern Australia. These forest types would have experienced frequent low- to moderate-intensity natural fire regimes. Forests that are chronically perturbed by high winds also may exhibit spatially heterogeneous structures of this type, as illustrated by the southern beech forests found in Tierra del Fuego (Rebertus et al. 1997) (Figure 7.7) and New Zealand. Silvicultural regimes utilizing selection, especially group selec-

FIGURE 7.7. Lenga forest in Tierra del Fuego illustrating the type of small gaps that are created by chronic wind disturbance within the stand; successful regeneration of the shade-intolerant lenga occurs within such gaps. Photo by J. Franklin.

tion, are sometimes viewed as models of this disturbance regime.

Episodic (stand replacing) and chronic disturbance regimes as portrayed here do represent extreme contrasts. Most natural disturbance regimes are actually mixtures of the two forms. For example, chronic disturbances and associated tree mortality create spatial heterogeneity within even-aged forests that are the result of stand-replacing disturbances, such as in coastal Douglas-fir forests of northwestern America (Franklin et al. 2002). Similarly, occasional high-intensity disturbances can cause stand replacement to occur in regions characterized by chronic regimes (Bowman 1999).

Traditional Forest Management Models

Traditional forest management approaches are generally are not well matched with natural disturbance patterns (Chapter 4). They are predominantly even-aged and usually based on clearcutting. Structural legacies are absent or limited, and management regimes of this type generally result in considerable within-stand homogeneity because uniformity in treatment is a specific goal. Moreover, these types of management regimes have the impact of an episodic (stand-replacing) disturbance regime but are often applied with the frequency of a chronic disturbance regime.

Considerations of patch size and location within managed landscapes have typically begun with the classic definition of a forest stand: a patch of forest distinct in composition or structure or both from adjacent areas (Helms 1998).

In practice, however, the goal was to create some form of regulated forest in which the natural arrangement would be reconfigured to achieve a particular (usually an even) flow of some commodities with a distribution of forest conditions as a byproduct. Although natural patches and patch boundaries might be considered in laying out harvest units, the creation of a new network of patches of different forest age classes was usually driven by other factors, including priorities for harvest, such as conversion of the oldest

and most decadent forest first or providing the type of logs (species, size, quality) currently needed by saw mills; accessibility; and technical considerations, such as logging methods, as well as approaches to slash disposal and stand regeneration.

Patch sizes (i.e., the size of harvest units) were typically a balance between operational factors and the topography and scale of local landforms. Distinct natural features, such as ridges and rivers, were often used as managed patch boundaries. Biodiversity conservation and other environmental issues were rarely considered, although creation of edge habitat for game species was sometimes recognized, whether it was actually a management goal or not.

Traditional forest management offered two contrasting approaches to imposing harvest units on the landscape: (1) dispersion of patches throughout the landscape (e.g., dispersed patch clearcutting), and (2) contiguous or continuous cutting.

Dispersed Model

The dispersed management model involves spatially dispersing harvest units of moderate size (e.g., 10 to 20 hectares) throughout a forest landscape (Franklin and Forman 1987) (Figure 7.8). This approach has often been favored by government agencies involved in timber management such as the U.S. Forest Service and a wide range of state government organizations in Australia (Florence 1996). It has been applied for a

FIGURE 7.8. Example of initial stage (first entry) using dispersed-patch clearcutting on the Tongass National Forest, southeastern Alaska, United States. Photo by J. Franklin.

variety of social and technical reasons including (after Franklin and Forman 1987)

- Opportunity to rapidly extend road systems in previously unroaded area
- Good match to technology and economics of logging
- Avoidance of large contiguous areas of slash after harvesting
- Potential for natural regeneration
- Dispersed impacts on hydrology and aquatic ecosystems
- Aesthetics and other social concerns

The concept of reducing the impacts of harvesting activities by dispersing them throughout a landscape is intuitively appealing as indicated by the old saying, "the solution to pollution is dilution." However, the reality is very different for at least some ecological functions, as noted below.

Dispersed-patch clearcutting can have significant negative impacts on forest biodiversity, disturbance regimes, and ecosystem processes (Franklin and Forman 1987). Dispersed harvesting typically leads to rapid forest fragmentation (Figure 7.9), large amounts of edge, reduced patch interior habitat, and, often, reduced overall patch size.

Since harvest unit (patch) sizes are typically based on economic and logging considerations rather than

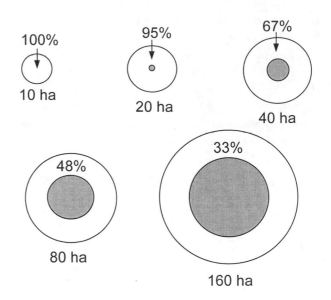

FIGURE 7.10. Amounts of interior and edge-affected (% shown) habitat associated with different sizes of circular forest patches surrounded by clearcuts in the Douglas-fir region of northwestern North America.

on biological considerations, they are often inappropriate for some forest-dependent biota. For example, in the Pacific Northwest, harvest unit sizes typically averaged 10 to 15 hectares, which can severely limit the amount of interior forest in the remainder of the landscape. Remaining uncut units need to be 80 to 100 hectares to achieve even modest levels of interior habitat within a clearcut landscape (Figure 7.10), and the northern spotted owl needs intact forest areas of 100 to 1,000 hectares (see Chapter 11). Similarly, consolidating clearcuts to limit fragmentation and provide large uncut areas suitable for use by the American marten (*Martes americana*) has been recommended in Maine, in the northeastern United States (Chapin et al. 1997). This was important because American marten typically occupy intact habitat patches of 150–250 hectares.

Dispersed-patch clearcutting also increases the potential for major damage from windstorms (Figure 7.11) (e.g., Gratkowski 1956; Alexander 1964; Sinton et al. 2000). This is a consequence of several factors, including increased (often rapidly increased) amounts of high-contrast edge between cutover and intact forest, placement of harvest unit boundaries in relation

FIGURE 7.9. Forest landscape in an advanced state of dispersed-patch clearcutting; note the highly fragmented state of the remaining intact forest and large amount of edge in this landscape (Warm Springs Indian Reservation, Oregon, United States). Photo by J. Franklin.

FIGURE 7.11. Dispersed-patch clearcutting and associated road networks create the potential for large-scale windthrow events, such as in the case of the Bull Run Watershed, Mount Hood National Forest, Oregon, United States. Photo by J. Franklin.

to soils and topography, and increased wind fetch (Franklin and Forman 1987).

The Continuous Model

The continuous cutting model of harvest has been the practice on many large industrial forest lands in northwestern America (Figure 7.12). Here, extensive areas of contiguous forest were progressively cut, resulting in large contiguous clearcut patches. Since clearcutting is the traditional approach, these large cutover patches typically exhibited little internal heterogeneity or structural complexity. Large-scale clearcuts can have significant negative impacts on many elements of biodiversity and ecosystem processes (e.g., Anderson et al. 1976). More recently, riparian buffers and other retained areas have provided low to moderate levels of internal (within-patch) structural heterogeneity.

Limitations on maximum clearcut patch sizes (based on public relations concerns) in some jurisdictions have required the separation of harvest units by narrow strips of trees, which are subsequently removed when tree regeneration reaches a designated height. Such "green-up" requirements that limit clearcutting of adjacent patches until some level of tree regrowth has occurred are primarily cosmetic and appear to provide few ecological benefits as noted below.

Fragmentation is usually not a problem on landscapes subject to continuous clearcutting, since no forested patches are retained, but the cumulative effects of timber harvesting on hydrological processes and aquatic ecosystems—as well as on forest biodiversity—can be profound (e.g., Anderson et al. 1976). Large contiguous blocks of harvested land are highly detrimental for some species, such as those with limited dispersal capabilities (Recher et al. 1980). Spence et al. (1996) believed that if the areas of harvested boreal forest in Canada were too large, then the distance between suitable habitat patches might not allow old growth–dependent carabid beetles to recolonize regenerating stands. Smith (2000) predicted that the rare Tasmanian carnivorous snail *Tasmaphena lamproides* would be sensitive to aggregated patterns of harvest disturbance.

Neither the dispersed nor the continuous even-aged approaches to forest harvest are based strongly on natural models. Both require an extensive road network, although there may be more opportunity for reduced densities of permanent roads under the continuous approach. As traditionally practiced, neither provide for within-harvest-unit (i.e., within-patch) heterogeneity (e.g., structural complexity). Continuous harvest results in larger harvest units (patch sizes) and, possibly, longer periods (between disturbances) for recovery in some stream drainages.

FIGURE 7.12. Continuous clearcutting of forest landscapes has been practiced on many large industrial forestlands. It can result in loss of structural complexity over large areas as well as cumulative impacts on hydrological and geomorphological regimes (industrial forest lands near Mt. Rainier, Washington, United States). Photo by J. Franklin.

It is possible to devise landscape-level approaches to forest harvest that both avoid many limitations of traditional even-aged management regimes, which are better matched to target organisms and processes, and are closer to natural models of disturbance. We will now consider important design attributes and the possible trade-offs among them, and then provide examples of several conservation-based landscape management designs for timber harvesting programs.

Key Variables in Designing Management Strategies

Important variables in designing forest management strategies (including timber harvesting) are patch size and shape, structural conditions within the recently disturbed patch (patch content), and rotation period.

Patch Size and Shape

Issues of patch size and shape are fundamental because of their influence on the availability of some contiguous habitat conditions (i.e., scale of contiguous patch types) or of some specific mosaic of habitat conditions; the amount of edge (i.e., boundaries between patches) within the landscape; and the extent of interior patch conditions (i.e., area free of edge effects) available within a patch and, collectively, within a landscape.

In addition, use of inappropriate patch sizes and shapes can have long-lasting consequences for forest landscapes because reversing the patterns that are created requires focused efforts over very long periods. There are also economic and aesthetic issues associated with selection of appropriate patch sizes and shapes.

Ecological considerations in the selection of an appropriate size for a management unit include

- Organisms of interest and their habitat requirements, such as dependence on interior patch or edge conditions
- Landscape context (or the condition of the forest surrounding a patch)—the silvicultural prescriptions, rotation period, extent of protected areas, and dominant landforms
- Natural range-of-variability in patch sizes

> **BOX 7.3.**
>
> **Woodland Caribou and the Value of Planning the Size of Harvest Units**
>
> *The value of planning the spatial arrangements of harvest units is illustrated by woodland caribou in Canadian forests (Thompson and Angelstam 1999) where the species apparently requires old forest patches covering hundreds of square kilometers. If harvest units are much smaller than the patch sizes created by natural disturbances, populations may be fragmented and susceptible to predation.*

- Silvicultural prescriptions to be utilized and their effects on habitat complexity within the harvested patches
- Watershed scale of interest

Selected patch sizes should be appropriate to provide for the needs of biota of interest, particularly organisms associated with interior forest conditions (see Box 7.3). Considerations should include potential edge effects, which are often extensive and occasionally immense, such as in the case of ungulate grazing in the forests of the Great Lakes region in the northern United States (8,000 meters.) (Alverson et al. 1994). The poor fit between harvest patch sizes (10 to 25 hectares) adopted for Douglas-fir forests in northwestern North America and some late-successional species has already been noted (Franklin and Forman 1987). In this case, patch size was grossly inadequate to provide unaltered interior forest microclimatic conditions because of edge effects in the high-contrast landscape of old-growth forest and clearcuts (Chen 1991); hence, species dependent on interior forest microclimate were at risk. The landscape also was unsuitable for species that required large contiguous intact habitat patches, such as the northern spotted owl throughout much of its range.

Both biological and social concerns may limit the size of harvest units that are acceptable, resulting in *adjacency constraints*. As defined by Davis et al. (2001): "The term *adjacent* refers to *what is beside* the feature of interest. When the term *adjacency constraint* is used, it generally means that we seek to *control the timing of activities that are beside (or near) one another*" (emphasis added).

A typical adjacency constraint is a "green-up" requirement that limits the harvesting of management units adjacent to recently cutover areas until regenerated trees reach a certain size. Green-up requirements are typically based on aesthetic rather than on ecological concerns. This approach results in the temporary retention of narrow strips of trees between harvest units that typically contribute little to biodiversity conservation.

Patch shape can influence patch and landscape attributes beyond the obvious effect it has on the proportion of edge-affected and interior environments. For example, accelerated mortality of trees is known to occur at the edges of patches (Chen et al. 1992; Esseen 1994; Laurance 2000). The shape and location of patch boundaries and the total amount of edge can profoundly influence the potential for wind damage. Complex patch boundaries, which are sometimes proposed in order to reduce the aesthetic impact of harvest patches, can focus and intensify air movement resulting in increased wind disturbance in adjacent forest patches (Miller et al. 1992). The total amount of edge (Franklin and Forman 1987) and the location of edges in relation to environmental variables such as topography, landform, and soil drainage (Moore 1977; Savill 1983) can influence windthrow in intact forest patches. Problem areas are patch boundaries on soils with high seasonal water tables or in specific topographic positions, such as gaps or low points along ridge tops (Gratkowski 1956; Alexander 1964; Ruth and Harris 1979). In southeastern Australia, tree mortality rates in retained forest strips within logged landscapes increased significantly with the increasing size of harvest units, suggesting a relationship with increased wind fetch (Lindenmayer et al. 1997a).

Internal Structural Complexity of Harvest Patches

Patch content matters. There has been a tendency to view patches as either habitat or nonhabitat (see Chapter 2), particularly in forest landscapes where harvested patches are clearcuts. Such cutover areas retained essentially no internal biological structures, thereby eliminating habitat for many organisms, and generated landscapes dominated by high-contrast edges and, hence, intense and extensive edge influ-

ences, including altered disturbance regimes (e.g., Franklin and Forman 1987).

However, if disturbed patches support moderate to high levels of structural complexity (i.e., biological legacies), they retain some capacity for sustaining forest organisms and reducing the extent of edge influences. Similarly, the social acceptability of management regimes may be increased as a result of reduced visual impacts; the effect of live tree retention and greater acceptability of partial cutting has been recognized for many years. Therefore, the structural content of harvested patches or the lack of such content is an important variable in determining appropriate harvest unit sizes, and the level of structural content is directly related to the silvicultural prescription adopted. Prescriptions that provide for significant retention of live trees and other structures (i.e., the retention harvesting approaches described in Chapter 8) can result in

- Increased capacity of the harvested patches to sustain biodiversity (refugia, lifeboating)
- Increased potential to accommodate species and processes as stand recovery proceeds (structural enrichment)
- Reduced edge impacts (including buffering of reserved areas)

Where compositional and structural heterogeneity is maintained on harvest units, such as by selective cutting and structural retention, larger harvest units are likely to be more acceptable socially and appropriate environmentally than under clearcutting regimes.

Natural Guides to Patch Mosaics

The types of patch mosaics created by natural disturbances and with which native flora and fauna have evolved can be used to guide the design of a forest management patch mosaic (including mean patch size) (see Chapter 4). We do not suggest that the historic landscape patterns should be precise templates for managed landscapes. In some cases it is impossible to replicate natural patterns, such as the large-scale catastrophic fires typical of the Douglas-fir forests in northwestern North America and the mountain ash forests of southeastern Australia (Hemstrom and Franklin 1982; McCarthy et al. 1999) (Figure 7.13). Such large-scale human disturbances are socially

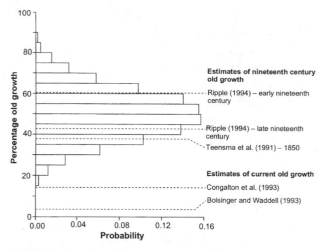

FIGURE 7.13. Probability distribution of the percentages of old-growth forest in the Oregon Coast Ranges over the past 3,000 years calculated from 100 independent simulations, referenced to published estimates of old growth currently and in the nineteenth century (from Davis et al. 2001).

unacceptable and probably ecologically inappropriate under modern conditions. Nevertheless, management regimes can be adopted to better approximate natural regimes, as discussed later in this chapter (see also Spies and Turner 1999; see Chapter 4).

Ultimately, non-ecological considerations are significant factors in the final selection of patch sizes, shapes, and internal content, including aesthetics and other aspects of social acceptability, economics, transportation systems, and logging methods. These will not be dealt with further here.

Rotation Period

Rotation period—traditionally defined as the length of time between regeneration harvests under even-aged management regimes—is an important variable in managing forests for biodiversity. This is primarily because it determines the rate at which the landscape is perturbed by harvesting, hence, the recovery period for harvest units, and, consequently, the proportion of a landscape in various age classes, including recently disturbed areas and late-successional conditions.

There are potentially strong interrelationships or trade-offs between rotation period and harvest prescriptions; high levels of structural retention (e.g.,

partial cutting rather than clearcutting) may make it possible to achieve some conservation objectives under short rotations and vice versa.

Rotations in managed forests have traditionally been set using economic criteria, specifically analyses based upon *net present value* (NPV)—the discounted value of the revenues and costs from use of forest resources over time (Davis et al. 2001). Calculation of NPV relies on the use of compound interest and (typically) a 4 to 6 percent discount rate. As a consequence, many forests are harvested on short rotations (e.g., 40 to 60 years), well before the culmination of mean annual increment (CMAI)—the time that maximizes annual growth per unit area of forest. Therefore, such rotation periods are set to maximize profits, not wood production.

Rotations based upon NPV and even (although to a lesser extent) on CMAI can create major problems for biodiversity conservation. At the landscape level they can result in

- Loss of late-successional forest stands (Spies and Turner 1999) because few if any old stands escape harvest (Seymour and Hunter 1999)
- Loss of large trees, snags, and logs if clearcutting is utilized (Franklin et al. 1981)
- Landscape dominance by cutovers and young forests with potential negative impacts on hydrological regimes, such as peak flood flows (Jones and Grant 1996), and geomorphological regimes, such as debris torrents (Anderson et al. 1976)
- Reduced average stand age (McCarthy and Burgman 1995)
- Reduced levels of carbon sequestration (Wayburn et al. 2000; Harmon 2001)

For all these reasons, rotation period is an important variable in matrix management. Habitat for species dependent upon late-successional forest conditions or structures is likely to be totally eliminated from landscapes managed by clearcutting on short rotations (Franklin et al. 1981; Stohlgren 1992; Lindenmayer 1994a). Similarly, species sensitive to the proportion of recently disturbed landscape will be negatively affected by short rotations (Milledge et al. 1991; Lamberson et al. 1994; Økland 1996).

Reducing the rate of timber harvest by adopting longer rotations is one powerful way to reduce some of the negative impacts of traditional economic rotations while still continuing to obtain forest products from a landscape (Curtis 1997; Thiollay 1997). Longer rotations can make positive contributions to biodiversity conservation by

- Increasing the time available for organisms to become reestablished in cutovers, although their ability to do so is strongly influenced by the amount, type, and spacing of retained vegetation on cutovers (Franklin et al. 1997; Lindenmayer and Franklin 1997a) (see Chapter 8).
- Increasing the number of age classes within a landscape to include late-successional stages and thereby contributing to stand- and landscape-level heterogeneity. Conversely, it decreases the total area of forest in young cohorts that are unsuitable for taxa sensitive to the effects of logging operations (Curtis 1997).
- Improving connectivity between patches of late-successional forest (Harris 1984; Norton 1999).
- Increasing opportunities for the recruitment of key structural and floristic attributes of stands that require a long time to develop, such as large trees, snags, and logs (Curtis 1997).
- Expanding the range of stand management options (silvicultural treatments) to improve habitat development (Curtis and Marshall 1993; Carey et al. 1996; Bergeron et al. 1999) (see Chapters 4 and 8).
- Buffering old-growth stands and other protected habitat in the landscape by reducing the extent of edges with high structural contrast (Harris 1984).
- Providing options for reductions in the size of permanent road networks (Franklin et al. 1997).
- Reducing the potential for negative cumulative effects on hydrological and geomorphological processes (e.g., peak flood flows)—and, consequently, reducing negative cumulative effects on aquatic biota (e.g., Harr 1986; Jones and Grant 1996; Jones et al. 2000).

- Increasing the array of forest management options by delaying some aspects of decision making (Curtis 1997). This has the additional advantage of allowing for more-informed decisions in the future based upon new and better data.

Curtis (1997) identified a range of other benefits from extended rotation periods, such as reduced visual impacts from harvesting. Longer rotation periods also will increase the quantity of carbon sequestered in the landscape (Pacific Forest Trust 2000; Harmon 2001), increase wood yields per unit area of land (Curtis 1994), and produce higher-quality forest commodities (Curtis 1997), including water (Langford et al. 1982) (see further benefits detailed in Chapter 8).

There are several ecological bases for selecting longer rotation periods. One approach is simply to determine the rotation period necessary to meet defined landscape-level goals for specified structures or structural conditions. For example, it is possible to select a rotation period sufficient to produce the minimal area of managed late-successional forest desired within a landscape or, conversely, to ensure that the area of recent cutover and regenerating stands is at (or below) the target threshold levels.

Natural forest cycles also are useful guides in selecting rotation age. Historic return intervals for stand-originating disturbances can be determined using various scientific approaches to provide insights into the proportion of the landscape that has been in different age and structural classes of forest, such as old growth. Dendrochronological approaches usually have good space and time resolution but have limited time depth, which is typically only two to three fire intervals. Paleobotanical methods, such as those that utilize pollen or charcoal, have weak time and space resolution but can provide a window on the past covering many millennia. Historical reconstructions of fire events are often possible for one or a few centuries. Stand reconstructions based on destructive analyses of stands can also provide useful insights on recent effects of recent disturbances. For example, the average interval between stand-replacing fires in mountain ash forests was determined to be 180 to 220 years by stand reconstruction methods (Lindenmayer 1999a). A logical inference is that rotations of at least

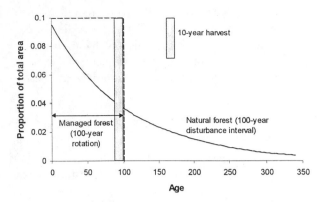

FIGURE 7.14. There can be large differences in the age class distribution of naturally disturbed and human-disturbed landscapes, even if the average frequency of disturbance events across the landscape is the same. A right-skewed age class distribution occurs in naturally disturbed forests with an average of 100 years between (stochastic) disturbance events, leaving some old stands in the 200- to 350-year age classes as shown by the solid curved line. This contrasts strongly with deterministic human disturbance regime (logging) with the same return interval (i.e., the rotation time is 100 years) where no stands exceed 100 years. Redrawn from Van Wagner 1978 and Seymour and Hunter 1999. Reprinted with the permission of Cambridge University Press.

half this period (90 to 110 years) would be much better for biodiversity conservation than the current thirty-five- to eighty-year rotations.

Natural return intervals for stand-replacing events occur at even longer intervals in many coastal forests of northwestern North America, and their occurrence may be related, at least in part, to long-term climatic fluctuations. A natural fire return interval of approximately 400 years was calculated for forests in Mount Rainier National Park, in Washington state (Hemstrom and Franklin 1982). Some stands in coastal British Columbia have return intervals of a millennium or more (Lertzman et al. 1996). Disturbance frequencies increase further south into Oregon and northern California (Agee 1993).

Variability in the frequency of disturbance (as well as the mean return interval of natural, stand-replacing events) can have significant implications for setting rotation times (McCarthy and Burgman 1995). As disturbances typically occur at irregular intervals over many centuries, there may be large variations in particular forest ages (such as old-growth forest) (Wim-

TABLE 7.1.

Proportion of the forest harvested at different rotation times to match the age class distribution created by inherent variability in fire frequency around a mean return interval of 100 years (after Seymour and Hunter 1999).

ROTATION TIME (YEARS)	PROPORTION OF FOREST HARVESTED (%)
300	10
200	15
150	20
100	35
50	20

berly and Spies 2001) (Figure 7.13). Such variability leads to long, right-skewed tails in age class distributions (Figure 7.14) with these tails representing very old forests that can be critical for the conservation of some organisms (Forest Ecosystem Management Assessment Team 1993; Henderson 1994; Marcot 1997). One approach to replicating such distributions is to allocate different proportions of a managed forest to successively longer rotations (Table 7.1) thereby mimicking the proportion of forest maintained in different stand ages around some average return interval, such

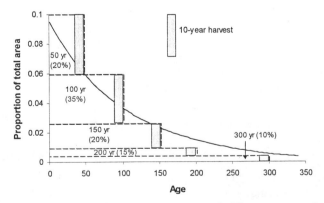

FIGURE 7.15. Multiple rotation ages of different areas of forest within the same landscape can help better mimic the variability in the frequency of natural disturbance regimes with a mean return interval of 100 years and can ensure the maintenance of some old forest cohorts—as would occur under natural fire regimes (see Figure 7.13). Redrawn from Seymour and Hunter 1999. Reprinted with the permission of Cambridge University Press.

as 100 years (Henderson 1994; Seymour and Hunter 1999) (see Figure 7.15).

As shown in Figure 7.15, a single rotation period may not be appropriate across an entire landscape. A range of rotation periods may be necessary not only to enhance biodiversity objectives, but also to achieve greater congruence between human disturbances and landscape patterns generated by natural perturbations, if that is a goal (see Chapter 4).

Limitations of Longer Rotations at the Landscape Level

Despite their numerous benefits, long rotations are not a universal solution for all conservation issues related to intensive forest management (Curtis 1997). For example, to develop trees with cavities on clearcut sites would require a substantial increase in rotation period in some eastern Australian eucalypt forests from the current 35 to 80 years to over 200 years. Some species require cavity trees exceeding 400 years in age (Ambrose 1982; Lindenmayer et al. 1991c). Even with 200-year rotations (which are unlikely to happen), only 25 percent of harvested stands (the stands greater than 150 years old) would support suitable nesting habitat for the more than 100 cavity-using species at any one time. Without structural retention at harvest, none of the stands would contain these critical old structures, so the only successful strategy over the long term is one based upon retention (see Chapter 8).

Maintaining very long-lived stands and their associated taxa cannot be resolved simply with long rotations. The main reason is our limited knowledge about the structure, function, and composition of very old forests and the high probability that, even if we were omniscient, human institutions and technology are not up to the task of re-creating such ecosystems. Even the Society of American Foresters (1984) has agreed that re-creating old-growth forests is not possible (Thomas et al. 1988). Therefore, maintenance of truly old forests requires that at least a portion of such stands be reserved from harvest in the form of large ecological reserves and midspatial-scale protected areas. Late-successional forests and old-growth trees are much older than even very long rotations (e.g., more than 300 years) (Figure 7.15). The natural life spans of many tree species, such as Douglas-fir and giant sequoia, exceed 1,000 years, and habitat development continues for many centuries following a disturbance (e.g., Franklin et al. 2002). There are apparently many other organisms, such as some species of lichens, that do not establish until trees are several centuries old (McCune 1993; Henderson 1994).

Concentrating or Dispersing Management Activities in Time and Space

The pattern in which management units are imposed on the landscape over time also matters. Davis et al. (2001) refers to these choices as "tactical" issues and foresters have traditionally selected one or two from what is actually an infinite array of possibilities. These are strategic choices from the standpoint of biodiversity conservation, because they can make the difference between survival and extinction for some taxa.

One fundamental choice is whether to favor dispersing or concentrating management activities in time and space. The most traditional approach in forest management has been to disperse activities so that active management (such as timber harvest) occurs in all parts of a landscape during every specified time period. The alternative is to concentrate activities in one or a few selected portions of a management unit (e.g., a drainage basin) during a given time period (e.g., decade). This has sometimes occurred de facto, such as when large forest units were progressively harvested.

Despite a tradition of dispersed approaches, the alternative needs to be seriously considered—concentrating activities in time and space. In fact, natural disturbance regimes in many forest landscapes often are highly concentrated, rather than dispersed in time and space. Many forest types are subject to infrequent, catastrophic disturbances at very large spatial scales, resetting entire landscapes and creating new patches of hundreds to thousands of hectares. Classic examples are the Douglas-fir forests of northwestern North America, the lodgepole pine forests of interior western North America, and mountain ash forests in Australia. Simi-

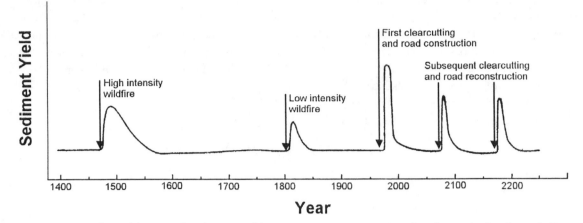

FIGURE 7.16. Hypothetical history of sediment yields in response to vegetation disturbance in the Cascade Range of Oregon, United States; infrequent catastrophic fire produced large pulses of sediment but allowed for extended periods of recovery, while current management results in chronically elevated levels of sediment (modified from Swanson et al. 1982).

larly, large, intense windstorms, such as the 1962 Columbus Day windstorm in western Oregon and Washington, can dramatically alter very large areas.

While such intense, large-scale disturbances and the immense patches associated with them are rarely acceptable to current human society (the large fires in Yellowstone National Park, in the United States, may be an exception), we note that the biota of these regions are adapted to such disturbance regimes. Of course, conditions within naturally disturbed patches differ greatly from those found on the clearcuts created by modern forestry; high levels of biological legacies, including islands of undisturbed vegetation, are characteristics of such disturbances (e.g., Delong and Kessler 2000).

Disturbance regimes comprising areas with long recovery periods may be particularly appropriate for aquatic organisms and ecosystems. Dispersed management regimes and the associated road networks have effectively created a regime of chronic disturbance, involving low- to moderate-intensity but frequent perturbations of aquatic ecosystems and lacking extended disturbance-free recovery periods (Swanson et al. 1982) (Figure 7.16). Effectively, a regime of chronically elevated sediment has been substituted for higher but episodic inputs of sediments resulting from catastrophic disturbances.

The importance of this contrast in disturbance regimes to anadromous fish and socially acceptable management approaches to aggregating management activities in time and space is described by Reeves et al. (1995):

Aquatic ecosystems throughout the [western North American] region are dynamic in space and time, and lack of consideration of their dynamic aspects has limited the effectiveness of habitat restoration programs. Riverine-riparian ecosystems used by anadromous salmonids were naturally subjected to periodic catastrophic disturbances, after which they moved through a series of recovery states over periods of decades to centuries. Consequently, the landscape was a mosaic of varying habitat conditions, some that were suitable for anadromous salmonids and some that were not. Life history adaptations of salmon, such as the straying of adults, movement of juveniles, and high fecundity rates, allowed populations . . . to persist in this dynamic environment. Perspectives gained from natural cycles of disturbance and recovery of the aquatic environment must be incorporated into recovery plans for freshwater habitats. In general, we do not advocate returning to the natural disturbance regime, which may include large-scale catastrophic processes, such as stand-replacing wildfires. This may be an impossibility given the patterns of human development within the region. We believe that it is more prudent to modify human-imposed disturbance regimes to create and maintain the necessary range of habitat conditions in space (100 kilometers) and time (10 to 100 years) within and among watersheds across the distributional range of an ESU [evolutionary significant unit]. An additional component of any recovery plan, which is imperative in the short-term, is the establishment of watershed reserves that contain the best existing habitats and include the most ecologically intact watersheds.

Longer intervals between harvest rotations would be another component of this new disturbance regime. In single basins in the central Oregon Coast Range, the desirable interval may be 150 to 200 years. . . . Concentrating rather than dispersing management activities could be another element of the new disturbance regime. This would more closely resemble the pattern generated by natural disturbances than does the current practice of dispersing activities in small areas. For example, if a basin has four subwatersheds, it may be better to concentrate activities in one for an extended period (50 to 75 years) than to operate in 25 percent of each at any one time.

Hence, concentrating activities in time and space may, to at least some degree, sustain some elements of biodiversity in some landscapes better than dispersing activities, particularly if concentration of activities will better meet the needs of targeted biota, allow for extended periods of recovery and reduce chronic impacts, allow for larger patches, and create options for reduced road densities.

Management approaches that concentrate activities in time and space are not going to be successful (either biologically or socially) if the practical result is large, contiguous clearcuts. Patch content, which ultimately depends upon the silvicultural regime adopted, is critical for acceptance of larger management units or harvested areas. Much higher levels of structural retention and, hence, internal patch heterogeneity are going to be necessary than would be the case with small dispersed patches. Longer rotations, of course, imply long recovery periods of disturbed areas and smaller percentages of a landscape in a recently disturbed state.

Examples of Conservation-Based Approaches

In this section, we present some conservation-based approaches to management that consider

- Natural models of disturbance within forest landscapes of interest (including recovery periods)
- Habitat conditions needed to sustain native biodiversity
- Trade-offs between patch size and internal heterogeneity and rotation age
- Transportation network requirements

The trade-offs among patch size, patch heterogeneity, and rotation age assume that (1) when treated patches have higher levels of internal structural complexity, patch size is less of an issue; (2) when treated patches have higher levels of internal structural complexity, rotation period is less of an issue; and (3) when rotation periods are longer, patch size is less of an issue (see also Chapter 9).

Landscape Management Using Historic Fire Regimes in the Douglas-fir Region

The scientific team centered at the H. J. Andrews Experimental Forest in the western Oregon Cascade Range (Luoma 1999) has pioneered an approach to landscape management based upon the historic catastrophic fire regimes characteristic of the conifer forests in this region. Two different planning areas were targeted for analysis (Augusta Creek [Cissel et al. 1998] and Blue River [Cissel et al. 1999]). Results from the two areas were similar and we use the Blue River planning area in our example.

Landscapes were first subdivided into noncontiguous areas representing three fire regimes that differed in fire-return interval, distribution of fire (patch) size, and severity (based upon overstory tree survival) based on empirical studies (Figure 7.17a). Silvicultural prescriptions were adopted for areas assigned to each of the fire regimes that reflected historic patterns of disturbances for those sites (Table 7.2). For example, in areas naturally subject to a fire regime of high frequency, low severity, and small size, a short rotation (100 years), small harvest unit sizes (less than 40

TABLE 7.2.

**Elements of the silvicultural prescriptions for the landscape areas with
differing natural fire regimes (from Cissel et al. 1999).**

AREA	ROTATION AGE (YEARS)	PERCENTAGE OF LANDSCAPE AREA			RETENTION LEVEL (% CROWN COVER IN OVERSTORY)
		Small block (less than 40 ha)	Medium block (40–80 ha)	Large (more than 80 ha) block	
1	100	60	20	20	50
2	180	40	40	20	30
3	260	20	40	40	15

hectares), and a high level of structural retention at harvest (50 percent crown cover in the overstory) were prescribed. Recognition of special area reserves (e.g., patches of old-growth forest for northern spotted owl nest sites) and aquatic conservation areas were also part of the landscape plan (Figure 7.17b).

The aquatic conservation strategy differed significantly from strategies based upon standard-width riparian corridors, such as the interim riparian reserve strategy in the Northwest Forest Plan (Figures 7.17b and 7.17c). The disturbance-based strategy was designed primarily to meet aquatic ecosystem objectives and secondarily to contribute to late-successional forest objectives. Aquatic reserves consisted of (1) small watersheds (50 to 200 hectares) that were reserved from any timber harvest and were

A.

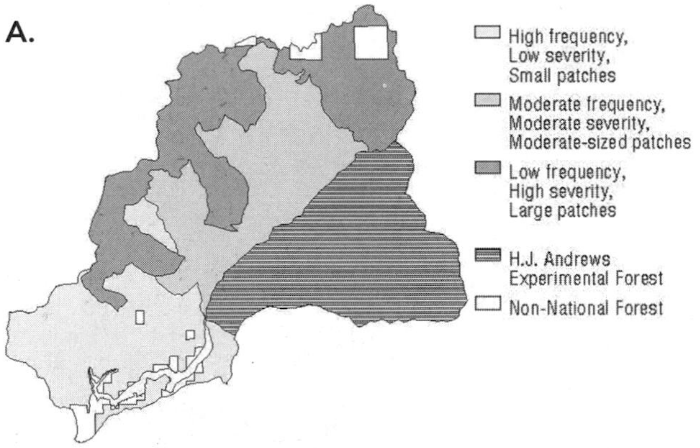

High frequency,
Low severity,
Small patches

Moderate frequency,
Moderate severity,
Moderate-sized patches

Low frequency,
High severity,
Large patches

H.J. Andrews
Experimental Forest

Non-National Forest

FIGURE 7.17. The 23,908-hectare Blue River landscape in the Cascade Range of western Oregon utilized in the development of a management plan modeled on natural disturbance regimes (adapted from Cissel et al. 1999). (*A*) Division of landscape into areas subject to different fire regimes. (*B*) Management areas differing in silvicultural regimes, special area reserves, and riparian reserves under the interim plan. (*C*) Management areas based upon area allocations of the Northwest Forest Plan. Reprinted with the permission of the Ecological Society of America.

B.

General Forest (Matrix)
Scenic (Matrix)
Special Area Reserves
Riparian Reserves

Interim Plan
Management Areas

C.

Landscape Area 1
Landscape Area 2
Landscape Area 3
Special Area Reserves
Aquatic Reserves

Landscape Plan
Management Areas

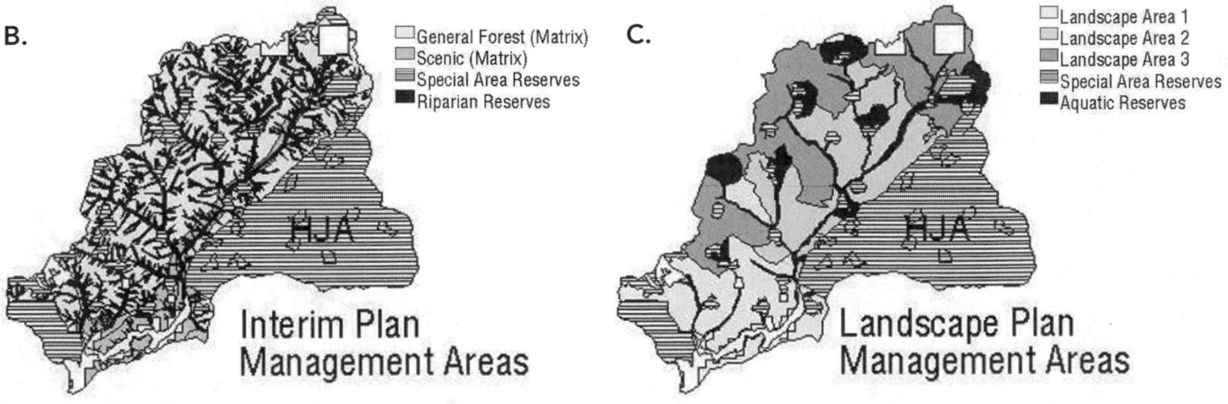

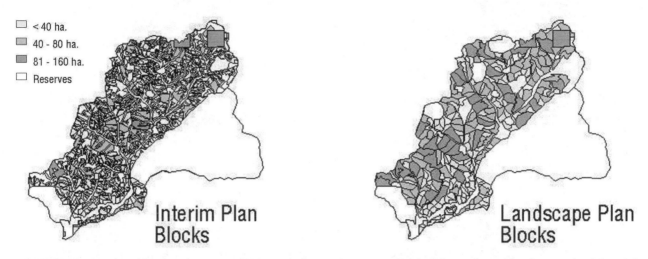

FIGURE 7.18. Simulated landscape patterns 200 years after implementing the disturbance-based landscape plan (*A*) and the Northwest Forest Plan (*B*) on the current Blue River landscape. Adapted from Cissel et al. 1999. Reprinted with the permission of the Ecological Society of America.

large enough to provide interior late-successional forest conditions, and (2) riparian corridors of varying width (approximately 70 to 200 meters of slope distance to either side of the stream), which were positioned around fish-bearing streams. In contrast with the riparian reserve strategy of the Northwest Forest Plan, riparian buffers were not established on non-fish-bearing permanent or intermittent streams. However, retention levels in the landscape plan can be varied across the landscape to provide for high levels of tree retention near streams.

The two approaches to landscape management—the disturbance-based plan and the Northwest Forest Plan (Forest Ecosystem Management Assessment Team 1993)—generated very different future landscape structures based on long-term simulations (Cissel et al. 1998, 1999) (Figure 7.18). Two hundred years in the future (2195 A.D.), when the Northwest Forest Plan was compared to the disturbance-based landscape plan, the landscape produced by the Northwest Forest Plan had (1) more area in patches of young (less than eighty years) forest due to higher harvest rates and shorter rotations within the matrix, (2) less-varied overstory and lower overall average retention levels in areas subject to timber harvest, and (3) a large gap in age classes due to near disappearance of mature stands. The disturbance-based plan pro-

duced larger overall patch sizes (and, consequently, more interior habitat), reduced high-contrast edge habitat, and resulted in more late-successional forest than did the Northwest Forest Plan.

There are other differences in both landscape structure and potential implications for biodiversity and ecosystem processes between the approaches of Cissell et al. (1998, 1999) and the Northwest Forest Plan (Forest Ecosystem Management Assessment Team 1993). Perhaps the most profound is that the disturbance-based model results in a landscape that better matches the patchwork and internal patch complexity found under the natural range of variability. The landscape strategy avoids the high and unnatural levels of landscape contrast that occur when an extensive fixed-width riparian buffer system (the riparian reserves of the Northwest Forest Plan) interacts with a harvested upland (the matrix of the Northwest Forest Plan) (Figure 7.18).

Minimum Fragmentation Approaches in Managed Landscapes in the Great Lakes Region

Many, if not most, managed forest landscapes already display substantial amounts of habitat fragmentation. As issues of patch size, fragmentation, and maintenance of interior forest conditions have emerged, federal land managers in the United States have begun to

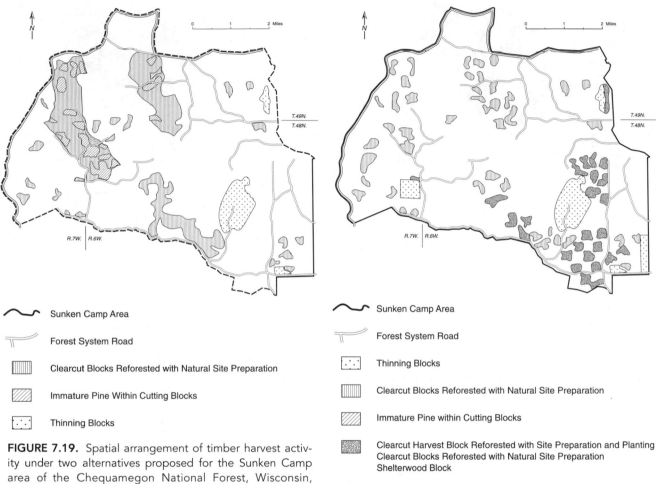

FIGURE 7.19. Spatial arrangement of timber harvest activity under two alternatives proposed for the Sunken Camp area of the Chequamegon National Forest, Wisconsin, United States. (*A*) Alternative designed to create large habitat units for nongame wildlife species.

(B) Alternative to dispersed timber harvest activity in small (less than 15-hectare) clearcuts (from Kick 1990).

develop and adopt landscape-level approaches that not only minimize additional fragmentation (so-called "min-frag" approaches), but also attempt to repair the effects of past management.

Projects in the Chequamegon National Forest (Wisconsin, United States) provide early examples of such efforts (Parker 1997). The selection of appropriate spatial and temporal patterns of harvest units was central to restoration programs in this forest, including the use of much larger patch sizes—whether for maintenance of interior forest conditions or for restoration. The need for larger blocks of old growth first emerged because of concerns over edge effects associated with ungulate herbivory on understory plants. Such unde-

sirable effects can extend for as much as 8,000 meters from the stand edge in this landscape (Alverson et al. 1994). However, restoration approaches designed to reduce fragmentation of forest landscapes have also been proposed as a part of other restoration and timber programs, such as on some of the pine barren habitats found in the Chequamegon (Parker 1997). Pine barrens under pre-European conditions had low tree densities with species-rich herbaceous and shrub layers that were maintained by recurrent natural fire. Logging and fire control programs modified patch mosaics and stand conditions. Restoration of these landscapes is carried out using management regimes more comparable to the natural conditions, such as a shift from managed patch sizes of 20 to 5,000 hectares, thereby simulating historical, fire-regenerated patch size. In addition, silvicultural regimes have been shifted from clearcuts to prescriptions that better simulate the structural complexity and plant species composition required by pine barren taxa.

A draft environmental impact statement for the Sunken Camp area on the Chequamegon National Forest provides an example of high and low fragmentation alternatives for a proposed jack pine restoration project (Kick 1990). Three action alternatives were proposed and analyzed:

1. Emphasis on creation of large units of habitat for plant and animal species
2. Emphasis on human recreation and creating large habitat units for nongame wildlife species
3. Emphasis on harvesting timber in clearcuts of 15 hectares or less

A comparison of areas of edge-affected and contiguous forest that would result from the various alternatives makes clear the contrast between a fragmentation-oriented approach (Alternative 3) and approaches designed to reduce edge effects and increase interior habitat (Alternatives 1 and 2) (Table 7.3) (Figure 7.19). Edge-affected habitat is much lower and contiguous interior forest area is higher under Alternatives 1 and 2 than under Alternative 3. Although fragmentation is reduced under Alternatives 1 and 2, not all potential conservation objectives are achieved for the landscape (see Alverson et al. 1994).

Timber Harvesting Modeled on Chronic Disturbance (Gap-Type) Regimes

Technically it is relatively easy to develop silvicultural regimes modeled on natural disturbance regimes in which forests are subject to chronic, low- to moderate-level disturbances. As noted earlier, such regimes produce forests with high levels of spatial complexity as a result of gap-scale disturbance events. From an ecological perspective, the stands consist of a fine-scale, low-contrast mosaic of structural units in which

essentially all stand structures and developmental processes are simultaneously present (Franklin et al. 2002).

Stands managed for timber production utilizing the gap-based disturbance model have a number of ecological advantages (Hammond 1991), foremost among them being retention of forest influence or dominance over essentially the entire managed landscape. Such approaches are often referred to as *group selection* by foresters. Care is necessary, however, to ensure that the prescribed sizes of harvested "groups" are comparable to natural gap sizes (e.g., Knight 1997) or at least sufficiently small that the microclimatic regimes are still dominated by forest rather than by open (clearcut) conditions. Other concerns are the extent of the road system required and the perpetuation of large-diameter trees, snags, and logs within the managed landscape.

There are few well-documented examples of gap-based timber harvesting, particularly on lands subject to large-scale forestry operations. Most applications are on relatively small, private forest holdings, such as those owned by individuals, family groups, indigenous peoples, and local communities. A gap-based approach to harvest of southern beech forests was implemented on public lands under timber license in New Zealand. Logging was accomplished using helicopters—limiting the negative economic and ecological consequences of road networks—and the program appeared to be successful both environmentally and economically. However, the project was terminated for political reasons.

Gap-based approaches to harvesting have been proposed for forests subject to chronic low-intensity fire, such as those in the southeastern United States. For example, group selection was adopted by the United States Congress as a part of an overall land-managed strategy for management on approximately 800,000 hectares of national forest land in the northern Sierra Nevada in California. This strategy, referred to as the "Quincy Library Group plan," has been incorporated into the current management plan for federal lands throughout the Sierra Nevada (USDA Forest Service 2001b).

Gap-based approaches do offer some outstanding opportunities to model forest harvesting on natural

TABLE 7.3.

Landscape attributes under three different management alternatives for the Sunken Camp area, Chequamegon National Forest, Wisconsin. Alternatives 1 and 2 are designed to create larger habitat units while Alternative 3 is designed to disperse activities in clearcuts of less than 15 hectares in size. Edge effects were calculated using a width of edge-affected region of 100 meters.

ALTERNATIVE	EDGE (KM)	AREA OF NEWLY CREATED EDGE (% OF LANDSCAPE)	AREA OF CONTIGUOUS INTERIOR FOREST (HA)
1	185	18.9	4884
2	156	15.9	5670
3	238	24.3	4788

disturbance regimes. Nevertheless, it is necessary to ensure that

- The scale of the harvested groups does not grossly exceed the scale of natural disturbances
- Levels of large trees, snags, and logs appropriate to achieve biodiversity and other ecosystem objectives are maintained within harvested stands
- Rotation periods or stand reentries allow for stand recovery
- Roads and skid trails do not unacceptably impact upon soils and aquatic ecosystems

It is useful to note that gap-based systems *do* constitute chronic perturbations of the forest landscape and that their potential negative ecological impacts will be related to their cumulative effect. In some respects they can be considered the ultimate expression of dispersing management activities in time and space.

Conclusions

How to impose management activities such as timber harvest in time and space is extraordinarily complex. Developing a functional multiscale approach to conservation of biodiversity and ecosystem processes in forests requires careful analysis. This vital issue has received little attention from either resource managers or conservation biologists, who often appear to assume that certain approaches (e.g., dispersion of

management in time and space, patch size, etc.) are obviously correct and can be applied uniformly in all places.

In fact, there are important alternatives. One approach—such as in regions subject to chronic disturbances—may be to manage more of the landscape extensively (but sensitively), thereby reducing contrast among the landscape patches and improving overall habitat levels and connectivity. In other landscapes, it may be better to concentrate activities in time and space and to utilize longer rotations, especially where there are aquatic ecosystems that evolved under episodic disturbance regimes. Smaller harvest units are not necessarily superior to larger harvest units, particularly when silvicultural prescriptions provide for significant retention of forest structures and internal patch heterogeneity following harvest. Interactions between time-space patterns of timber harvest and transportation networks need careful consideration, particularly as the impacts of midslope roads in mountainous topography become increasingly unacceptable. Ultimately, the interactions amongst patch size and shape, patch content, rotation length, and transportation networks and decisions as to whether to concentrate or disperse activities within the landscape must be considered in the context of the biological and environmental goals, economic and technical constraints, and social acceptability. Uncertainties and risks will have to be a part of such analyses. Trade-offs among goals are certain (see Chapter 9). But alternatives and consequences need to be openly considered and analyzed rather than simply adopting past patterns and preferences.

Other Landscape-Level Considerations

Fundamental to matrix management is the protection of critical habitats and sensitive areas within managed forests (see Chapter 6). However, in many forest landscapes, human activities have substantially reduced the original extent of some habitat types (e.g., late successional forest) so that the viability of taxa dependent upon them is threatened (e.g., Berg et al. 1994). This is especially true for forest regions that have long been settled, such as Europe and southcentral Asia. It also can be a problem in regions that have only recently been a forest frontier, as illustrated by the Siuslaw National Forest in the Oregon Coast Ranges (Harris 1984). Old-growth forests occupy only 3.3 percent of that national forest as a result of human-caused fires in the mid-nineteenth century and high rates of timber harvest during the last forty years. Furthermore, with an average old-growth patch size of approximately 28 hectares, only eight stands exceed 140 hectares, and the majority (61 percent) cover 16 hectares or less (Harris 1984). Such dramatic changes negatively and significantly impact the array of species known to be closely associated with late-successional forests (Forest Ecosystem Management Assessment Team 1993; Marcot 1997). Another example is old-growth eucalypt forest in Australia. Past human disturbances (such as logging) have made late-successional forests of many eucalypt types rare (Recher 1985; Woodgate et al. 1994). Only small areas of old growth remain in the wood production montane ash forests of the Central Highlands of Victoria—the largest patch is 57 hectares and most patches are less than 20 hectares in size. The limited amount of old growth jeopardizes the long-term persistence of populations of other species such as the sooty owl, yellow-bellied glider, and Leadbeater's possum in wood production areas (Milledge et al. 1991; Lindenmayer et al. 1999a; see Chapter 12).

Landscapes that have been subjected to extensive human modification, such as the Siuslaw National Forest and the Central Highlands of Victoria, may be places where increasing the total area of old growth, as well as conserving existing remnant old growth, should be a primary conservation objective. Some young forests could be reserved from future timber harvest and allowed to redevelop late-successional conditions (Lindenmayer and Possingham 1995a; Økland 1996). In some cases, a decision to actively facilitate development of old growth may be desirable or necessary (Forest Ecosystem Management Assessment Team 1993; Mladenoff et al. 1994). Examples of such activities include

- Use of prescribed fire and mechanical treatments to reduce the potential for catastrophic fires and to reintroduce periodic fire into forest types characterized by low- to moderate-intensity fire regimes but which have been subject to fire suppression (Sierra Nevada Ecosystem Project 1996; Bowman 1998; Miller and Urban 2000; USDA Forest Service and USDI Bureau of Land Management 2000).
- Thinning young stands that were either naturally regenerated or planted following harvest. Dense, uniformly stocked stands, such as those created for timber production purposes, may require long periods to develop structural features characteristic of late-successional forests (e.g., large trees, snags, logs, and spatial heterogeneity). Thinnings and other stand and tree treatments can accelerate development of these structures.
- Managing riparian buffers to enhance structural development, such as by thinning dense young stands and planting desirable species that will produce large, decay-resistant woody debris (Forest Ecosystem Management Assessment Team 1993).

Chapter 8 provides further detail on stand management activities that can enhance biodiversity conservation and ecosystem processes.

Landscape restoration programs can sometimes be assisted by catastrophes such as wildfires and floods (Gregory 1997). As outlined in Chapter 4, these events leave large quantities of biological legacies and can help restore some of the stand structural complexity, landscape heterogeneity, and aquatic ecosystem integrity lost as a result of forest management practices. Therefore, it is important for agencies responsible for forest management to develop salvage harvest policies regarding the retention of biological legacies following catastrophes so as to promote landscape restoration.

Finally, landscape restoration efforts can often benefit from taking advantage of existing areas of conservation value in the matrix. For example, existing areas of old-growth forest can be useful nodes around which to expand the size of the (often limited) old-growth estate (McCarthy and Lindenmayer 1999a). Similarly, places that have been subject to lower-intensity harvesting operations in the past and which support larger numbers of biological legacies can be good locations for focused restoration efforts. The value of landscape maps to assist such exercises is obvious.

Integrating Matrix Management Strategies at the Landscape Level

Landscape management of matrix forests needs to be guided by explicit objectives. Relevant questions might include: What are the largest patch sizes that need to be maintained and where will they be located? How many wildlife corridors should there be per unit area of forest? How are riparian and upland objectives to be balanced? The checklist in Table 6.1 in Chapter 6 incorporates a range of items for consideration as part of an integrated approach; of course, not all items apply equally to all landscapes. For example, the size of riparian buffers will depend partially on the amount of other vegetation retained and the silvicultural systems employed in adjacent harvested areas. Different numbers and types of midspatial-scale protected areas within the matrix may be required depending on the rotation time and/or the extent of stand-level structural retention (see Chapter 8).

Goals for wood production from the matrix must be integrated with the conservation goals within the landscape of interest. Wood production goals have often been set at levels that are not consistent with maintaining desired levels of biodiversity and ecosystem processes. Wood production goals often failed to account for uncertainties and risks associated with natural disturbances, such as wildfires, floods, and windstorms; events that occur at the scale of the 1962 Columbus Day windstorm or the 1980 Mount St. Helens eruption (both in the Pacific Northwest region of the United States) can have profound consequences for both commodity and conservation objectives.

Concepts such as "landscape resource accounting" can be useful by giving consideration to relationships among forest biodiversity, forest commodities, and forest services (e.g., timber yields and water production) at the landscape level. For example, in the mountain ash forests of southeastern Australia, there are direct and positive relationships between increased production of water, the conservation of area-sensitive species, and the proportion of old-growth forest in the landscape. Landscape-level forest planning needs to account for these different values within the bounds of logical and clearly stated objectives and use models that simulate future spatial changes in landscapes under different management regimes to compare and assess effects. Models can help managers conceptualize and visualize long-term consequences of near-term management decisions, including cumulative changes.

Final Comments

Currently, four broad categories of approaches for landscape-level matrix management are recognized:

1. Identification and appropriate management of mesoscale reserves, including buffers for aquatic ecosystems, specialized habitats, biological hotspots, and remnants of late-successional or forest within the matrix

2. Identification of landscape-level goals for the retention or maintenance of particular habitats or structures and limitations or thresholds on other conditions

3. Design, construction, and maintenance of transportation systems (primarily road networks) so as to minimize impacts on species, critical habitats, and ecological processes

4. Decisions regarding placement of harvest or other treatment units in time and space—including considerations of patch size, shape, and content, rotation length, and the interactions of these elements—and of dispersion versus concentration of activities

This chapter has focused on Categories 2, 3, and 4 and highlighted the diverse ways in which these aspects of matrix management can enhance biodiversity conservation. Generic and "cookbook" approaches do not work because each landscape differs in its physical and biological characteristics (including those imposed by history), social and economic constraints, and specific management objectives, including species assemblages targeted for conservation.

Many aspects of landscape-level matrix management represent a significant challenge for forest managers, in part because of the difficulty in conceptualizing key processes (and major problems) over large spatial scales and long temporal scales. Computer-based visualization techniques can help meet some of these challenges and assist managers to, for example, simulate cumulative changes in landscape patterns accruing from different spatial arrays of many sequential harvest units. Weyerhaeuser Company has recently employed such visualization techniques to model interrelationships between different harvesting scenarios and landscape patterns for its lands on Vancouver Island (British Columbia). As with all modeling tools, it is essential that users are well aware of the assumptions and limitations of such techniques; otherwise "virtual reality can quickly descend into real stupidity."

We have only two generic recommendations: First, do not manage every landscape in the same way. Rather, spread the risks by creating a range of conditions and spatial patterns in different landscapes as described by the risk-spreading approach in Chapter 3. Second, consider the principles for conserving biodiversity (Chapter 3) and, with those in mind, work systematically through the hierarchically structured checklist of items for consideration in matrix management (Table 6.1 in Chapter 6). These principles and the strategies to address them should be considered at a range of spatial scales, including the stand level, which is the topic of Chapter 8.

C H A P T E R   8

Matrix Management in the Harvested Stand

Carrying out our local and global responsibilities to protect biological diversity in forests does not end at the landscape level. Building on the solid foundation, we must ensure that all our activities at the stand level . . . are carried out in an ecologically responsible way.

—HAMMOND (1991)

Stand-level matrix management encompasses three broad strategies: (1) structural retention at the time of regeneration harvest, (2) management of regenerated and existing stands to create specific structural conditions, and (3) long rotations or cutting cycles.

Each strategy contributes uniquely to the maintenance of biodiversity within managed stands so that the strategies can be effectively combined to address a broader range of objectives. For example, the advantages of long rotations are enhanced by structural retention at the time of harvest. Active management of dense regenerating stands for structural complexity can prevent the development of dense young forests with limited habitat value for biodiversity that might occur if only structural retention is applied.

Innovative silvicultural systems that address both commodity production and biodiversity conservation can be achieved by integrating the above three strategies. Information about natural disturbance regimes, including the types, numbers, and spatial patterns of biological legacies, can be used to guide the development of these systems. New silvicultural systems that conserve biodiversity are emerging, such as variable retention harvesting (Franklin et al. 1997), the biodiversity pathways model of active stand management (Carey et al. 1996), and modifications of traditional clearcutting (Fries et al. 1997). In tropical forests, more ecologically sensitive "reduced impact logging" (RIL) systems are being developed (Putz et al. 2000). The ultimate goal is the development of comprehensive conservation-based silvicultural systems that incorporate all aspects of management from structural retention at regeneration harvest through to active ecologically oriented management of stands that are grown on rotation times sensitive to environ-mental considerations and are applied over multiple rotations.

The implementation of any strategy and associated silvicultural system is driven by management objectives and physical and biological conditions present in a landscape. Matrix-based strategies must be developed on a case-by-case basis (see Part III).

Although each of the three strategies—as they emerge in newly developed silvicultural systems—can contribute significantly to biodiversity conservation, a holistic approach to forest management is needed where the combined contribution of an array of matrix-based strategies (and large ecological reserves) is considered. Potential deficiencies in any single approach and a lack of information about long-term effectiveness do underscore the importance of dedicated large ecological reserves. The real challenge then is the development of sophisticated plans that integrate considerations at the stand, landscape, and large ecological reserve levels.

Developing and maintaining structural complexity in managed stands is central to any forest management program that has the serious intent of maintaining forest biodiversity and ecosystem processes. Unfortunately, a common premise in conservation biology is that a forest landscape is divided into suitable habitat (the reserves and corridors) and nonhabitat (everything else), effectively ignoring the potential contribution of managed stands. Of course, this is not the situation, even in landscapes totally dominated by intensive forestry practices, such as plantations (see Chapter 10).

TABLE 8.1.

Checklist of factors for matrix management.

LARGE ECOLOGICAL RESERVES (CHAPTER 5)

CONSERVATION STRATEGIES WITHIN THE MATRIX:
LANDSCAPE-LEVEL CONSIDERATIONS
(CHAPTERS 6 AND 7)

*Protected habitat within the landscape matrix: protected
areas at intermediate-spatial scales*

Special habitats
    Cliffs, caves, rockslides, thermal features, meadows,
       vernal features
Remnant patches of late-successional forest
Biological hotspots
    Calving areas, source areas for coarse woody debris,
       populations of rare species
Fire, wind, and other disturbance refugia
Aquatic ecosystems and riparian buffers
    Springs, seeps, vernal features, lakes, ponds, wetlands,
       streams and rivers, and associated buffers
Wildlife corridors
Culturally sensitive areas within the matrix

Other landscape-level considerations

Transportation systems (e.g., roading networks)
*Landscape-level goals for specific structural features (e.g.,
   large snags)*
Spatial and temporal patterns of timber harvesting
    Dispersed versus aggregated
    Size of harvest units
    Rotation lengths
*Restoration and re-creation of late-successional forests or
   other habitat features*

CONSERVATION STRATEGIES WITHIN THE MATRIX:
STAND-LEVEL CONSIDERATIONS (CHAPTER 8)*

*Habitat within management units or stands**

*Retention of structures and organisms at time of
   regeneration harvest**
*Creation of structural complexity through stand
   management activities**
*Lengthened rotation times**

Strategies addressed in this chapter are indicated by an asterisk (*).

The internal structure and composition of harvest units and managed stands is a primary determinant of the degree to which a managed forest landscape will sustain biodiversity and maintain ecosystem processes. The objective of matrix management at the stand level is to purposefully increase the contribution of harvest units to the conservation of biodiversity. As noted in Chapter 3, this is accomplished by enriching the structural and compositional complexity of managed stands. Stands can be managed to sustain species, increase habitat diversity, improve connectivity, buffer sensitive areas, and sustain ecosystem processes, including site productivity.

Although traditional silvicultural practices sometimes contribute to these goals, they can also detract from them. For example, traditional commercial thinning, which is designed to produce large and evenly distributed trees, may actually simplify and homogenize stand structure. Hence, the need for new approaches to thinning.

In this chapter, we emphasize approaches that will enhance structural and compositional complexity in managed stands with an emphasis on landscapes in which both timber production and conservation of biodiversity are management goals. The checklist of strategies for matrix management from Chapter 6 is reproduced in Table 8.1 along with the related guiding principles from Chapter 3 (Table 8.2).

Stand Management and Biodiversity Conservation Goals

Virtually all primary forests around the world are structurally complex—even those in the early stages of succession following a natural disturbance. This structural complexity ranges from the forest floor with its array of types and patterns of coarse woody debris (Franklin et al. 1981; Harmon et al. 1986; Maser et al. 1988) to complex, vertically integrated forest canopies (Parker 1997; Brokaw and Lent 1999; Parker and Brown 2000; Schütz 2001).

Intensive timber management practices, such as clearcutting followed by establishment of densely stocked plantations, simplify and homogenize stand structure (Halpern and Spies 1995; Lindenmayer and

TABLE 8.2.

Matrix management and the achievement of the general principles outlined in Chapter 3.

PRINCIPLE	STRATEGY
Principle 1. Maintenance of connectivity	• Riparian and other corridors • Protection of sensitive habitats with the matrix • Vegetation retention on logged areas throughout the landscape • Careful planning of roading infrastructure • Landscape reconstruction
Principle 2. Maintenance of landscape heterogeneity	• Riparian and other corridors • Protection of sensitive habitats within the matrix • Midspatial-scale protected areas • Spatial planning of cutover sites • Increased rotation lengths • Landscape reconstruction • Careful planning of roading infrastructure • Use of natural disturbance regimes as templates (see Chapter 4)
Principle 3. Maintenance of stand complexity	• Retention of structures and organisms during regeneration harvest • Habitat creation (e.g., promotion of cavity-tree formation) • Stand management practices • Increased rotation lengths • Use of natural disturbance regimes as templates
Principle 4. Maintenance of intact aquatic ecosystems	• Riparian corridors • Protection of sensitive aquatic habitats within the matrix • Careful planning of and maintenance of roading infrastructure
Principle 5. Risk-spreading	• Adoption of array of strategies critical to principles listed above • Ensuring that strategies are varied between different stands and landscapes ("don't do the same thing everywhere")

Franklin 1997a). These impacts are accentuated by harvesting over multiple rotations. In fact, almost all traditional regeneration harvest approaches (clearcut, seed trees, shelterwood, and group selection) create simplified forest conditions. The Silvicultural System Project conducted in the mountain ash forests of Victoria, Australia, is illustrative (Squire et al. 1987). Several silvicultural prescriptions—including shelterwood and group selection—were incorporated into the experiment, but all prescriptions provided for eventual removal of all living trees after eighty years and differed only in the proportion of the stand removed at each entry. Hence, all of the traditional silvicultural approaches tested had negative long-term impacts on cavity-dependent fauna that occur in natural mountain ash forests (Lindenmayer 1992).

A challenge for forest managers is to develop and implement management practices that restore stand structural complexity (Principle 3 from Chapter 3) as well as facilitate connectivity and maintain the integrity of aquatic ecosystems (Table 8.2). Some of these goals can be achieved by creating and maintaining the structural complexity and compositional diversity (see Table 8.3) that will, in turn, sustain targeted elements of biodiversity.

Management Approaches for Enriching Managed Stands

Three broad strategies can be used to manage stands for biodiversity conservation, each enriching structural complexity and thereby also enriching the diversity of habitats in managed stands. These strategies are (1) structural retention at the time of

TABLE 8.3.

Components of compositional and structural diversity in managed stands that contribute significantly to maintenance of biodiversity and ecosystem processes.

COMPOSITIONAL DIVERSITY (both overstory and understory) THAT PROVIDES KEY HABITAT (e.g., nesting sites, substrate for epiphytes):

- Forms critical structures (e.g., large persistent woody debris)
- Sustains symbiotic partners (e.g., mycorrhizal-forming fungi)
- Important food sources (e.g., fruits and protein-rich herbage)
- Key processes (e.g., nitrogen fixation)
- Species diversity (e.g., deciduous trees in an evergreen coniferous forest)

STRUCTURAL DIVERSITY (INDIVIDUAL STRUCTURES), INCLUDING:

- Large-diameter trees
- Decadent living trees (e.g., stems with decay, branch brooms, and dead tops)
- Standing dead trees (snags)
- Logs and coarse woody debris on forest floor
- Large-diameter branches
- Complex bark (e.g., exfoliating or deeply furrowed bark)
- Thick litter layers

STRUCTURAL DIVERSITY (SPATIAL PATTERNS):

- Gaps (canopy openings)
- Anti-gaps (heavily shaded patches)
- Vertically continuous or multilayered canopies

regeneration harvest, (2) management of established stands to maintain or create structural and functional complexity, and (3) long rotation periods or cutting cycles.

A fourth approach is the introduction of particular plant species to increase the compositional diversity of a stand (Tappeiner et al. 1997), although this is effectively covered under strategy 2, discussed later in this chapter. All three strategies will be needed if biodiversity is to be effectively conserved in production forests (e.g., Hammond 1991; McComb et al. 1993; Carey et al. 1996; see Chapter 9).

Structural Retention at the Time of Regeneration Harvest

Retaining structures from the original stand—such as large decadent trees, snags, and logs—at the time of regeneration harvest is an emerging strategy for maintaining and enhancing the structural complexity of stands in temperate production forests (Recher et al. 1980; McComb et al. 1993, Franklin et al. 1997; Fries et al. 1997) (see Figure 8.1). Structural retention strategies are modeled on the biological legacies that characterize most natural forest disturbances (see

FIGURE 8.1. Large decadent trees, large snags, and large logs are features of forest stands that are difficult to create but which can be maintained within managed forests by incorporating structural retention in regeneration harvest prescriptions. Such a strategy is highlighted in this area of harvested Douglas-fir forest on the Willamette National Forest, Oregon, United States. Photo by J. Franklin.

Chapter 4) (Franklin et al. 2000b; Franklin and MacMahon 2000). A review by Bunnell (1999b) showed that many species from a wide range of groups can benefit from the retention of biological legacies in a stand.

Structural retention can make substantial contributions to biodiversity conservation by

- Maintaining or "lifeboating" biota on a harvested site during and following logging by conserving their essential habitat, such as snags or logs
- Structurally enriching the new stand, thereby allowing organisms to more quickly recolonize harvested sites
- Modifying post-logging habitat conditions, such as microclimate, making it suitable for particular species
- Facilitating movement of organisms through harvested areas
- Buffering protected zones, such as riparian corridors, within the matrix

These roles are described more fully in Chapter 3 and also in Franklin et al. (1997).

Structural Retention—What to Retain? How Much to Retain? What Spatial Pattern to Retain?

Identifying and prioritizing management objectives, which defines the trade-offs between economic and conservation goals, must precede the development of a silvicultural prescription (Gibbons and Lindenmayer 1996; Franklin et al. 1997). Once management objectives are defined and the relevant information assembled, silvicultural prescriptions that provide for structural retention can be developed. These prescriptions must address three key issues:

1. What structures are to be retained?
2. How much of each structure should be retained?
3. How should these structures be spatially arranged?

Empirical data that can be used to address these three issues is often limited. However, useful information is provided by studies of structural and floristic attributes of natural stands (e.g., old-growth forests)

and of the types, numbers, and patterns of biological legacies that remain in stands following natural disturbance (e.g., Delong and Kessler 2000; see Chapter 4). Large structural features that are difficult or impossible to re-create in managed stands under a given management regime (including long rotation periods) are prime candidates for retention, since they often fulfill pivotal roles in forest ecosystems. For example, forest ecosystems in the Northern Hemisphere need very large decadent trees and snags for large primary cavity excavators, such as woodpeckers (e.g., Thomas 1979; Brown 1985), as well as for other species of birds and mammals (e.g., Wesolowski 1996). Similarly, large logs have greater ecological value and persist longer as structures in forest, stream, and river systems than do small logs (Maser et al. 1988). Large logs in advanced states of decay had the richest bryophyte flora of any substrate in a study in Douglas-fir forests in western Oregon (United States); structural retention was recommended as the best way to maintain bryophyte diversity (Rambo 2001).

Types of Retained Structures

The types of structural features that should be considered for retention in logged forests include large living trees—particularly those exhibiting decadence and other age-related features, such as large branches; large snags; overstory and understory plant species, which contribute to multiple vegetation layers and vertical heterogeneity; large logs; and intact areas of forest floor (i.e., the organic accumulations on the soil surface).

These structural features play valuable roles in biodiversity conservation (Franklin et al. 1997, 2002), although specifics regarding roles vary with forest type and species assemblages.

Large Living Trees

Large living (decadent) trees have many important ecological values (Figure 8.2). In addition to their numerous roles as living trees, when they die they become sources of snags, logs, and other coarse woody debris in forests and associated aquatic ecosystems (Harmon et al. 1986; Maser et al. 1988; Cascade Center for Ecosystem Management 1995; Recher 1996; Franklin et al. 2002). Some authors combine partially

FIGURE 8.2. Large-diameter trees can play many roles in biodiversity conservation. Large-diameter trees can provide suitable attachment sites for mud-nest-building birds such as the fairy martin (*Hirundo ariel*) in Australia. Photo by D. Lindenmayer.

FIGURE 8.3. Cavities are a critical resource for numerous species of vertebrates and invertebrates worldwide. In Australia alone, more than 300 species of vertebrates depend on cavity-trees. Photo by E. Beaton.

dead and decayed trees and snags into a single category in their discussions of wildlife habitat requirements, recognizing that dead and dying trees are related developmentally and are often equally suitable as habitat (e.g., Thomas 1979; Brown 1985; Lindenmayer et al. 1997a). Live trees with significant decay do tend to last longer than totally dead trees and provide cavities that are useful to a broader array of vertebrates; they may also facilitate partitioning of resources by potential competing species. Both categories of trees can be used for foraging, nesting, and as roosting sites for many species of vertebrates and invertebrates (McClelland and Frissel 1975; Saunders et al. 1982; Scotts 1991; Wesolowski 1995; Gibbons and Lindenmayer 2002) (Figure 8.3). Ohmann et al. (1994) identified over 100 species of vertebrates in the Pacific Northwest region of the United States that require cavities in living trees and/or standing snags as primary habitat.

Although some organisms are apparently insensitive to whether cavities occur in living or dead trees, others require cavities within living trees. The red-cockaded woodpecker (*Picoides borealis*) within the longleaf pine (*Pinus palustris*) forests of the southeastern United States is an example (Jackson 1978). This species excavates nests in large living trees that have significant wood decay caused by brown rot (*Fomes pini*) but which still provide significant sap flows that deter predation by snakes and other predators.

Large living trees provide distinctive architectural features that often warrant explicit recognition in structural characterizations of forests (e.g., Franklin et al. 2002). These include large-diameter branches, complex branching systems and multiple tops, brooms (or multilayered branch thickets resulting from mistletoe damage), and bark habitats.

Large-diameter branches provide habitat for many organisms simply by virtue of their size and longevity (Burgman 1996; Peck and McCune 1997). For example, they are foraging sites for several species of woodpeckers in the *Nothofagus* spp. forests of central Chile in South America (Willson et al. 1994). A suite of invertebrate taxa depends upon wounds and bark-free areas on large branches in some European forest types. The conservation of many species, such as saproxylic insects, can be promoted by allowing dead limbs to accumulate in production forests in Europe (Schiegg 2001). In the Douglas-fir forests of the Pacific Northwest, the occurrence of many species of lichens and mosses is associated with networks of large branches, which are found primarily on old-growth trees (Clement and Shaw 1999). Large branches are also the primary nesting habitat for the endangered marbled murrelet (*Brachyramphus marmoratus*) in these forests (Ralph et al. 1995). Some Douglas-fir develop elaborate epicormic branch systems that are effectively "perched" ecosystems with organic soils, complex communities of microorganisms and invertebrates, and nests of vertebrates (Figure 8.4). Epicormic branches develop from dormant buds laid down in the axils of branches and twigs and may be important in many coniferous forest types (Franklin et al. 2002).

The extensive and complex canopies of large old trees provide diverse habitats for invertebrates, including spiders and predatory and parasitic insects (Schowalter 1989). These predators and parasites help control insect herbivores and also contribute to the prey base for vertebrates. The canopies are also habitat for the adult stages of many soil and aquatic insects—taxa that further enrich canopy food webs and sustain key processes in soils and aquatic ecosystems (Hooper et al. 2000; Palmer et al. 2000).

The distinctive bark habitats of large living trees often support many invertebrates and epiphytes. The rough bark characteristic of old-growth western North American conifers is habitat for a suite of insects and spiders. In Australian mountain ash forests and forests dominated by other eucalypt species, extensive bark streamers are primary habitat for a group of flightless crickets as well as many other invertebrates that are, in turn, prey species for arboreal marsupials (Smith 1984) and birds (Loyn 1985a). Bark production in mountain ash forests can exceed 1.5 tons per hectare, especially in mature and old-growth stands; this amount is five times greater than in sixty-year-old stands (Lindenmayer et al. 2000c).

Retention of large living trees is a powerful strategy for lifeboating species and for structurally enriching stands that develop after logging. In Sweden, aspen (*Populus tremula*) trees retained on harvest units

FIGURE 8.4. Epicormic branch systems on an old-growth Douglas-fir tree. Such large, complex branching systems provide outstanding locations for the development of epiphytic plant communities, nesting sites for many vertebrates, and accumulations of organic soils. Photo by J. Franklin.

help maintain populations of lichens and bryophytes that are otherwise sensitive to the effects of timber harvesting (Hazell and Gustafsson 1999). This also can allow species to persist through two or more cutting cycles (Kaila et al. 1997; Gundersen and Rolstad 1998; Price and Hochachka 2001; Rolstad et al. 2001). The ability of residual old-growth trees to sustain in situ populations of sensitive lichen and moss species has also been documented in young coniferous forests in the northwestern United States (Neitlich and McCune 1997; Peck and McCune 1997).

Management prescriptions that facilitate the structural enrichment of harvest units can promote much earlier recolonization of young forests than would otherwise be possible (Carey 1995; Carey and Curtis 1996).

Retained living trees, as well as other plants, contribute significantly to maintaining the integrity of the below-ground ecosystem following timber harvesting. High levels of energy are required to maintain below-ground structures (e.g., fine root and mycorrhizal hyphal systems), organisms, and processes (Hooper et al. 2000). Living trees are the primary source of this energy in forest soils; in many forests, more than half the photosynthetic production is utilized below the ground (Allen et al. 2000). High levels of microbial, invertebrate, and fungal diversity and complex food webs are also associated with the death and decay of networks of roots as well as the materials exuded and leached from them in the living state. Hence, for all these reasons, below-ground organisms and processes must be sustained, typically by maintaining significant numbers of above-ground trees and shrubs.

Maintaining populations of fungi capable of forming mycorrhizal associations is a particularly important forest management goal, even in production forests. This may require attention to the individual plant species and species groups to be retained (see Amaranthus et al. 1994). For example, some fungi form symbiotic relationships with species in the Pinaceae and Ericaceae families while others are associated with many hardwood species (angiosperms) or some groups of gymnosperms (e.g., members of the Cupressaceae family). Therefore, different above-ground tree symbionts are necessary to maintain the full spectrum of mycorrhizal-forming fungi.

In areas of highly unstable soils, such as the steep headwall sections of streams, living trees and other woody plants may be retained specifically to maintain the extensive, intact root systems that stabilize soils.

Large Dead Standing Trees and Logs

The value of dead standing trees (snags) and logs for the maintenance of biodiversity was discussed in general terms in Chapter 3. Their value is now well understood by many forest ecologists and forest managers (Harmon et al. 1986; Franklin et al. 1987; Gibbons and Lindenmayer 1996; Bunnell et al. 1999) (Figure 8.5). Indeed, some authors estimate that 20

FIGURE 8.5. Coarse woody debris is a key structural component of all unmanaged forests and plays numerous roles in biodiversity conservation and ecosystem function. This stand of *Nothofagus* spp. forest near Te Anau in southern New Zealand highlights the large amount of coarse woody debris that can occur in natural forest. Photo by D. Lindenmayer.

percent of all forest biodiversity is associated with dead wood (Hunter 1990; Grove 2001). However, dead trees and logs are still viewed as wasted wood, fire and safety hazards, and management impediments in traditional forest management paradigms, often making their conservation a challenge.

Some of the numerous functions of snags and logs are as

- Substrate for the germination and development of plants (nurse logs) (Harmon and Franklin 1989; Barker and Kirkpatrick 1994)
- Habitat for rich assemblage of detritivores and decay organisms (Maser and Trappe 1984; Berg et al. 1994)
- Shelter for a wide variety of forest-dependent vertebrates (Tallmon and Mills 1994; Wilkinson et al. 1998)
- Protected runways for the movement of terrestrial animals (Maser et al. 1978; McCay 2000)
- Hunting and resting perches (Thomas 1979; Brown 1985; Backhouse and Manning 1996)
- Basking sites for reptiles and mammals (Cogger 1995; Webb 1995)
- Foraging sites for wildlife (Thomas 1979; Maser and Trappe 1984; Smith et al. 1989), including species-rich invertebrate assemblages (Taylor 1990; New 1995)

- Hiding cover and habitat for fish and other aquatic organisms (Harmon et al. 1986; Sedell et al. 1988; Koehn 1993; Gregory 1997; Naiman and Bilby 1998).
- Influences on hydrologic and geomorphic processes in streams and rivers (Harmon et al. 1986; Sedell et al. 1988; Naiman and Bilby 1998; Gippel et al. 1996; Naiman et al. 2000).
- Long-term sources of energy and nutrients in forest and aquatic ecosystems (Harmon et al. 1986). Swift (1977) noted there were two key consequences of the high lignin content of wood and consequent slow composition rates. First, it ensured a slow release of nutrients giving a buffering effect to nutrient cycles. Second, the humification of wood residues is thought to contribute disproportionately to the formation of soil organic matter. This has important consequences for the long-term productivity of a site, since organic matter has important influences on soil structure, water holding capacity, and nutrient storage.
- Sites for nitrogen fixation (Harmon et al. 1986).
- Potential fire fuels (Luke and McArthur 1978).
- Providing mesic refugia for an array of organisms during drought and/or fire. Large fallen trees may act as micro fire breaks, not only because of their diameter and length but also because of the moisture they contain and the fact that the litter adjacent to them also has relatively high moisture levels (Andrew et al. 2000).
- Contributing to heterogeneity in the litter layer and patterns of ground cover.

Harmon et al. (1986) and Maser et al. (1988) provide extensive reviews on the functions of woody debris. Graham et al. (1994) discuss the quantities of logs needed to be retained per hectare of logged forest to maintain forest productivity.

Issues associated with the role and retention of snags are reviewed by Gibbons and Lindenmayer (1996, 2002) and, briefly, below.

The importance of snags for cavity-dependent birds and mammals has been known for many decades (e.g., Fischer and McClelland 1983; Raphael and White 1984; BC Environment 1995; Imbeau et al. 2001), but different species need different tree species, tree sizes, and states of snag decay (Thomas 1979; Gibbons and Lindenmayer 2002). Large-diameter snags often occur in low densities (Spies and Franklin 1988; Grief and Archibold 2000) and are irregularly distributed within forest stands and landscapes (Spies and Franklin 1988; Lindenmayer et al. 1991b). The abundance of snags often limits populations of snag-dependent species (Thomas 1979; Newton 1994), particularly in even-aged and/or intensively managed forests (reviewed by Newton 1998; Gibbons and Lindenmayer 2001). It is clear that the needs of cavity-dependent taxa are typically not met in production forests (Hope and McComb 1994; Gibbons and Lindenmayer 1996).

Maintaining populations of snags is a challenge for natural resource managers. This is because of

- Their importance for wildlife (Raphael and White 1984; Steeger et al. undated).
- Their relatively short life span—few snags remain intact and standing for more than a few decades (Graham 1982; Maser et al. 1988; Lindenmayer et al. 1997a). There are tree species that have exceptional snag life spans, such as more than seventy years in some South African and Australian forests and one to two centuries in the case of western redcedar (*Thuja plicata*) and other Cupressaceae species throughout the world (in part because of chemicals that make the heartwood decay resistant).
- The difficulty of developing management programs to perpetuate snag populations because of other management activities, such as slash burning to reduce logging debris (Horton and Mannan 1988; Lindenmayer et al. 1990a).
- Other management problems associated with snags, such as their contribution to the spread of wildfires (Crowe et al. 1984) and the hazards they pose to forest workers involved in management, including logging, fire suppression, planting, thinning, and aerial treatments of fertilizers and herbicides (Styskel 1983; Squire et al. 1987; Hope and McComb 1994). For example, in British Columbia, regulations of the Health and Safety Board require cutting all snags within work areas and roadsides.

Multiple Vegetation Layers and Understory Vegetation

Many retention strategies focus on overstory trees. However, retaining understory vegetation can be critical for biodiversity conservation, particularly in forests subjected to intensive harvesting practices such as clearcutting and broadcast slash burning (Hansen et al. 1991). Besides being biota, understory trees and shrubs have many roles (Carey et al. 1996; Bunnell et al. 1999) (Figure 8.6), which include

- Providing food resources for vertebrates (Seebeck et al. 1984) and invertebrates
- Contributing to nutrient cycling (Ashton 1976; Adams and Attiwill 1984), particularly in the case of species with special attributes such as an association with nitrogen-fixing bacteria
- Serving as nursery sites for other plants (Howard 1973), such as tree ferns in Australian eucalypt forests (Duncan and Isaac 1986)
- Acting as nesting and perching sites and movement routes for birds and other animals
- Providing habitats for invertebrates (Woinarski and Cullen 1984)

FIGURE 8.6. Tree ferns are a key understory component of wet eucalypt forests and rainforests throughout eastern Australia. They are nursery sites for many other plant taxa, nesting places for birds and some mammals, and a food source for large vertebrates such as the mountain brushtail possum. Tree ferns are also sensitive to intensive harvesting practices such as clearcutting, and special consideration needs to be given to their retention during regeneration harvesting. Photo by D. Lindenmayer.

- Providing microhabitats for the development of fungi, which are food resources for animals (Maser et al. 1977; Claridge 1993)
- Contributing to vertical canopy diversity (Franklin et al. 2002), thereby increasing the range of foraging substrates for birds (Woinarski et al. 1997) and bats (Brown et al. 1997)
- Acting as sediment traps and, in turn, promoting soil development processes

Long-lived and disturbance-sensitive understory species can be negatively affected by intensive forestry operations (Ough and Ross 1992; Halpern and Spies 1995; Mueck et al. 1996). Retention strategies that focus on small islands of vegetation, or aggregates, which are kept free of logging and slash fire, are particularly valuable for maintaining these communities and their associated animal biota (Ough and Murphy 1998).

Intact Areas of Forest Floor

Maintaining intact areas of the forest floor, including patches supporting coarse woody debris, promotes the conservation of litter invertebrates and hypogeal fungi. Besides being biotic components, these organisms are part of the diet of some vertebrates (Maser et al. 1978; Harvey et al. 1987; Claridge 1993; North 1993 in Franklin et al. 1997). They carry out important ecosystem functions as part of the detrital food chain. Hypogeal fungi are often symbionts that maintain forest productivity. And, in some types of forest, up to 50 percent of the nitrogen and nitrogen fixation occurs on the forest floor (Graham and Jain 1998).

Numbers of Retained Structures

The simple answer to the question of "How many structures should be left?" is "Enough to achieve specified management goals." However, structural retention strategies differ widely with the management objectives, which may be to maintain biota on site, enable organisms to quickly recolonize the area following logging, or facilitate movement through cutovers to areas of more suitable habitat; or all three.

Even for extensively studied forest ecosystems, quantitative data are limited on levels of retained structures that are needed to achieve particular goals. Calculating appropriate numbers of structures is complicated because of the array of issues that must be taken into account (Box 8.1). For example, selecting unmerchantable stems for retention can be inappropriate because they do not persist as long as trees with sound wood, which limits their long-term value for cavity-dependent taxa (Gibbons and Lindenmayer 2002) and other fauna. Several studies are currently under way that should provide empirical data on the effectiveness of different levels of retention for the conservation of specific species (e.g., Halpern and Raphael 1999).

Crude estimates of the appropriate number of retained trees and logs for harvest units have been made by summing values required for each species targeted for conservation (e.g., DeGraaf and Shigo 1985) and linking this number to information on the recruitment rate of new structures (Gibbons and Lindenmayer 1997). McComb and Lindenmayer (1999) used this approach to calculate the total number of living and dead cavity trees and logs in various diameter classes required for all vertebrates in a maturing forest in northwestern Washington state. They estimated that 6.4 logs and 27.3 snags and living cavity-bearing trees (in varying diameter classes and states of decay) were required per hectare to provide optimum habitat for all species. Caveats associated with such estimating procedures include the following:

- Knowledge about the quantitative relationships between log attributes and species requirements is limited for most forest types.
- Values will be different in stands of different ages.
- Many organisms require structural features (such as canopy cover) in addition to adequate log, snag, and cavity-tree levels.

In Swedish production forests, a standard prescription for live tree retention (part of a national standard for forest certification) is retention of sufficient stems to sustain at least ten large old trees per hectare during the next rotation (Hazell and Gustafsson 1999). Of course, forest managers and timber workers must be trained to avoid cutting these retained green trees during post-regeneration thinning operations (Hazell and Gustafsson 1999).

Many guidelines have been developed for specified densities of wildlife trees and logs on harvested sites. These are based primarily on expert opinion (e.g.,

Anonymous 1992) and/or some arbitrary proportion of the numbers of these structures measured in unlogged forests. We endorse such approaches as logical starting points where no other information is available, although their adequacy in sustaining

---

BOX 8.1.

The Number of Retained Trees in Harvested Mountain Ash Forests in the State of Victoria, Australia

Data from the well-studied mountain ash forest ecosystem in Victoria illustrate the complex task that calculating numbers of trees for retention in harvest units involves. Tentative numbers have been calculated for conservation of the endangered arboreal marsupial Leadbeater's possum. Based on a validated statistical relationship of the habitat requirements of this cavity-dependent species, Lindenmayer et al. (1991a) estimated that a perpetual supply of five to ten cavity trees per hectare was needed to achieve a 50 to 80 percent probability of occurrence of the species in logged and regenerated forests. Significantly more cavity trees are needed to meet the requirements of the more than 100 other cavity-dependent species that occur in these forests (Gibbons and Lindenmayer 1996). Moreover, maintaining cavity-bearing stems representing a range of developmental stages is a major challenge in managed mountain ash forests. Cavity-dependent species have different cavity-tree requirements. The greater glider requires tall, intact cavity trees whereas Leadbeater's possum typically utilizes short highly decayed stems (Lindenmayer et al. 1991b). The latter type of cavity-trees takes more than 200 years to develop as they progress through several preceding stages of decay (Lindenmayer et al. 1993a). Therefore, management of cavity-bearing trees actually involves the creation and maintenance of cohorts that represent a developmental sequence of increasing senescence. Tree mortality and collapse further increase the initial number of stems required for retention (Ball et al. 1999). So, to ensure the continued presence of a set number of cavity-trees in a stand, significantly more trees need to be retained at the time of regeneration harvest.

This example shows that calculating required levels of retained trees on logged areas can be very complex. It requires integrating information on longevity of retained trees, the rates at which trees are degraded, and the array of organisms that are of management interest and their cavity-tree requirements (Gibbons and Lindenmayer 2002).

FIGURE 8.7. Contrasting spatial patterns for structural retention in harvest units: (*A*) Dispersed retention in Sierra Nevada mixed conifer forest (California, United States). (*B*) Aggregated retention of live trees and other structures in aggregates or small forest patches (BC Coastal Forest Project of Weyerhaeuser Company, Vancouver Island, British Columbia, Canada). (*C*) Mixture of dispersed and aggregated retention (Cedar River watershed, Washington, United States). Photos by J. Franklin.

specific species needs to be subsequently validated (Gibbons and Lindenmayer 2002).

Spatial Patterns of Structural Retention

The spatial arrangement of structures within stands is as important to structural complexity as the diversity and density of individual structures (Franklin et al. 1997, 2002). Two contrasting spatial patterns for retention are *dispersed retention*, in which structures are dispersed uniformly over a harvested area; and *aggregated retention*, in which structures are concentrated in aggregates or small forest patches within a harvested area (Figure 8.7).

We reiterate that structural retention applies to logs, areas of intact forest floor and other important structural components, as well as to trees, as noted earlier in this chapter.

Both the dispersed and the aggregated patterns of structural retention have particular advantages and disadvantages. These depend upon such variables as the objectives of structural retention, the biology of taxa targeted for management, potential loss of retained structures (such as to windthrow), and operational constraints, such as worker safety and logging costs (Table 8.4). There are many variants on the dispersed and aggregated approaches, including combinations of both approaches.

TABLE 8.4.

Contrasts between dispersed and aggregated structural retention (from Franklin et al. 1997).

OBJECTIVE ON HARVEST UNIT	PATTERN OF RETENTION	
	Dispersed	*Aggregated*
Microclimate modification	Less, but generalized over harvest area	More, but on localized portions of harvest area
Influence on geohydrological processes	Same as above	Same as above
Maintenance of root strength	Same as above	Same as above
Retain diversity of tree sizes, species, and conditions	Low probability	High probability
Retain large-diameter trees	More emphasis	Less emphasis
Retain multiple vegetation (including tree) canopy layers	Low probability	High probability
Retain snags	Difficult, especially for soft snags	Readily accomplished, even for soft snags
Retain areas of undisturbed forest floor and intact understory community	Limited possibilities	Yes; can be as extensive as aggregates
Retain structurally intact forest habitat patches	Not possible	Possible
Distributed source of coarse woody debris (snags and logs)	Yes	No
Distributed source of arboreal energy to maintain below-ground processes and organisms	Yes	No
Carrying capacity for territorial snag- and/or log-dwelling species	More	Less
Windthrow hazard for residual trees	Average wind firmness greater (strong dominants), but trees are isolated	Average wind firmness less, but trees have mutual support
Management flexibility in treating young stands	Less	More
Harvest (e.g., logging) costs	Greater increase over clearcutting	Less increase over clearcutting
Safety issue	More	Less
Impacts on growth of regenerated stand	More; generalized over harvest area	Less; impacts are localized

Dispersed Retention

A dispersed pattern of structural retention in a harvest unit may be desirable or necessary to achieve specific conservation goals (Table 8.4). Examples are maintenance of below-ground biota and processes (e.g., inocula of hypogeal mycorrhizal fungi); maintenance of root strength; and provision for well-distributed populations of snags and logs (Table 8.4). Dispersed retention is also superior in achieving modified environmental conditions (e.g., temperature, relative humidity, and insolation) throughout a harvested area, which can be important to the survival of some organisms, such as salamanders (e.g., Naughton et al. 2000). Some vertebrates require dispersed cavity trees, such

as arboreal marsupials (Lindenmayer et al. 1990b) and large parrots (Rowley and Chapman 1991; Nelson and Morris 1994; Krebs 1998). This is necessary to accommodate the social behavior of these territorial species. The benefits of retaining dispersed trees also have been documented for species of old-growth-selected epiphytic lichens (Sillett et al. 2000).

Aggregated Retention

Aggregated retention involves the retention of small, intact areas of forest in harvest units. The appropriate size of aggregates depends upon many variables, including forest and conservation management objectives, the silvicultural system being applied in the harvest unit, and the forest type. Some variability in aggregate size is probably also appropriate. Aggregates currently being retained in clearcut harvest units in northwestern North America typically vary in size from 0.4 to 1.5 hectares (Franklin et al. 1997). Aggregates of 0.2 hectare in size have been used in shelterwood-harvested *Nothofagus* forests in Tierra del Fuego, South America (see Chapter 15).

Aggregates are intended to be *a part of—not apart from*—the harvested stand in which they are located. The objective of aggregated retention is not to create large forest islands that provide true forest-interior conditions (which is the function of large ecological reserves). Aggregates provide small distributed refugia within harvest units that function as lifeboats for biodiversity and structurally enrich the managed stand throughout the rotation. In some forest ecosystems, aggregates may mimic the remnant (unburned) patches left by wildfire (British Columbia Ministry of Forests 1995; Delong and Kessler 2000).

Aggregated retention offers a number of environmental and practical advantages over dispersed retention (Franklin et al. 1997) (Table 8.4). Assuming that aggregates are fully protected from logging, slash disposal, and site preparation operations, aggregated retention is designed to retain a wide range of structures, including different sizes and conditions of living trees, snags, and logs; multiple or continuous canopy layers; undisturbed understory and forest floor conditions; and small areas where environmental conditions come close to those of an intact forest, even though they are not true forest-interior conditions.

Aggregated retention allows the goals of understory and overstory vegetation retention to be achieved simultaneously, as in the "understory island" strategy developed by Ough and Murphy (1998).

With regard to management and operational considerations, aggregated retention typically

- Provides safer working conditions by creating no-work zones, which makes it possible to retain hazardous structures, such as soft snags (Hope and McComb 1994; Hickey et al. 1999). Some authors view aggregated retention, such as wildlife tree patches in British Columbia (Fenger 1996), as one of the only ways to protect snags without increasing risks to timber workers. In these cases, aggregates

may need to be of sufficient size to overcome worker safety issues; in other words, collapsing snags and decadent trees cannot fall on workers operating in the harvest zone outside retained patches.
- Allows for efficient logging operations.
- Limits the area in which tree growth and reproduction is suppressed by retained overstory trees.
- Makes aerial treatment of managed stands, such as with fertilizers and herbicides, more feasible.

Aggregated retention is an efficient strategy for lifeboating many elements of biodiversity within a logged area. Stand islands (aggregates) provide more niches and, consequently, more species are likely to persist using aggregated rather than dispersed structural retention (Berg et al. 1994; Cascade Center for Ecosystem Management 1995; Scientific Panel for Sustainable Forest Practices in Clayoquot Sound 1995). Some animals depend upon the adjacency of overstory and understory structures, such as using overstory trees for nesting and understory plants for foraging (Lindenmayer et al. 1991a, 1997b). Aggregated retention also provides for the intact soil and organic layer conditions needed by many mycorrhizae-forming fungi—organisms that are essential to the sustained productivity of harvested stands (Perry 1994).

Loss of retained living trees and snags to windthrow and exposure may be less with aggregated than dispersed retention under some circumstances, particularly if treatments, such as topping or pruning, are used to make trees less susceptible to blowdown. Aggregated retention of "wildlife habitat clumps" was adopted in harvested forests in the Australian state of Tasmania because of concerns about windthrow (Forest Practices Board 1999b; Duhig et al. 2000). Other approaches to minimizing windthrow include laying out aggregates with streamlined shapes, locating them in topographically sheltered parts of harvest units, and incorporating sound (decay-free) dominant trees (Gibbons and Lindenmayer 1996; Franklin et al. 1997). When dispersed retention consists primarily of dominant, sound trees, losses to windthrow may be as low or lower than that encountered with aggregated retention.

In prescribing the location, size, and shape of aggregates, many factors have to be considered including (1) local site conditions, (2) stand conditions (including whether aggregates are to be a representative cross section of the stand being harvested), (3) spatial distribution over the harvest unit, and (4) biology of the taxa targeted for management.

The degree to which the aggregates are representative of the stand (point 2 above)—is an important issue. Structural features, such as large soft snags or large logs, can be used as focal, or anchor, points for aggregates. Biasing aggregate locations toward parts of a harvest unit that are nonforested (e.g., wetlands) or of low productivity, are occupied primarily by unmerchantable trees, or are disease centers is *not* appropriate if the goal is to lifeboat forest biodiversity and structurally enrich the managed stand. In some forest types it may be inappropriate to locate, for example, retained aggregates in riparian zones because midslope and upslope tree species are not represented and the shallower root systems of trees in wet or moist bottomlands can make them highly susceptible to windthrow. In other cases, it may be appropriate to locate aggregates around biotically rich nonforested features, including streams, rock outcrops, or wetlands, particularly if the aggregates also incorporate the structural complexity of the surrounding stand. In the Queen Charlotte Islands of British Columbia, Canada, culturally modified trees such as those used in the construction of canoes by indigenous people are often used as anchor points in aggregated retention.

The biology of target taxa can be particularly important (point 4 above). Taylor and Haesler (1993) believed that limiting retained trees to riparian buffers might exacerbate patterns of territorial behavior among Tasmanian forest birds and thereby reduce occupancy rates of cavity trees. Increased stand complexity through structural retention may not affect all species equally (Newton 1994). In the case of cavity-dependent vertebrates, Gibbons and Lindenmayer (2002) speculated that the retention of living trees and snags in Australian forests might benefit larger, more aggressive generalist taxa than smaller species with more specialized nest and den requirements. There are large differences in the response of the northern flying squirrel (*Glaucomys sabrinus*), Townsend chipmunk (*Tamias townsendii*) and Douglas' squirrel (*Tamiasciurus douglasii*) to retention management strategies in the Pacific Northwest (Carey 2000).

Concluding Remarks on Structural Retention

A combination of both aggregated and dispersed approaches to retention will typically be desirable, since it is usually impossible to achieve all the objectives of a retention harvest strategy by depending exclusively on either approach. The combination of dispersed and aggregated approaches also may be essential to meet particular objectives. A good example is the case of arboreal marsupials in the mountain ash forests of Victoria. Many species require spatial adjacency of understory and overstory components (e.g., to help young move from nest sites to neighboring foraging sites), but they also require nest sites (in large cavity trees) to be well spaced throughout the forest because of social behavior (such as territoriality). Dispersed aggregates throughout a harvest unit will best meet the requirements of these species.

It is important to recognize that past harvesting practices—such as where trees were left behind for seed or other reasons (Isaac 1943)—may be poor guides to responses expected under a designed retention strategy.

Some organizations have significant experience in retention harvesting. For example, the BC Coastal Project of Weyerhaeuser Corporation is currently one of the foremost practitioners of retention harvesting in North America (Weyerhaeuser 2000); its prescriptions are specifically designed to balance ecological and economic objectives. Presently, there are relatively few data available to assess the effectiveness of newly emerging retention strategies, but some valuable studies are under way (e.g., Chambers et al. 1999; Carey 2000).

The DEMO (Demonstration of Ecosystem Management Options) project in the Douglas-fir forests in western Oregon and Washington is a major experimental test of alternative retention approaches. DEMO is a replicated cutting experiment designed to assess the consequences of different levels (15 and 40 percent) and patterns of retention (dispersed and aggregated) on biodiversity and ecosystem processes (Franklin et al. 1999b; Halpern and Raphael 1999).

As with all the matrix management strategies discussed in this book, there is no prescription that can be applied generically and uncritically to all stands. There is an infinite variety of forest, environmental, and social conditions, and the most effective structural retention design will be based on considerations of the natural history and management experiences of the particular location, as well as on common sense. Moreover, extensive stand retention may not be necessary on every hectare of production forest; it may not even be possible in some locations for reasons such as worker safety. Structural retention prescriptions can be guided by

- Lessons from patterns created by natural disturbance regimes and spatial variation in quantities of biological legacies across landscapes (see Chapter 4)
- Management objectives
- Organisms targeted for management
- Types and intensities of other matrix-based management strategies (e.g., riparian buffers and midspatial-scale protected areas (see Chapters 6 and 7)
- Variation in stand conditions needed as a part of adaptive management studies and experiments (see Chapter 16)

Applying retention uniformly on all harvest units may homogenize landscapes just as with widespread clearcutting. Dispersed retention will be most appropriate for some species and ecological processes and aggregated retention (or a combination of both) for others. Hence, spatial variation in structural retention is appropriate and another technique for risk-spreading (see Chapter 3). Variability in stand conditions increases the chance that suitable habitat will occur for most species in at least part of a landscape.

The Variable Retention Harvest System

The variable retention harvest system (VRHS) is a systematized approach to structural retention that is defined as

an approach to harvesting based on the retention of structural elements or biological legacies (trees, snags, logs, etc.) from the harvested stand for integration into the new stand to achieve various ecological objectives. . . . Major variables . . . are types, densities, and spatial arrangements of retained structures. (Helms 1998)

Implicit in the VRHS is acceptance of the idea that some of the productive capacity and economic value of the stand will be devoted to maintenance of biodiversity (and other values such as the maintenance of ecosystem processes) rather than maximizing the regeneration and growth of commercial tree species (Franklin et al. 1997).

VRHS first emerged as a concept in the recommendations of the Scientific Panel for Sustainable Forest Practices in Clayoquot Sound (1995) on the west coast of Vancouver Island, British Columbia. The Scientific Panel recommended that the government

replace conventional silvicultural [clearcut] systems in Clayoquot Sound with a "variable retention silvicultural system." The purpose of this system is to preserve, in managed stands, far more of the characteristics of natural forests. The variable retention harvest system provides for the permanent retention after harvest of various forest structures or habitat elements, such as large decadent trees, snags, logs, and downed wood from the original stand that provide habitat for forest biota.

The Scientific Panel recommended the retention of at least 15 percent of the forest. This retention was to occur primarily as 0.1- to 1-hectare aggregates representative of the forest conditions within and well dispersed throughout the cutting units. Furthermore, aggregates were to be spatially distributed so that all parts of the harvest unit were within two tree heights of an aggregate or stand edge.

Subsequently, MacMillan-Bloedel Corporation (the largest wood products company in Canada at that time) announced that it was phasing out clearcutting and adopting the VRHS as its primary silvicultural system (Dunsworth and Beese 2000). Because government cutting permits could be issued only for officially recognized silvicultural systems, a legal definition of VRHS in British Columbia was adopted:

> Retention system means a silvicultural system that is designed to: (a) retain individual trees to maintain the structural diversity over the area of the cutblock for at least one rotation, and (b) leave more than half of the total area of the cutblock within one tree height from the base of a tree or group of trees, whether or not the group of tree or group of trees is within the cutblock.

The BC Coastal Forest Project of Weyerhaeuser Corporation (the successor to MacMillan-Bloedel) has continued to refine its application of structural retention (Figure 8.7).

Structural retention is mandated for essentially all regeneration harvest units on federal lands within the range of the northern spotted owl in the northwestern United States (USDA Forest Service and USDI Bureau of Land Management 1994b). The direction is to

retain at least 15 percent of the area associated with each cutting unit. . . . As a general guide, 70 percent of the total area to be retained should be aggregates of the oldest live trees . . . and hard snags occurring in the unit. Patches should be retained indefinitely.

The VRHS concept continues to evolve. The consensus in North America is that

- A minimum level or threshold of retention is necessary for the practice to be socially credible and ecologically effective
- At least some of the retention has to be of large structures (e.g., dominant trees and large snags)
- The spatial distribution of retention is important (i.e., retention cannot be concentrated along the edges of a harvest unit)
- Structures must be retained for at least one rotation—in other words, structures that are retained only temporarily, such as a shelterwood overstory, do not meet the goal of structural retention

Beyond this broad consensus, VRHS encompasses a broad continuum of silvicultural prescriptions (see Figure 8.8). It is flexible in terms of levels of stand retention and the array of structural conditions that can be created (e.g., even-aged, multi-aged, or all-aged)

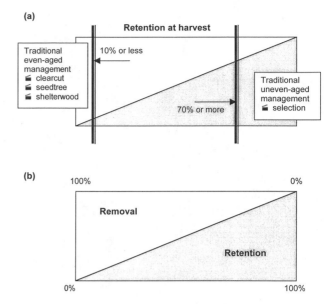

FIGURE 8.8. The variable retention harvest concept (modified from Franklin 1993b; Franklin et al. 1997)

(Figure 8.8). VRHS is not simply a modification of traditional regeneration harvest systems (cf. Florence 1996; Smith et al. 1997); it is based upon a much broader array of objectives and structural objectives than traditional regeneration harvest systems. Most particularly, it focuses upon what is left behind rather than what is removed.

VRHS avoids many difficulties associated with attempts to incorporate new ecological objectives into traditional tree regeneration–oriented harvest systems (i.e., clearcut, seed tree, shelterwood, and selection). It allows foresters to communicate clearly how the stand is to be treated and how it will appear following harvest, whereas attempts to adapt old terminology to new objectives can lead to poor communication. For example, to describe a cutover with 30 to 40 percent of the trees retained as a "clearcut" or "clearcut with reserves," because the trees were retained for objectives other than the regeneration of a commercial tree species, can be misleading. Hence, use of VRHS terminology—in which the specifics are provided regarding what and how much is to be retained and in what spatial arrangement—can lead to much clearer communication among professionals as well as with the interested public.

FIGURE 8.9. Structural profile of a forest stand that is a fine-scale mosaic and adapted to small group selection with structural retention harvest methods—mixed conifer forest in the Sierra Nevada range of California, in the United States (Aspen Valley, Yosemite National Park). Drawing courtesy of Robert Van Pelt.

BOX 8.2.

The Warra Long Term Ecological Research Experiment in Tasmania— Testing New Silvicultural Systems

The silvicultural systems trial at the Warra Long Term Ecological Research (LTER) site in the southern Australian state of Tasmania is a useful example of a major study designed to develop new silvicultural systems that better integrate wood production with biodiversity conservation and other environmental values (Hickey et al. 1999). Clearcutting has been the traditionally applied harvesting system in the wet forests of southern Tasmania (Figure 8.10). There have been concerns about the negative environmental impacts of clearcutting, including detrimental effects on biodiversity. The organization responsible for harvesting on public land (Forestry Tasmania) instigated a major cutting trial to explore new methods to manage the wet eucalypt forests of southern Tasmania. The range of treatments tested (and their potential benefits) include the following (modified from Hickey and Neyland 2001):

TREATMENT	POTENTIAL BENEFITS
Clearcut, burn, and sow (traditional harvest system)	Economically and operationally efficient, effective regeneration
Clearcut, burn, and sow with understory islands	Increased biodiversity values
Cable harvested 300×80-meter strips and low-intensity burn	Natural seedfall, low soil damage, protection of rainforest
Cable harvested in 300×240-meter patch, and low-intensity burn	Natural seedfall, low soil damage, protection of rainforest
Dispersed retention (10 percent basal area retention, and low-intensity burn)	Natural seedfall, more cavity trees, supply of large logs
Aggregated retention (30 percent basal area retention, "fairways" one log width either side of a skid trail, aggregate strips of 0–1 hectare in size)	Natural seedfall, more cavity trees, increased worker safety
Single tree/small group retention (permanent skid trail, repeat cutting every twenty years, site scarification)	Natural seedfall, enhanced biodiversity values, protection of rainforest.

Some entirely new methods of harvesting are showing considerable early promise, such as the 30 percent aggregated retention or "fairway" system with skid tracks two tree widths wide, with trees retained between the "fairways" (J. Hickey personal communication).

The Warra silvicultural systems trial is an excellent example of a proactive approach to integrate multiple uses in matrix forests through testing new harvesting methods that are not constrained by traditional silvicultural paradigms.

The potential for innovative approaches using VRHS is extensive. For example, a form of group selection with structural retention could be utilized in managing forest types that are naturally subject to frequent, light-to-moderate-intensity disturbances that create small gaps (see Franklin and Fites-Kaufmann 1996; Sierra Nevada Ecosystem Project 1996). Examples would include the western North American pine forest types that are chronically perturbed by wildfire (Figure 8.9) and the wind-disturbed *Nothofagus* forests of Tierra del Fuego (Rebertus et al. 1997).

Longleaf pine forests in the southeastern United States provide another potential application of a VRHS approach. Conservation of the red-cockaded woodpecker (*Picoides borealis*) is a significant issue in these forests (Sharitz et al. 1992; Rudolph and Conner 1996). One proposed harvesting system to create habitat for the species was an "irregular shelterwood" with a wide range of tree diameters (Rudolph and Conner 1996). This was criticized by Engstrom et al. (1996), who, in turn, recommended some alternative harvest techniques. An approach for habitat creation that clearly qualifies as variable retention harvesting in longleaf pine forests has been described by Mitchell et al. (2000).

The focus in applying VRHS should be on ensuring the creation of appropriate stand conditions for particular management objectives, including the conservation of biodiversity, while still allowing for commodity production (Seymour and Hunter 1999) (Box 8.2). The flexibility of VRHS allows foresters to adapt to altered societal and stand management objectives and ensures its continuing evolution (DeBell et al. 1997) (Figure 8.10).

In conclusion, the overarching goal of VRHS is to develop structurally complex managed forests that meet explicitly defined management objectives. Each prescription is expected to be a unique solution to such key questions as the type, density, and spatial arrangement of retained structures.

Reentries into VRHS units will be a continuing challenge—should retained structures be left in perpetuity or rotated at each harvest cycle? and so forth. Again, specific treatments will depend on management objectives. Computer visualizations may assist managers by providing images of likely future stand

FIGURE 8.10. Aerial and ground views of a subset of types of structural retention in the Warra silvicultural systems trial in Tasmania. Photos by J. Hickey.

conditions under various management alternatives (Ball et al. 1999). However, empirical data from VRHS applications, monitoring, and experiments are likely to provide the most useful information. This is why adequate record keeping on applications of VRHS, such as why particular treatments were applied within a harvest unit and what additional operations were conducted (e.g., pruning to reduce the susceptibility of trees to windthrow), is so important. This information is essential to guide subsequent generations of forest managers and wildlife managers in evaluating the success of various prescriptions, including responses of species to stand conditions.

Managing Stands for Biodiversity

The active management of young stands to enhance compositional and structural diversity as they regenerate and grow is a second important approach to managing the matrix for biodiversity. Regardless of the valuable contribution of a structural retention, active management of young regenerated stands can dramatically increase diversity of individual structures, levels of spatial heterogeneity within a stand, and the rate at which structural complexity develops (McComb et al. 1993; Carey et al. 1996) (see also Carey et al. 1999b in the section Biodiversity Pathways, later in this chapter). The dense young stands that will typically develop following timber harvest and subsequent natural or artificial regeneration are likely to be slow in developing structural complexity and can undergo extended periods in a dense, relatively sterile state without active management. Many simplified stands already exist that would benefit from active management to enhance structural development and biodiversity conservation. Tens of thousands of hectares of young forests incorporated into the late-successional reserves described in the Northwest Forest Plan are excellent examples (USDA Forest Service and USDI Bureau of Land Management 1994a). These stands were created by clearcutting and planting with timber production as the primary management objective. The plan has changed the goal for management of these stands to

restoration of old growth or, at least, structurally complex conditions. Appropriate silvicultural treatments can be used to speed structural development within many of these stands.

Stand Management Techniques for Enhancing Structural Complexity

Many techniques can be used to create and maintain structural complexity and biodiversity within young managed stands (Table 8.5). Creative management (such as the biodiversity pathways approach described later in this chapter) typically combines many different techniques (McComb et al. 1993). Most of the practices outlined below are not part of traditional stand management practices, which focus primarily on wood production. Many of these practices aim at not only promoting the development of particular stand structural attributes, but also encouraging key processes such as crown-class differentiation, stratification of the canopy, tree decadence, and understory development (Carey et al. 1999b).

When conservation of biodiversity is a principal aim, re-creating and perpetuating aspects of decadence that are characteristic of natural forests (particularly older stands) are one of the most difficult tasks. Not only is the concept of creating decadence and decay foreign to many forest managers, but also the processes that contribute to such features are often poorly understood. As noted by Carey et al. (1999b):

> Managing decadence is the most challenging aspect of intentional ecosystem management. . . . Decadence is more than snags and logs; it is a process that is influential in multiple aspects of ecosystem development from providing cavities for wildlife, to creating gaps in the canopy, to altering forest floor climate and structure. Active management is necessary to maintain decadence in the first 150 years of ecosystem development. Thinnings without active management for decadence could result in dimunition [*sic*] of decadence, decline in coarse woody debris, and a change in trajectory of forest development away from complexity and resiliency.

TABLE 8.5.

Some techniques for managing forest stands to create or maintain structural complexity and compositional diversity.[a]

Precommercial and commercial thinning to grow large-diameter trees

Variable density thinning ("skips and gaps") to create structural heterogeneity within stand

Thinning "from above" (selectively removing dominant trees) or branch pruning to sustain or release

    Shade-tolerant understory trees

    Understory shrubs and herbs

Conservation of tree or other plant species that fulfill different structural and functional roles such as

    Deciduous hardwood species in evergreen coniferous stand

    Species hosting nitrogen-fixing bacteria

    Species with high capacity to host epiphytes

    Species with distinctive bark or branching habits

    Species with edible fruits

Creating decadence

    Creating logs and coarse woody debris

    Stimulating development of decadence in living trees

    Creating artificial cavities

Installing nest boxes or similar artificial structures

Prescribed burning

Planting desired tree or understory plant species

Introducing or enriching populations of desired animal species

[a]These states need not be mutually exclusive.

We agree that active management is often desirable to maintain or re-create decadence as well as other structural features in managed forest stands. Methods useful in developing structural complexity are presented below.

Thinning to Create Populations of Large-Diameter Trees

Large-diameter trees are absent from many young forests. Traditional thinning practices—such as "thinning from below"—can accelerate the production of the large-diameter trees needed to achieve many objectives. Traditional thinning can be used to reduce overall stand density, eliminate inferior species and defective stems, favor large dominant trees, and favor trees of desirable form and condition. Such thinnings are typically applied uniformly through a stand. Considerable adaptation of traditional thinning practices is often necessary when structural complexity and biodiversity is a primary goal. Uniform thinning operations can produce simplified forest stands with a limited range of tree sizes and conditions and an understory dominated by one or a few aggressive plant species (Carey et al. 1999b). Uniform thinning can also result in reduced patterns of spatial heterogeneity in plant species composition, function and structure—the reverse of most natural stand development processes (Franklin et al. 2002). Finally, some structural elements may be eliminated, such as slower-growing shade-tolerant plants that contribute to plant diversity and the diversity of animal habitats.

The creation of large trees is, in itself, valuable. Such trees are the sources for large snags, large woody debris for stream and river ecosystems, and suitable trees for large cavity-excavators, such as large woodpeckers. However, other stand attributes are essential to meet many habitat objectives, and a singular focus on production of large-diameter trees should not override other structural objectives.

Variable-Density Thinning to Create Structural Heterogeneity

Variable-density thinning regimes in which thinning intensity and tree marking rules are varied within the stand of interest (Carey and Johnson 1995; Carey and Curtis 1996) are a useful approach to increasing heterogeneity in stand density and canopy cover. Variable-density thinning is sometimes referred to as the "skips-and-gaps" approach. In such a prescription, some portions of a stand are left lightly or completely unthinned ("skips") providing areas with high stem density, heavy shade, and freedom from disturbance while other parts of the stand are heavily harvested ("gaps"), including removal of some dominant trees providing more light for subdominant trees and understory plants (Carey et al. 1996). Intermediate levels of thinning are also applied in a typical variable-density prescription.

Variable-density thinning addresses a variety of stand development objectives, although it is generally more difficult to apply than uniform thinning. However, tools, such as global positioning systems, can make spatially variable stand management relatively straightforward and cost effective (G. Schreuder personal communication).

Physical removal of trees that are felled or girdled may not be necessary in thinnings aimed at enhancing biodiversity conservation. Some or all of the thinned material may be retained to contribute to stand structural complexity and organic matter. However, where trees have commercial value and are physically accessible, they probably will be removed; this can provide financing for additional stand treatments to further enhance conservation of biodiversity.

Releasing Understory Trees and Other Plants

Thinning can be used to sustain or stimulate the development of desirable understory trees and other plants that are needed to achieve specific ecological objectives (Carey et al. 1996; Bunnell et al. 1999). As an example, thinning dense tree canopies can provide for increased light penetration needed to sustain or redevelop understory shrub and herb communities that may provide important functions, such as browse for ungulates, small-animal habitats, or nitrogen fixation. The relationship between increased light and understory response can be very complex, however, particularly with regard to understory species composition, which may be critical for achieving conservation objectives. Hence, simply making light available may not achieve a desired goal; ecological knowledge of the existing stand and the autecology of understory species is crucial.

A good example of both the importance and the complexity of managing forest canopies to sustain understory communities is found in the old-growth Sitka spruce (*Picea sitchensis*) western hemlock (*Tsuga heterophylla*) forests of southeastern Alaska. Understory plants are very important as winter foraging habitat for the Sitka black-tailed deer (*Odocoileus hemionus sitkensis*), especially evergreen forbs (Hanley et al. 1989; McClellan et al. 2000). Hence, management of the overstory to maintain these valuable components is appropriate but may not be accomplished simply by uniform thinning in young stands, which typically have an understory dominated by shrubs and coniferous regeneration (Alaback 1982, 1984); uniform thinning can result in dense, sterile, monospecific understories. Effects of canopy manipulation on the accumulation and persistence of snow is also an important variable influencing the availability of browse in these forests, creating endless additional considerations in silvicultural prescriptions (Hanley and Rose 1987; Hanley et al. 1989).

Shade-tolerant tree species (which are sometimes slow to establish) are critical components in the structural development of stands dominated by shade-intolerant pioneer species. In the Douglas-fir forests of northwestern North America, the development of a midstory of western hemlock, western redcedar, and true firs contributes significantly to multilayered or vertically continuous canopies that are characteristic of old-growth forests (Franklin et al. 2002). Silvicultural treatments can enhance the survival and growth of these species, provided the stems are vigorous and well foliated and capable of responding to improved light and moisture conditions (i.e., vigorous healthy trees albeit in the "suppressed" canopy classification). Latitude and sun angles are an important consideration in developing prescriptions to release such understory trees by thinning from above. At mid- and high latitudes, the sun is never overhead and canopy openings need to be spatially displaced to account for the sun angle in order to stimulate growth of understory plants (Van Pelt and Franklin 1999), although thinning around the plant(s) of interest may also be needed to reduce competition for soil moisture.

Conservation of Trees and Other Plants with Special Roles

Thinning regimes have traditionally focused on the release of potential crop trees—species with little or no commercial value and individuals with defects or poor form were routinely removed. The ecological role of many species and stems that would have been removed in earlier times needs to be carefully considered in thinning prescriptions designed to enhance structural complexity.

Conservation of tree species that enhance the environmental, structural, and functional complexity of forest stands can be enhanced by thinning, as well as by underplanting (see below). Conserving deciduous hardwood trees within forests that are dominated by evergreen conifers is an excellent example of a practice that can enhance stand complexity (BC Environment 1993; Bunnell et al. 1999). Maintaining a deciduous hardwood component in a conifer stand provides areas within the forest that contrast in (1) light and temperature conditions (especially in winter), (2) quality and quantity of forage for ungulates, (3) litter quality, (4) soil chemical and physical properties, (5) mycorrhizal and other fungi, and (6) habitat structure (including improved conditions for epiphytes).

Large, old decadent trees have many special roles, and thinning can free them from the competitive effects of younger trees and help extend their life span (Sierra Nevada Ecosystem Project 1994). In Sweden, thinning around large-diameter oak (*Quercus* spp.) trees benefits invertebrates and cryptogams. Prescribed fire can also be used to free large decadent trees from competition, although risks to the health and survival of trees from such treatments must be carefully considered.

Species that add or expand the functional capabilities of a forest stand are obvious candidates for retention during thinning. These include plants that have distinctive architectural features, such as unusual branching habitats, bark characteristics, or

wood decay patterns (e.g., cavities); value as sources of food for wildlife; and significant positive impacts on the chemical, physical, and biological properties of soils and on nutrient cycling. For example, red alder (*Alnus rubra*) and ceanothus species (*Ceanothus* spp.) have nitrogen-fixing root symbionts. Many other species accumulate cations and improve soil chemical and physical properties, such as representatives of the families Cupressaceae and Taxodiaceae (Zinke 1962; Zinke and Crocker 1962; Alban 1969; Turner and Franz 1985; Kiilsgaard et al. 1987).

In summary, maintaining tree species and other plant taxa that enhance structural complexity and ecosystem processes is an essential consideration in prescribing thinnings and other intermediate stand treatment activities in managed stands. Directly or indirectly, biodiversity conservation can be greatly enhanced with stand treatments or, alternatively, negatively impacted if only commercial species and potential crop trees are considered.

Creating Snags, Logs, and Other Woody Debris

Coarse woody debris (CWD), including snags, logs, and other large wood structures, is essential for biodiversity conservation in forests around the world (Fischer and McClelland 1983; Harmon et al. 1986; Maser et al. 1988; Johnson and O'Neil 2001). Levels of CWD have been seriously depleted in most managed forests by timber harvesting and associated slash disposal and site preparation practices. As a result, managed stands typically have low levels of coarse woody debris (Berg et al. 1994; Lindenmayer et al. 1999g) and few large-diameter snags and logs (Linder and Östlund 1998). In fact, many management programs were designed specifically to eliminate snags in perpetuity (e.g., Crowe et al. 1984). Any legacies of CWD that do survive initial harvest and related treatments decline rapidly with the second and third rotations.

Accelerating the development of decadence in trees and snags can be valuable for enhancing biodiversity conservation. Under natural conditions, some types of snags and cavity-bearing trees require long periods to develop. Australia provides an extreme example because cavity-excavating species—such as woodpeckers—are absent (Saunders 1979; Perry et al. 1985); termites,

FIGURE 8.11. Topping live green trees is one approach to creating snags in recently harvested forest areas where snags are either absent or eliminated for safety reasons. An alternative is to remove most of the canopy but leave some live branches so that natural decay processes can occur prior to death. Eliminating most of the foliage dramatically reduces the potential for windthrow of the structure (Willamette National Forest, Oregon, United States). Photo by J. Franklin.

fungi, and bacteria require several centuries to develop cavities in eucalypts that are suitable for wildlife use (Saunders et al. 1982; Mackowski 1984).

Size, species, location, and current condition of trees are important considerations in snag creation. Small snags have short life spans (Graham 1982; Maser et al. 1988; Brown 1985; Lindenmayer et al. 1997a). Some cavity-dependent species need a number of snags in close proximity (Lindenmayer et al. 1996) while others will utilize widely spaced snags (Gibbons and Lindenmayer 2002). Existing decay can

be an asset for some objectives but can also contribute to a reduced snag life span.

Creating snags is sometimes more practical than protecting existing snags (e.g., Rose et al. 2001). Killing live trees to create snags is an increasingly common silvicultural practice (Figure 8.11) (Hutto 1995). Fire can be used to kill live trees retained at harvest (Cascade Center for Ecosystem Management 1995; National Board of Forestry 1996b). Pheromones can be applied to attract bark beetles and other decay-promoting insects to trees (Bull and Partridge 1986). Mechanically girdling trees at their base is another technique (Baumgartner 1939; Moriarty and McComb 1983; Hennon and Loopstra 1991; National Board of Forestry 1996b), although tree death may not occur for several years following mechanical girdling. Removing the entire canopy from live trees (topping) produces instant snags (Carey 1993). Topping can be accomplished with explosives or chainsaws (Bull et al. 1981; Carey and Sanderson 1981; Lindner and Östlund 1995; Chambers et al. 1997; Rose et al. 2001). Girdling is problematic, however, because (1) sap rot occurs before heart rot, and (2) treefall can occur before there is sufficient top and heart rot to make the snag useful for cavities.

We favor the removal of most, but not all, of the canopy to create snags; this leaves some live branches behind and allows the tree to undergo more natural decay processes and to provide cavity habitat over longer periods as both a living and a dead structure. Rose et al. (2001) provide comparative information on effectiveness of different methods of snag creation in coastal Oregon forests in the northwestern United States.

Selecting the most appropriate method to create snags and decadence in living trees depends upon the tree species (which differ in wood properties and decay organisms and processes), the array of species targeted for conservation, and the presence or absence of primary cavity-excavating species (McComb and Lindenmayer 1999). Artificial approaches are most likely to be successful in achieving ecological goals when the decay organisms and decomposition patterns are comparable to those that occur naturally (Highley and Kent Kirk 1979; Manion and Zabel 1979; Wilkes 1982a,b).

FIGURE 8.12. Forests in Sweden have been severely depleted of important structural features. In that country, 2-meter-high cut stumps (or "eternity trees") are retained in an attempt to substantially increase dead wood volumes. Photo by D. Lindenmayer.

Trees can be felled to provide logs and other woody debris on the forest floor (e.g., Rose et al. 2001). Generally, large logs persist for longer periods and provide higher-quality habitat than small ones (Ashton 1986; Harmon et al. 1986; Maser et al. 1988). Logs on harvested sites have many conservation uses (e.g., Sverdrup-Thygeson 2000). Where size or the abundance of logs or potential logs (i.e., trees) is insufficient to meet specified objectives (e.g., Kaila et al. 1997), other approaches may be needed. In Swedish production forests, a management goal has been to substantially increase the amount of woody debris in managed forests by retaining stumps approximately 2 meters tall (sometimes termed "eternity trees") (Figure 8.12) and by using prescribed burning to create large quantities of burned dead wood (National Board of Forestry 1996b). The potential to bunch or bundle small logs into larger structural units is being explored where larger trees are not available (Carey et al. 1999b). Small log structures and brush piles (McCay 2000) may have the potential to provide sheltering habitat needed by some animals and can be easily created from residual products of logging.

Trees can be mechanically uprooted in some forests if the creation of root wads and associated depressions in the forest floor is a management goal. Logs with root wads attached are particularly valuable as components of aquatic ecosystems (streams, rivers, and estu-

aries) because of their stability and structural complexity (Harmon et al. 1986; Maser et al. 1988). Scientists at Harvard Forest in the eastern United States actually uprooted two entire stands of trees to simulate the effects of a hurricane and examine the feasibility of the approach for adding structural complexity (Cooper-Ellis et al. 1999).

Stimulating the Development of Decadence in Living Trees

Deliberate efforts to create decadence in trees, such as decay columns and cavities, can promote habitat diversity within managed stands. One straightforward method is to create entry points for decay by breaking branches and scarring tree boles and root systems (Lindenmayer et al. 1993a), either during logging or as a separate operation. A variety of techniques, from creation of cavities to extensive linear scarring of tree boles (to simulate scarring associated with falling boles), have been pioneered by Tim Brown (personal communication) in northwestern North America. Active introduction of decay organisms can be used to supplement the mechanical scarring processes or can be conducted separately. A variety of techniques have been tried, especially in North America (Sanderson 1975; Backhouse and Lousier 1991). These include inoculating stems with fungi and growth hormones (Conner et al. 1981; Parks et al. 1995), shooting projectiles with rot into the tree bole, inserting plugs of decaying wood, and directly inoculating with cultures of decay organisms.

Although there are many ways to stimulate development of decadence in trees, only a few have been tested, but some of the preliminary results are encouraging (Carey et al. 1996).

Creation of Artificial Habitats

The provision of artificial cavities such as nest boxes is another way to create habitat in managed stands depleted by clearing and logging (Newton 1998) (Figure 8.13). This technique can be particularly useful where natural decadence takes a long time to develop. Nest boxes added to forests in Germany throughout the 1950s resulted in a five- to twenty-fold population increase in some bird species (Bruns 1960). Artificial cavities have resulted in other spectacular population recoveries of birds, such as the pied flycatcher (*Ficedula hypoleuca*) in Finland (von Hartman 1971), three species of bluebirds (*Sialia* spp.), and the wood duck (*Aix sponsa*) in North America (Haramis and Thompson 1985). Nest box dimensions have been deliberately chosen to minimize competition with other

FIGURE 8.13. The loss of cavity-trees is a substantial problem in many managed forests around the world. One approach that attempts to provide artificial nesting habitat for cavity-dependent vertebrates is the provision of nest boxes—a strategy used widely throughout the world. Successful nest box programs often require an extensive effort to establish and then monitor and maintain on a regular basis, as in the case of the large-scale nest box project under way in managed mountain ash forests in Victoria. Photo by E. Beaton.

hollow-using species in some cases (Miller 1970). It is also possible to create artificial cavities in live trees, snags, and logs using chainsaws. Several techniques have been developed to create cavities for the red-cockaded woodpeckers (Copeyon 1990; Allen 1991; Taylor and Hooper 1991).

Nest boxes and other artificial structures have significant limitations, such as the costs of construction, installation, and maintenance (McKenney and Lindenmayer 1994). Extensive and long-term nest box programs are likely to be rare because of the costs and the need to regularly maintain and replace them. In addition, they often are directed toward the conservation of a limited set of species because many taxa have narrowly defined cavity-tree requirements (Gibbons and Lindenmayer 2002). Therefore, nest box programs are more likely to be used as adjuncts to other forest and conservation management strategies. Moreover, establishment of a nest box program should not obviate the need to address the underlying reasons for the shortage of important natural structural elements in a forest, such as the modification of silvicultural systems to create and maintain suitable numbers of cavity trees (Lindenmayer 1996).

Use of artificial structures needs to be closely monitored to determine if they are effective and actually used by the species targeted for conservation. Some species may not use nest boxes (e.g., some types of woodpeckers) because cavity excavation is part of their breeding biology (Backhouse and Lousier 1991). Artificial cavities may weaken trees making them susceptible to wind storms (Hooper et al. 1990). There is also the potential to lose such structures to cavity-using pests, such as the introduced honey bee (*Apis mellifera*) and feral birds (e.g., the European starling [*Sturnus vulgaris*]) (Gibbons and Lindenmayer 2002). A study of the tawny owl (*Strix aluco*) in northern England found that when artificial cavities (nest boxes) were established, birds used them preferentially and ceased occupying naturally occurring nest sites (Petty et al. 1994).

Prescribed Burning

Prescribed burning is a useful tool to achieve a broad variety of stand management objectives, including tree thinning, reduced fuel levels for high-intensity wildfires, spatial heterogeneity within stands, desired spa-

In such cases, it can be used to reduce fuels, eliminate fuel ladders, and reduce densities of undesirable plant species. Prescribed burning is increasingly used to restore desired structural, compositional, and fuel conditions altered by past fire-control programs (Figures 8.15 and 8.16). Prescribed fires, including slash fires to reduce logging debris, need to be carefully managed to limit impacts on retained structures, such as snags and cavity trees (Horton and Mannan 1988) and understory islands. Direct impacts of repeated prescribed burns on elements of biodiversity (such as

FIGURE 8.14. A prescribed burn in a longleaf pine (*Pinus palustris*) stand. Regular burning programs are used to maintain desired stand densities, eliminate competing hardwood species, and stimulate understory plants (Joseph E. Jones Ecological Research Center, Georgia, United States). Photo by J. Franklin.

FIGURE 8.15. (*A*) Prescribed fire is being used in retained forest aggregates in managed Swedish forest stands to enhance habitat conditions for fungi and invertebrates that require burned wood, including many red-listed taxa. (*B*) Many red-listed species in Scandinavian forests can be relatively abundant on burned dead wood substrates in Russian forests as shown in this stand south of Moscow. Photos by D. Lindenmayer.

tial mosaics of forest- and grasslands or meadow, reproductive stimulation of selected plants, and reduced densities of particular plant taxa. Burning may also fulfill specific habitat needs, such as providing large quantities of burned wood needed as habitat for threatened or red-listed insects in Sweden (National Board of Forestry 1996b).

Prescribed fire is a particularly valuable tool in forest types where the natural fire regime includes frequent light- to moderate-intensity fires (e.g., ponderosa pine [*Pinus ponderosa*] forests in the southwestern United States and the longleaf pine forests in the southern United States) (Figure 8.14).

FIGURE 8.16. Prescribed fire has been used extensively as a stand management tool to reduce fuel loads, stimulate reproduction of the giant sequoia, and eliminate competing tree species. This area has dense regeneration of giant sequoia and has been subjected to several prescribed burns (Sequoia National Park, California, United States). Photo by J. Franklin.

ants) also need to be considered (York 2000). In the forests in southeastern Australia, high-intensity regeneration fires lead to very high rates of loss of retained trees (Lindenmayer et al. 1990a; Gibbons et al. 2000; see Chapter 12).

High-intensity fires are not usually part of a management regime for an established stand, but they can be useful as stand-initiating events for conservation purposes or to dispose of slash and create seedbeds in a variety of forest types. An example of the use of catastrophic fire is in regeneration of jack pine (*Pinus banksiana*) stands for Kirtland's warbler (*Dendroica kirtlandii*) in North America.

Some managers favor mechanical treatments rather than prescribed burning within established stands. Thinning stands can reduce fuel loadings and the potential for catastrophic fire, but mechanical treatments obviously do not replicate all of the effects of fire. Impacts of mechanical treatments on understory plants also need to be considered (Ough and Ross 1992). Fire is less manageable and needs more preparation. In addition, mechanical treatments can provide income from harvested wood.

Prescribed fire is sometimes combined with mechanical treatments (Agee 1993). An example is the natural process–based management (NPM) regime

proposed for western North American pine and mixed conifer forests (Fiedler et al. 2001).

Underplanting Desired Tree and Other Plant Species

Particular trees or other plant species may be viewed as important or desirable understory components in managed stands. Underplanting these species may be the best approach to restore desired levels in managed stands if propagules for these species are scarce or absent or conditions for germination and early growth are unfavorable. Good examples are shade-tolerant tree species such as western hemlock and western redcedar in young and mature Douglas-fir stands in the Pacific Northwest (Tappeiner et al. 1997; Keeton 2000).

Translocating or Enriching Populations of Desired Animal Species

Introducing or supplementing populations of particular animal species, often those threatened or endangered, may help achieve desired population levels in some managed forests. The success of this approach depends on many factors, including the presence of suitable habitat for survival and reproduction. Translocation often involves high risks and costs, and historically many translocation efforts have been unsuccessful (Serena 1995; Fischer and Lindenmayer 2000).

The Biodiversity Pathways Concept

Ecological and economic techniques for creating structurally complex forests to better conserve biodiversity are the focus of many current research projects and land management planning processes. The Washington Forest Landscape Management Project (WFLMP) is one of these and has as its goal development of integrated management approaches that provide alternatives to land allocation as a means of resolving conflicts among forest values (Carey et al. 1996, 1999b). WFLMP focused on the

> development and analysis of silvicultural treatments, direct wildlife habitat improvements, and conservation of biological legacies . . . that either maintain organisms, structures, and functions associated with late-successional forests or accel-

erate their reestablishment following timber harvest. (Carey et al. 1996)

The "biodiversity pathways" concept ultimately emerged from WFLMP.

Biodiversity Pathways Model and Its Assessment

The biodiversity pathways model management regime includes active stand management, structural retention at harvest, and rotation times longer than those currently in use. In addition to precommercial thinning, several types of variable-density commercial thinning are used to reduce overall stand density, open portions of the canopy to stimulate the growth of understory plants, provide coarse woody debris, and create cavity trees. These structural objectives were identified and discussed earlier.

Four measures were used for quantitative comparisons of the ecological effectiveness of different management strategies:

1. The capacity of the forest to support vertebrate diversity (based upon the occurrence of up to 130 native species)
2. The maintenance of forest floor function based on the integrity (completeness) of the mammal community inhabiting the forest floor
3. The maintenance of ecological productivity based on the abundance of arboreal rodents (a surrogate measure of the reproductive activity of fungi and plants in the ecosystem)
4. The maintenance of populations of black-tailed deer and North American elk

Structural attributes were used as the basis for ranking various management strategies in regard to these ecological measures. The economic measures used in the study included the net present value of wood, sustained timber revenues, total and sustainable volume of wood products, and premiums for wood products.

Seven alternative management regimes were simulated that represented variations on the following three primary approaches:

1. No manipulation pathway (NMP)—forest protection only

2. Timber fiber pathway (TFP) with an emphasis on maximizing present net value
3. Biodiversity pathway (BDP) emphasis with the goal of maintaining 30 percent of the landscape as late-successional forest

The study focused on a 7,800-hectare landscape dominated primarily by young (less than eighty-year-old) closed-canopy forests of western hemlock located on the Olympic Peninsula (Washington, United States). Management regimes were simulated for 300 years. The BDP option provided the best mix of benefits and met the goal of maintaining at least 30 percent of the landscape in late-successional forest structure. The chief elements of the BDP prescription were

- Clearcutting with the retention of biological legacies
- Planting widely spaced Douglas-fir with natural regeneration of western hemlock and other species
- Precommercial thinning designed to favor multiple plant species at fifteen years

Commercial thinning regimes and the rotation age alternated in each management cycle: variable-density thinnings at thirty and fifty years with a final harvest at seventy years and variable-density thinnings at thirty, fifty, and seventy years with final harvest at 130 years.

Achievements of the Biodiversity Pathways Model

Active management of stands for biodiversity (the BDP strategy) resulted in a more rapid development of late-successional forest conditions than the passive strategy that involved no stand manipulations (NMP) (i.e., forest protection only) (Carey et al. 1999b). In fact, simply protecting the landscape (NMP) slowed ecological restoration, while management for net present value (timber fiber production [TFP]) prevented ecological restoration. Excluding stand management after timber harvesting delayed the development of complex structures characteristic of later successional stages. The time required to achieve 30 percent late-successional coverage was eighty years for BDP and 180 years with NMP. The late-successional goal of 30 percent was never achieved under a management regime for net present value (TFP) unless very wide riparian buffers were established, in which case it took 240 years to achieve.

Management strategies had profound consequences at the stand and landscape levels (Carey et al. 1996, 1999b). A landscape managed under the TFP emphasis had low to very low values for all four ecological measures examined, comparing very poorly with both BDP and NMP on these criteria. TFP eliminated habitat for fourteen species of vertebrates and provided for only 32 percent of potential biodiversity compared with 98 percent for the BDP strategy. Management by BDP captured more than 90 percent of the capacity of the landscape to support vertebrate diversity, about 80 percent of the potential forest floor function, and almost 70 percent of the ecological productivity.

Managing to maximize present net value (the TFP strategy) put at least twenty-five terrestrial vertebrate species at risk (Carey et al. 1996). Although the TFP strategy produced the highest economic value, net present value of the BDP approach was 82 percent of the maximum present net value, and it produced the greatest sustainable income among all the alternatives tested (Carey et al. 1999b).

An additional notable finding was that wide riparian reserves can have negative consequences for development and continuity of late-successional forest habitats, assuming that the total area in such a condition (e.g., 30 percent) is a management constraint (Carey et al. 1996, 1999b). Such potentially negative implications of wide riparian buffers echoes problems with riparian reserves in the Northwest Forest Plan that were discussed in Chapter 6.

Long Rotations and Cutting Cycles

Adoption of long rotations is the third of the general approaches to the management of stands to conserve biodiversity. As noted earlier (and in Chapter 7), the use of long rotations can have direct and significant consequences on biodiversity conservation at both landscape and stand levels. Rotation length affects the proportion of a landscape that is in a recently logged condition—a significant issue because of its potential effects on ecosystem processes, such as hydrological responses (Chapter 7). In the simulation study by Carey et al. (1999b), even moderately long rotations (70–130 years) resulted in 72 percent fewer clearcuts per decade.

Rotation length is a critical variable that determines the proportion of a landscape in older, more developed forest stands as well as in dense young stands undergoing competitive exclusion (e.g., Franklin et al. 2002). In comparison with the short rotations used under many intensive timber management regimes, long rotations provide the potential to grow larger trees, accumulate more organic matter, and develop other structural features associated with more advanced successional conditions.

What is a long rotation? Rotations in forestry have traditionally been based upon either economic or biological criteria worldwide. Typically, rotation periods are determined primarily by traditional investment economics (i.e., the discounted present net value of a forest stand). Such rotation times vary from twenty to sixty years, depending upon site productivity and species.

Biologically based rotations are usually based on the culmination of the mean annual increment (CMAI) in wood production. Rotation lengths based on this criterion are typically two to three times longer than economic rotations for the same tree species and site conditions. Lengthening rotations to at least CMAI increases stand- and landscape-level wood production, since economic rotations invariably harvest stands far short of their peak productivity.

CMAI actually equates to achievement of stand maturity. For example, CMAI may be 90 to 150 years in the Douglas-fir forests of northwestern North America. This is at least a century short of the development of structural conditions typical of old-growth stands (Franklin et al. 2002).

Here we define long rotations as those that are significantly longer than the economic rotation time for a site. Rotation times will have to be significantly longer than those based on present net value if stands are to be allowed to develop complex stand structures. Rotations longer (perhaps much longer) than those based upon CMAI may also be necessary, although stand treatments such as thinning may delay culmination of CMAI for many decades (Curtis 1995; Curtis and Carey 1996).

One generic forest management proposal involves the use of long rotations to develop structurally complex managed forests with large-diameter trees (Wiegand et al. 1994). Such proposals usually include a series of silvicultural treatments during stand development to help create specific structural elements. Rotation times may be extended by 50 to 300 percent. Thus, stands typically managed on an 80-year rotation would be grown on a cycle up to 240 years. Of course, as rotation lengths increase, the ability of landowners and societies to maintain such long-term commitments becomes an issue. There is also a risk of losing stands to disturbances such as wildfires.

Opportunities for Stand Development with Long Rotations

Long rotations provide the explicit opportunities for

- Growing structures that are essential habitats for many elements of biodiversity but which require a long time to develop, such as large-diameter trees, large snags and logs, and within-stand spatial and vertical heterogeneity.
- Reestablishment of organisms, although this can be greatly influenced by the amount, type, and spacing of residual vegetation (Lindenmayer and Franklin 1997a; Foster et al. 1998; Turner et al. 1998).
- Expansion of a range of silvicultural options, including thinning and other post-logging site treatment prescriptions (Curtis and Marshall 1993; Carey and Curtis 1996). Multiple harvesting options within Canadian black spruce forests illustrate the stand-level benefits of long rotations (Bergeron et al. 1999; see Chapter 4).
- Accumulation of more organic matter (and hence greater carbon acquisition) and restoration of nutrients and soil organic matter lost during timber harvesting.

Limitations of Long Rotations at the Stand Level

Long rotations have serious limitations if they are adopted without the additional tools of structural retention at the time of harvest and active stand management to promote structural complexity.

The first limitation of long rotations is that some structural elements (and related species and processes) are completely lost from harvested stands or, at least, are present in a much smaller percentage of a landscape. Obviously, the total population of those structures in a managed landscape will be much higher under management systems where they are maintained (retained) as elements of harvest units than in landscapes managed under long rotations. Large-diameter, moderately decayed snags provide an example. The re-creation of such snags takes at least 100 years in a typical Douglas-fir stand in northwestern North America or several centuries in Australian mountain ash forests. Large trees must first be grown and then die or be killed. A period of time is needed to develop suitable levels of stem decay. Under a hypothetical example, a managed Douglas-fir stand might support large decayed snags for only 20 years of a 120-year rotation. Hence, only one-sixth of the landscape would have these structures at any one time. In mountain ash forests, it is not possible to create these structures at all under a 120-year rotation, so they would virtually disappear from large parts of managed landscapes (Lindenmayer and Possingham 1995b).

Structural retention at harvest can provide for large and well-distributed populations of large moderately decayed snags and large-diameter living trees in contrast with conditions under long rotations and clearcutting. The loss of structural attributes from forests managed on a long rotation but without retention could be critical for many cavity-dependent vertebrates.

A second limitation of moderately long rotations (up to 200 years) is that the establishment of some organisms is not possible (Duffy and Meier 1992; Stohlgren 1992; Lindenmayer and Franklin 1997a). Some species simply do not establish themselves in early- or midsuccessional stands (Marcot 1997). Species of epiphytic bryophytes and foliose lichens provide examples of species that appear to need more than 200 years to develop in conifer forests in northwestern North America (McCune 1993), at least partially because of the large-diameter branches that are required (Clement and Shaw 1999).

In summary, there are both theoretical and practical limitations to management strategies based upon long rotations, even when the particular organisms and structures (and the processes needed to sustain them) have been recognized and quantified.

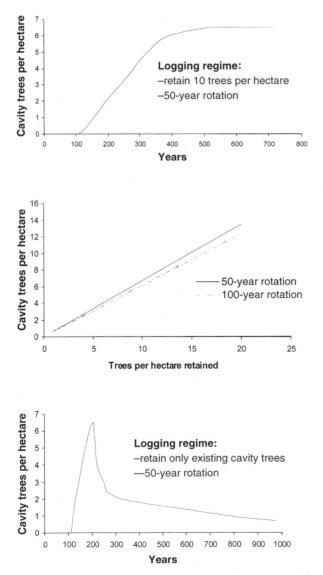

FIGURE 8.17. Forecast numbers of cavity-trees under different tree retention scenarios and/or rotation times in the mountain ash forests of the Central Highlands of Victoria (modified from Ball et al. 1999).

Use of Multiple Rotations

Many limitations of long rotations can be overcome by developing and retaining desired structures over multiple rotations. This approach can accommodate organisms that depend upon structures that require several centuries to develop. The large cavity-bearing trees used by arboreal marsupials in Australian mountain ash forests are an example (Lindenmayer et al. 1993a). Another example is the red-listed fungi that inhabit large decaying logs in Norwegian forests (Sverdrup-Thygeson 2000). Siitonen et al. (2000) has recommended the retention of old-growth characteristics, such as large logs at the time of harvest, to ensure that critical stand attributes are maintained in Scandinavian forests. An added advantage of long-term structural retention through multiple rotations is that it can promote the development of multicohort stands—a forest condition most suitable for some species.

These long-term approaches require planning and visualization of complex forest stand structures for very long periods and over several rotations, a capability promoted by current computer software (Figure 8.17) (e.g., Ball et al. 1999). However, long-term approaches still need to be translated to actions on the ground, and, as noted by John Wells (in Hunter 2001), "we'll have to plan to produce desired habitat features just as carefully as we plan to produce certain quantities of wood volume per unit area."

Selection Harvest Practices

In this chapter, we have focused primarily upon forests managed for some level of timber production using silvicultural systems that lead to a one-, two, or three-aged stand. However, the principles of stand management for the maintenance of structural complexity and conservation of biodiversity apply equally to forests managed selectively (to provide a many-aged condition). In these cases, trees are removed as individuals or small groups (individual tree and group selection). These management regimes maintain forest cover and, therefore, a forest-controlled environment over essentially the entire landscape—an outcome that has positive benefits for many ecological processes and forest-dependent organisms.

Unfortunately, descriptions of traditional selection cutting methods make no reference to the maintenance of decadent trees, snags, and logs as structural components of managed stands (e.g., Smith et al. 1997). Therefore, objectives related to individual structural components, spatial heterogeneity in stand structure, and the provision of habitat for specific organisms and assemblages have rarely been explicitly incorporated into prescriptions for selective management, although there are exceptions.

In fact, there is an infinite array of approaches to incorporating structural complexity into selection management. For example, the concept of group selection with structural retention offers the potential to create enhanced stand conditions for the California spotted owl in the Sierra Nevada pine and mixed conifer forests in the western United States (Franklin et al. 1997; see Chapter 11). An alternative is to select areas of a forest that are reserved from any harvest in order to maintain structural complexity.

Harvesting by group selection can be very closely modeled on natural disturbance regimes in forest ecosystems that are chronically perturbed by fire or wind (Coates and Burton 1997). Structural retention, active stand management, and long rotations are all activities that can be readily incorporated into group selection prescriptions. The size and frequency distributions of natural disturbances can be used as a silvicultural guide. Unfortunately, many prescriptions for group selection prescribe harvest on areas that are much larger than most openings created by natural disturbances (Knight 1997).

The topic of selection management deserves much greater exploration than we have provided here. However, silvicultural approaches that are essentially even-aged dominate many temperate and boreal forest regions and hence were our emphasis in this chapter.

CHAPTER 9

Revisiting a Multiscaled Approach to Forest Biodiversity Conservation

Management for diversity calls for diversity of management.

—EVANS AND HIBBERD (1990)

There is immense potential for positive synergies when an array of matrix management strategies are simultaneously employed at multiple spatial scales, just as negative impacts can accumulate in managed landscapes if the need for multiscaled strategies is ignored (see Chapter 5). The implication is that different conservation management strategies need to be considered jointly rather than in isolation in order to meet specified objectives as part of a comprehensive plan for biodiversity conservation.

Trade-offs are possible between management strategies. Comparable objectives for biodiversity conservation can sometimes be achieved from different combinations of, or different emphases on, matrix management strategies. The need for a varied approach to conservation management and potential trade-offs between different strategies are conceptualized in a simple model in this chapter.

The conservation of forest biodiversity requires a comprehensive multiscaled approach ranging from reserves to landscape-level and stand-level matrix management strategies. This is essential to meet the diverse multiscaled requirements of individual species as well as the varying needs of different taxa (Chapter 3). In Chapter 6 (and again in Chapter 8), we presented a checklist of factors that need to be considered as part of a comprehensive plan for forest biodiversity conservation. However, these different multiscaled approaches do not act in isolation.

There can be positive synergistic benefits for biodiversity from embracing several strategies concurrently, such as implementing stand retention over multiple lengthened rotations (Chapter 8). For exam-ple, a combination of strategies that retain snags and overstory hardwoods within cutover areas while strategically positioning mature forest adjacent to harvest units was found to be appropriate for reducing logging impacts on the southern flying squirrel (*Glaucomys volans*) in the forests of Arkansas, in the United States (Taulman et al. 1998).

Compounding (or cumulative) negative impacts for biodiversity can also occur and are usually the result of ignoring the need for an array of different strategies (Chapter 3). For example, the loss of structural complexity within stands can accumulate over many cutover sites and deplete landscape heterogeneity.

The value of integrating different strategies is illustrated in Chapter 8 by the evolution of new silvicultural systems such as the variable retention harvest system (Franklin et al. 1997) and the biodiversity pathways concept (Carey et al. 1996, 1999b). In this chapter, we look again at a comprehensive plan for biodiversity conservation by outlining a theoretical framework that integrates different elements of forest management to meet specified objectives.

Maintaining habitat at multiple spatial scales is the overarching goal of any comprehensive plan for enhanced biodiversity conservation (Chapter 3). It will sometimes be possible to meet this broad objective (as well as more specific biodiversity conservation objectives associated with it) using different combinations of forest management strategies. One way of conceptualizing this is to visualize management regimes as the integration of various individual approaches. For example, the subcomponents of a harvesting regime at the stand level might incorporate structural retention, stand management, and appropriate cutover sizes and rotation periods (see Chapter 8). The implementation of each of these subcomponents can vary; there are gradients from low to high levels of stand retention, extensive stand management to no intervention, small to large cutover sizes, and short to long rotation times. Integration of each of these subcomponents gives rise to a hypothetical harvesting regime (Figure 9.1). There may be trade-offs among different approaches while still ensuring that designated objectives are achieved with a given management regime. For example, increased levels of stand retention at the time of regeneration harvest may allow for larger cutover units or reduced levels of stand management. Similarly, increased levels of stand retention may make it possible to reduce the rotation period. In some cases, a harvesting regime will need to be modified to meet desired biodiversity goals. Such a transition can be made via a shift of emphasis on different subcomponents of that regime, such as increased levels of stand retention during regeneration harvest.

Just as there can be variation in applying any subcomponent of the harvesting regime (Figure 9.1), there is also a gradient in the intensity of application of landscape-level strategies for matrix management. Wildlife corridors can be wide or narrow, the road system can be extensive or limited, and the size of midspatial-scale protected areas within the matrix can be large or small (Figure 9.2).

Management activities at the stand level cannot be divorced from considerations at broader spatial scales,

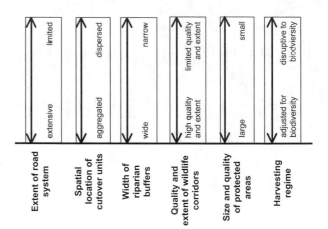

FIGURE 9.2. Graphical presentation of the various components of a comprehensive plan for forest biodiversity conservation at the landscape level. Note that each component is itself composed of a range of subcomponents—in other words, the harvesting regime is a function of stand level retention, rotation time, cutover size, and the overall level of stand management (see Figure 9.1).

as was made clear in Chapters 6, 7, and 8. Harvesting regimes are intimately linked with other considerations, such as the width of riparian buffers and wildlife corridors and the protection of specialized habitats and biodiversity hotspots in the matrix. This is illustrated diagrammatically in Figure 9.3, which shows the juxtaposition of a harvesting regime (depicted as an emergent property of its subcomponents in Figure 9.1) in relation to other matrix management strategies.

The interrelationships between a harvesting regime and other considerations can be illustrated by a few hypothetical scenarios. For example, if there are increased levels of stand retention as part of a modified harvesting regime, then it might be possible for the width of wildlife corridors or riparian buffers to be reduced and yet still meet specified conservation objectives. Similarly, the levels of protection required for some types of specialized habitats may be relaxed when the selected harvesting regime limits the levels of physical and structural contrast between cutover units and adjacent protected areas. Clearly, many other permutations are possible. Lengthened rotation times with stand retention may allow requirements for strictly protected wide wildlife corridors to be relaxed. As in the case of the altered harvesting regimes out-

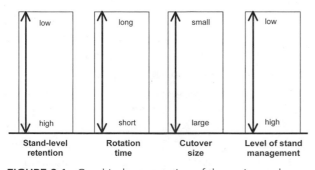

FIGURE 9.1. Graphical presentation of the various subcomponents of the harvesting regime employed at the stand level. Each of these components can be changed independently or in conjunction with one another, depending on the management objectives.

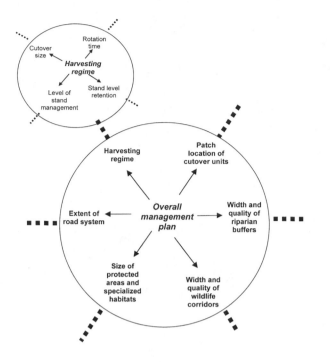

FIGURE 9.3. Graphical presentation of a hierarchical model that can be used to conceptualize how a detailed multi-scaled plan for forest biodiversity conservation may be implemented. The overall management plan is derived from several components at the landscape level, each of which is composed of subcomponents at smaller spatial scales. For simplicity, in this diagram only the subcomponents of the harvesting regime have been shown. But implicit in this hierarchical model is the fact there will be a suite of subcomponents of each of the key areas presented in the inner wheel of the diagram. Understanding the interactions between the different scales and management options is an important aspect of the overall approach.

lined above, redirected emphasis on particular matrix management strategies may be required to achieve some biodiversity conservation goals.

Any comprehensive plan for biodiversity conservation requires taking account of strategies at multiple spatial scales—from large ecological reserves to individual stands. The key points of this chapter are that (1) there can be cumulative benefits (or, alternatively, negative impacts) of different strategies (e.g., Taulman et al. 1998), (2) different combinations of strategies can achieve the same objectives for biodiversity conservation that will sometimes (but not always) allow for trade-offs between such strategies, and (3) different emphases on particular components of a management regime can have implications for other attrib-

utes of the forest estate. A classic example might be where the aim is to reduce the size of cutover areas, which may then lead to a need for more cutover areas in a landscape and an associated increase in the road network and number of stream crossings (see Chapter 7). The inference from such outcomes is that in order to meet specified conservation objectives, different approaches to matrix management need to be considered simultaneously, and not in isolation. The simple model outlined in this chapter is a way of conceptualizing this.

As discussed in previous chapters in this book, the combination of strategies to best achieve desired biodiversity conservation outcomes will vary between landscapes as a function of specified objectives and the suite of organisms or ecological processes targeted for management. It is essential for forest managers to have a clear vision about both the forest and the stands that compose that landscape. An explicit statement of these objectives can be complemented by the development of strategies and practical tools needed to achieve these objectives. Some of the strategies and tools for use by forest and wildlife managers are discussed in Chapters 5 through 8.

Matrix Management in Plantation Landscapes

Studies show that the potential apparently exists to minimize negative impacts on biodiversity in plantations.

—MOORE AND ALLEN (1999)

Plantation landscapes (as well as near-natural forests) can benefit from the application of the principles outlined for matrix management in Chapter 3 and discussed further in Chapters 6, 7, and 8. Plantations are rarely valued for the conservation of biodiversity and there are certainly many species that cannot be conserved in them. Nevertheless, an increasing number of studies show that plantations—at both the landscape and the stand levels—can conserve at least some components of biodiversity. Some (often relatively minor) changes in plantation management can enhance their conservation value and ensure that they are not biological deserts devoid of diversity.

A variety of approaches for matrix management at the landscape and stand levels have been outlined in Chapters 6, 7, and 8. Although the primary focus was on natural forests, matrix-based strategies can also be successfully applied to plantations (defined here as planted forests of commercially important tree species). Planted forests may be composed of exotic species, as are the conifer plantations in the United Kingdom, South Africa, New Zealand, and Australia or the extensive areas of eucalypts in Brazil, China, Spain, and almost 100 other countries worldwide (Doughty 2001). Alternatively, they can be planted areas of trees native to a given continent such as loblolly pine (*Pinus taeda*) in the southeastern United States (Moore and Allen 1999) and blue gum (*Eucalyptus globulus*) and shining gum (*Eucalyptus nitens*) in Australia (Eldridge et al. 1994). Often the trees used in these plantations have been subject to programs of tree breeding and genetic modification. Intensive site preparation (e.g., ripping and mounding), thinning, and application of herbicides and pesticides (Peterken 1996; Moore and Allen 1999) are often characteristic of plantation forestry, further modifying site conditions. Indeed, plantations share many attributes with intensive agricultural systems, except that they are managed on longer rotations.

Roughly 75 percent of the world's plantations are in temperate zones and the remainder in the tropics and subtropics (World Commission on Forests and Sustainable Development 1999). In 1996, the combined area of plantations worldwide was estimated to exceed 130 million hectares (Cubbage et al. 1996). In 2000, the estimated area of plantations had increased to 187 million hectares, or about 5 percent of the world's total forest area (Food and Agriculture Organization of the United Nations 2001). There are almost 20 million hectares of eucalypt plantations alone, and plantations of *Pinus* species represent 20 percent of all plantings. Managing plantations to contribute to biodiversity is, therefore, important because of their extent. In countries such as the United Kingdom and New Zealand, where much of the original forest cover has been removed, a large proportion of the existing forest is plantations (Spellerberg and Sawyer 1997; Peterken 1999). In Australia, plans are well advanced to triple the extent of plantations in the next two decades (Department of Primary Industries and Energy 1997).

TABLE 10.1.

Matrix management strategies in plantations (modified and expanded from Spellerberg and Sawyer 1997).

Retain snags, woody debris of indigenous tree species

Retain some plantation trees at the time of harvest

Maintain some native understory elements

Protect areas of native vegetation within plantation boundaries

Protect, restore, and buffer riparian areas

Increase the range of plantation tree species used

Manage some stands on long rotations

Plan the location of plantation establishment to limit impacts on conservation values

The primary goal of plantation forestry is the efficient production of large quantities of timber and pulp. However, there are also important opportunities for designing and managing plantations to enhance their value for biodiversity (Clout 1984; Hanowski et al. 1997) (Table 10.1). Inferences from several studies indicate that changes in plantation management regimes will increase biodiversity conservation, sometimes significantly (see the case study in Chapter 13). In other cases, even the limited structural complexity of plantations provides suitable habitat for some forest species (see Chapter 13) or enhances connectivity for others (Renjifo 2001), or both.

Landscape-Level Matrix Management in Plantations

Three broad categories of approaches for landscape-level matrix management were highlighted in Chapters 6 and 7: (1) protection of midspatial-scale protected areas within the matrix such as biodiversity hotspots, specialized habitats, and aquatic areas; (2) planning the road network; and (3) planning the temporal pattern of timber harvesting.

The value of these strategies in plantations has been well documented worldwide. Preservation of small and intermediate-sized protected areas of native forest within plantation estates has been shown to

provide valuable habitat for many species of vertebrates in Australia (Goldingay and Kavanagh 1991; Fisher and Goldney 1998), New Zealand (Clout and Gaze 1984), Europe (Peterken 1999), Brazil (Zanuncio et al. 1997), and Chile (Estades and Temple 1999). Clout and Gaze (1984) identified several species of New Zealand birds that were absent from radiata pine monocultures but which occurred in plantation landscapes that incorporated some native vegetation. Retained patches do not always have to be large to be useful. In southeastern Australia, Lindenmayer et al. (2001a) showed that within extensive plantations of radiata pine, remnant patches of eucalypt forest as small as 1 hectare had value as habitat for forest birds, reptiles, frogs, and mammals (see Chapter 13).

In the case of aquatic ecosystems, retaining attributes such as native riparian vegetation provides habitat for many species within plantation landscapes (Suckling et al. 1976; Recher et al. 1987; Estades and Temple 1999). Riparian vegetation also can act as a dispersal corridor for terrestrial fauna thereby contributing significantly to connectivity (Lindenmayer and Peakall 2000; see Chapter 13).

There can be positive benefits for biodiversity conservation arising from changes in the spatial and temporal patterns of harvesting of plantation forests. In southeastern Australia, in the pine-eucalypt remnant system at Tumut (see Chapter 13), Lindenmayer and Pope (2000) believed that functional connectivity between populations of the greater glider residing in eucalypt patches could be maintained by staggering the spatial configuration of clearcut radiata pine stands adjacent to remnants. Radio-tracking of the animals revealed they use the radiata pine forest to move between eucalypt habitat patches. Therefore, at any one time, some areas of advanced regrowth radiata pine in the matrix should link eucalypt remnants to maintain connectivity.

Stand-Level Matrix Management in Plantations

Three broad strategies for stand-level matrix management were outlined in Chapter 8: (1) structural retention at the time of harvesting, (2) stand management

FIGURE 10.1. Early stages of establishment in a blue gum (*Eucalyptus globulus*) plantation in southwestern Western Australia. Plantation establishment is taking place on semi-cleared former grazing, and the retention of even small areas such as the paddock trees in the foreground and the remnant patches of native forest in the mid-ground (on the left-hand side of the photograph) will have some value for biodiversity conservation. Photo by D. Lindenmayer.

activities to create structural complexity, and (3) altered rotation times to meet specified stand-level objectives.

These strategies can be combined in some cases to develop entirely new silvicultural systems, such as the variable retention harvest system proposed by Franklin et al. (1997). Each of these broad strategies and altered silvicultural systems can be useful in plantation forest management.

Maintaining some stand structural attributes within plantation forests can significantly enhance the value of plantation landscapes for biodiversity. When large living trees and snags are retained at the time of regeneration harvest they are used by many species of birds (Land et al. 1989; Kavanagh and Turner 1994). Observations in the Tumut Fragmentation Experiment in southeastern Australia have shown that some bird species are strongly associated with windrows of eucalypt logs—the remains of the original native forest cleared to plant radiata pine stands (see Chapter 13). These findings and other studies in Australia have highlighted the importance of such windrows for native fauna within softwood plantations (e.g., Friend 1982; Curry 1991). Recommendations have been made to reduce the rate of disappearance of these important biological legacies and ensure that it is not ac-

celerated by damage from machinery during harvesting operations (Lindenmayer and Pope 2000).

Modifying silvicultural systems in plantations is another approach to enhance biodiversity conservation at the stand level. Moore and Allen (1999) described the benefits of a wider spacing between planted rows of trees to increase native plant diversity. They also highlighted the value of exempting thickets of understory vegetation from site preparation activities. Work in the Tumut Fragmentation Experiment (see Chapter 13) has shown that threatened bird species such as the olive whistler (*Pachycephala olivacea*) often occur within stands of radiata pine that support intact thickets of native understory vegetation (Lindenmayer et al. 2001a). Appropriate treatment of understory vegetation within these stands during harvesting operations could make an important contribution to the conservation of this species. Notably, such understory vegetation often occurs in gullies or the wet seepage areas associated with the margins of swamps—places that could be readily captured within protected riparian zones. Other workers have found that plantations can provide habitat for rare or unusual species. In New Zealand, nationally threatened birds such as the brown kiwi and kakapo (*Strigops habroptilus*) have been recorded in plantations (Colbourne and Kleinpaste 1983; McLaren 1996; D. Norton personal communication). Radiata pine plantations in that country are also known to support many species of vascular plants (Allen et al. 1995).

Other aspects of plantation management at the stand level can have positive impacts on biodiversity. Planting a wider range of tree species can result in a greater diversity of habitats and their dependent taxa (Spellerberg and Sawyer 1997). This can be useful in amenity areas within and adjacent to plantations (such as campgrounds and picnic areas) where such vegetation can be valuable for groups such as birds (Clout 1984) and invertebrates.

Altered rotation times also can be valuable for biodiversity. Peterken et al. (1992) showed that managing some parts of the British conifer plantation estate on long rotations would allow them to develop old-growth characteristics needed by some wildlife taxa (see Box 10.1).

BOX 10.1.

Plantation Forestry and Biodiversity Conservation in the United Kingdom

*Some of the most innovative approaches to biodiversity conservation in plantation landscapes have come from the United Kingdom. The forests there are dominated by introduced plantations of Sitka spruce (*Picea sitchensis*)—now the most commonly occurring tree in that country (Peterken 1996). Extensive areas of conifer plantations have been established in Wales and, more recently, in Scotland. It has been increasingly recognized that plantation forests have considerable value for biodiversity, not only for vertebrates (Peterken 1996), but also for a wide range of other groups such as fungi (Humphrey et al. 2000) and invertebrates (Ozanne et al. 2000). The Forestry Commission in Great Britain recognizes the important role of its plantations in contributing to biodiversity conservation (C. Ozanne personal communication).*

In a detailed discussion about British plantations, Peterken (1996) outlined a wide range of landscape- and stand-level matrix management strategies to promote biodiversity conservation. These included (among others) the following:

- *Setting aside riparian areas, wetlands, and bogs.*
- *Ensuring that a range of habitats is incorporated within the boundaries of plantations, such as seminatural woods, moorland, and former farmland.*
- *Maintaining attributes of old plantation forests*

within newly regenerated stands such as large logs, and decadent trees and snags.

- *Increasing the rotation times on lower (more sheltered) slopes to allow stands to develop "old growth" features. The rotation time for the remaining areas of the plantation is then shortened to limit impacts on wood production. Decreased rotation times are applied on upper slopes that are more prone to wind damage—the major form of natural disturbance in British forests (Savill 1983).*
- *Adopting silvicultural systems other than clearcutting in some stands (e.g., selection or large-group selection methods) to promote shrub and understory growth and the development of vertical heterogeneity.*

Native forests are comparatively scarce in Great Britain, and plantations compose approximately 85 percent of the forest estate. These plantations have been established using trees from other parts of the world, primarily North America. Nevertheless, it is clear that these areas have considerable value for biodiversity conservation—values that can be enhanced by adopting many of the principles and approaches for matrix management (Peterken 1996) discussed in Chapters 3, 6, 7, and 8.

Long-Established and Newly Established Plantations

The principles of matrix management apply both to areas proposed for new plantation developments and to long-established plantations. An important aspect of plantation establishment is their location (Hanowski et al. 1997). Spellerberg and Sawyer (1997) noted that many plantations in the United Kingdom had replaced valuable habitats such as lowland heathlands. The retention of patches of remnant native vegetation and riparian areas is fundamental to plantation design criteria in the semi-cleared grazing lands of southeastern Australia that are increasing targeted for conversion to widespread radiata pine plantations (Lindenmayer and Pope 2000). Similarly, Peterken (1996) discusses the importance of conserving a range of habitat types within the boundaries of new areas

proposed for British plantation expansion to produce a landscape mosaic rather than a monoculture.

Matrix management practices are also relevant to long-established plantations. Rotation times can be altered (see Box 10.1), the spatial pattern of harvest units can be modified, riparian areas planted with plantation trees can be rehabilitated with native vegetation, and structural features can be retained at the time of transition to the next rotation.

In summary, plantation forests can make a positive contribution to the conservation of biodiversity with some modifications of management regimes (Dyck 2000). Importantly, biodiversity conservation has become a component of plantation forest policy in many countries. Plantations are also being considered under certification agreements, such as those developed under Forest Stewardship Council guidelines (Cauley et al. 2001). Biodiversity provisions were included as the document *Principles for Commercial Plantation For-*

est Management in New Zealand, which was signed by many stakeholder groups in 1995. Similarly, the extent of native vegetation within the exotic plantation forests is used as a measure of sustainability in the Australian state of New South Wales. Spellerberg and Sawyer (1996) recommended that the conservation of biodiversity should be a key part of ecological stan-dards developed for plantations. Indeed, it is clear that even intensely managed forest landscapes (including plantations) can support elements of native biodiversity if managed according to even a small subset of matrix-based principles. Plantations will typically support more biodiversity than urbanized or agricultural lands (Putz et al. 2000).

PART III Case Studies in Developing Multiscaled Plans for Biodiversity Conservation

A recurring theme in Chapters 6, 7, 8, and 9 is that there is no formula for biodiversity conservation in the matrix that is universally applicable. Rather, there will be idiosyncratic, multiscaled solutions for individual landscapes, forest types, and species assemblages crafted on a case-by-case basis. Working from this premise, Part III is a set of case studies that illustrate how multiscaled biodiversity planning and matrix management have been applied, providing real-world examples of the principles outlined in Chapter 3 and the various approaches described in Chapters 5, 6, 7, and 8. Three of the five case studies (Chapters 11, 12, and 15) not only illustrate issues associated with matrix management, but also contain lessons about the value of comprehensive multiscaled approaches to forest biodiversity conservation.

The first case study relates to the management of the northern, California, and Mexican spotted owls in western North America. For the northern spotted owl, a matrix-based stand-retention strategy was incorporated into the Northwest Forest Plan (Tuchmann et al. 1996) to facilitate the dispersal of the species between large habitat reserves that incorporated old-growth Douglas-fir forest. Although large ecological reserves and matrix retention are still the primary strategy, subsequent work in the southern part of the subspecies' range has shown that it is actually favored by a diverse landscape rather than one dominated by closed forests. An interim conservation strategy based entirely on matrix-based conservation was adopted initially for the California spotted owl, but it was subsequently modified to include large ecological reserves. Both subspecies demonstrate that management strategies should be dynamic—not

static—as they evolve in response to new information and adjustments in objectives, including acceptable levels of risk. They provide real-world examples of the need for adaptive management.

The second case study is the conservation of Leadbeater's possum in the montane ash forests of Central Victoria (southeastern Australia). It highlights the need for an integrated array of conservation strategies at multiple scales—large ecological reserves, intermediate scale reserves and wildlife corridors within the logged matrix, and fine-scale stand-retention strategies on cutover sites.

The third and fourth case studies are large-scale ecological "experiments" in which we explore various themes concerning interrelationships between the conservation value of the matrix and remnant patches within it. These experiments include the Tumut Fragmentation Experiment in southeastern Australia and the Biological Dynamics of Forest Fragments Project in Brazil. The Tumut study reinforces the ideas outlined in Chapter 10 that matrix management principles are applicable to biodiversity conservation in plantations as well as in natural forests.

The final case study is from the southern beech (*Nothofagus* spp.) forests of Tierra del Fuego in Chile and Argentina. This is a classic example of an approach embracing the maintenance of habitat at multiple scales and all associated guiding principles (the maintenance of connectivity, landscape heterogeneity, stand complexity, and aquatic ecosystem integrity) of matrix management. The various reserve and within-matrix strategies adopted in Tierra del Fuego match the checklist of approaches suggested in Table 6.1 and Table 8.1 (see Chapters 6 and 8).

Case Study 1: Northern, California, and Mexican Spotted Owls

Key Points

- Scientifically credible forest species conservation programs must include strategies for biodiversity conservation in the matrix, even for old growth–related species.
- Matrix management, whether for single or multiple species, must include strategies at multiple spatial scales.
- Closely related organisms (e.g., different populations of the same subspecies or different subspecies of the same species) may have different requirements not only for ecosystem and autecological reasons, but also for societal reasons. A flexible approach to matrix management needs to allow for a response to such challenges.
- The general principle of providing critical habitat at multiple spatial scales (see Chapter 3) was fundamental to developing conservation strategies for spotted owls but also led to the maintenance of populations of other species and of ecosystem processes.
- Two examples of the adaptive and evolutionary nature of conservation plans are evident in the spotted owl case studies: (1) the addition of large ecological reserves to an interim matrix-based conservation strategy for the California spotted owl, and (2) recognition that the northern spotted owl appears better adapted to a heterogeneous forest landscape structure than to one dominated solely by closed canopy forests in a portion of its range.

Many of the principles presented in earlier chapters have been vital to the development of conservation strategies adopted for the spotted owls of western North America. The strategies adopted for all three subspecies illustrate (1) the critical role of the matrix in conserving wide-ranging interior forest birds, and (2) the necessity of planning and working across a range of spatial scales, from regions to stands.

It is notable that the approaches to the conservation of the three subspecies differ in their relative emphasis or dependence on the reserve and matrix components of the landscape. This case study also illustrates how conservation plans that begin with a focus on a single species often evolve into more comprehensive efforts involving multiple species and even multiple ecosystems.

The spotted owl is a medium-sized nocturnal owl that occurs in forested areas of western North America (Forsman et al. 1984) (Figure 11.1). The species consists of three recognized subspecies: California spotted owl (*Strix occidentalis* ssp. *occidentalis*), northern spotted owl (ssp. *caurina*), and Mexican spotted owl (ssp. *lucida*). Although all three subspecies occupy forests, they occur in distinctively different biotic regions (Figure 11.2) and forest types that vary from the dense, continuous humid coniferous forests of northwestern North America to the fragmented forests of conifers and hardwoods found in arid southwestern North America. Environmental and habitat conditions vary greatly over the species range and, hence, there are substantial differences in the autecology of

FIGURE 11.1. Spotted owls are medium-sized nocturnal forest-inhabiting birds; California spotted owl in Sequoia–Kings Canyon National Park, California, United States. Photo by C. Halpern.

the subspecies of the spotted owl, such as differences in prey species.

Northern Spotted Owl

The northern spotted owl occurs primarily in old-growth and other structurally complex forests characteristic of northwestern North America. Forests characteristic of the western hemlock (*Tsuga heterophylla*) zone are probably the most common habitat but the subspecies has a range that also includes the mixed evergreen and mixed conifer forests of southwestern

Oregon and northwestern California (Franklin and Dyrness 1988). Long-lived forests dominated by Douglas-fir (*Pseudotsuga menziesii*), western hemlock, western red cedar (*Thuja plicata*), Sitka spruce (*Picea sitchensis*), and true firs (*Abies* spp.) are characteristic of the western hemlock zone, while a significant hardwood component (e.g., tanoak [*Lithocarpus densiflorus*], Pacific madrone [*Arbutus menziesii*], and canyon live oak [*Quercus chrysolepis*]) are characteristic in the subspecies southern range.

The natural disturbance regime in northwestern coniferous forests is primarily large-scale, stand-replacing events (wildfires or, less often, windstorms) that occur at intervals of one to many centuries. These disturbances result in the creation of extensive, largely even-aged stands, but such naturally regenerated stands have high levels of structural complexity due to

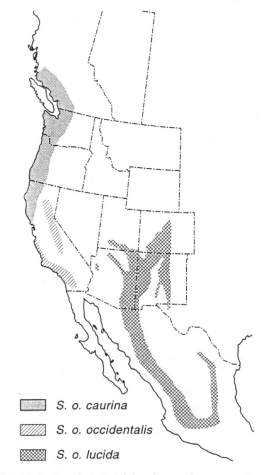

S. o. caurina

S. o. occidentalis

S. o. lucida

FIGURE 11.2. Distribution of the three subspecies of spotted owl in North America. Redrawn from Forsman et al. 1984.

biological legacies that remain after disturbance events. The natural range of variability in the proportion of suitable old-growth forest habitat within the range of the owl is estimated to have been between 30 and 70 percent for the last several centuries (see Figure 7.13 in Chapter 7).

Habitat conditions within the range of the northern spotted owl have changed dramatically during the past century as most old-growth forests were converted to young managed forest stands. Accelerated logging in the last major reservoir of such forests (federal government lands) occurred using a dispersed patch clearcutting system between 1950s and the 1980s—a strategy that rapidly fragmented the forest landscape (Franklin and Forman 1987). This resulted in increasing concern over the loss of suitable habitat for the owl in the 1970s and 1980s (Forsman et al. 1984; Yaffee 1994). In 1990, the northern spotted owl was listed as a threatened species throughout its range by the U.S. Fish and Wildlife Service (Yaffee 1994).

The conservation strategy ultimately adopted for the northern spotted owl was one of the first to explicitly recognize the importance of the matrix in an overall regional strategy. Initial proposals for conservation of the northern spotted owl were built around a system of *spotted owl habitat areas* (SOHAs)—areas that would provide habitat for one to three pairs of owls. Despite the considerable efforts expended in developing the SOHA strategy by the management agencies (federal Forest Service and Bureau of Land Management), this approach was not scientifically credible (Thomas et al. 1990) and, consequently, not legally acceptable (Yaffee 1994).

Subsequently, the U.S. Forest Service appointed the Interagency Scientific Committee to Address the Conservation of the Northern Spotted Owl (Thomas et al. 1990—often referred to as the Thomas Committee) to develop a credible conservation strategy for the owl. After extensive scientific review, the Thomas Committee proposed the creation of a system of *habitat conservation areas* (HCAs) over the entire range of the species (Thomas et al. 1990). Whenever possible, the HCAs were designed to provide sufficient habitat for at least twenty pairs of owls, the spacing between them to be no more than 12 miles (approximately 19 kilometers), which was two-thirds of the observed average distance covered by dispersing owls.

The Thomas Committee recognized that connectivity among HCAs was an important issue despite the large ecological reserve network proposed. The committee considered creating corridors between HCAs to facilitate dispersal of juvenile owls. However, a corridor-based strategy was inappropriate since, based on radio-tracking data, juveniles were known to disperse randomly with little or no tendency to follow corridors of suitable habitat. Furthermore, juvenile owls dispersed through a wide variety of habitat conditions, and the amount of forest fragmentation appeared unrelated to either dispersal distances or number of days they survived. Consequently, the committee concluded that an alternative strategy was needed to facilitate dispersal (Thomas et al. 1990).

The Thomas Committee proposed a matrix-based strategy to enhance connectivity for the northern spotted owl, noting that the "validity of the proposed strategy depends as much on condition of the habitat between the HCAs as it does on the status of the HCAs themselves" (Thomas et al. 1990).

The matrix strategy was based on maintaining a level of forest cover within the landscape between the HCAs that would facilitate successful dispersal of spotted owls (Thomas et al. 1990). The committee provided a guideline that was described as the *50-11-40 rule*. This rule provided that at least 50 percent of the matrix should be in forest stands that were at least 11 inches (27.5 centimeters) diameter at breast height and had a 40 percent canopy cover. This condition of forest was considered to be sufficient to provide some protection from predation as well as marginal foraging habitat for dispersing owls (although assumptions regarding the adequacy of such forests for dispersal and foraging continue to be debated).

Subsequent policy analyses commissioned by the U.S. Congress (Johnson et al. 1991) and by President Bill Clinton (Forest Ecosystem Management Assessment Team 1993) reiterated the Thomas Committee's concerns about the importance of conditions in the matrix. However, participants in these policy analyses were directed to move beyond the conservation of the northern spotted owl to include other conservation issues (Thomas 2000). The Scientific Committee on Late Successional Forest Ecosystems was sometimes referred to as the "Gang of Four," although over 125 scientists and resource specialists were ultimately involved in this effort. This scientific team expanded matrix considerations from connectivity for owls to providing habitat and facilitating connectivity—including maintenance of high-quality aquatic ecosystems—for a wide range of late-successional animal and plant species. Three increasingly conservative approaches to matrix management were added to a continuum of reserve levels (from none to all remaining late-successional forests). Potential costs and benefits of each of thirty-four alternatives (various reserve and matrix combinations) were assessed. The matrix-based variations involved increments of lengthened forest rotations and increased levels of structural retention at harvest (Johnson et al. 1991).

Building on the Gang of Four's report, the Forest Ecosystem Management Assessment Team (Forest Ecosystem Management Assessment Team 1993) conducted further analyses of the matrix, its ecological role, and management alternatives. This information was incorporated into the Northwest Forest Plan that was adopted for management of federal lands within the range of the northern spotted owl (Tuchmann et al. 1996).

Reserves within which timber harvest activities were excluded or severely constrained were the central elements of the Northwest Forest Plan (Tuchmann et al. 1996). These included late-successional reserves (approximately 3.4 million hectares) and riparian reserves (approximately 1.2 million hectares). Along with other areas already reserved by congressional or administrative action, areas effectively reserved from timber harvest covered 8.6 million hectares, or 77 percent of the 11.1 million hectares of federal forest land within the range of the northern spotted owl.

Despite the emphasis on reserves, habitat conditions in the matrix received significant attention in the Northwest Forest Plan. Approximately 16 percent of the federal land base was explicitly identified as "matrix" in the Northwest Forest Plan—timber can be harvested on some of these areas but there are specific constraints on harvesting techniques (USDA Forest Service and USDI Bureau of Land Management 1994a,b). Clearcutting is prohibited: "a minimum of 15 percent of the trees and forest on each harvest area must be permanently retained to provide for biodiversity and other ecological functions" (USDA Forest

Service and USDI Bureau of Land Management 1994b).

The general guideline is that two-thirds of the retention is to be in the form of aggregates of moderate to large size (0.2 to 1 hectare or more) with the remainder as dispersed structures. In addition, "as a minimum, snags are to be retained with the harvest unit at levels sufficient to support species of cavity-nesting birds at 40 percent of potential population level" (USDA Forest Service and USDI Bureau of Land Management 1994b).

The objectives of structural retention in harvested areas are essentially the same as those described by Franklin et al. (1997) for other variable retention harvest practices. These include providing refugia for elements of biodiversity ("lifeboating"), structurally enriching the next generation of forest, and improving connectivity in the managed landscape. The ultimate goal is, of course, maintenance of higher levels of ecological function and of biodiversity in the managed landscape.

Monitoring programs have been designed and implemented to assess the effectiveness of the Northwest Forest Plan in providing for both the northern spotted owl (Lint et al. 1999) and old-growth forest (Hemstrom et al. 1998).

Research on the ecology of the northern spotted owl is continuing (another part of the adaptive management strategy) and the results strongly suggest that conservation strategies need to evolve. A study on habitat quality and fitness in northern spotted owl populations in northwestern California provides an important example of how fundamental premises can be altered by more research (i.e., a form of validation monitoring) (Franklin et al. 2000b). The effects of landscape characteristics and climate on survival, reproductive output, and recruitment were considered in a ten-year population study of marked owls on ninety-five territories in northwestern California. Landscapes that were mosaics of older forest conditions and other vegetation types (including very young forests and openings) proved to be superior habitat compared with those dominated by continuous mature or old forest cover. One obvious reason is that the primary prey species of spotted owls in this region is the dusky-footed woodrat (*Neotoma fuscipes*), which occupy areas of young forest and brush.

California Spotted Owl

In some cases, the importance of the matrix for maintaining populations of forest organisms has led to conservation strategies based primarily on the management of the matrix rather than on the establishment of large ecological reserves. The initial or interim policy adopted for management of the California spotted owl provides an example in which the USDA Forest Service policy was more complex—involving both matrix and reserves—than that adopted for the northern spotted owl.

The California spotted owl is a declining subspecies that may be vulnerable to extinction (LaHaye et al. 1994; USDA Forest Service 2001b). Currently, its primary habitat is forest on federal government land in the Sierra Nevada in California. Many of these forests are dominated by pine, pine-oak, and mixed conifers.

In 1992, an interim conservation strategy was developed and adopted for the California spotted owl by the U.S. Forest Service that provided for the management of the entire forest matrix to maintain well-distributed populations (Verner et al. 1992). A specific provision of the strategy was the retention of all trees exceeding 85 centimeters in diameter on harvested lands. This approach was selected in preference to the establishment of a system of reserves, which biologists predicted would result in a significant reduction in the total number of owls—an outcome viewed as unacceptable.

The interim strategy guided the management of national forests in the range of the California spotted owl for nearly a decade during which an environmental impact analysis process was conducted to amend land-use plans in the Sierra Nevada and Modoc Plateau regions of California. That process was controversial, lengthy, and exhausting, and we will not elaborate on it further. The Sierra Nevada Ecosystem Project (1996, 1997) contributed substantially to the scientific knowledge base used in development of the final plan, playing a role somewhat comparable to that of Forest Ecosystem Management Assessment Team (1993) in the development of the Northwest Forest Plan.

The *Sierra Nevada Forest Plan Amendment Final Environmental Impact Statement* was adopted in January 2001 and dramatically expanded the matrix-based interim strategy to include reserves. It provides direction on the management of national forest lands for the California spotted owl as well as on many other conservation issues, such as old-growth forests and fur-bearing mammals (USDA Forest Service 2001a,b). This plan was reviewed by the administration of President George W. Bush and reaffirmed.

Regional Forester Brad Powell discussed in his record of decision (USDA Forest Service 2001b) the objectives of the selected alternative with regard to the California spotted owl, and noted that

The objective of the conservation strategy is to provide a high likelihood of maintaining viable populations of California spotted owl, well distributed across the national forests within the Sierra Nevada planning area. This strategy seeks to maintain habitat capable of supporting existing owl populations, stabilize current population declines, and provide increases in owl habitat over time. This strategy is based on providing and improving fundamental components of spotted owl habitat such as: a high foliage volume and complex vegetation structure at nest sites; a high percentage of home ranges in forests with moderate to high cover that are concentrated near nest sites; and habitat for primary prey species, especially the northern flying squirrel. This is accomplished through a *multi-scale landscape strategy* [emphasis added] to: (1) Protect and manage old forest emphasis areas to provide large area reserves of high-quality spotted owl habitat [i.e., reserves]; (2) Protect and manage individual spotted owl home range core areas located in the general forest matrix [i.e., meso-scale reserves]; (3) Manage the general forest outside of core areas to maintain and increase the amount of suitable spotted owl habitat [i.e., matrix stand prescriptions]; and (4) Address fire hazard and risk by reducing surface and ladder fuels within strategically placed area treatments focusing upon the urban wildland intermix zone and in old forest emphasis areas of high hazard and risk.

More specifically, the adopted strategy for conservation of the California spotted owl on national forest lands in the Sierra Nevada and Modoc Plateau includes the following components:

- *Old forest emphasis areas* (OFEAs) are established on approximately 1.6 million hectares to maintain remaining high-quality old-growth forests, which are also prime California spotted owl habitat. Management direction for OFEAs is protecting existing old forest ecosystems, expanding the amount of such forests by restoration, and reintroducing fire.
- *Protected activity centers* (PACs) for the California spotted owl are to be identified and protected. The 120-hectare PACs are foci for larger, *home range core areas* (HRCAs), which range from 240 to 960 hectares and are to be managed according to the guidelines for OFEAs. Fire fuel treatments are limited on PACs to no more than 10 percent of known PACs during any decade; moderate to high levels of tree canopy are to be maintained.
- *General forest areas* (about 1.5 million hectares) are to be managed to maintain and increase the amount of suitable California spotted owl habitat. All large live trees are to be retained as a part of vegetative and fuel treatment programs, including timber harvest. Specific direction for retention by species groups and areas are all trees greater than 76.2 centimeters for westside conifers, 61.0 centimeters for eastside conifers, 30.5 centimeters for westside hardwoods, and 20.3 centimeters for hardwoods in blue oak woodlands. Prescriptions are not to reduce canopy cover below 50 percent on westside forests and 30 percent on eastside forests, except in the urban-wildland interface. Prescriptions also provide for retaining minimal levels of snags exceeding 38 centimeters in diameter. Finally, all forests with stands of large trees (as defined by a California structural classification) are to be managed to perpetuate those conditions.

The relationship between these categories and the land allocations that we have identified in this book should be obvious. The large ecological reserves in this regional strategy (Chapter 5) include OFEAs, federally designated wilderness areas, national parks and monuments, and some California state parks. Total large reserved areas preserved on the national forests alone, including OFEAs, congressionally reserved lands (e.g., wilderness), areas administratively reserved from timber harvest, and aquatic refuges, is 2,256,000 hectares—nearly 50 percent of the national forests in the affected region—without including national or state parks. The PACs and OFEAs are the small reserved areas within the matrix (Chapter 6); riparian reserves are included in the Sierra plan as well. The management direction for the general forest areas is an example of managing stands within the matrix for biodiversity (Chapter 8).

Mexican Spotted Owl

The recovery plan for the Mexican spotted owl is less detailed than those for the other two subspecies (USDI Fish and Wildlife Service 1995). The recommendations in the plan are for a combination of

(1) protection of both occupied habitats and unoccupied areas approaching characteristics of habitat. And, (2) implementation of ecosystem management within unoccupied but potential habitat. The goal is to protect conditions and structures used by spotted owls where they exist and set other stands on a trajectory to grow into replacement nest habitat or to provide conditions for foraging and dispersal.

The Mexican spotted owl recovery plan does provide for the protection of all known activity centers. These protected activity centers (PACs) are to be 243 hectares each. The harvesting of trees more than 22.4 centimeters in diameter is proscribed in the PACs, although the necessity of managed fuelwood harvest is accepted. Road or trail building in PACs is to be avoided. In fuel treatments of PACs, a core area of 40 hectares around the nest site is to be excluded. It is noted in the recovery plan that "the intent of these guidelines is not to preserve these PACs forever, but rather to protect them until it can be demonstrated that we can create replacement habitat through active management" (USDI Fish and Wildlife Service 1995).

A second element in the recovery plan is the protection of currently unoccupied but potentially suitable nesting and roosting habitat for the Mexican spotted owl. Guidelines are provided for the identification of such areas, including important structural features such as a high basal area of trees, large trees, multistoried canopy, high canopy cover, and decadence in the form of downed logs and snags. A significant hardwood tree component is typically present. Target and threshold conditions are established by geographical recovery units and vary from 10 to 25 percent of the target area. Such areas are to be managed with several structural goals and with an emphasis on use of uneven-aged harvest systems, long rotations if the even-aged harvesting is used (more than 200 years), and retention of all large trees (exceeding 61 centimeters in diameter), hardwood trees, large logs, and snags.

The approach adopted is obviously a matrix-based strategy that includes (1) identification and sensitive management of important biotic areas (PACs) (see Chapter 7), and (2) direction regarding the management of potential habitat (Chapter 8).

Case Study 2: Leadbeater's Possum and Biodiversity Conservation in Mountain Ash Forests

Key Points

- A conservation strategy based solely on reserves is inadequate for Leadbeater's possum, even at reservation levels of 20–30 percent.
- Investigations of old forest, coupled with known habitat requirements of arboreal marsupials, such as Leadbeater's possum, have clarified the essential structural features that need to be retained and perpetuated as part of matrix-based stand management strategies.
- Studies of logging effects on stand structure indicate how many key habitat attributes are lost and what modifications to silvicultural systems are needed to enhance species conservation in the matrix.
- Developing a comprehensive conservation plan for Leadbeater's possum requires taking multiscale factors and multiscale ecological processes within mountain ash forests into account—the habitat requirements of the species, connectivity, metapopulation dynamics, and key ecosystem processes such as fire, logging, and climate change. Therefore, implementing a range of strategies at different spatial scales is fundamental. They should include large ecological reserves, midspatial-scale reserves within the matrix, wildlife corridors, and stand retention in harvested areas. Strategies have been implemented at each of these scales except stand retention in harvested areas. Modification of existing

clearcutting practices remains an important unaddressed problem.
- Restoring forest landscapes is critical to conservation management of Leadbeater's possum in order to expand the old-growth component of the existing managed forest estate.
- The implementation of monitoring programs to assess the effectiveness of management strategies is a fundamental part of any comprehensive biodiversity conservation plan. A monitoring program in mountain ash forests has been designed and recently implemented to examine the effectiveness of tree retention strategies on logged sites and assess whether midspatial-scale protected areas within the matrix actually support the populations of the arboreal marsupials they were designed to conserve.

This case study focuses on the outcomes of almost twenty years of empirical research and computer modeling on Leadbeater's possum and the mountain ash forests that it inhabits in the Central Highlands of Victoria, southeastern Australia (Figure 12.1). We discuss research on logging impacts and the matrix-based forest management strategies needed to mitigate these impacts. The work clearly demonstrates the need for a comprehensive plan for Leadbeater's possum based on integrated conservation strategies at multiple scales—large ecological reserves, midspatial-scale protected areas and wildlife corridors within the matrix, and fine-scale stand retention strategies on logged sites.

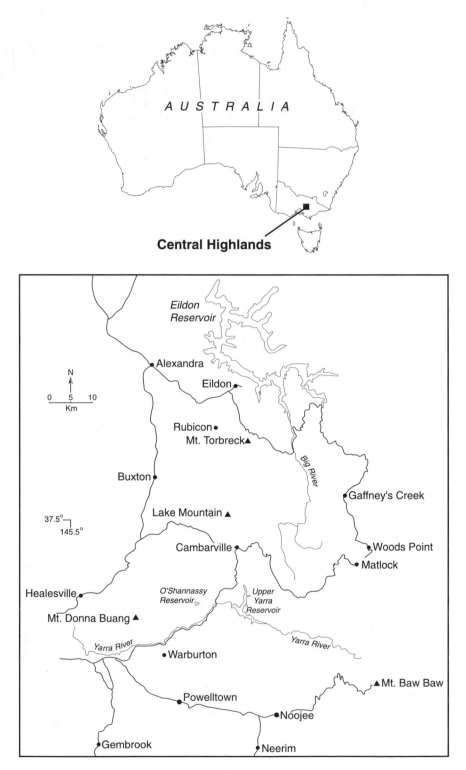

FIGURE 12.1. The location of the Central Highlands of Victoria.

Background on Mountain Ash Forests and Leadbeater's Possum

The mountain ash forests in the Central Highlands of Victoria include stands of mature and old trees with heights approaching 100 meters, making these trees the tallest flowering plants in the world (Ashton 1976) (Figure 12.2). Mountain ash forests support many forest-dependent taxa (Lindenmayer 1996), including more than 100 species of birds (Loyn 1985a, 1998), several hundred plant species (Ashton 1986; Mueck 1990), and over thirty mammal taxa (Lumsden et al. 1991). They also provide the majority of known habitat for the endangered arboreal marsupial Leadbeater's possum—a species patchily distributed within its 60×80-kilometer range (Lindenmayer 1989) (Figure 12.3).

The major management issue in mountain ash forests is the impact of widespread clearcutting operations on biodiversity, including Leadbeater's possum (Lindenmayer 1989, 1996). Mountain ash forests are among the most valuable stands for production of timber and pulpwood in eastern Australia (Macfarlane and Seebeck 1991), and a large forest industry has developed around this important resource (Government of Victoria 1986; Gooday et al. 1997). Clearcutting of 15 to 40 hectare areas during a single operation is the traditional form of harvesting and it leaves few live standing trees (Squire et al. 1991; Lutze et al. 1999).

FIGURE 12.3. Leadbeater's possum. Photo by D. Lindenmayer.

Codes of Forest Practice allow for up to three adjacent 40-hectare cutovers (Department of Natural Resources and Environment 1996). A high-intensity slash fire is used to burn logging debris (e.g., bark, tree crowns, and branches), remove fuels, and create a nutrient-rich ash seedbed for regeneration of new stands (Campbell 1984). The planned interval between clearcutting is reportedly eighty years (Government of Victoria 1986), but extensive areas of much younger (thirty-five- to fifty-year-old) ash forests have been logged in the past few decades.

Impacts of Clearcutting in Mountain Ash Forests

Contrasts between the vegetation structure and plant species composition of old-growth stands and young stands recovering after logging have been assessed to measure the impacts of clearcutting in mountain ash forests. Statistical analysis of extensive empirical field data (see Figure 12.4) shows that old-growth mountain ash stands are characterized by

- Numerous large-diameter logs that provide high volumes of coarse woody debris on the forest floor (350 to more than 1,000 cubic meters per hectare) (Lindenmayer et al. 1999g)
- Numerous large living and dead cavity-trees that vary considerably in characteristics such

FIGURE 12.2. A stand of old-growth mountain ash forest within the protected water catchments where there has never been timber harvesting. The figure in the mid-ground highlights the size of the trees in this stand. Photo by E. Beaton.

a. Probability of occurrence of tree ferns

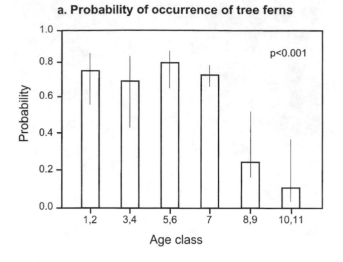

b. Log of cavity trees

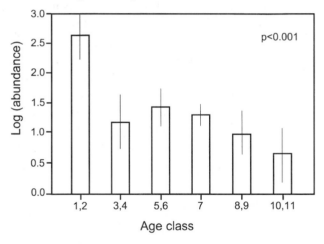

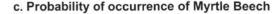

c. Probability of occurrence of Myrtle Beech

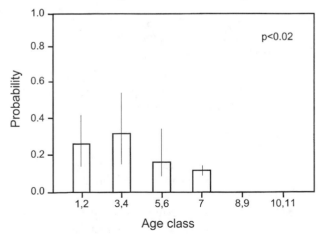

FIGURE 12.4. Some structural differences between young and old mountain and alpine ash forests. The age classes vary from those denoted 1+2 (old-growth stands more than 250 years old) through to age classes 10+11, which correspond to young post-logging regrowth forests (stands less than 20 years old) (from Lindenmayer et al. 2000c). Three features are shown: (A) occurrence of tree ferns, (B) logarithm of the abundance of cavity-trees, and (C) presence of rainforest (myrtle beech) in the understory.

as diameter, height, and stage of senescence and decay (Lindenmayer et al. 1993b, 2000c)

- An abundance of tree ferns and understory rainforest trees (Lindenmayer et al. 2000c,d)
- Trees of markedly different ages within the same stand resulting in a multi-aged forest (Lindenmayer et al. 1999f)

In addition, old-growth stands are some of the few places that support features such as clumps of mistletoe and associated vertebrate taxa such as the mistletoe bird (*Dicaeum hirundinaceum*) (Lindenmayer et al. unpublished data).

Clearcutting has serious negative consequences for a variety of taxa (see Figure 12.5):

- Cavity trees are significantly reduced in abundance (Lindenmayer et al. 1991b). These trees are nesting and denning sites for arboreal marsupials, including Leadbeater's possum. Large areas of forest are rendered unsuitable for cavity-dependent animals, and the recurrent application of clearcutting on a fifty-year rotation ensures these areas will never again become suitable for the entire suite of cavity-dependent fauna such as Leadbeater's possum.
- Tree fern populations are severely depleted (Ough and Ross 1992; Ough and Murphy 1996; Lindenmayer et al. 2000c). Tree ferns are important foraging sites for mammals such as the mountain brushtail possum (Lindenmayer et al. 1994b).
- Thickets of long-lived fire-resistant understory plants are lost (Mueck et al. 1996; Ough and Murphy 1998), including understory rainforest trees (Lindenmayer et al. 2000d) that are nesting sites for birds such as the pink robin (*Petroica rosea*; Loyn 1985a).

FIGURE 12.5. The principal stages in a clearcutting operation in mountain ash forests. (*A*) Cutting with snags left standing on site. (*B*) High-intensity slash burning. (*C*) Reseeding on the broadcast burnt site. Photos by D. Lindenmayer.

- Landscape composition is altered and the limited remaining areas of old-growth forest (now reserved from logging) are isolated among extensive stands of young forest recovering after harvesting. These changes have negative effects on some wide-ranging vertebrates such as the sooty owl (*Tyto tenebricosa*) and yellow-bellied glider, which are strongly associated with large areas of old-growth forest (Milledge et al. 1991; Lindenmayer et al. 1999a; Incoll et al. 2000). Old-growth forests are also important habitat refugia for the mountain brushtail possum and the greater glider, and populations of these species in late-successional stands that are fragmented by widespread clearcutting may not be viable in the medium to long term (Possingham et al. 1994; Lindenmayer and Lacy 1995a,b; McCarthy and Lindenmayer 1999a).

Field-validated regression models of nighttime count data confirm Leadbeater's possum is typically found in patches of regrowth and old-growth mountain ash forest characterized by both numerous large cavity-bearing trees (used as nest sites) (Lindenmayer and Meggs 1996) and a dense understory of wattle (*Acacia* spp.) trees, which are a foraging resource for the species (Lindenmayer et al. 1991a, 1994a). Colonies of Leadbeater's possum are totally dependent on large trees with cavities that require 200–400 years to develop (Lindenmayer et al. 1991c, 1993a)—a period five to eight times the length of current clearcutting rotations.

Strategies for the Conservation of Biodiversity in Mountain Ash Forests

Impacts of clearcutting on the long-term conservation of Leadbeater's possum have been an issue for over forty years (e.g., Rawlinson and Brown 1980; Lindenmayer and Norton 1993). Warneke (1962) and Dempster (1962) strongly recommended conservation efforts to reduce its risk of extinction. More recently the Victorian government has made a commitment to multiple forest use (Government of Victoria 1986) and the conservation of native wildlife in timber production forests (Government of Victoria 1988).

The Victorian Flora and Fauna Guarantee Act (1988) is also an attempt to better conserve Leadbeater's possum. The Act states that resource management must ensure that "Victoria's native flora and fauna . . . can survive, flourish and retain their potential for evolutionary development in the wild" (Government of Victoria 1988).

A range of legislative processes can be instigated under the Flora and Fauna Guarantee Act, including (1) adding a given species, community, or threatening process to the schedule of the Act; (2) preparing management plans and action statements for the conservation of a species or community or mitigating a threatening process; and (3) invoking an interim conservation order for habitat protection.

Notably, the loss of trees with cavities is listed as a threatening process in Victoria (Wilson and Clark 1995). This is an important consideration given that Leadbeater's possum is dependent on cavities, the number of trees with cavities is declining in mountain ash forests (Lindenmayer et al. 1997a), and current forestry operations significantly deplete cavity-bearing trees (Gibbons and Lindenmayer 1997, 2002).

This leads to the undesirable situation that the powers provided in the Act have never been utilized despite the good intentions of the legislation and the fact that clearcutting is legally a threatening process.

A draft management strategy for the conservation of Leadbeater's possum was completed in the early 1990s (Macfarlane and Seebeck 1991). Later, a recovery plan for the species was published outlining management actions that need to be put into effect (Macfarlane et al. 1998). Proposed strategies in these two documents included large ecological reserves and various forms of matrix management such as the establishment of midspatial-scale protected areas, wildlife corridors, and altered cutting regimes (Macfarlane and Seebeck 1991). These are discussed below.

Large Ecological Reserves

The Yarra Ranges National Park was established in the mid-1990s. It is a large reserve encompassing three large water catchments that have been closed to logging for up to 100 years (Land Conservation Council 1994). The reserve contains the most extensive areas of old-growth, mature, and multi-aged mountain ash forest remaining in the Central Highlands of Victoria (Lindenmayer and Possingham 1994). The Yarra Ranges National Park supports about 20 percent of the existing total area of 170,000 hectares of ash-type forest in the region (Macfarlane et al. 1998) and contributes significantly to the conservation of Leadbeater's possum. Yet, if it is the only element in a conservation strategy, populations in the area are at risk of extinction from high-intensity wildfires that could affect the entire park (Lindenmayer and Possingham 1995b). Populations of Leadbeater's possum in the national park are also at risk from the effects of global warming; the range of the species is predicted to contract significantly as a result of altered climatic conditions (Lindenmayer et al. 1991d; Brereton et al. 1995). Although the existing national park spans a wide range of elevational gradients, climatic conditions provided in existing wood production forests presently occupied by Leadbeater's possum are unique and could be vital in the future (Mackey et al. 2002).

The potential impacts of catastrophic wildfires and climate change highlight the need for the conservation of Leadbeater's possum in the 80 percent of its range that is outside of the Yarra Ranges National Park—within matrix landscapes subject to timber harvesting. Recommendations for the management of Leadbeater's possum in matrix lands have been made by the government of Victoria (Macfarlane and Seebeck 1991; Macfarlane et al. 1998). There are also codes of forestry practice that attempt to limit the impacts of mountain ash logging on biodiversity values (Department of Conservation, Forests and Lands 1989; Department of Natural Resources and Environment 1996).

Protected Areas within the Matrix

Midspatial-scale approaches within wood production areas are also important for the conservation of Leadbeater's possum. These include the provision of wildlife corridors, streamside reserves, and unloggable forests on steep and rocky terrain (Macfarlane et al. 1998). Although these areas are important, they also have limitations. For instance, Leadbeater's possum rarely inhabits narrow wildlife corridors located between cutover areas (Lindenmayer et al. 1993b), although they may use them in moving between reserved areas (Lindenmayer and Possingham 1996).

The linear shape of corridors is apparently not conducive to their use, possibly because of the species' complex social behavior and a diet composed of widely dispersed food (saps from understory *Acacia* spp. trees and bark-dwelling insects) (Lindenmayer and Nix 1993).

In addition, excluding logging from up to 30 percent of the area of individual 4,000- to 10,000-hectare timber management units is still inadequate to support long-term viable populations of Leadbeater's possum (Lindenmayer and Possingham 1994, 1995a). This is because Leadbeater's possum typically avoids steep terrain (Lindenmayer et al. 1991a) and riparian areas are dominated by cool temperate rainforest (*Nothofagus cunninghamii*; Lindenmayer et al. 2000d) that is unsuitable foraging and nesting habitat for the species. There is a very high risk of extinction of Leadbeater's possum if the remaining 70 percent of the forest is clearcut, thereby precluding the development of suitable matrix habitat (Lindenmayer and Possingham 1995b).

Protection of areas of old-growth forest within timber production landscapes is an essential midspatial-scale strategy for the long-term conservation of Leadbeater's possum (Lindenmayer and Lacy 1995a; Macfarlane et al. 1998). A simulated experiment in which old-growth patches were sequentially deleted showed that particular old-growth stands were pivotal for retention; their loss led to a very high predicted loss of Leadbeater's possum in a given forest block (Lindenmayer and Possingham 1996).

Expansion (restoration) of old-growth patches is an additional conservation strategy for Leadbeater's possum because current areas are inadequate. Existing 4,000- to 10,000-hectare forest blocks that are devoted to wood production presently contain less than 5 percent old growth and are dominated by mountain ash stands less than twenty to sixty years old (Lindenmayer and Possingham 1995a; Commonwealth of Australia and Department of Natural Resources and Environment 1997). One way to increase the area of old growth would be to withdraw about 600 hectares of regrowth forest from timber harvesting within each forest block and to protect them as multiple reserves each 50 to 100 hectares in size. These areas could be allowed to develop into suitable habitat for Leadbeater's possum (Lindenmayer and Possingham

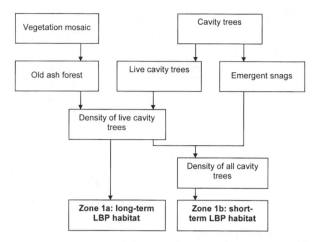

FIGURE 12.6. A model for the selection of midspatial-scale reserves set aside in the matrix for the conservation of Leadbeater's possum (LBP). Redrawn from Macfarlane et al. 1998.

1995a), although this is a long process since the large cavity-bearing trees occupied by the species take 200 to 400 years to develop (Lindenmayer et al. 1991c). This approach has recently begun to be adopted (S. Smith personal communication). When restoring the landscape, existing areas of old-growth forest should be the nodal points for future expansion; arboreal marsupials benefit more from regrowth forest reserved adjacent to existing old-growth patches than from regrowth stands set aside far from these areas (McCarthy and Lindenmayer 1999a).

Management zoning is another midspatial-scale conservation strategy that has been implemented. The zoning partitions wood production forests by use: Zone 1, where the conservation of Leadbeater's possum is a priority; Zone 2, where wood production is a priority; and Zone 3, where joint land use is a priority.

Identification of Zone 1 is based on a broad understanding of the habitat requirements of the species (Macfarlane and Seebeck 1991; Macfarlane et al. 1998), particularly the abundance of cavity-trees (Figure 12.6). Although the zoning system has some advantages, it also has problems (Lindenmayer and Cunningham 1996):

- The zoning system is not based on all key habitat components of Leadbeater's possum (such as the abundance of understory plant species) with the result that some areas suitable for the species will inevitably be lost.

- The zoning system is temporary. Cavity-trees are lost in mountain ash forests as a result of natural collapse (Lindenmayer et al. 1997a), and areas of existing Zone 1 forest revert to Zone 2 or 3 and become available for clear-cutting when the number of cavity-bearing trees falls below a threshold number (Macfarlane et al. 1998).
- The application of clearcutting within former areas of Zone 1 forest on a fifty- to eighty-year rotation permanently precludes the development of suitable new habitat for Leadbeater's possum; therefore, new Zone 1 habitat cannot be added in matrix lands.
- On-ground mapping errors have resulted in accidental logging of Zone 1 habitat (Lindenmayer 1996).
- The zoning system may result in two broad categories of forest in wood production areas—cutover areas between 0 and 50 years (the rotation time) and old-growth stands greater than 300 years in age. This not only limits the range of age classes in the forest but also could limit the recruitment of new areas of old growth to mountain ash landscapes.

The zoning system for Leadbeater's possum has recently been incorporated within a more general forest zoning approach for mountain ash forests. High conservation value sites are specified as Special Management Zones (SMZs) and the remainder of the forest, where intensive forestry can be practiced, is designated as General Management Zone (GMZ). Despite the renaming of the zones, the five problems listed above still remain.

Stand-Level Matrix Management

Retaining trees on logged sites is another important element of an overall conservation strategy for Leadbeater's possum (Figure 12.7). This tactic helps deal with the potential problems of catastrophic fires in the Yarra Ranges National Park, limitations of wildlife corridors, riparian strips, zoning, and the restricted current extent of old-growth patches. Leadbeater's possum has been recorded in post-logging regrowth forest where numerous large cavity-trees have been retained (Smith and Lindenmayer 1992). However,

FIGURE 12.7. Tree retention strategies in logged mountain ash forest. Photo by D. Lindenmayer.

while the approach is valuable, in isolation it is insufficient because

- Retained trees often are destroyed or badly damaged by high-intensity slash fires, and trees that do remain standing often have poor survival rates (Lindenmayer et al. 1990a, 1997a).
- Tree retention strategies, even if increased by 100 percent over those presently recommended, will still leave significantly fewer cavity-trees in logged areas than occurred in unmanaged stands (Ball et al. 1999).
- Numbers of retained trees may be insufficient to meet the habitat requirements of a wide range of other cavity-dependent taxa in addition to Leadbeater's possum (Gibbons and Lindenmayer 1997, 2002).

The Need for a Risk-Spreading Strategy

Each strategy discussed above has important potential benefits and limitations for Leadbeater's possum. However, the long-term conservation of Leadbeater's possum depends upon implementation of *multiple* conservation strategies covering a range of spatial scales (Lindenmayer 2000). A risk-spreading strategy (see Chapter 3) also requires variation in the way any practices at a particular spatial scale are implemented on the ground. The value of variable on-ground prescriptions has recently been highlighted in modeling studies showing that the initial 50–100-hectare reserves set aside within the matrix to conserve Leadbeater's possum (Lindenmayer and Possingham 1994) would be inadequate. More recent analysis that takes into account patterns of spatial correlation in fire regimes have shown that patches close together are more likely to burn than patches far apart (McCarthy and Lindenmayer 2000). As in the case of the California spotted owl (Chapter 11), this highlights the dynamic nature of matrix-based management strategies and their need to evolve in response to new information and insights.

Alternative Silvicultural Systems in Mountain Ash Forests

Concerns over the impacts of traditional forms of clearcutting led to the establishment of the Silvicultural Systems Project in mountain ash and other eucalypt forest types. This was instigated to test and develop economic methods of timber harvesting that also consider other non-commodity values (Squire et al. 1987; Squire 1990). The broad objectives of the Silvicultural Systems Project were "to identify and develop silvicultural systems with clear potential as alternatives to the clearfelling (clearcutting) system and model those systems against clearfelling in terms of the long-term balance between socio-economic and environmental considerations" (Squire 1990).

A range of types of forest logging treatments were investigated in the Silvicultural Systems Project, including clearcutting, shelterwood, small-gap selection, large-gap selection, seed tree, and strip cutting. In addition, various methods of seedbed preparation were examined. Squire et al. (1987) and Squire (1990) give details of these various treatments.

Although the Silvicultural Systems Project was laudable in exploring alternatives to clearcutting, the study was limited by its use of a restricted set of traditional silvicultural methods (Lindenmayer 1992). Each treatment resulted in the removal of all stems in a given area on a fifty- to eighty-year rotation. For ex-

ample, one of the shelterwood systems removed retained trees only three years after the regeneration felling (Saveneh and Dignan 1998). Given removal of all stems, coupled with the requirement for nest sites in large old cavity-trees by virtually all species of arboreal marsupials (including Leadbeater's possum), all silvicultural practices tested have detrimental long-term on-site impacts (Lindenmayer 1992). It has been established that adequate regeneration can be obtained by the use of harvesting regimes other than clearcutting (Campbell 1997) and high-intensity slash fires (Squire 1993). In addition, recent work has clearly demonstrated that mountain ash seedlings can regenerate in gaps as small as 0.1 hectare (compared with the typical cutover sizes of 10–40 hectares) (Van der Meer et al. 1999). This indicates that it is certainly possible to implement modified harvesting regimes using a wider range of cutting practices than is currently applied (Lindenmayer and McCarthy 2002a). However, such alternative methods have not been widely embraced in mountain ash forests; the vast majority of harvested sites (more than 95 percent) are still logged using clearcutting and high-intensity slash-burning methods (Lutze et al. 1999), which means these threatening processes have yet to be adequately addressed.

In summary, implementing a range of strategies at different spatial scales is fundamental for the conservation of Leadbeater's possum. Strategies should include large ecological reserves, midspatial-scale reserves within the matrix, wildlife corridors, and structural retention in harvested areas. Strategies have been implemented at each of these scales except for structural retention, and the modification of existing clearcutting practices remains an unresolved problem.

Clearcutting and Natural Disturbance Regimes in Mountain Ash Forests

The importance of biodiversity conservation strategies that use knowledge from natural disturbance regimes was described in Chapter 4. The traditional view of disturbance in mountain ash forests is that of high-intensity stand-replacing wildfires that produce even-aged regrowth forests (Attiwill 1994). This view of wildfire and clearcutting as being "ecologically equivalent" has been used to justify the widespread application of clearcutting (National Association of

Forest Industries 1989; O'Neill and Attiwill 1997). However, high-intensity stand-replacing fires are only one disturbance pathway in mountain ash forests. Lower-intensity fires that lead to only partial stand replacement also occur, resulting in complex multi-aged forests composed of overstory ash-type eucalypt trees of several age cohorts (Ambrose 1982; Chesterfield et al. 1991; Lindenmayer et al. 1991b; McCarthy and Lindenmayer 1998). Landscape-level factors influence variation in fire intensity in mountain ash forests; multi-aged stands are most likely to occur in parts of forest landscapes characterized by low levels of incident solar radiation (Lindenmayer et al. 1999f). Biological legacies in mountain ash forests will vary between stands in response to fire intensity, which is, in turn, influenced by stand location in the landscape.

Logging and regeneration methods should more closely resemble natural disturbance regimes and promote structural complexity in harvested stands to enhance wildlife habitat values (Lindenmayer and McCarthy 2002a). New silvicultural systems need to provide for (1) retention of more snags as well as more living trees that will remain standing through several cutting events and eventually develop cavities (Gibbons and Lindenmayer 2001), (2) protection of intact thickets of logging-sensitive understory vegetation, (3) protection of existing large logs and recruitment of new ones to the forest floor, and (4) creation of more truly multi-aged stands (an inevitable outcome if 1 and 2 above are adopted).

Monitoring as Part of Ecologically Sustainable Forest Management in Mountain Ash Forests

Without rigorous monitoring, there is no way to determine whether management strategies implemented over the last decade are effective and whether mountain ash forests are being managed in an ecologically sustainable way. A long-term monitoring study has commenced for populations of Leadbeater's possum and other species that are threatened with decline or extinction as a result of timber harvesting. This study uses a retrospective approach to resurvey populations of arboreal marsupials on sites first surveyed in 1983 (Lindenmayer and Incoll 1998). A pool of 161 sites has been established and each site has had a history of repeated field surveys over the past two decades. All 161 sites were resurveyed in 1997–1998 and a randomly selected subset of 50 of the 161 sites were resampled for animals in 1999, 2000, and again in 2001.

The monitoring program has been designed to provide strong statistical inferences regarding population trends. Year-to-year partial sampling allows short-term temporal fluctuations in population dynamics to be separated from long-term trends such as population declines (Welsh et al. 2000). The 161 sites encompass a wide range of stand conditions that vary in the history of human and natural disturbance. Repeated surveys of the vegetation structure on the sites (Lindenmayer et al. 1997a) coupled with the known history of disturbance patterns allow testing of hypotheses that attempt to account for observed population declines or increases (if any). Response of arboreal marsupials to tree retention strategies on logged sites are being closely examined, as is the abundance of animals on midscale protected areas within the matrix. The monitoring work is specifically designed to evaluate strategies proposed for enhanced matrix management in mountain ash forests.

The two government natural resource agencies responsible for the management of mountain ash forests (Parks Victoria and the Victorian Department of Natural Resources and Environment) co-sponsor the monitoring program (administered from The Australian National University). Information from the program will be made available to both organizations. If this information leads to altered management practices (such as modified methods of timber harvesting), this could be viewed as adaptive management (see Chapter 16). However, the present lack of variation in the current cutting regimes in mountain ash forests (95 percent of cutovers are clearcut) dramatically limits the range of stand conditions available for study as demanded in a true adaptive management approach.

Aspects of monitoring in the mountain ash forests conform to what has been termed an "adaptive monitoring approach" (Ringold et al. 1996; see Chapter 16)—monitoring protocols have been adjusted since the commencement of the program on the basis of new information. The original protocol for the selection of the subset of sites (N = 50) from the total monitoring pool (N = 161 sites) in any given year was based on targeting sites found to support high numbers of arboreal marsupial (irrespective of the species). Analyses of data found limited patterns of persistence of Leadbeater's possum on such selected sites from year to year. As Leadbeater's possum is the species of primary concern for managers in mountain ash land-scapes, new site selection protocols have been adopted. Protocols are now based, in part, on ensuring that all sites that support Leadbeater's possum in a given year are monitored in the following year. This increases the chance of identifying population trends of Leadbeater's possum, without jeopardizing the quality of data gathered for other species (Lindenmayer et al. unpublished data).

CHAPTER 13

Case Study 3: The Tumut Fragmentation Experiment

Key Points

- Landscape context is important for many species— conditions in the matrix (here extensive plantations of exotic radiata pine) considerably influence their presence and abundance in remnant eucalypt patches.
- Landscape context differences between eucalypt remnants and large continuous areas of native eucalypt forest were not significant for those species of arboreal marsupials that occurred in the matrix.
- Small and intermediate-sized protected areas within the matrix (i.e., remnant eucalypt patches) have considerable value for forest vertebrates and are used by many taxa for shelter and breeding. Larger remnants supported more species of arboreal marsupials, small mammals, and birds. However, even small and intermediate-sized remnants contributed significantly to the maintenance of forest biodiversity.
- The interactions between the remnant eucalypt patches and the radiata pine matrix were significant. For example, many bird species occurred in the matrix because of the spatial juxtaposition of the two landscape context classes.
- Structural complexity within the radiata pine matrix (e.g., large eucalypt logs and native understory vegetation) strongly influenced the ability of some birds and small mammals to persist.
- Genetic analyses indicated that gullies and riparian vegetation were an important dispersal route for small mammals through the landscape.

- Incorporating the matrix as a dispersal sink or as additional area for foraging significantly improved the predictions from computer-based simulation models for metapopulation dynamics on presence and abundance of animals in remnant eucalypt patches.
- Some species may not be adequately conserved by matrix-based strategies in the plantation estate (e.g., the yellow-bellied glider) even if large remnant patches are set aside, streamside reserves are established, and structural features in the matrix are retained. Large ecological reserves are required to protect such species.
- Most theories associated with fragmentation are derived from agricultural landscapes where habitat remnants are surrounded by a matrix that is hostile to biota. More complex responses to landscape conditions occur where the surrounding matrix is not totally inhospitable.
- Even though the remnant eucalypt patches at Tumut will never be part of a reserve system, they make important contributions to biodiversity conservation in the region.
- Plantation estates around the world are usually considered to have limited conservation value. However, the implementation of some matrix-based conservation strategies can significantly increase the contributions to biodiversity.

The statistically designed, large-scale Tumut Fragmentation Experiment is so named because it incorporates many features required of a classical experiment, including (1) randomized selection of sites in different treatment classes, (2) extensive replication of

229

sites in the different treatments, and (3) controls, which are located in large areas of unfragmented forests whose environmental conditions compare to those in the eucalypt remnants.

Demographic and genetic studies, as well as computer simulation modeling, are integrated within the experiment and used to examine the influence of landscape context, habitat fragmentation, and matrix conditions on arboreal marsupials, small mammals, birds, and plants. Unlike the case studies in Chapters 11 and 12, the Tumut Fragmentation Experiment does not focus on the resolution of a set of forest management and species conservation problems. Rather, it is an extensive scientific investigation designed to provide a better understanding of (1) interrelationships between the matrix and habitat fragments and their influence on population dynamics, and (2) the role of the matrix in biodiversity conservation. Nevertheless, important inferences can be drawn from the study about matrix-based strategies to enhance biodiversity conservation within plantation landscapes (Lindenmayer and Pope 2000).

Background to the Tumut Fragmentation Experiment

The Tumut Fragmentation Experiment is located in the Buccleuch State Forest, about 100 kilometers west of Canberra in southern New South Wales (NSW), southeastern Australia (Figure 13.1). It is located in a 50,000-hectare plantation of exotic radiata pine established in areas that formerly supported native eucalypt forest. Clearing of native forest for radiata pine commenced in the mid-1930s and continued for about fifty years; subsequently, new softwood plantations were initiated on previously cleared grazing land. A total of 192 patches of remnant *Eucalyptus* spp. forest of varying size, shape, and vegetation types were not cleared during plantation establishment and are now surrounded by extensive stands of radiata pine. Large continuous areas of native eucalypt forest exist beyond the boundary of the pine forest in the Kosciuszko and Brindabella national parks as well as in the Bondo and Bungongo state forests.

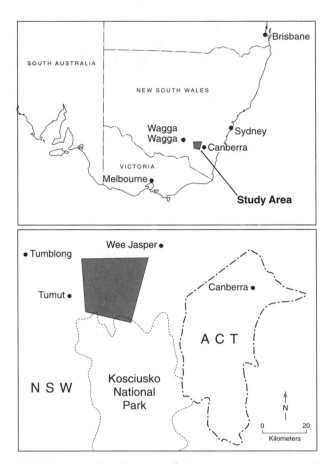

FIGURE 13.1. The location of the Tumut region in southeastern Australia.

Three broad categories of sites, each differing in landscape context, have been sampled for vertebrates at Tumut:

1. Sites in remnants or fragments of native *Eucalyptus* forest located within the boundaries of the radiata pine plantation
2. Sites dominated by radiata pine
3. Sites in the large areas of continuous *Eucalyptus* forest adjacent to the plantation (Lindenmayer et al. 1999b,c, 2001a) (Figure 13.2 and Figure 13.3)

Of the 192 eucalypt remnants at Tumut, 86 were selected for sampling using a randomized and replicated statistical procedure. These remnants varied in size (1–124 hectares) and shape (long and narrow versus elliptical or round), as well as in other features. A set of 40 sites in the matrix of radiata pine were se-

FIGURE 13.2. Aerial view of eucalypt patches embedded within the radiata pine matrix at Tumut. Photo by D. Lindenmayer.

lected and matched to the 86 eucalypt remnants on the basis of environmental, climatic, and terrain conditions. Another 40 sites were selected within large continuous areas of *Eucalyptus* forest. These were matched to the eucalypt remnants on the basis of environmental and climatic conditions, terrain, and vegetation cover in the Tumut region.

Actual field data for patch occupancy by birds, small mammals, and arboreal marsupials were compared with predictions of the same measures derived by spatially explicit computer simulation models for metapopulation dynamics (Lindenmayer et al. 1999d, 2000a; McCarthy et al. 2000). Finally, genetic analyses examined patterns of genetic variability among small mammal populations in relation to patch structures and landscape connectivity (Hewittson 1997; Lindenmayer et al. 1999h; Lindenmayer and Peakall 2000).

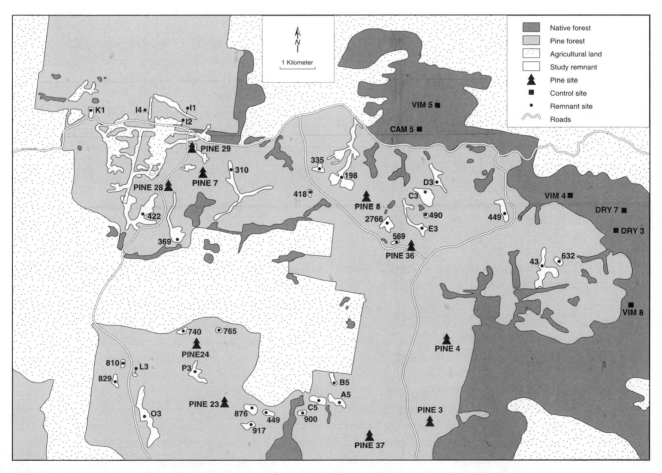

FIGURE 13.3. A subset of the three broad types of sites in the Tumut Fragmentation Experiment.

An important feature of the Tumut Fragmentation Experiment is the extent of continuous eucalypt forest that contains the forty sites matched to the eucalypt remnants. These control areas cover tens of thousands of hectares and are large enough to support viable populations of all species that occur in the study region. In an extensive review of fragmentation studies, Andrén (1994) noted that only 10 percent of investigations included habitat areas larger than 1,000 hectares in size. Much fragmentation research has concentrated on very small fragments (less than 10 hectares), which may extrapolate poorly to larger fragments and reserves and limit the inferences that can be drawn (Zuidema et al. 1996).

Another important feature of the Tumut Fragmentation Experiment is the sample of forty sites in the radiata pine matrix. Many fragmentation studies have sampled only habitat fragments and ignored the potential habitat value (occupancy) of the surrounding matrix (Simberloff et al. 1992; Beier and Noss 1998).

Field Survey Results and the Matrix

An objective of the Tumut Fragmentation Experiment was to test effects of specified landscape (matrix) conditions on vertebrates: arboreal marsupials, small mammals, and birds. Fragments of eucalypt forest surrounded by radiata pine forest supported a different faunal assemblage than stands embedded within extensive continuous eucalypt forest. In the case of arboreal marsupials, the large context effects were partly a consequence of the complete absence of most species from the radiata pine matrix. Two species—the yellow-bellied glider and the squirrel glider (*Petaurus norfolcensis*)—were lost from all the remnants, including relatively large ones (more than 120 hectares). Possible reasons for loss of these species are that the squirrel glider was rare at the time of fragmentation and the yellow-bellied glider has a very large home range (bigger than the size of most patches). Additional analyses were focused only on eucalypt remnants and continuous eucalypt forests (i.e., the sites dominated by radiata pine forest were ignored).

Significant landscape-context effects were observed for the two species absent from the matrix (the common brushtail possum [*Trichosurus vulpecula*] and the greater glider) (Lindenmayer et al. 1999b). These effects were not found for two species that persisted at small population sizes in the radiata pine matrix (the common ringtail possum [*Pseudocheirus peregrinus*] and the mountain brushtail possum; Lindenmayer et al. 1999b). Indeed, the data suggested that both these species *increased* in abundance in the remnants relative to the control species—an effect not anticipated in island biogeographic or metapopulation theory (see Chapter 2).

Subsequent analysis of the amount of eucalypt cover in the matrix surrounding each remnant highlighted very strong matrix effects on the dynamics of populations in the remnants. Arboreal marsupial populations in remnants were *higher* in isolated remnants than in those with neighboring eucalypt patches in the surrounding radiata pine matrix. These results suggest the possibility of a *fence effect* (sensu Wolff et al. 1997). That is, animals attained higher densities in isolated remnants because the surrounding matrix was hostile and animals were reluctant to disperse from their natal patch. Whatever the cause, the findings clearly demonstrate that strong interrelationships between matrix conditions and habitat fragments exist, and that information on both is needed to interpret population dynamics in such complex landscape mosaics (Lindenmayer et al. 2001a,b).

Strong interrelationships between the matrix and habitat fragments also were recorded for birds. The forest mosaic provided habitat for many birds (more than ninety species), and complex landscape context, habitat fragmentation, and matrix effects were identified. Some birds were ubiquitous, such as the grey shrike thrush (*Colluricincla harmonica*). Another group of taxa was recorded most often in the remnants (e.g., little raven [*Corvis mellori*], superb fairy wren [*Malurus cyaneus*], and shining bronze cuckoo [*Chyrysococcyx lucidis*]). Of these, some were more likely to be recorded in small eucalypt remnants (the golden whistler and Australian magpie). Others such as the eastern yellow robin (*Eopsaltria australis*) and superb lyrebird (*Menura superba*) were more likely to be detected in the intermediate-sized eucalypt remnants. The sacred kingfisher (*Todiramphus sanctus*), leaden flycatcher (*Myiagra rebecula*), and white-naped honeyeater (*Melithreptus lunatus*) were among the group detected most frequently in larger remnants. Other birds, such as cicada bird (*Coracina tenuirostris*), gang-gang cockatoo (*Callocephalon fimbriatum*), and olive-backed oriole

(*Oriolus sagittatus*), favored large continuous areas of native forest. A number of native species were significantly more likely to occur in the matrix (e.g., rufous whistler [*Pachycephala rufiventris*] and the brown thornbill [*Ancathiza pusilla*]) than in the eucalypt remnants or large continuous areas of native forest.

A combined measure of species presence and abundance (termed a *bird frequency profile*) formed the basis of analyses of the composition of the bird assemblage at each site. There was no fixed community of birds in the three broad landscape context classes or a climax bird community but rather a complex reassembling of bird species in relation to landscape context, remnant area, and conditions in the landscape surrounding a given site. There was strong empirical evidence for a gradient in the bird frequency profiles between radiata pine stands and large continuous areas of native eucalypt forest (Figure 13.4). Changes in the bird frequency profiles along this continuum encompassed changes both in the identity of the taxa in the assemblage and relative abundance of each species between the two broad radiata pine and continuous eucalypt landscape context classes. The remnants connected the bird frequency profiles of these two forest types, and the nature of the gradient depended strongly on remnant size (Lindenmayer et al. 2001a). However, the change in the bird frequency profile for the various types of sites was strongly influenced by the landscape conditions that surrounded the field sites. For example, the occurrence of many bird species in the radiata pine matrix was significantly related to the amount of eucalypt forest adjacent to these radiata pine sites (Lindenmayer et al. 2001a). Pine sites where the surrounding landscape contained some eucalypt remnants had a different bird frequency profile from pine sites where the surrounding landscape was pure radiata pine. Radiata pine sites with adjacent eucalypt patches had a bird frequency profile similar to that of small and intermediate-sized eucalypt remnants (Lindenmayer et al. 2001a). Thus, the matrix influenced the occurrence of birds in the eucalypt remnants, but the eucalypt remnants also influenced the patterns observed in the matrix. Therefore, simultaneous consideration of the landscape mosaic (i.e., the interplay between eucalypt remnants *and* the surrounding landscape matrix) was critical for determining the response of birds in the experiment.

The results of the Tumut Fragmentation Experiment highlighted why it is important not to assume that what appear to be patchy landscapes from a human perspective automatically correspond to patchy wildlife populations with dynamics conforming to metapopulation processes. The red wattlebird (*Anthochaera carunculata*) provides a good example; the species was significantly more abundant in large areas of continuous forest than in radiata pine stands and had intermediate levels of abundance in the remnants and the radiata pine matrix. Extensive analyses of patterns of spatial dependence in bird distribution (e.g., Koenig 1998) showed that the patterns observed for the red wattlebird and many other species did not conform to that predicted by classic metapopulation models (sensu Hanski and Gilpin 1991) with particular taxa restricted only to a "mainland" (here the extensive native forest areas) or to certain patches or types of patches (see Chapter 2). Rather, the radiata pine matrix surrounding the remnants provided suitable or partially suitable habitat for many species, including the red wattlebird (Figure 13.4).

Studies of birds at Tumut revealed a range of interesting matrix effects:

- Small and intermediate-sized patches of eucalypt forest supported a wide range of species—and contributed substantially to the regional conservation of biota.
- There were strong spatial interrelationships between the radiata pine matrix and remnant patches of eucalypt forest. Many taxa oc-

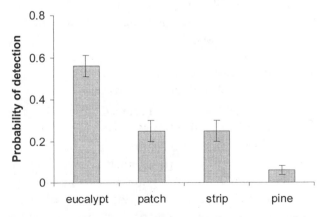

FIGURE 13.4. Gradient of response for the detection of the red wattlebird for the pine-eucalypt remnant-eucalypt control landscape context classes.

curred in the radiata pine matrix because of its proximity to eucalypt forest—they would have been rare or absent if the landscape were a plantation monoculture. Hence, the landscape mosaic contributed strongly to high levels of species richness observed.

- Some of the responses to landscape condition and habitat fragmentation observed in the study were quite different from those seen in other fragmentation investigations (e.g., the preferential use of small and intermediate-sized patches by some native taxa).
- Some of the novel results were not consistent with paradigms such as nested subset theory (see Chapter 2) in which new taxa would be added to an original (minimal) assemblage of birds in an ordered and progressive fashion in response to increasing remnant size (Patterson 1987). Rather, there were substitutions in the species represented in the different bird assemblages across the gradient from small to large remnants and the controls.

Many fragmentation concepts were developed in agricultural landscapes where the matrix surrounding habitat remnants is often considered to be hostile. More complex responses, such as those derived from the Tumut Fragmentation Experiment, may occur where the surrounding matrix is not totally inhospitable—as reported from a plantation environment in South America (Estades and Temple 1999). Such findings have implications for investigations of fragmentation because important effects may be overlooked if (1) the use of the landscape matrix is ignored (Beier and Noss 1998), (2) only a limited range of fragment sizes are studied (Andrén 1994; Zuidema et al. 1996), or (3) control sites (the large continuous areas of native forest at Tumut) are either unavailable or not examined (Margules 1992).

Stand Structural Features in the Matrix and Animal Response

Conditions in the matrix strongly influenced the ability of birds and small mammals to persist there. For example, piles of windrowed eucalypt logs remaining after clearing of the original native forest by machinery provided critical habitat for small mammals in radiata pine stands. These were the

only locations in the radiata pine matrix where these taxa were recorded (Lindenmayer et al. 1999c).

Understory conditions also were significant predictors of the presence and abundance of many species of diurnal forest birds within radiata pine stands in the Tumut Fragmentation Experiment (Lindenmayer et al. 2001a). For example, occurrences of native understory plants such as dogwood (*Cassinia aculeata*) and bracken fern (*Pterideum esculentium*) provided cover for taxa such as the brown thornbill (*Ancathiza pusilla*) and white-browed scrub-wren (*Sericornis frontalis*). The presence of these structural features often meant that overall differences in species richness were less than expected among radiata pine stands, remnants, and large continuous areas of native forest (Lindenmayer et al. 2001a).

Simulation Modeling and the Matrix

The contribution of the matrix to population dynamics at Tumut was highlighted by computer simulation modeling. When the effects of matrix on dispersal mortality were incorporated in model specifications, better congruence was obtained between predicted and actual values for the occupancy of remnant eucalypt patches and the overall abundance of arboreal marsupials (Lindenmayer et al. 2000a) (Figure 13.5). Similarly, for one native bird—the white-throated treecreeper—model predictions were greatly improved if the model included the ability of the species to forage up to 500 meters from the remnant eucalypt patches and into the surrounding radiata pine matrix (McCarthy et al. 2000).

Integrated Genetic and Demographic Work and the Matrix

The bush rat (*Rattus fuscipes*) occupied remnants of eucalypt forest surrounded by extensive stands of radiata pine at Tumut and were significantly more likely to occur in larger patches (Lindenmayer et al. 1999c). Genetic analyses of patch populations and those in large continuous areas of native forest revealed interesting patterns of genetic variability. No correlation between genetic distance and geographic distance was found (Hewittson 1997). Rather, populations of the

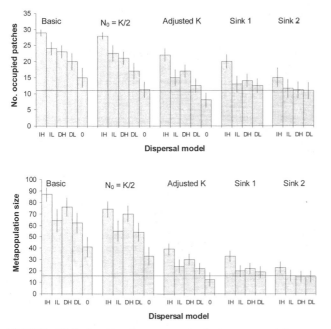

FIGURE 13.5. Levels of congruence between model predictions and actual values for patch occupancy by the greater glider when the effects of the matrix on dispersal are (Sink 1 and Sink 2 scenarios—see below) and are not included in the simulations (modified from Lindenmayer et al. 2000a). The solid horizontal line shows the actual number of occupied patches. The dispersal models are *IH* (island model with a high migration rate), *IL* (island model with a low migration rate), *DH* (patch size/interpatch distance^2 with a high migration rate), *DL* (patch size/interpatch distance^2 with a low migration rate), and *O* (no interpatch migration). The scenarios are *Basic* (all patches are at full carrying capacity at the start of simulations), $N_0 = K/2$ (simulations initialized at half carrying capacity), *Adjusted K* (variation in habitat quality according to patch quality), and *Sink 1* and *Sink 2* (the negative effect of the matrix conditions on interpatch migration included in the simulations; see Lindenmayer et al. 2000a for further details). Reprinted with the permission of the Ecological Society of America.

bush rat located considerable distances apart were more closely related if they occurred in patches connected by a drainage line than spatially adjacent patches not connected by a watercourse (Hewittson 1997). An explanation for this result was that animals use watercourses as dispersal routes—there was limited movement among patches not linked via riparian vegetation, even where they are spatially adjacent (Lindenmayer and Peakall 2000). This result confirms the importance of riparian vegetation and intact

aquatic ecosystems as habitat for some species and emphasizes the value of riparian protection within matrix lands for connectivity (General Principle 1 in Chapter 3).

The Nanangroe Experiment

A lack of knowledge about the status of different species prior to habitat fragmentation is a problem common to almost all fragmentation studies (Margules 1992)—and it applies to the Tumut Fragmentation Experiment. A new study recently begun in the Tumut region, the Nanangroe Experiment, avoids this limitation. It is a long-term (longitudinal) direct study of changes in vertebrate fauna inhabiting woodland fragments as the surrounding grazed landscape matrix is transformed into radiata pine plantation. It is uniquely focused on landscape changes in the matrix surrounding habitat fragments (Figure 13.6). Groups targeted for study are birds, terrestrial mammals, arboreal marsupials, reptiles, and frogs.

The Nanangroe Experiment is located on a grazed woodland landscape approximately 15 kilometers northeast of the boundary of the Tumut study area. The results of this experiment will provide an interesting contrast with those from the Tumut Fragmentation Experiment. Already, after the initial three years of work, the study shows that responses of several species in the remnant-mature

FIGURE 13.6. A patch of remnant woodland being surrounded by newly planted stands of radiata pine plantation in the Nanangroe Experiment. Photo by D. Lindenmayer.

pine plantation system in the Tumut Fragmentation Experiment are entirely different from those in the cleared and grazed remnant woodland system at Nanangroe (e.g., the common ringtail possum and the sacred kingfisher; Lindenmayer et al. 2001c). These findings not only highlight the importance of the matrix in examining species' responses, but also demonstrate the need to gather empirical data to track such responses.

Conclusion

Results from the Tumut Fragmentation Experiment reinforce the concept that active planning of plantation landscapes can enhance biodiversity conservation (Chapter 10). Many guiding principles outlined for matrix management in Chapter 3 and discussed further in Chapters 6, 7, and 8 can be applied to plantation landscapes.

Case Study 4: The Biological Dynamics of Forest Fragments Project

Key Points

- Matrix conditions significantly influenced the population dynamics of many groups in fragments demonstrating interrelationships between within-fragment dynamics and the matrix.
- Differences in plant species dominating regrowth vegetation in the matrix influenced the recolonization patterns of fragments by some birds (e.g., insectivores). Matrix conditions also influenced both the magnitude of edge effects (such as rates of annual tree mortality) and how far the effects penetrated forest fragments.
- Many bird, frog, and small mammal taxa moved through regrowth moist forest matrix and recolonized habitat fragments.
- Matrix habitats can be broadly classified according to the level of contrast with habitat fragments; the most favorable matrix for biota (and least negative in its effects on remaining habitat fragments) is one most similar to unmodified primary forest. Levels of similarity span attributes such as floristic and structural composition as well as microclimatic conditions.
- Loss of primary forest led to the loss of some primary forest–dependent bird species. Nonetheless, the modified landscape mosaic comprising habitat fragments and regrowth moist forest had significant value for the conservation of biodiversity.

This case study focuses on the Biological Dynamics of Forest Fragments Project (BDFFP) established more than two decades ago in tropical moist forests of north-central Brazil (Figure 14.1). The project's initial goal was to address what were then very controversial issues flowing from the theory of island biogeography (Macarthur and Wilson 1967), such as the SLOSS debate—will more species be conserved by a single large or several small reserves? (Wilson and Willis 1978; Simberloff and Abele 1982; see Chapters 2 and 5).

The BDFFP commenced during a period of mass clearing and destruction of tropical rainforests throughout the Amazon basin (a process that continues today: Whitmore 1997; Laurance et al. 2000). One objective was to determine the minimum size of fragments needed to conserve the extraordinary biodiversity of rainforest ecosystems. To address this issue, the proposed study design was a large-scale experiment with replicated forest fragments in logarithmically increasing size classes (1 hectare, 10 hectares, 100 hectares, 1,000 hectares).

Due to a variety of circumstances, not all size classes of fragments were created, and for those size classes that were created, replication was limited. The extent of clearing (and subsequent cattle grazing) in the surrounding matrix was variable, resulting in matrix conditions adjacent to the fragments ranging from pasture to advanced regrowth moist forest. Perhaps fortuitously, this modified design has forced the BDFFP to focus more closely on the interrelationships between the landscape matrix and dynamics within fragments (Bierregaard et al. 1992) than on the SLOSS question (Gascon et al. 1999). Information about matrix-fragment interrelationships from the BDFFP have generic value for increasing understanding of biota response in the complex landscape mosaics that characterize many regions around the world (Forman 1995).

Project Background

The Biological Dynamics of Forest Fragments Project began under a different name—the "Minimum Critical Size of Ecosystems Project" (Lovejoy 1980; Lovejoy et al. 1986). This was because the initial goal of the experiment was to provide empirical data on the minimum area needed to conserve rainforest ecosystem processes and biodiversity (Lovejoy and Oren 1981). This was done by tracking the loss of species from rainforest fragments of different sizes after they were isolated by clearing of the surrounding vegetation for cattle grazing (Bierregaard and Stouffer 1997).

The Minimum Critical Size of Ecosystems Project began in 1979, north of Manaus in north-central Brazil (Figure 14.1). It was initially funded by Brazil's National Institute for Research in Amazonia (INPA) and the World Wildlife Fund-U.S. Brazilian law required landowners to leave 50 percent forest cover on any land targeted for development. The Minimum Critical Size of Ecosystems Project planned to take advantage of this by cooperating with owners of cattle ranches in creating a replicated set of twenty-four rainforest fragments of 1, 10, 100, and 1,000 hectares in size. Each remnant was to be located 200–300 meters from adjacent forest.

Fragments of tropical moist forest were identified in consultation with owners of cattle ranches at the start of the experiment and prior to clearing. Detailed baseline data were gathered for an array of taxonomic groups at these sites. Two fragments (one of which was 10 hectares and the other 1 hectare) were isolated in 1980. An additional five fragments were isolated in 1983 (two 1-hectare fragments, two 10-hectare frag-

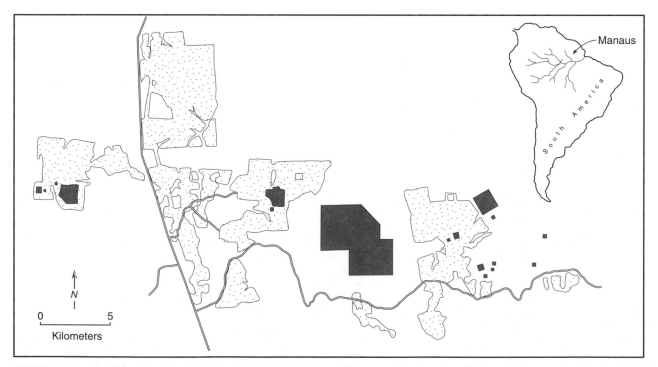

FIGURE 14.1. The location of the Biological Dynamics of Forest Fragments Project. The black areas correspond to the locations of the rainforest remnants targeted for detailed study. The stippled areas are consolidated blocks of rainforest. Redrawn and modified from Bierregaard and Stouffer 1997. Reprinted with the permission of the Gustav Fischer Verlag.

ments, and one 100-hectare fragment) and three in 1984 (two 1-hectare fragments and one 10-hectare fragment). However, problems with the Brazilian economy and a reduction in subsidies to establish new areas for cattle grazing ended forest clearing in 1984 before other fragments could be added to the experiment (Bierregaard and Stouffer 1997), although one more 100-hectare fragment was added in 1990. Thus, the final experiment had eleven fragments ranging in size from 1 to 100 hectares. In addition, the degree of isolation was 70–100 meters for three remnants and 150–1,000 meters for the others.

The clearcut areas around the two fragments isolated in 1980 were burned, but a new forest quickly established in the surrounding matrix. These re-growth stands were again cleared after five years. The matrix surrounding the fragments established in 1983 was not burned, and new stands also quickly regenerated in these areas. The matrix surrounding the fragments established in 1984 was burned and pasture was established. Therefore, the matrix surrounding the eleven fragments in the experiment varied from pasture to advanced regrowth moist forest—a wide variation due to abandonment of grazing in most pastures surrounding the fragments and to regrowth type that depended on past grazing intensity and fire history (Bierregaard and Stouffer 1997).

Although the initial aim of the project was to examine species-area relationships, the enforced changes to the experimental design meant the project subsequently focused more intensively on other aspects of landscape modification as reflected in its name change. These biological dynamics included edge effects (Kapos 1989) and interrelationships between the biota of fragments and matrix conditions (Bierregaard and Stouffer 1997; Gascon et al. 1999).

Despite design problems with the BDFFP, the study has yielded many valuable results, some challenging theories such as island biogeography (Gascon and Lovejoy 1998) and others highlighting the importance of the matrix for interpreting fragmentation effects (Tocher et al. 1997; Gascon et al. 1999). Although the study has "before" and "after" components, the large between-site variations in fragment size, isolation distance, and matrix conditions make it more like an observational study than a true experiment.

The Matrix and Responses of Selected Groups in the Project

The range of groups studied in the BDFFP includes (among others) birds (Stouffer and Bierregaard 1995; Bierregaard and Stouffer 1997), butterflies (Brown and Hutchings 1997), leaf-litter insects (Didham 1997), amphibians (Tocher et al. 1997), nonflying mammals (Malcolm 1997), and canopy trees (Rankin-de Mérona et al. 1992). In addition, a number of studies of edge effects have been completed (e.g., Kapos 1989; Mesquita et al. 1999; Sizer and Tanner 1999). Obviously, the methods and designs of field investigation vary between these different groups and types of studies, with varying numbers of control sites in continuous forest and matrix sites in the pasture and/or regenerating forest in the areas surrounding the remnants (see the Methods section of Gascon et al. 1999).

Much has been published on the BDFFP (see Bierregaard et al. 1992; Laurance and Bierregaard 1997), and it is not possible to review even a small fraction of the work here. But we briefly outline some findings for several groups as they relate to fragment-matrix interrelationships.

Birds

Initial changes in the matrix that isolated fragments in the BDFFP had large impacts on individual species of birds and on the overall bird community. Isolation did not immediately lead to extinctions. Conversely, there was an initial flux of individual birds into fragments as they were displaced from the surrounding clearcut matrix—an ephemeral "crowding effect" (Bierregaard and Stouffer 1997). Capture rates of individual species and overall species richness eventually declined below pre-fragmentation levels (Bierregaard and Lovejoy 1989). A large change in species composition was also documented relative to the assemblages characteristic of pre-fragmentation primary forest (Bierregaard et al. 1992; Bierregaard and Stouffer 1997).

Following experimental fragmentation, the bird recolonization patterns in fragments were influenced by matrix regrowth vegetation; this was especially noticeable in groups of birds such as insectivores (Stouffer and Bierregaard 1995).

TABLE 14.1.

Variations in responses of different taxonomic groups in the Biological Dynamics of Forest Fragments Project (modified from Gascon and Lovejoy 1998).

GROUP	BEFORE AND AFTER SPECIES DIVERSITY	CHANGE IN SPECIES COMPOSITION	INVASION OF MATRIX-ASSOCIATED SPECIES
Birds	Decrease	High	Low
Frogs	Increase	Low	Medium
Small mammals	Increase	Low	Medium
Ants	Decrease	High	Low
Butterflies	Increase	High	High

Frogs, Small Mammals, and Other Groups

Complex and contrasting fragmentation effects were recorded for frogs, small mammals, and other groups such as butterflies and ants (Table 14.1). In the case of frogs, few taxa were lost in comparison with species richness levels prior to fragmentation. Conversely, species diversity actually increased (irrespective of fragment size), primarily because of the influx of species capable of exploiting new, more open conditions created by changes in the matrix (Tocher et al. 1997; Malcolm 1997). In addition, the fragments in the BDFFP did not appear to be truly isolated for many species of frogs because they could inhabit and move through the matrix, indicating that frogs were less susceptible to fragmentation than other groups examined in the BDFFP (Tocher et al. 1997).

Few species of small mammals were lost as a consequence of fragmentation. There was no evidence of a change in community composition (Table 14.1). Indeed, recently isolated forest fragments supported more individuals and more species of small mammals than control areas in continuous forest did (Malcolm 1997). The increase in species richness was associated with invasions of species that could use and move through matrix habitats (Gascon and Lovejoy 1998).

Ants and butterflies, like birds, showed high levels of turnover in composition compared with the original taxonomic assemblages. However, overall species richness declined among ants, but increased among butterflies (for the same reasons that influenced the post-fragmentation diversity of frogs and small mammals). Fragment size and isolation appeared to have limited direct impacts on butterflies. But indirect factors associated with landscape change, such as edge effects and the creation of altered conditions in the matrix, strongly influenced the composition of butterfly communities (Brown and Hutchings 1997). For example, light-loving butterflies were common at the edges of forest fragments.

The Matrix and Edge Effects

The strong influence of the matrix on the magnitude of edge effects was demonstrated in the BDFFP. For example, when edges were created at the boundary between primary forest and adjacent pasture, light penetrated the forest both vertically and horizontally, altering microclimatic conditions such as temperature regimes and relative humidity (Kapos 1989; Sizer and Tanner 1999). Wind speeds also changed dramatically at forest-matrix edges (Lovejoy et al. 1986). The extent that edge affects penetrated varied substantially depending on the measured attributes (see Figure 2.8 in Chapter 2) such as leaf fall, tree mortality, seedling recruitment, or, ultimately, plant species composition (Laurance et al. 1998). Laurance et al. (2001) reported dramatic impacts on the liana community structure—increased ecophysiological stresses on trees through altered light and nutrient regimes led to heavy infestations of lianas and contributed to significantly increased tree damage and mortality, especially near the edges of rainforest fragments.

Edge effects in the BDFFP varied depending on whether the surrounding matrix was cleared pasture or regrowth forest recovering from clearing. For example, annual tree mortality in fragments was significantly higher if the surrounding vegetation was pasture compared with regrowth forest (Mesquita et al. 1999). To summarize, edge effects led to substantial alterations in the physical environment at the boundaries between forest fragments and the surrounding matrix (e.g., Laurance 2000) with consequent changes in the biota.

Conclusion

Species varied substantially in their response to landscape change and habitat fragmentation in the BDFFP (Gascon and Lovejoy 1998) as has been observed in many other studies (e.g., Robinson et al. 1992). Larger forest fragments always supported more species than smaller ones—a result consistent with species-area theory (Preston 1962; Rosenzweig 1995) and island biogeography theory (Macarthur and Wilson 1967). However, many other results from the BDFFP contradict predictions from island biogeography theory. In this sense, the project achieved its goal of testing aspects of this widely discussed theory and has demonstrated its limited utility (Gascon and Lovejoy 1998). BDFFP has also highlighted the fact that predicting responses of biota to landscape modification requires an understanding of interrelationships among *all* landscape components—habitat fragments, the varied states of the matrix within which they occur, and areas of continuous unmodified (or primary) forest (Gascon et al. 1999). This is a valuable outcome because tropical landscapes are often complex mosaics (Mesquita et al. 1999). Contrary to island biogeography theory, species diversity of groups such as frogs, small mammals, and butterflies *increased* in abundance following fragmentation as a result of influxes into the fragments by taxa that can exploit modified conditions in the matrix (Gascon and Lovejoy 1998). Although butterflies showed increased species diversity after fragmentation, they did undergo a large turnover in community composition: 40 percent of the original butterfly species were replaced by open-country species that respond positively to increased available light (Hutchings 1991 in Gascon and Lovejoy 1998). Obviously, focusing only on species diversity (i.e., species numbers) would have masked important details about the effects of landscape change on the biota.

A meta-analysis of the responses of several groups showed strong interrelationships between matrix conditions and within-fragment population dynamics. Gascon et al. (1999) assessed the responses of several groups targeted for study in the BDFFP (ants, small mammals, birds, and amphibians) using the approach employed by Laurance (1991a) (see the Extinction Proneness section in Chapter 2). Small mammals, frogs, and birds all exhibited strong positive relationships between abundance in the matrix and vulnerability to habitat fragmentation (Table 14.1) (i.e., species that did not use or persist in the matrix tended to disappear from fragments). Conversely, populations of those taxa that utilized the matrix increased or remained stable in the fragments (Gascon et al. 1999). Understanding how biota use (or do not use) all components of a landscape was essential to understanding how species are influenced by landscape modification and habitat fragmentation.

CHAPTER 15

Case Study 5: The Rio Condor Project

Key Points

The Rio Condor Project exemplifies integrated, multiscale approaches that can be taken in previously undeveloped landscapes. Measures taken to maintain biodiversity include:

- Creation of a system of large ecological reserves
- Identification and protection of sensitive areas within the matrix, including riparian and wetland buffers; sensitive sites, such as steep slopes and high-elevation forest; important biological sites or hotspots; and archeological sites
- Maintaining and enriching structural conditions and habitat diversity in harvested areas by aggregated retention and conservation of coarse woody debris
- Implementation of a comprehensive environmental monitoring and research program to obtain new information as a part of the adaptive evolution of the project

The Rio Condor Project is the first industrial forestry project that incorporates a comprehensive program to conserve biodiversity at the regional, landscape, and stand levels (although others are emerging). The project is also unusual in the extensive involvement of academic experts in design and implementation of the project, thereby providing a high degree of scientific credibility. The monitoring and research program is exceptional in its scale and sophistication. Furthermore, the requirement for an annual independent environmental audit helps ensure environmental standards and commitments will be achieved.

Despite the above, significant impediments to the evolution of the project continue:

- Bureaucracies create impediments to adaptive management
- Failure to scale monitoring and other environmental requirements to levels of forest harvest created financial penalties for reductions in timber harvest levels
- Opposition from some environmental groups despite the potential for significant conservation benefits

The Rio Condor Project is a sustainable forestry project developed for a large industrial forestry property in Tierra del Fuego, South America. The project was developed by Trillium Corporation (a North American company) and associated legal entities (hereafter collectively referred to as Trillium Corporation) working collaboratively with a group of Chilean scientists and Chilean government agencies. The project exemplifies an approach that integrates the conservation of biodiversity and ecosystem processes with commercial forest production. It also illustrates an ideal situation in which project designers begin with an essentially intact landscape. In preparing this case study we relied heavily on information from the document *Toward an Ecologically Sustainable Forestry Project: Concepts, Analysis and Recommendations*, prepared by the Independent Scientific Commission for the Rio Condor Project (Arroyo et al. 1996).

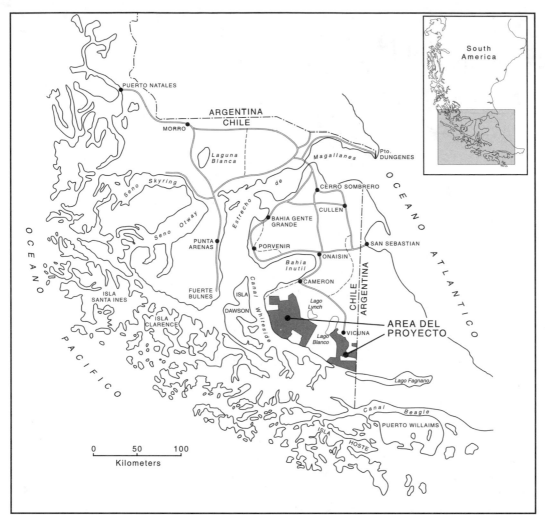

FIGURE 15.1. Location of the Rio Condor Project in Tierra del Fuego, South America.

Physical and Biological Features of Rio Condor

The Rio Condor Project covers approximately 272,000 hectares of land in two large blocks located at approximately 54°S latitude in southwestern Tierra del Fuego, Chile (Figure 15.1). It is a mountainous region with elevations typically ranging up to 1,000 meters (Figure 15.2). Topography is generally moderate, however, with localized steep slopes (more than 50 percent); there are extensive areas of flat to gently undulating land associated with major river valleys and broad areas of peat bogs, locally known as *turba* (Figure 15.3).

We provide the reader with more background information on the biophysical setting for this case

FIGURE 15.2. Forested mountainous topography characteristic of much of the Rio Condor Project. Photo by J. Franklin.

FIGURE 15.3. Extensive flat to gently sloping topography characteristic of the intermountain valleys and peat bogs (turba) within the Rio Condor Project. Photo by J. Franklin.

study than for other chapters in Part III because the characteristics of this remote region are not well known and are highly relevant to some of the conservation measures ultimately selected.

Climate

Rio Condor has a cold temperate maritime climate as would be expected of an island situated at the southern tip of South America. Precipitation is well-distributed seasonally, but annual values decline from 900 to 1,000 millimeters on the Pacific Coast to around 500 millimeters in eastern portions of the property. Mean annual temperatures are around 3 to 5°C at lower elevations with mean winter minimums (July) around –0.5 to –4°C. Strong persistent winds are characteristic of the region, especially during the spring and early summer. Snow commonly occurs during the winter, but a significant persistent winter snowpack is not characteristic.

Low clouds and fog are typical of the forest environment. This means atmospheric inputs of moisture and nutrients, through condensation and precipitation on tree canopies, are potentially important to forest ecosystems. For example, in one study within the Magellenic region, it was estimated that cloudwater deposition could be up to 800 millimeters annually

and inputs of nitrogen from cloud moisture could be as much as 8 kilograms per hectare per year (Weathers and Likens 1997).

Vegetation

The Rio Condor property is a mosaic of forested and nonforested vegetation at an approximate ratio of 60:40. The forests consist primarily of three species of southern beech (*Nothofagus* spp.), and they vary in species composition along the west-east climatic gradient. The nonforested vegetation consists primarily of turba (32 percent) and alpine areas (5 percent) with minor amounts of wet steppe, coastal scrub, and rock outcrops. Although large continuous areas of forest and of turba do occur, complex landscape mosaics of forested and nonforested patches are common.

Forests

Rio Condor forests are variously composed of the evergreen coigue de Magallanes (*N. betuloides*, hereafter referred to as "coigue") and the deciduous species lenga (*N. pumilio*) and nirre (*N. antarctica*). Coigue is found primarily in western areas of the property in varying mixtures with lenga. Pure lenga forests dominate the eastern (drier) half of the property. Nirre is primarily a small species occupying ecotones such as the upper timberline and the boundaries between forest and turba or steppe. Community structure of the forests is often simple with understories in lenga forests often consisting only of herbaceous species (Figure 15.4); a shrub layer composed of *Berberis ilicifolia* is common in forests along the west coast (Figure 15.5).

The forests are predominantly multi-aged old-growth forests. Dominant trees on sites at lower elevations typically range from 20 to 25 meters tall, 28 to 42 centimeters dbh (diameter at breast height), and are as old as 250 years. The primary disturbing agent in these forests is wind, which generally creates small gaps (e.g., 0.1 to 0.3 hectare) within which new tree cohorts are regenerated, most commonly from an existing seedling bank. High winds may occasionally blow down large patches of forest (Figure 15.6).

FIGURE 15.5. Mixed lenga and coigue stand typical of the western portions of the Rio Condor property; a shrubby understory of *Berberis ilicifolia* is characteristic of these forests. Photo by J. Franklin.

FIGURE 15.4. Pure lenga stand typical of the central and eastern portions of the Rio Condor property; the depauperate understory consisting of herbaceous species and large amounts of coarse woody debris is characteristic. Photo by J. Franklin.

FIGURE 15.6. Extensive wind-damaged lenga stand in the Rio Condor Project area. Photo by D. Lindenmayer.

Turba

Wetlands are extensively represented on the Rio Condor property and consist primarily of three types of peat bogs (sphagnum-dominated, grass-dominated, and cushion-plant-dominated). They occur at all elevations, although the most extensive tracts are found at low to moderate elevations in the southern and western parts of the property. Most are raised peat bogs in which the center of the wetland is higher than the surrounding land. Wetlands, primarily peat bogs, completely dominate the headwaters and lengthy channel segments of several rivers, including the Rio Grande and the Rio Condor.

Aquatic Ecosystems

Large rivers, streams, and numerous lakes and ponds are significant elements of the Rio Condor landscape (Figure 15.7). The stream drainage network is much denser in the wetter western and southern sections of the property. Major river systems include the Rio Grande, which runs into the Atlantic Ocean near Ar-

FIGURE 15.7. Segment of the Rio Condor River illustrating the strong influence of wetlands, primarily peat bogs, in the river drainages within the Rio Condor Project. Photo by J. Franklin.

berline. Estimated densities for beaver colonies are 6.6 to 8.5 colonies per kilometer of stream in Rio Condor. Except for only some of the smallest (first order) tributaries, essentially all aquatic ecosystems have been significantly modified, and 8 to 10 percent of the forest in Rio Condor has been destroyed by beaver activity.

A peculiar and important feature of the Rio Condor landscape is the effective "decoupling" of forested areas from larger streams and rivers throughout many of the watersheds. Wetlands—primarily peat bogs—and wet steppe completely dominate the headwaters of the Rio Grande and Rio Condor Rivers and major segments of the river channel. Such nonforested ecosystems also dominate the stream reaches that are intermediate between forested mountain slopes and major river valleys. They are also sites of extensive beaver activity (Figure 15.8) and former current sites for grazing by domestic livestock. The chemical and physical properties (e.g., sediments) of the major rivers and streams are strongly influenced by peat bogs and beaver and are effectively isolated from influences of human activities on forested uplands. The decoupling of the forested uplands from the major rivers is an important consideration when designing meaningful monitoring programs to assess the impact of forestry activities on aquatic ecosystems.

gentina, and the Rio Condor, which drains to the Pacific Ocean. Several very large lakes are adjacent to and fed by tributaries from project lands.

Significant sport fishing—an important economic and social consideration in Tierra del Fuego—based on introduced sea-run brown trout (*Salmo trutta*) and rainbow trout (*Oncorhynchus mykiss*) exists in river and lake ecosystems in the region.

The aquatic ecosystems are significantly modified from their natural condition by naturalization of the beaver (*Castor canadensis*), which was introduced to create a fur industry in Tierra del Fuego in 1946. The beaver lacks any significant natural or human predation and has spread to occupy essentially all suitable and much marginal habitat, such as sites at upper tim-

FIGURE 15.8. Wetlands and wet steppe dominate stream reaches located between forested uplands and major river and stream systems; these reaches are also sites of extensive activity by introduced beaver. Consequently, chemical and physical properties of larger streams and rivers are dominated by peat bogs and beaver rather than processes within the forested areas, an important consideration in monitoring. Photo by J. Franklin.

FIGURE 15.9. Two mammals of special management interest in Rio Condor are the (*A, left*) red fox, an uncommon forest species subject to poaching, and (*B, right*) guanaco, a large herding browser that is abundant throughout the property. Photos by J. Franklin.

Biodiversity

Species diversity of most groups is low to moderate in Rio Condor—not surprising on a high-latitude island. But some groups and some species are very abundant and the flora and fauna exhibit moderate levels of endemism. There is moderate diversity of forest vascular plants, with most of the endemics concentrated in coastal forests. Two plant groups with high levels of diversity are forest lichens and fungi.

Vertebrate diversity, while modest in terms of total number, include a number of interesting species. Forest bird species diversity (twenty-four species) includes 92 percent of all birds known from Tierra del Fuego and includes a forest owl (*Strix rufipes*), a large woodpecker (*Campephilus magellanicus*), and the austral parakeet (*Micropsittaceae ferruginea*). In the wet-

lands and steppe, there are large numbers of waterfowl (geese, ducks, swans, and wading birds) and buff-necked ibis (*Theristicus caudatus*), the latter nesting in the canopies of old-growth beech forests. The Andean condor (*Vultura gryphus*) is a notable inhabitant of the skies, and nests on rocky crags within the Rio Condor Project.

Mammal diversity is low (thirteen native species), but it includes two species of special management interest—guanaco (*Lama guanicoe*) and red fox (*Pseudalopex culpaeus lycoides*). The red fox is considered endangered in this region of Chile (Figure 15.9a) and is a species that could be affected by forestry programs, particularly because of the increased access for poachers. The guanaco is a very abundant, large herding herbivore (Figure 15.9b) that could potentially influence the success of tree regeneration through its browsing activity. It occurs throughout the project area from the steppe to alpine habitats and numbers thousands of animals. Unlike the situation on mainland South America, on Tierra del Fuego the guanaco has no natural predators. Nevertheless, it is considered an endangered species because of its status on the mainland, which complicates approaches to population control and management on Tierra del Fuego.

Human Influences

Native Americans occupied Tierra del Fuego prior to western settlement in the late nineteenth century but were effectively extirpated. Doubtlessly they influenced conditions and animal populations within the Rio Condor region, particularly in coastal regions and, perhaps, at the steppe-forest ecotone at the eastern edge of the property. There is substantial archaeological evidence that they used the forest (Arroyo et al. 1996).

Coastal portions of Rio Condor have been heavily impacted in the past by grazing, logging, and burning (some dating back to western settlement over 100 years ago). Evidence remains in the condition of the landscape and the presence of several small abandoned communities. Much of the early logging was of a selective nature that removed individual trees but left forest stands largely intact. Grazing by domestic livestock (sheep and cattle) continued around the margins of the property until 1999 and feral animals (e.g., cattle and horses) still occur throughout the property, although they are being eliminated whenever possible.

Fires of human origin have occurred at the margins of the property, particularly at the forest-steppe ecotone—some caused by sport fishermen. Forest regeneration on several of these burned areas has been very slow.

Objectives of the Management Program

The Rio Condor Project is intended to be a sustainable forestry project in which the production of value-added wood products is effectively integrated with conservation and ecotourism. The goal is to maintain biodiversity and ecological processes at both the landscape level and within the harvested blocks while also achieving an adequate financial return. The goal of the previous landowners was to maximize short-term economic returns through aggressive timber harvest to produce woodchips for the export market. Trillium Corporation has adopted a different philosophy for its management with the intention of developing a sustained yield of high-value hardwood timber and prod-

ucts. Its plan is to set a new global standard for environmentally responsible industrial forestry, demonstrating the compatibility of economical and ecological goals. To achieve this standard, Trillium

- Developed and adopted a set of stewardship principles to guide all aspects of project development and implementation
- Chartered an independent scientific commission composed of a team of Chilean academics to provide information and advice
- Voluntarily committed the company to an environmental impact assessment and permitting process under a new law administered by CONAMA (Comision Nacional de Medio Ambiento), the Chilean equivalent of the U.S. Environmental Protection Agency

Stewardship Principles

The development and adoption of a set of stewardship principles was one of the earliest steps in the development of the Rio Condor Project. President D. Syre of Trillium Corporation signed these principles on 11 October 1993 as the basis upon which project development would proceed, making the commitment in recognition of the "vastness, location and quality of the land and forests of the Rio Condor Project" (Arroyo et al. 1996).

The principles addressed the issues of environmental quality ("The Project will be operated so that, in the long term, the aggregate environmental quality of the land, water and forests of the Project will meet or surpass baseline conditions") and sustainability ("The Project's forests will be responsibly managed for indefinite sustainable hardwood production, using the best available scientific knowledge to assure the protection of . . . the ecosystem") (Arroyo et al. 1996).

Related to these principles were commitments to specific practices, including exclusive use of native tree species (no exotics), primary dependence upon natural regeneration, and primary wood processing within the region (no export of raw wood in the form of logs or woodchips).

Trillium Corporation also voluntarily created an ombudsman-like land-steward position to provide independent oversight of its activities in Rio Condor, particularly with regard to adherence to the stewardship principles. A North American lawyer, R. Jack, was designated the first land steward and was succeeded in 1996 by J. F. Franklin.

Independent Scientific Commission

Trillium Corporation recognized the importance of incorporating the best scientific information available when developing its management plan for the Rio Condor Project. For this reason, it collaborated with the Chilean Academy of Sciences in identifying an independent group of active Chilean academic scientists representing a breadth of relevant disciplines. This team of six (eventually seven) scientists was chartered in 1994 as the Independent Scientific Commission (ISC) under the chairmanship of M. T. K. Arroyo of the University of Chile. Activities carried out by the ISC between 1994 and 1999 included

- Completion of baseline studies on the ecosystems and organisms of the Rio Condor Project area
- Development of recommendations for management practices
- Development of recommendations regarding the location, size, composition, and management of biological reserves
- Development of recommendations for monitoring and research programs at Rio Condor

Environmental Analysis and Permitting Process

Trillium Corporation could have initiated harvesting in its forest lands immediately upon obtaining title to the land in 1993 since it held valid harvesting permits issued by the Chilean National Forestry Agency (CONAF). However, Trillium volunteered to be the first forestry project to develop and submit an environmental impact statement (EIS) under a new Chilean environmental law administered by CONAMA. Development of the EIS was contracted to an environmental consulting firm, Dames and Moore, but the documents incorporated much of the information and most of the recommendations provided by ISC.

Two EISs were ultimately produced. The first was completed in 1996 and resulted in a permit issued by CONAMA in 1997. But the legal validity of this permit was challenged by some environmental organizations, and the Chilean Supreme Court ruled that the permit was invalid because the government had failed to adopt regulations necessary for valid implementation of the new law. A second EIS was completed by Trillium in 1998 and resulted in a new permit from CONAMA. The legality of the second permit was upheld through a new series of court challenges and became the basis for initiating forestry operations.

The permit issued by CONAMA incorporated an extensive series of requirements for environmental protection by Trillium Corporation and its operating subsidiaries. A long list of mandated monitoring activities was a major condition of the permit; these incorporated, but substantially expanded and specified details of, the voluntary monitoring and research program that the company had developed with the ISC and submitted as part of the EIS.

The engagement of an independent environmental auditor (IEA) also was a condition of the CONAMA permit. The IEA is a scientific and technical team that audits the performance of the company with regard to the conditions of the permit and subsequently reports their findings directly to CONAMA. Trillium pays all of the costs for the IEA and its activities. CONAMA selected the auditing organization from three finalists offered by Trillium that, in turn, made its selection from proposals submitted by interested organizations. In 1999, CONAMA selected Geotechnica Consultores, an environmental consulting firm based in Santiago, Chile, as the IEA.

The decision to voluntarily participate in the new environmental permitting process proved to be extremely expensive for Trillium Corporation in terms of both direct costs and five years of delay before harvesting activities could begin. Furthermore, the conditions laid down in the environmental permit were not scaled to the size of the project. Initiation of timber harvest, even at the level of a single tree, effectively brought into force all the conditions of the permit, including the very extensive mandated monitoring program (designed to assess impacts of a large harvesting program) and engagement and funding of the IEA.

Key Elements of the Rio Condor Management Plan

The conservation objective in managing the Rio Condor Project is to maintain biodiversity and ecosystem processes at both the harvest unit and the landscape level. Therefore, both reserve-based and matrix-based approaches are utilized. Indeed, the planning and management at Rio Condor address essentially all of the elements identified in the landscape checklist in Chapter 6 (Table 6.1). These include creation of large ecological reserves, conservation of specialized habitats and critical structures within the matrix, and an adaptive management process to develop critical new knowledge through monitoring and research (Chapter 16).

Reserves

Large ecological reserves are a major conservation element of the Rio Condor Project—of course incorporating significant reserves as part of an industrial forest management plan is unprecedented. However, as part of its commitment to stewardship, Trillium recognized the importance of large ecological reserves in its plan because of the large scale of its landholdings in this region, the poor representation of regional ecosystems in existing reserves, and the lack of significant Chilean government lands in the region from which public reserves could be created.

The ISC strongly recommended establishment of ecological reserves on Rio Condor prior to any harvesting. They suggested that these be selected so as to incorporate the full range of ecosystems present, including commercial forest; environmental variability of the property (east-west and altitudinal gradients); and viable populations of endemic, rare, and endangered species (Arroyo et al. 1996).

Trillium concurred with these recommendations and asked ISC to identify the locations for a series of permanent ecological reserves that would cover 25 percent of the property and incorporate up to 10,000 hectares of commercial forest. ISC made an initial proposal for five large reserves in 1997. Subsequently, there were careful technical evaluations of boundaries and the extent of forest and other vegetation types.

TABLE 15.1.

Total area and area of commercial forest in the permanent biological reserves within the Rio Condor Project.

NAME	AREA (HA)	COMMERCIAL FOREST (HA)
Canal Whiteside	2,216	1,144
Rio Caleta	4,812	2,096
Lago Escondido	17,908	3,967
Lago Blanco-Kami	43,162	6,988
Total	68,098	14,195

In March 1999, ISC and Trillium Corporation agreed upon the final boundaries for a set of four large ecological reserves covering a total area of 68,098 hectares and incorporating 14,195 hectares of commercial forest (Table 15.1; Figure 15.1). Two of the originally proposed reserves were connected and incorporated into the Lago Blanco-Kami Biological Reserve. The smallest reserve (Canal Whiteside) was established primarily to provide for representation of the species-rich coastal forest. With the agreement of Trillium Corporation, the final reserve system encompassed nearly 50 percent more commercial forest than originally stipulated.

Matrix-based Conservation Strategies

Matrix-based conservation strategies in the Rio Condor Project included

- Buffers for riparian areas, wetlands, and coastal sites
- Protection of steep slopes and high-elevation sites
- Permanent structural retention within harvested sites in the form of aggregates
- Conservation of coarse woody debris within harvested forests
- Protection of sites with special plant populations
- Identification and conservation of archeological sites

These measures were proposed by Trillium Corporation as part of its management plan based largely upon recommendations of the ISC. Subsequently, most of these measures were mandated by CONAMA as conditions of the environmental permit.

Buffers around Aquatic Features and Wetlands

The Rio Condor Project incorporates protected (unmanaged) forest buffers around all sensitive landscape features, including streams and rivers, lakes and ponds, and at the ecotones between forested and nonforested vegetation, such as wetlands or steppe.

Riparian buffer widths vary with the aquatic feature. The Rio Condor and its floodplains are considered to be particularly sensitive and biologically important; there is a minimum 250-meter no-harvest forest buffer on either side of the river, and the buffer must include all of the first terrace. The buffer on other rivers and larger streams is 50 meters on both sides as measured from the high-water mark, or 30 meters of continuous forest; if the latter is chosen, then only limited harvesting is allowed within the next 20 meters. Banks of all small streams (0.5 to 2.0 meters wide) are protected.

All turba have a no-harvest buffer of at least 10 meters and limited harvest for an additional 20 meters. However, since no harvest of nirre is allowed, and the wetland-forest ecotones are often composed of pure nirre, the buffers are typically wider than the minimums.

The effect of the protective buffers is to provide an extensive system of connected riparian reserves within the Rio Condor landscape and to protect aquatic and wetland features from the direct impacts of forest harvesting. Approximately 8,520 hectares of additional commercial forests are in riparian, wetland, and other types of buffers.

Protection of Forests on Steep Terrain

Harvesting activities are restricted on all slopes greater than 45 percent and above 450 meters in elevation. Any proposal for harvesting beyond these limits requires the development of a special plan that must be approved by CONAMA. However, the company has never proposed harvests on such sites and is unlikely to do so. Harvesting on sites where dominant trees average less than 9 meters in height is also proscribed by the management plan. These restrictions provide for the protection or reservation of substantial amounts of closed forest in the more rugged mountainous regions of Rio Condor. They also provide for

the protection of most areas of shallow and unstable soils. The percentage of the landscape protected under these provisions is estimated to be in excess of 10 percent.

Aggregated Retention within the Harvest Units

Aggregated retention within the harvest units is the primary strategy that has been adopted to maintain biodiversity within managed forest stands. When structural retention occurs as small intact patches within the harvest unit, it is referred to as *aggregated retention*, because the aggregates are intended to be a part of—and not apart from—the harvested and managed stand.

The purpose of structural retention is detailed in Chapter 8. Briefly, it involves retaining structural elements from the harvested stand in order to provide refugia for elements of biodiversity, structurally enrich the subsequent managed forest stand for purposes of biodiversity and ecosystem function, and facilitate movement of organisms through the managed landscape (Franklin et al. 1997, 2000b).

The primary silvicultural harvest system approved by CONAF for use at Rio Condor is a shelterwood with natural regeneration. This is an even-aged management system in which an initial entry removes approximately 60 percent of the basal area, leaving a protective overstory (shelterwood) with approximately 50 percent canopy cover. Regeneration is allowed to develop for about fifteen years (or to a height of 5 to 7 meters) at which point the remaining overstory is removed, releasing the young even-aged cohort. Up to 200 hectares of continuous harvested area is allowed. Intermediate cuts (commercial thinnings) are also anticipated as part of the management regime. The proposed rotation age for managed stands is 110 years. Notably, silvicultural approaches approved by CONAF and its academic advisors for lenga forests (shelterwood) are very traditional. Harvesting using group selection with structural retention would come much closer than a shelterwood in terms of emulating the natural disturbance regime. Application of an even-aged system to forests that are naturally uneven-aged results in major alterations of the structure (e.g., spatially homogenizing the stands) and function of the native forests. Structural retention can reduce the simplification of the stands under a shelterwood system. Unfortunately, CONAF appears to be uncomfortable with retention of large, decadent trees as part of the shelterwood overstory and generally requires that they be removed or at least killed.

The Rio Condor management plan requires a minimum of 10 percent permanent structural retention as small aggregates that are kept free of logging disturbance. While a 10 percent retention level may seem low, it is important to remember that the primary harvest prescription is a shelterwood in which approximately 50 percent canopy cover is maintained for about fifteen years following the initial entry. Therefore, the context for retention is very different than it would be under a clearcut harvest regime. Furthermore, although harvest removal of the shelterwood overstory is programmed, it rarely occurs in practice and substantial additional dispersed retention of trees is a result.

Aggregate size and spacing have evolved during the first year of forest harvest at Rio Condor. The initial prescription for retention was the identification and marking of a 0.1-hectare aggregate on each hectare of harvest area prior to commencement of logging. Observations in the first harvest units suggested that the size and integrity of 0.1-hectare aggregates was inadequate, so aggregate size was increased to 0.2 hectare—with appropriate adjustments in spacing—in subsequent harvest units. Tree-marking crews are directed to position aggregates on or around structural features that are important for conservation of biodiversity (such as nesting trees for the austral parakeet) or that are difficult to grow or retain where logging occurs (such as concentrations of large decadent trees, large snags, and accumulations of large logs).

Marking in most harvest units also has provided for some dispersed retention, primarily in the form of individual large decadent trees that are retained for wildlife conservation and other purposes. Such trees are considered unacceptable as part of the shelterwood overstory by CONAF under its marking rules—they do not count toward the 50 percent canopy cover objective—so they must be identified and marked separately. Only a limited number are allowed because they can substantially increase canopy cover over acceptable levels. As noted earlier, CONAF prefers to see all decadent trees removed or at least killed.

Conservation of Coarse Woody Debris

Coarse woody debris, particularly large decaying logs, is abundant in natural forests of lenga and coigue. Although traditional forestry practices attempt to minimize levels of such material by intensive utilization and slash disposal practices, the importance of this material in maintaining biodiversity and important ecosystem processes, such as nitrogen fixation, is well understood (Harmon et al. 1986; Maser et al. 1988). Most ecosystem-oriented forestry practices include measures to maintain levels of coarse woody debris.

Silvicultural prescriptions at Rio Condor recognize the ecological values of coarse woody debris and attempt to conserve any slash and existing woody debris that cannot be readily utilized, a policy recommended by the ISC. Unfortunately, such efforts run counter to forestry tradition and approved CONAF practice, which means that the company is under pressure to utilize or dispose of as much coarse woody debris material as possible.

Special Management Areas

Populations of unusual plant species, communities, and habitats are scattered throughout the area of the Rio Condor Project. The company is committed to the identification and protection of these types of sites. Plant communities associated with Tertiary-period rock outcrops in the Vicuna region are a specific example that was identified by ISC. The environmental permit requires the identification and protection of these outcrops and their protection with 50-meter buffers. Another example is the provision for protection of the Puerto Arturo region and delta of the Rio Condor. Other important biological areas, such as nesting sites for the Andean Condor, are being identified and protected as part of the evolving management plan.

Finally, sites are being selected and committed to long-term experimentation and monitoring as described below. These include a set of four small adjacent watersheds in the Vicuna region that are reserved and will undergo calibration for future watershed experiments.

Archaeological Sites

There were seventy-seven archaeological sites within the Rio Condor Project area identified during baseline studies by ISC. Such sites are protected under Chilean law and require varying degrees of protection and conservation (i.e., excavation by qualified archeologists). The appropriate treatment of each of these sites is stipulated in the environmental permit issued by CONAMA and varies from complete protection of some sites to excavation of others. In addition, for the purposes of protecting archaeological resources, there is a 75-meter buffer along the entire sea coast and a 500-meter buffer along Seno Almirantazgo between Canal Whiteside and the Lago Blanco-Kami Biological Reserve.

Recreational Sites

There are provisions in the management plan and the environmental permit for identification and protection of important recreational sites, including those that could contribute to the development of ecotourism.

Adaptive Management

Adaptive management is the final important element of the conservation strategy for the Rio Condor Project. Central to the adaptive management component is a comprehensive monitoring and research program mandated by the conditions of the environmental permit and subject to an annual independent audit.

Monitoring and Research Program

A proposal for a comprehensive monitoring program was originally developed by the ISC, company staff, and the land steward and included as part of the EIS submission. The program addressed issues identified by ISC in its report (Arroyo et al. 1996) and the Santiago Accord as well as data management and evaluation and the critical issue of adaptive change in the monitoring plan itself (Ringold et al. 1996; Franklin et al. 1999a).

Two research projects critical for evaluating the effects of management were referenced but not included in the proposed monitoring plan. One of these was a small watershed experiment based on the Hubbard Brook model (Bormann and Likens 1979) to evaluate long-term sustainability based on nutrient balances.

The second was a replicated harvest cutting experiment to evaluate the success of the silvicultural system (shelterwood with 10 percent retention) at sustaining biodiversity, especially of relatively unknown organisms such as fungi, lichens, mosses, and invertebrates.

The monitoring program volunteered by the company—and the two research projects—is mandated and substantially specified by conditions of the permit issued by CONAMA, which also imposed additional monitoring requirements identified by CONAMA. As noted earlier, the performance of this massive program is not scaled to the actual level of timber harvest but, rather, must be fully implemented with initiation of any timber harvest. There is also a requirement for annual evaluations of the monitoring and research by the independent environmental auditor.

The monitoring program at Rio Condor is large and complex (see Table 15.2). Establishment of permanent sampling sites or stations for repeated long-term measurements is a basic element of the program. Much of the monitoring also involves comparisons of paired treated (harvested) and untreated (control) sites. This greatly increases the potential to distinguish management effects from the natural dynamics of populations and processes. Scientific leadership is provided by academics from Chile and Argentina with many of the field measurements and installations conducted by company staff. Data management is the responsibility of the company.

An investigation of the silvicultural responses of coigue-dominated forest to shelterwood harvesting is also a part of the Rio Condor adaptive management program (see Table 15.2). Experience in harvesting coigue-dominated forests has been very limited, so the potential for natural regeneration is not known. Some experimental studies that demonstrate successful regeneration of coigue is possible are required prior to initiation of timber harvesting in these stands.

However, several problems exist with CONAMA's mandated and precisely specified monitoring and research program:

- The high degree of specificity makes it difficult for the program to be adaptive.
- Some government decisions regarding monitoring appear to be arbitrary and based upon political rather than technical considerations. An example is the expensive requirement for

TABLE 15.2.

Major components of the mandated monitoring program for the Rio Condor Project.

- Meteorological monitoring at three locations (western, central, and eastern) within the project area
- Hydrologic monitoring (flow, temperature, sediment) on five major rivers within the project area
- Aquatic ecosystem monitoring (stream flows, chemistry, and physical properties; habitat structure; biota, including fish and invertebrate populations) on permanent paired stream reaches (control and harvested) at eight locations within the project area
- Harvest impacts on physical and chemical properties of soils (detailed studies at three locations)
- Growth and mortality of residual stand and regeneration of trees, including browsing impacts of Guanaco, on harvested sites (permanent treated and control areas)
- Harvest impacts on mammal populations (including the red fox) and bird populations through the establishment of permanent treatment and control sites
- Succession, population dynamics, and growth of untreated natural forests
- Assessment of Guanaco population dynamics
- Assessment of invasive plant species in roaded and harvested areas
- Harvesting experiment (replicated) to assess effectiveness of selected silvicultural system in sustaining biological diversity, including fungi, lichens and mosses, and invertebrates
- Watershed experiment to assess impacts of harvesting regime on overall long-term nutrient balance of forest ecosystem and on structure and function of associated small stream ecosystems
- Coigue harvest study to determine regeneration requirements and capability of coigue-dominated forest

monitoring hydrology and water quality in five large rivers in a landscape in which turba and beaver are dominant river influences. These effectively uncouple the large streams and rivers from activities in, and influences of, upland forests.

- Mandating the two scientific experiments forces the company to take direct responsibility for them rather than delegating this work to an independent academic group.
- The lack of scaling in monitoring requirements results in high overhead costs that penalize the company if it chooses—as it has—

to limit rather than maximize allowable timber harvests; the direct costs of the monitoring, research, and auditing activities are currently estimated at about $US3 per hectare, or $US850,000 per year.

Adaptability in a monitoring and research program is critical (Ringold et al. 1996; Franklin et al. 1999a). Interpreting the ecological significance of the data is a major challenge because there is little scientific basis for identifying important thresholds of change. Elements of the monitoring program need to be continually evaluated for their relevance and sensitivity. Hence, an interdisciplinary, inter-institutional assessment of the monitoring program will be carried out every five years.

Evolution of the Rio Condor Project

The Rio Condor Project has been a continuing study in adaptation since its inception. It has responded to many stimuli, including potential commercial markets for wood products and carbon, financial considerations, a long and evolving environmental analysis and permitting process (culminating in a permit, albeit one with a large and highly specified number of conditions), and challenges from some environmental organizations.

On-the-ground implementation of the Rio Condor Project began in October 1999 with simultaneous initiation of the logging and the environmental monitoring and auditing program. It continues to evolve, with major proposals made for carbon sequestration and for the value-added manufacture of wood products. Unfortunately, the carbon proposal, which was focused primarily on preservation of old-growth forests, is not eligible for carbon credits under the current guidelines (Marrakesh meeting in 2001) for Kyoto Treaty implementation.

Environmental Controversies

One would assume that environmental organizations would strongly support a proposal for an innovative industrial forestry project such as Rio Condor. This has not been the case. The Rio Condor region is beautiful and a relatively wild and unmodified landscape dominated by natural old-growth forests. For these reasons, there has been significant orchestrated opposition to the Rio Condor Project (including legal challenges) from some environmental organizations. These organizations and individuals are opposed to any timber harvest activity in the region, regardless of its design and environmental safeguards. Many are committed to ending all timber harvesting south of 44°S latitude and creating an international Gondwandaland National Park. Clearly the issue here is preservation of large wildland landscapes rather than conservation of forest biodiversity. No credible scientific challenge has been raised to the Rio Condor Project regarding conservation of forest biodiversity, particularly in view of the comprehensive multiscale approaches that are part of the project design. The establishment of 68,000 hectares of permanent ecological reserves is unprecedented and provides the only ecologically significant lenga reserves in Tierra del Fuego. Furthermore, a recent scientific analysis of forest conservation needs in South American temperate forests (Armesto et al. 1998) shows the major threats to forest biodiversity are in central Chile and northern Argentina and not at high latitudes and altitudes, including Tierra del Fuego.

Some environmental organizations have remained neutral with regard to the Rio Condor Project, recognizing the value of an innovative sustainable forestry project, the significant contributions already made to conservation of forest biodiversity (and the potential for more), and the locations where there are truly critical threats to Chilean forest biodiversity. CODEFF (National Committee for the Defense of Fauna and Flora) is one such organization.

Initiation of Harvesting and Environmental Monitoring Program

Timber harvesting began in October 1999 at very modest levels. Approximately 1,500 hectares of forest were treated with the first entry of the approved silvicultural system of shelterwood with 10 percent aggregated retention. The harvest was divided between the western region near Puerto Arturo and Vicuna in the eastern portion of the property. No additional harvest cutting is planned for at least two years, although logs felled along roads and within the forest may be uti-

lized in milling operations. To illustrate the contrast between current activities and the original plans, annual harvest levels approved in the environmental permit would eventually reach 2,368 treated hectares and 543,000 cubic meters of harvested wood annually; the monitoring program was actually planned for this level of activity.

The environmental monitoring and auditing program also began in October 1999. The first audit by the independent auditing team was concluded with a positive report in April 2000. The monitoring and research program will continue to develop as negotiated between Trillium and CONAMA. Design work is under way on the first replicate of the harvesting-biodiversity experiment. Topographic and boundary surveys have been completed for the experimental watersheds in Vicuna.

Development of Carbon Proposal

Development of a proposal for a carbon offset project under the Kyoto Treaty is a recent and potentially profound evolutionary stage of the Rio Condor Project. Approximately 5.2 million metric tons of carbon are being marketed based upon carbon emissions reduction through ecosystem preservation and reduced impact forest management. Under this proposal, the area of permanent ecological reserves at Rio Condor would double (to approximately 136,000 hectares) and the amount of included commercial forest would increase from 14,000 to 35,000 hectares. Additional carbon credits would accrue by increasing the level of aggregated retention in harvested areas from 10 to 20 percent.

This project has been approved by responsible agencies in both Chile and the United States. SGS, an environmental auditing firm, has certified that the project will prevent emissions that would otherwise have occurred and has verified the accuracy of the carbon calculations. Unfortunately, the proposed project is not of the type approved for recognition under the current agreement regarding forestry carbon projects allowed under the Kyoto Treaty, even though it would preserve native old-growth forest. The carbon could be marketed outside the provisions of the Kyoto Treaty, and Trillium is strongly committed to carbon marketing on the property, regardless of what happens regarding the Kyoto Protocol.

PART IV Adaptive Management and the Human Aspects of Matrix Management

Making resource management decisions, including developing and adopting policies for the conservation of biodiversity, is a complex social science offering challenges of comparable or greater magnitude than those associated with the natural sciences. Society makes all of the fundamental decisions with regard to conservation policies. Directly or indirectly, consciously or unconsciously, society sets the goals, determines their relative priorities, defines acceptable levels of risk (or certainty) regarding various outcomes, and determines the importance of maintaining options. This is accomplished through a diversity of mechanisms, including tradition, economic policy and markets, adoption of various political and legal positions, and development of international agreements. In some democratic societies, the legal and social contracts are often adjudicated by the courts, even to the degree that courts may decide what constitutes a scientifically credible conservation plan, an acceptable standard of certainty or risk in efforts to sustain species, and a sufficient monitoring program. Society also influences policies by the level of financial support it provides for resource management, monitoring, and scientific research and by the degree that it chooses to enforce laws and agreements that it has adopted. Hopefully, society has available to it comprehensive scientific information relevant to the questions at hand rather than an abridged information base and that it utilizes such information in making informed decisions regarding policies. Scientists and resource managers have an important role in providing society with the relevant knowledge, including the potential and limitations of the relevant ecosystems and organisms and sources and levels of uncertainties and risks.

Resource management involves informing and implementing societal decisions. Central to resource management is the recognition that adopted management approaches—from the level of the large regional plan to specific management prescriptions for a single, local project—are working hypotheses. Knowledge is always limited, and unpredicted events (human and natural) will intervene so that there are always substantial uncertainties about the outcomes. The reluctance of resource professionals—be they foresters, fisheries biologists, wildlife managers, or conservation biologists—to acknowledge the substantial uncertainties associated with their proposals has obscured the reality that adopted policies and practices are working hypotheses rather than firm statements about outcomes. This is highly relevant to conservation of biodiversity because, fundamentally, *conservation biology is theoretical and the effectiveness of most conservation policies and programs is largely unproven.*

Adaptive management acknowledges

- uncertainties inherent in resource management
- our intuitive understanding that we are going to learn more about the ecosystems and organisms of interest
- the value of systematically engaging in the collection and incorporation of that new knowledge

In the remainder of this book, we broadly address the application of adaptive management to the management of biodiversity in forest landscapes. This is, of course, only one subset of the social issues related to development and implementation of policies with regard to conservation of biodiversity and ecosystem processes. Elements of adaptive management include a review of the basic philosophy of adaptive management and the role of long-term monitoring (Chapter 16) and of scientific research (Chapter 17) in the continual evolution of improved (more effective) management. The development and implementation of meaningful monitoring programs is particularly challenging for managers, policy makers, scientists, and stakeholders worldwide. Statistically valid forest monitoring programs are currently very rare, especially in view of the central role of monitoring in assessing the sustainability and effectiveness of any management program in achieving its goals.

Additional social dimensions in matrix management are considered in Chapter 18, including fundamental challenges to the notion that matrix management is an essential component of a comprehensive strategy for conservation of forest biodiversity. If matrix-based approaches are to be broadly adopted, many stakeholder groups (including those from both NGOs [nongovernmental organizations] and resource-based corporations), resource managers, and scientists will need to modify strongly held positions. General conclusions about the importance of the matrix for the conservation of biodiversity are provided in the concluding chapter (Chapter 19).

CHAPTER 16

Adaptive Management and Long-Term Monitoring

Adaptive management . . . "has a ruthless hold on uncertainty."

—DAVIS ET AL. (2001)

Management for the conservation of forest biodiversity must be based on collection of new data to bridge gaps in the existing knowledge base. Adaptive management is needed to ensure that conservation strategies continue to evolve and progress. Monitoring will be an integral part of any adaptive management framework—it is a fundamental part of ecologically sustainable forestry. All stakeholders, managers, and decision makers need to appreciate the critical role of monitoring and work to ensure that it is adequately and consistently funded. It is impossible to systematically assess whether management goals are being achieved without adequate monitoring, which, in turn, ensures that the effectiveness of policies, legal obligations, and social commitments to sustainable forest practices can be assessed.

Adaptive management is an acknowledgement of the uncertainties inherent in management of biological resources, the certainty that more will be learned about these resources and their management responses, and the value of systematizing the learning process. It is adaptive because "it acknowledges that there will always be unpredictability and uncertainty in managed ecosystems, both as humans experience new situations and as these systems change through management" (Davis et al. 2001).

Adaptive management is directed toward acquiring new knowledge from experience, monitoring, and research, and integrating that information into improved (more effective) management practices. Adaptive management involves a systematic approach to the processes of acquiring and incorporating new in-

formation, but there are wide differences of opinion (as illustrated below) regarding the degree to which the process is formalized and, especially, the requirement for focused experimentation. We present both the broader perspective and the more formalized perspective on adaptive management.

Monitoring is an essential element of any adaptive management program and, more broadly, the basis for any objective assessment about the level of success achieved by a management program. Although it might be assumed that management organizations would want to have objective measures of the effectiveness of their programs, remarkably little effort has been expended on monitoring. Government agencies, companies, and other organizations managing lands typically have not acknowledged the importance of monitoring nor provided sufficient, sustained support for monitoring.

Quantitative long-term data on the effectiveness of the general concepts underlying conservation programs, including matrix management, are lacking (McComb et al. 1993; Hobbs 1997; Lindenmayer 1999a). Yet, effective management of any ecosystem requires at least a rudimentary understanding of major components, processes underlying ecosystem function, and ecosystem responses to disturbances. For most forests, such knowledge is limited. Additional information is needed to better inform and improve matrix management (Carey 2000). In this chapter, we outline an adaptive management approach to enhance the collection of data to better inform conservation management.

Adaptive Management: Gathering and Applying New Knowledge to Managing Forest Landscapes for Biodiversity

Adaptive management is the acquisition of additional knowledge and the utilization of that information in modifying programs and practices so as to better achieve management goals. Potential sources of information include management experience, monitoring, and research. Adaptive management requires an active, planned, and systematic effort to acquire and utilize information, regardless of its specific source; implicit is a comprehensive, well-documented approach to the management of records and other data.

Analyses of the major uncertainties associated with management plans and prescriptions are major components of adaptive management. As described by Davis et al. (2001):

> The process of adaptive management includes highlighting uncertainties, developing hypotheses around a set of desired system outcomes, and structuring actions to evaluate or test these ideas. Although learning occurs regardless of the management approach, adaptive management attempts to make that learning more efficient.

Management activities are viewed as experimental tests of hypotheses about ecosystem responses or species responses under adaptive management. One difference between broader and narrower interpretations of adaptive management is that the narrow definition (see below) requires an active program of rigorously testing alternative management proposals, while a broader interpretation allows more passive approaches in which a variety of data is collected and utilized to inform decisions.

Four key elements of adaptive management are identified by Davis et al. (2001):

1. A ruthless hold on uncertainty . . . [starting] with an acknowledgement of the uncertainties associated with proposed management policies.

2. The description of key management policies as testable hypotheses.

3. The search for, and use of, information that will enable testing the hypothesis or hypotheses . . . [which] can range from informal observations of foresters and other specialists . . . to formal replicated experimental design, but does require a conscious attempt to assess the validity of the hypothesis or hypotheses in question.

4. An institutional mechanism that ensures that the hypotheses will undergo periodic, fairminded review and that management policies can change as a result of the review.

The first of these elements—a "ruthless hold on uncertainty"—is one that makes adaptive management threatening to many stakeholders. While uncertainties associated with management plans may be obvious and the acquisition of additional relevant information intuitive to most stakeholders, accepting adaptive management also means accepting the impermanence of decisions with regard to both management approaches and specific resource outputs. This is disconcerting to stakeholders who would like assurances that timber harvest levels will be sustained or ecological reserve boundaries maintained in perpetuity and not constantly reassessed as the result of new information. Hence, while adaptive management appears logical and desirable in prospect, stakeholders often find it difficult to accept the practical consequences of an adaptive management policy. The struggle to maintain the flexibility needed for adaptive policies is very difficult in the face of stakeholders' desire for assured outcomes, as noted by Davis et al. (2001), who propose the use of outside review as one mechanism to overcome the natural tendency of people and organizations to defeat attempts to embrace adaptive management.

A Formalized Approach to Adaptive Management

Several scientists have defined a highly formalized approach to adaptive management (Holling 1973; Walters 1986, 1997; Walters and Holling 1990). They propose a highly integrated approach involving research, monitoring, and management designed to assess (test) and improve the effectiveness of resource management prescriptions (Shaw et al. 1993), which obviously encompasses programs and projects designed to maintain biodiversity on forest landscapes (including the matrix).

An adaptive management system is defined by Holling (1973) as one that "can absorb and accommodate future events in whatever unexpected form they may take." Experimentation is viewed as a core element in the formalized approach to adaptive management, essential to an improved understanding of a system that will make improved management possible. In these experiments with the system, the goal is to learn as much as possible from both successes and mistakes (Taylor et al. 1997). The adaptive management framework does allow natural disturbances and human activities to be utilized as experimental opportunities (for example, analysis of conditions created by and responses of taxa to logging; Lindenmayer and Franklin 1997a) and findings incorporated into management.

However, in the formalized approach to adaptive management, the method is a rigorous process and not a trial-and-error approach (Walters 1986). Trial-and-error management is problematic because policy alternatives are not properly specified. Formalized adaptive management has strict requirements for the documentation of objectives, assumptions, policy options, and outcomes. It is based on clear hypotheses stemming from real policy options informed by previous experience and understanding; however, these hypotheses are not constrained by a requirement that the approaches being tested in the field must work (Walters 1997). This formalized approach to adaptive management increases the likelihood that new knowledge will be generated and subsequently embraced in on-the-ground management (Taylor et al. 1997). Rigorous monitoring is a fundamental requirement of

adaptive management—something that has rarely been accomplished in resource management (including biodiversity conservation) programs (see below).

In implementing a formalized adaptive management program there are a series of logically linked steps (Figure 16.1):

Step 1. Gather all available information about the system. Based on that information, create alternative models regarding management of that system and clarify policies on approaches that will meet management goals, possibly using simulation models (Walters 1986).

Step 2. Create a small set of testable hypotheses for different management options. It may be necessary to stipulate the degree of difference between the several options (Taylor et al. 1997). Sometimes, this step involves consideration of entirely new management approaches (even paradigm shifts) that are outside existing procedures and policies. The potential contribution of scientists at this step should be obvious.

Step 3. Develop an experimental design and monitoring program. The design must specify which system components are to be used as response variables (i.e., measured to assess the success of different management options). A pilot study may be required (Silsbee and Peterson 1993; Urban 2000). A robust experimental design is necessary to avoid the limitations of trial-and-error management.

Step 4. Implement management changes—such as altered reserve designs or silvicultural systems in managed stands—based on the results of the experiments. Monitoring and continuing assessment of the data stream continue with regard to the modified management strategies. Thus, continued and iterative field research is coupled with result-driven management actions.

Step 5. Carefully document the adaptive management program, including detailed information about all the steps of the process (Taylor et al. 1997).

The success of any adaptive management study depends upon two important contingencies: (1) management actions implemented now must maintain as many future management options as possible, and (2) tight linkages and feedbacks must be maintained be-

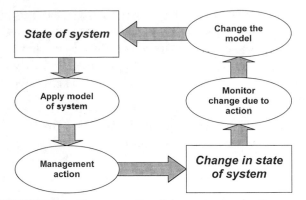

FIGURE 16.1. The connecting linkages in an adaptive management framework. Redrawn from Walker (1998). Reprinted with the permission of CSIRO Publishing.

tween scientists and managers (i.e., scientific research and monitoring) and also with policy decision makers (Dovers and Norton 1994).

The linkage to policy is fundamental, ensuring that the results of research are not lost (e.g., Underwood 1995). If new information is not considered and appropriately acted upon, the linkages in an adaptive management framework and knowledge hierarchy are broken. As noted by Hilborn (1992): "If you cannot respond to what you have learned, you really have not learned at all."

Adaptive Management and Stand- and Landscape-Level Studies

Along with Davis et al. (2001) and other authors, we take the position that relevant knowledge for adaptive management can be acquired from a variety of sources and can incorporate less-formalized approaches. Sources of information can include routine management treatments (especially if they incorporate randomly selected control or untreated sites), routine (but well-designed) monitoring programs, and retrospective studies of responses to past human and natural disturbances. However, well-designed experiments (see Box 16.1) can have extraordinary value in adaptive management, although the costs and other logistical difficulties associated with large-

BOX 16.1.

A Hypothetical Example of an Adaptive Management Natural Experiment

This box contains a hypothetical example of an adaptive management program for wood production areas in which plausible options for different silvicultural systems are tested. After progressing through the initial steps outlined above and defining the entities targeted for measurements (such as particular elements of the biota), a range of logging regimes is identified. Some areas might be clearcut while others are selectively harvested, and varying levels of stand retention could be employed in various other places subject to logging (e.g., the variable retention model described in Chapter 8). In addition, "control" areas with no timber harvesting (e.g., those in large ecological reserves) might be necessary to interpret the results.

Variation in logging regimes is critical because it provides the variation in stand conditions needed to test different silvicultural systems. It is also important because the effects of natural disturbances like wildfire on landscape pattern and structural conditions can be variable (Lorimer and Frelich 1994, Peterson and Pickett 1995; see Chapter 4). A major problem for adaptive management studies in many parts of the world is that the same method of logging is deployed throughout a given forest type leaving little or no variation in silvicultural systems to examine. For example, more than 98 percent of Victorian alpine ash (*Eucalyptus delegatensis*) forest is clearcut (Lutze et al. 1999) and alternate methods of harvest are very rarely contemplated.

Although variation in cutting regimes is essential, it is also important to have replicates of these "treatments" to provide some statistical power in the experimental design and to allow for valid contrasts between them. Data on the intensity of disturbance and the extent of vegetation retention on logged sites have to be carefully recorded and stored in a database. This is because activities such as site preparation, the retention of biological legacies, and regeneration methods can significantly influence the trajectory of stand recovery and the organisms associated with such processes (Hazell and Gustafsson 1999; see Chapter 4).

At all times, information on target species or key ecosystem processes gathered from monitoring studies would be used to assess the efficacy of logging prescriptions (e.g., the extent of canopy retention needed across the landscape, the width of wildlife corridors, or the number and spacing of trees retained within cutover sites). New sites where modified prescriptions and strategies are applied (based on the results of the adaptive management program) are then added to an expanded logging and monitoring study.

FIGURE 16.2. Experiments are important elements in adaptive management, as exemplified by the MASS (Montane Alternative Silvicultural Systems) study established cooperatively by industry, government, and academic institutions on Vancouver Island, British Columbia, Canada. Photo by J. Franklin.

scale, long-term experiments are always going to limit their number.

Some large, long-term forest management experiments are under way in production landscapes around the world (Figure 16.2), including in North America (Carey 1994; Arnott et al. 1995; Halpern and Raphael 1999) and Patagonia (Pickett 1996). Excellent examples of such experiments that incorporate different timber harvesting regimes include

- Demonstration of Ecosystem Management Options (DEMO) experiment in the Pacific Northwest region of the United States
- Montane Alternative Silvicultural Systems (MASS) experiment on Vancouver Island in British Columbia, Canada (Phillips 1996)
- Sicamous Creek Silvicultural Systems Project in interior British Columbia (Hollstedt and Vyse 1997)
- Warra Silvicultural Trial in Tasmania (Hickey et al. 1999) (see Box 8.2 in Chapter 8)

Advantages of these large-scale, long-term experiments are that they provide practical examples of how silvicultural systems can be altered and function as pilot studies for silviculturalists as well as provide empirical knowledge about ecological responses (Harris and Farr 1974; Beese and Bryant 1999). Unfortunately, long-term experiments of this type are often

very difficult to sustain as political and organizational emphases and budget allocations shift. An example is the Silvicultural Systems Project (SSP) in Victoria, southeastern Australia (Squire et al. 1987), which was significantly downsized when funds to maintain and measure the experiment were reduced. The treatments imposed in SSP were also confined to traditional silvicultural approaches, which has limited the application of the results, although this was not a factor in its demise (Lindenmayer 1992; see Chapter 12).

Although most adaptive management studies have been at the stand level, there have been calls for rigorous landscape-scale experiments (Wiens 1992). A few large-scale experiments have commenced (e.g., Schmiegelow and Hannon 1993; Schmiegelow et al. 1997) but more would be desirable (Simberloff 1998). For landscape-scale adaptive management experiments to work, it is essential that, as in the case of stand-level experiments, (1) each landscape pattern be replicated to provide statistical power, and (2) experiments include contrasting landscape patterns or treatments.

Adaptive Management and the Lack of Certainty

The implementation of iterative learning by adaptive management and monitoring has no defined endpoint. There is no single best (correct) silvicultural model or final set of prescriptions for matrix management—management actions will continue to change in response to new information and insights. This is fundamental to the concept of ecologically sustainable forest management. Adaptive management in forests is a long-term process in resource management; as noted by Walters (1986), "there is no point in learning about something you intend to destroy shortly."

As noted earlier, the concept of continuous change is a difficult one for forest managers, conservationists, governments, and other parties with interests in forest management. These groups typically seek certainty in management outcomes (Walters 1986). Governments are also eager to have certainty because they see this it as a way to resolve difficult and often socially divisive forestry and conservation issues. Similarly, some conservationists criticize adaptive management principles as merely a strategy by commodity interests to prolong high harvest levels and delay much-needed changes to silvicultural prescriptions. Taylor et al. (1997) correctly noted that adaptive management is not an excuse to maintain the status quo of some management practices, particularly ones with potentially negative impacts.

Certainty in management outcomes and access to timber resources can strongly influence private and government financial investment. Calculations of sustained timber yields assume some level of long-term access to exploit the forest estate. The need for certainty is one of the main reasons many forest management disputes are resolved by land allocation—assigning some areas reserve status and allowing wood production in others (see Chapter 18).

The tension between the perceived need for certainty and the need for continuous change in management is a challenge to the adoption of adaptive management as an integral part of modern forestry. Another challenge is the development of protocols or frameworks for making changes in prescriptions once they are found to be deficient. Changing traditional or interim prescriptions can be difficult and often controversial.

Adaptive Management and the Precautionary Approach to Forest Management

Adaptive management requires a more (rather than less) cautious approach to the use of forest resources because of the need to maintain future options and the difficulty of re-creating these resources quickly once they are gone. Adaptive management in forest management also requires that stakeholders accept and cultivate a philosophy of change as a natural part of resource management and conservation. Managers must not be penalized for trying approaches that are later shown to fail (Taylor et al. 1997). As noted by Williams and Johnson (1995), "ultimately, success in adaptive management harvest regimes requires, more than anything else, an institutional

framework that embraces patience, persistence, and commitment."

Although the adaptive management model is a potentially valuable one, it remains largely untested (Dovers and Lindenmayer 1997; Simberloff 1998), with few examples of its successful application (but see Taylor et al. 1997; Innes et al. 1999). Of course, this may be due to the relatively recent emergence of the concept; certainly, some very serious efforts to implement adaptive management are under way by both government and private organizations (see below). Some authors have recommended alternative frameworks to better interface science, management, and policy (e.g., Rogers 1997). Nevertheless, future experiences will determine if adaptive management can be truly effective.

Governments and management agencies need mechanisms to respond to new information and to institute changes rapidly and, often, frequently. Adoption of static forest agreements (such as the Regional Forest Agreement process in Australia) does not resolve issues regarding whether or not forests are being managed in an ecologically sustainable way (see Box 5.5 in Chapter 5).

There are now numerous and significant efforts to implement adaptive management strategies by a wide variety of forest management organizations, including government agencies, industrial forestry organizations, and other institutions. Adaptive management is a central element in the Northwest Forest Plan adopted for federal lands in the northwestern United States (Forest Ecosystem Management Assessment Team 1993; Tuchmann et al. 1996). One specific land allocation in this plan is *adaptive management areas*—locales (totaling more than 500,000 hectares) where experimentation and innovation are emphasized. Research and monitoring are also central elements in the Northwest Forest Plan (e.g., Haynes and Perez 2001). In addition, failure to conduct adequate monitoring programs would be a basis for challenging the legal validity of the plan.

Some comprehensive adaptive management programs have also been developed in connection with industrial forestry projects. For example, adaptive management is central to the Rio Condor Project in Tierra del Fuego (see Chapter 15), where it is mandated as a part of the government environmental permit and is central to eventual green certification. The financial obligations associated with monitoring, research, and other adaptive processes at Rio Condor are significant.

The BC Coastal Forest Project of Weyerhaeuser Company has developed (and is implementing) an extraordinary adaptive management program for forest resource management in connection with a shift from clearcutting to variable retention harvesting. This program includes extensive monitoring, research, and large-scale experimentation. There are also major interactions with scientists and stakeholders, including regular scientific reviews. Collaborative efforts with employees have also been an important part of the process.

Long-Term Monitoring: A Fundamental Part of Responsible Management

The major reason for monitoring is to assess the effectiveness of any resource management program and, ultimately, improve management. Most fundamentally, monitoring is necessary to generate the empirical data that are the definitive measure of the degree to which a management program is achieving its objectives. Although monitoring is a fundamental part of any adaptive management strategy, it is necessary for many other reasons, such as to comply with requirements in government plans, permits, and agreements (e.g., habitat conservation plans), meet certification requirements, or fulfill other legal obligations. For example, monitoring is one of the key elements for certification under the standards of the Forest Stewardship Council (1996). It is also an integral element of any responsible business plan, particularly with regard to periodic assessments of the condition and value of assets.

There is no way to determine whether conservation-oriented management strategies are effective without rigorous monitoring, hence the interrelationship between monitoring and adaptive management. Despite the fundamental importance of monitoring, relatively few credible long-term forest monitoring programs are actually under way anywhere in the world. For example, almost half of the monitoring programs initiated in New Zealand were unreported or not completed, suggesting a high rate of failure of monitoring projects (Norton 1996). We do not consider traditional continuous forest inventories, which have focused primarily on timber growth and wood volumes, to be credible ecosystem monitoring programs, although significant expansions to incorporate other ecological parameters have occurred in many inventory programs in recent years.

There are many reasons why adequate monitoring programs are difficult to develop and to successfully and continuously implement.

First, meaningful (credible) and practicable monitoring programs are difficult to design (see below) and expensive to implement. Many monitoring programs are poorly designed and, consequently, their outcomes have limited value for conservation management. For example, the field methods used often fail to address the questions posed (Norton 1996).

Second, objectives of monitoring programs are often poorly defined, explicit statements regarding expected outcomes are lacking, and the relevance to forest management is unclear (see Roberts 1991; Macdonald and Smart 1993). Poorly planned and unfocused monitoring programs are ineffective and often fail completely (Orians 1986).

Third, monitoring is often considered a routine activity that does not merit significant management effort (i.e., an activity that does not contribute to achieving immediate goals) or scientific input—viewpoints often held by both managers and scientists. There may be fears that results of monitoring will put favored management objectives or programs at risk.

Fourth, data management commonly receives grossly inadequate attention. Adequate management of long-term environmental datasets is very challenging and requires substantial technical expertise and significant financial support (Michener and Brunt 2000). Few monitoring programs have employed adequate data management procedures, such as appropriate standards for data documentation, and few have provided adequate financial support, which probably should average 20 to 25 percent of the monitoring program budget. Again, neither managers nor scientists generally appear to have recognized the technical challenges, costs, or critical role of data management, preferring to utilize financial resources for other priorities. As a consequence, immense amounts of important environmental data, including long-term records of management activities in forest landscapes, have disappeared or effectively have been lost due to inadequate documentation and quality control.

Fifth, adequate and sustained funding for monitoring programs is rarely available. Since monitoring is directed toward assessing rather than carrying out management programs, funds for monitoring typically have low priority. Even if adequate funds are provided to plan and initiate a monitoring program (a rare occurrence), most programs have subsequently been starved for financial and logistical resources. It is very difficult to assure that adequate financial resources will be continuously available. Most government agencies and corporations have annual budget cycles that make monitoring programs, which are necessarily long term, vulnerable to budget cuts as short-term shifts occur in organizational priorities and objectives. Development of trust funds is one way to deal with this problem, but we know of only one organization that has adopted this approach (The Nature Conservancy).

And, finally, failure of an organization to institutionalize a monitoring program can also be a problem when the initial momentum for its establishment is provided by a single individual or a small group. When the dedicated individual or cadre disappears from the organization, the program may languish.

Some of these impediments to monitoring programs are being overcome as monitoring moves from an optional to a required activity, such as to fulfill legal and market-based requirements for credible assessments of the effects of forest management activities on biodiversity and other environmental variables. In effect, the question is increasingly not one of whether monitoring will take place but, rather, how programs will be designed and implemented. Monitoring *will* be carried out, and the courts and the marketplace often will be the ultimate judges of its credibility.

Design of Monitoring Programs

Much of the following section has been adapted from the article "Complementary Roles of Research and Monitoring: Lessons from the US LTER Program and Tierra del Fuego" (Franklin et al. 1999a).

There are many ways of categorizing monitoring programs. In the United States, three major types of monitoring have been recognized:

1. *Implementation monitoring*, used to determine whether the types and levels of activities stipulated under a management plan are actually conducted, such as leaving x numbers of live trees and snags behind on y cutover areas
2. *Effectiveness monitoring*, used to determine whether or not the management plan has accomplished its resource goals, such as sustaining viable populations of species z
3. *Validation monitoring*, used to determine whether the accomplishment of specific goals was actually a consequence of the management activities that were undertaken

These categories of monitoring involve very different levels of complexity, although all three are essential elements of an adaptive management program, and the first two categories are typically legally mandated or market-mandated monitoring programs. Implementation monitoring is very straightforward: observing and recording whether or not you did what you said you were going to do. Effectiveness monitoring is much more challenging: ascertaining whether or not you accomplished your resource goal. Most of what we have to say about the design of monitoring programs later in this section is focused on effectiveness monitoring. Validation monitoring is the most challenging category of monitoring because it involves establishing causal relationships between some management action(s) and an environmental response. Most validation monitoring will be indistinguishable from classical scientific research since it involves hypothesis-based and rigorous experimentation.

The Role of Science in the Development and Implementation of Monitoring Programs

Traditionally, monitoring has been viewed as a management activity unrelated to scientific research. Although for many years scientists have stressed in numerous scientific articles and books the importance of monitoring (e.g., Goldsmith 1991; Noss and Cooperrider 1994; Spellerberg 1994), they have often refused to involve themselves directly in monitoring, viewing it as the routine collection of data for nonscientific purposes. This prejudice of the scientific community is based on a variety of factors (see Goldsmith 1991; Hellawell 1991), particularly the failure of academic reward systems to recognize and credit involvement in adaptive management experiments and monitoring programs (Taylor et al. 1997). The fact that many resource managers do not consider science or scientists to be essential participants in development and operation of their monitoring programs is another factor.

Monitoring programs must be collaborations between managers and scientists in our view. Extensive scientific involvement is necessary in the development, operation, and interpretation of credible natural resource monitoring programs (Franklin et al. 1999a). Results of scientific research and scientific expertise are needed in at least four major aspects of monitoring (Goldsmith 1991; Sparrow et al. 1994; Morrison and Marcot 1995; Franklin et al. 1999a): (1) design of monitoring programs, including selection of parameters and development of the sampling design (where, when, and how to sample as well as details of the statistical design), (2) quality control, (3) interpretation of results, and (4) periodic assessments of the effectiveness of the monitoring program.

In addition, direct and extensive involvement of scientists is going to be necessary in almost all validation monitoring efforts. The need for scientific rigor in monitoring programs is actually very high in order to produce robust and defensible results that provide stakeholders (including decision makers) with clear evidence as to whether or not changes in management practices or policies are needed. Our views on the relationship of science to monitoring differ strongly with Hellawell (1991), who contends that research is not relevant to monitoring. In contrast, Franklin et al. (1999a) cite the example of monitoring long-term nutrient balances and changes in soil productivity— complex mainstream research issues—which are identified as fundamental elements for monitoring under such sustainability protocols such as the Montreal Process.

The interpretation of results from monitoring programs can be extremely challenging. For example, separating long-term population declines from annual

FIGURE 16.3. Monitoring is a critical element in adaptive management, providing stakeholders with major challenges in design, implementation, and interpretation; significant long-term investments are typically necessary, such as installations for continuous hydrological monitoring and water sampling (stream gauging station at H. J. Andrews Experimental Forest, western Cascade Range, Oregon, United States). Photo by J. Franklin.

and cyclical fluctuations in population size is a nontrivial task, particularly since populations of some species can recover relatively quickly following human disturbances, such as logging. Franklin et al. (1999a) noted that carefully collected monitoring data could be extremely valuable for research (e.g., documenting long-term trends in ecosystem dynamics). For example, long-term monitoring data coupled with new analytical methods helped identify the cumulative effects of road networks and wood production on hydrological regimes in the Pacific Northwest (Jones and Grant 1996) (Figure 16.3). Indeed, rather than treating monitoring as second-rate science, Walters (1997) believes that researchers should be attracted to the prospect of conducting large-scale, well-designed, and replicated studies of direct relevance to natural resource management.

General Observations about the Design of Monitoring Programs

Designing a monitoring program involves difficult decisions about what parameters are to be monitored and how, when, and where measurements are to be made. We believe that those who suppose that these critical questions will be resolved for them, perhaps with some prescriptive guidebook, are going to disappointed. Standardized monitoring programs may be characteristic of large-scale, national, and international programs intended to assess environmental conditions and trends related to development of national and global policies. However, most monitoring programs are directed to assessing specific accomplishments and impacts of smaller-scale plans or programs. Examples include monitoring associated with plans for individual resource management units (national forests, national parks, corporate tree farms, etc.), with habitat conservation plans, or for regional collections of resource management units, such as the Northwest Forest Plan for management of U.S. government lands within the range of the northern spotted owl. The monitoring may be related to a specific organism of interest or overall environmental impacts of a specific management activity or project.

Our experience is that monitoring programs focused on regional and local projects and management units are likely to be highly idiosyncratic and not amenable to design using textbook models (Goldsmith 1991; Franklin et al. 1999a). Although the broad categories or topical areas of monitoring may be similar in many of these programs, the appropriate selection of parameters may vary with local circumstances, sometimes widely. This variation may be related to ecological, social, or economic circumstances as well as to legal and policy demands. Similarly, appropriate answers to the questions of when, where, and how to sample will often vary among projects even when the same parameter has been selected.

Design issues are further complicated by the fact that monitoring programs are not likely to have a hierarchical structure. While the concept of a nested monitoring program is very appealing—in other words, one with a design in which sampling for various parameters uses a common design (such as a grid of sampling points)—such designs are rarely feasible. Most operational monitoring programs incorporate parameters that are measured on widely contrasting spatial and temporal scales and, consequently, using different methodologies or technologies. For example, a monitoring program may include periodic assessments of population dynamics of individual species (annual), stream flow (continuing), land-use patterns (periodic), and natural disturbances (periodic or episodic).

The Specifics of Designing Monitoring Programs

Major challenges in the design of a monitoring program involve decisions about the parameters (what), spatial and temporal scales of sampling for each of the parameters (where and when), and issues of statistical design and selection of sampling methods (how), such as specific instrumentation, plot design, and sensors. Scientific research and expertise are key to all of these aspects of design (Franklin et al. 1999a).

Selection of parameters is the first of the challenges. Almost any stakeholder can come up with a shopping list of what should be monitored. Such lists are only useful as starting points in planning; they do not reflect realities of operating or financing a monitoring program. In making the selection of parameters, it is important to determine which are likely to be sensitive indicators of important ecological conditions (i.e., which are ecologically meaningful). There are also important practical realities to consider, such as whether a candidate parameter can be readily measured using existing technologies and, if so, what it will cost to make these measurements. Monitoring programs must, ultimately, focus on a subset of attributes rather than attempt to monitor everything (Zeide 1994; Lindenmayer 1999b).

Practicality will often lead to the selection of surrogate parameters, such as structural attributes of forest stands, as substitutes for difficult and expensive species-by-species inventories. This approach is justified by the strong ecological links that have been established in many forest ecosystems between a wide range of aquatic and terrestrial taxa and ecosystem processes and structural features. As an example, the level of coarse woody debris is a relatively easily measured structural parameter that is related to many biotic components (Harmon et al. 1986; Maser et al. 1988). This makes it a valuable candidate for monitoring. In some cases, the response of taxonomic groups for which such structural objectives are set could be included in the monitoring programs. For example, Carey (2000) proposed that monitoring responses of squirrel assemblages is a valuable way to assess effectiveness of management regimes. Particular species may be inappropriate for monitoring if they are rare, however, because data will be too limited to detect trends; more common taxa for which large datasets can be readily gathered sometimes reveal more about the state of an ecosystem.

It should be noted that some parameters will be selected because the organisms or processes are of special interest to society regardless of whether they are useful environmental indicators (e.g., populations of some large, rare, and charismatic vertebrates). This is because society as well as technicians has a say in selection of parameters and, further, society's say is often legally mandated.

Once a monitoring parameter has been selected, the next challenge is to develop a sampling design—formalizing the answers to where, when, and how in a statistically robust design. Developing sound estimates of parameters with sufficiently low error terms that a statistically significant change might actually be identified raises difficult questions. Where will the sampling take place (which may have the sub-question of where are sensitive locations within the landscape?) Answers to such questions are not always as obvious as they might appear (see Franklin et al. 1999a and Chapter 15 for an example with regard to effects of forestry activities on water quality in Tierra del Fuego). Temporal issues involve decisions about the sampling intervals that are to be used—continuous, daily, annual, multiyear; a broad range of sampling intervals are likely to be used in programs that are focused partially or totally on the conservation of biodiversity.

The where and when of a quantitative monitoring program must ultimately be formalized into a sampling design that provides the basis for statistically valid measures of change. This can be difficult and may in fact be the decisive issue in choosing among proposed monitoring parameters (Hinds 1984). Ultimately, many of the data are going to be subjected to rigorous statistical analyses and thus sufficient statistical power is needed to make inferences about trends, such as population declines, possible (Mac Nally 1997; Burgman and Lindenmayer 1998; Strayer 1999). Related issues of Type I errors (a false declaration of impact when none exists) and Type II errors (a false declaration that no change has taken place or that an observed change is random) are also fundamental to the design of monitoring programs (Fairweather 1991; Mapstone 1995). On this basis, monitoring designs should include wherever possible and appropriate:

- Controls in untreated forest and treatment replicates to take account of spatial heterogeneity and random variation and to allow estimates of error
- Pretreatment monitoring to establish natural trends and any pretreatment differences between plots
- Environmental stratification to provide opportunities to detect interactions between treatments and environmental variables
- Replication at more than one location to avoid location-specific phenomena or geographic bias

Long periods of monitoring may be necessary to distinguish treatment effects from effects of climatic fluctuations or episodic or stochastic events. It is also important to set clear objectives for monitoring and to put protocols in place to ensure that they are achievable. As noted by Roberts (1991), too often monitoring has been "planned backwards on the collect now (data), think-later (of a useful question) principle."

Given that monitoring programs will often be costly and must run for prolonged periods, their design needs to be subject to extensive and rigorous peer review (Macdonald and Smart 1993).

Interpreting the Results of Monitoring Programs

Interpreting the ecological significance of a change in a monitored parameter may be the greatest challenge in a monitoring program. "In a well designed monitoring program, a statistical change may be observed—but is it ecologically significant?" (Franklin et al. 1999a). We will rarely have the level of knowledge on cause-effect relationships in ecosystems that allow for quick and clean inferences from monitoring results. Scientists have a major role to play in this aspect of monitoring because their specialized knowledge can assist significantly in interpretation of results (although, as noted below, all stakeholders must also be a part of the interpretative process and resulting decisions about subsequent changes in management and monitoring). Examples of scientific involvement may be in conducting new research to provide additional information relevant to interpretation of the monitoring program, synthesis of existing information, and participation as consultants and members of expert panels.

When changes are observed in monitored parameters, many questions need to be addressed before decisions are made about managerial or regulatory responses (Franklin et al. 1999a):

- Is the observed change (e.g., in soil nitrogen or organic matter content) real, or is it a sampling artifact?
- If the observed change is judged to be real, then is this change permanent or temporary, and, if temporary, what is the probable rate of recovery?
- What are the potential environmental consequences of a change of the magnitude observed?
- Were similar changes observed in undisturbed (control) environments or only in managed areas?

Many responses to management will be temporary, with recovery processes commencing almost immediately. In such cases, the rate of recovery may be more important that the initial change itself. If the observed change is judged to be real, the next question concerns the potential consequences of that change. For example, what effect will a 2 percent reduction in soil organic matter or a 10 percent increase in bulk density of soil have for long-term site productivity? As noted above, information that allows such interpretations is limited for many parameters and even conflicting in others. Actual thresholds—points at which there are major changes in the relationship between the parameters and the response—may exist for some parameters but not for others (see Chapter 2). High levels of buffering also may allow ecosystems to tolerate short-term shifts without long-term damage. Compensating mechanisms or processes may also come into play.

Monitoring of untreated (control) ecosystems is essential for assessing whether the observed changes are actually responses to treatments or are part of a larger pattern of environmental change. This is particularly critical for parameters that have high levels of year-to-year variability, such as populations of small mammals.

In the preceding paragraphs we have tried to make clear the large challenges associated with *interpreting* the results of monitoring programs. An evaluative process is usually required when monitoring is being conducted to identify patterns and rates of change (trends). There is rarely a scientific basis for establishing thresholds for changes in parameters although it may be appropriate to stipulate standards or thresholds in some parameters, such as density of tree regeneration following harvest (Franklin et al. 1999a).

Monitoring as an Adaptive Process

Monitoring *must* be viewed as an adaptive process; indeed, it may ultimately be the most evolutionary component of an adaptive management program given its central importance both technically and socially. Much is going to be learned with each annual increment of monitoring effort—about the merits of the parameters and sampling schemes selected and about interpretation of data. Associated with that learning will be decisions about managerial responses to monitoring results and modifications (additions, deletions, and changes) in monitoring parameters and protocols. Stakeholders as well as resource managers and scientists must be involved in many of these decisions. Indeed, monitoring organizations need to be aware that regulatory agencies and stakeholder groups will likely insist on access to basic monitoring data; hence, formally incorporating them into the evaluation of both the data and monitoring program is appropriate.

As noted by Franklin et al. (1999a):

Any monitoring program should be viewed as a series of approximations which will be modified periodically as: (1) Initial parameters fail to adequately fulfill our objectives or improved designs and measurement technologies for these parameters emerge; (2) New and improved parameters are identified through empirical or theoretical research or become feasible due to availability of new technologies; and (3) Monitoring objectives change.

Therefore, monitoring programs must actively evolve rather than remain static. Third-party reviews can be an important part of that process (Palmer 1987). At the same time, it is imperative that a monitoring program maintain long-term consistency in data quality and comparability, a concept that has been termed *adaptive monitoring* (Ringold et al. 1996).

Monitoring—Long Time Frames and the Potential Contribution of Retrospective Studies

Many components of forest monitoring programs will be long-term because the processes of stand development and maturation (and associated recovery of some taxa) can take many hundreds of years (Franklin et al.

1981; Ruggerio et al. 1991; Botkin and Talbot 1992; Franklin and Fites-Kaufmann 1996). Some important ecological trends also will be impossible to detect without long-term data that are rigorously gathered on a repeated basis (Hinds 1984). However, retrospective studies often make it possible to quickly gather information to guide adaptive management without waiting the many decades (or even centuries) required for one or more management cycles (rotations) or more to start some integrated adaptive management-and-monitoring studies. For example, with careful study design valuable information on forest disturbance effects can be gathered from sites that were previously logged (e.g., Carey 2000). Unfortunately, the potential for retrospective studies is often limited by lack of records on silvicultural treatments.

Adaptive management does put an increased emphasis on adequate, spatially explicit documentation of management activities. Tools such as global positioning systems and geographic information systems are extraordinarily valuable in this documentation process. Properly maintained, such databases can be useful even if monitoring programs are temporarily suspended, because subsequent evaluative or monitoring activities will not have to start from scratch.

Funding and Other Issues for Forest Monitoring Programs

Ensuring continued financial support for monitoring programs appears to be one of the most difficult challenges in monitoring. The long-term nature of monitoring programs does not fit with the annual budget cycles of government and granting agencies so that innovative approaches often will be necessary to maintain monitoring programs. Approaches can include trust funds or endowments dedicated to monitoring programs, an approach pioneered by The Nature Conservancy in the United States. Monitoring funds could be provided by levies on timber products (Lindenmayer and Recher 1998)—a strategy that could be linked with the forest certification process. Given the central role of monitoring as the definitive test of the effectiveness of resource management as well as in assessing compliance with regulatory and market-based goals, solutions to the funding issue must be found.

Successful Monitoring Programs

There are increasing numbers of examples of successful efforts to design and implement monitoring programs even though these are scarcer than one would like and most have been operative for only a few years. An outstanding example of large-scale monitoring of a variety of parameters is the monitoring program designed and installed in Channel Islands National Park (California, United States) that covers such diverse biotic elements as pinnipeds, seabirds, the rocky intertidal zone, kelp forests, terrestrial vertebrates, land birds, terrestrial vegetation, fishery harvest, and weather (Davis and Halvorson 1988). The Nature Conservancy has established meaningful monitoring programs on many of its ecological reserves within the United States and has pioneered the establishment of trust funds to ensure long-term funding for such activities.

Monitoring programs are being adopted as a part of the large regional ecosystem management plans on federal lands in the United States, such as the Northwest Forest Plan (Mulder et al. 1999). Objectives and protocols have been developed and published for effectiveness monitoring of this plan with regard to late-successional and old-growth forest (Hemstrom et al. 1998), the northern spotted owl (Lint et al. 1999), and the marbled murrelet (Madsen et al. 1999b). Substantial research is also under way as part of the validation monitoring for the plan (Haynes and Perez 2001).

Industrial forestry programs are also beginning to incorporate meaningful ecological monitoring, often as a part of regulatory requirements, such as those that are associated with environmental permits or habitat conservation plans. An extensive monitoring program was a part of the permitting process for the Rio Condor Project (Chapter 15). Plum Creek Timber Company has undertaken extensive monitoring as a part of a multispecies habitat conservation plan for its lands in the northwestern United States. A major monitoring program has been instituted as a part of Weyerhaeuser Company's BC Coastal Forest Project on Vancouver Island, British Columbia (Canada), and it addresses many aspects of ecosystem structure as well as species responses to treatments. This is a part of a much larger adaptive management program that includes significant formalized experimentation. Programs of this type provide models of useful approaches and critical elements of monitoring programs.

Knowledge Gaps in Forest and Biodiversity Management: Areas for Future Research

Although precise solutions to ecological problems are rarely possible, science has developed practical and efficient ways to cope with our inability to know everything. . . . In the longer term, the success or failure of policies for ecologically sustainable development will depend on continuing strategic research to identify and quantify the key areas of risk, to identify strategies for forest management that are likely to achieve sustainability and can be tested by active adaptive management, and to develop procedures for routine monitoring and assessment of whether the objectives of the policies are being met.

—CORK (1997)

Matrix management for the conservation of forest biodiversity must be strengthened by new data. Obtaining these data will sometimes require innovative approaches that integrate methods from different scientific disciplines, such as landscape ecology and molecular genetics. Experiments and other forms of investigation, such as retrospective studies and modeling, are essential elements of this process. Investigations into how new scientific findings are adopted as part of natural resource policy and on-the-ground management are also needed to ensure that matrix-based conservation strategies continue to evolve within an adaptive management framework.

There are important gaps in the information needed to support biodiversity conservation in matrix lands, some of which are discussed in this chapter. Tackling these deficiencies will require rigorous research and monitoring (which can be indistinguishable; see Chapter 16) if conclusions drawn from it are to be defensible and convincing to resource managers, policy makers, and other stakeholders. We outline approaches to bridging existing knowledge gaps through experiments, observational studies, and modeling and discuss the strengths and limitations of each method. Finally, we discuss why research is required on how new findings are accepted and adopted, since such results need to be incorporated into on-the-ground operations.

Topics for Future Ecological Study

Few ecological studies are specifically designed as examinations of the importance of the matrix in biodiversity conservation. This will undoubtedly change, especially as adaptive management is accepted and widely adopted. We outline below some areas of research that may contribute to a better understanding of the matrix as a landscape component and in turn to enhanced management of the matrix for biodiversity conservation.

Retention Strategies and Silvicultural Systems

Developing and assessing new silvicultural techniques and systems that will better sustain biodiversity and ecosystem processes are a logical place to begin this chapter. Three generic approaches to, and many specific examples of, silviculture for biodiversity conservation are presented in Chapter 8. However, exploration of generic approaches has scarcely begun. Quantitative research on levels of structural complexity required to attain specific conservation objectives is particularly needed. As noted earlier, there are few quantitative answers to such questions as "How many logs are needed and how should they be arranged spatially to facilitate movement of a particular species (e.g., the California redback vole) into harvested openings?" or "How does the level of live tree retention affect productivity in a regenerated stand?" Information of this type is also needed as a basis for analyses of cost-benefit ratios of various strategies and levels of structural retention.

The effectiveness of various alternative silvicultural strategies, such as structural retention, is not well known (Franklin et al. 1997; Gibbons and Lindenmayer 1997; Halpern et al. 1999; Carey 2000). Of course, the effects of traditional forestry practices on biodiversity and long-term productivity have typically not been fully assessed, either. Nevertheless, the lack of information about alternative management approaches is surprising in view of the numerous government and professional reports that have called for more basic information on ecological processes and effects of natural and human disturbances in forest ecosystems (e.g., National Research Council 1990). In Australia, more than seventy-five inquiries to the timber industry since World War II have highlighted the need for such information (Resource Assessment Commission 1992).

There is an urgent need for carefully designed research to (1) identify and quantify relationships between structural and floristic features of forest stands and the requirements of forest-dependent biota, and (2) evaluate the effectiveness of specific retention and stand management prescriptions in achieving goals.

Data are generally lacking that demonstrate linkages between species habitat and vegetation structure and composition (Morrison et al. 1992). Even where such relationships have been well established and vigorously field tested, such as between cavity trees and arboreal marsupials in Victorian mountain ash forests (Lindenmayer et al. 1994a), the relative effectiveness of different spatial patterns of retention is unknown (Lehmkuhl et al. 1999). Information needs of this type are immense and cannot be met by academic scientific efforts, although these are important; much of the data must be generated as a part of adaptive management programs.

Areas of study should include research to reevaluate or modify traditional silvicultural approaches as well as to create new ones. For example, there is merit in revisiting seed tree systems previously applied in wet sclerophyll forests of Australia (Cunningham 1960) and selective harvest approaches applied in northwestern America (Curtis 1998). The need for research on silvicultural systems is given further emphasis by the fact that many (if not most) traditional forms of cutting are not based on rigorous scientific—including replicated—experiments. For example, prior to the establishment of the DEMO project (described in Chapter 8), there had never been a large-scale, replicated regeneration-harvest experiment in Douglas-fir forests (Franklin et al. 1999b). Of course, objectives in matrix management are much broader than simply that of wood production (Ford 1999).

Silvicultural research also needs to consider other aspects of forestry operations and ways that they can be modified to enhance biodiversity and ecosystem conservation, or at least to minimize negative impacts. Transportation and logging methods are important areas for investigation. For example, the effects of practices such as closing and revegetating road systems after harvesting, and alternative modes of log transport, such as helicopters, need to be analyzed for their potential value in conserving biodiversity.

Future silvicultural research will often need to be designed so that the effects of individual silvicultural practices can be separately evaluated. Silvicultural researchers have a strong tendency to test entire "systems" with multiple treatment components rather than to isolate effects of specific practices. Most silvicultural treatments actually involve several activities, such as roading, logging technique, cutting prescription, site preparation, fuel treatments, and regeneration technique. Each of these activities has consequences for biodiversity and ecosystem processes. Studies need to be designed so that the individual components of the silvicultural system can be altered and evaluated. This has been termed the *disaggregation of logging* by Putz et al. (2000), who discuss it in the context of cascading effects on biodiversity of logging roads in tropical forests; roads lead to massive hunting pressure on wildlife with corresponding negative impacts on animal-related ecological processes, such as seed dispersal. Populations of feral predators that use logging roads to gain access to harvest units provide a similar example from southeastern Australia (May and Norton 1996).

It is critically important that future silvicultural research be designed to allow clear inferences about effects of varying important stand variables, such as amounts and spatial patterns of retention or effects of size of retained aggregates. A general failing of much silvicultural research has been its focus on comparisons of regeneration harvest systems—such as shelterwood, seed tree, clearcut, and selection—rather than on systematic manipulation and evaluation of important stand parameters, such as the density or spatial arrangement of residual trees. The "system" comparisons invariably result in experimental designs in which effects of individual stand components are confounded. Simultaneously varying several variables strongly limits the ability to draw inferences about effects of manipulating forest structures. Hence, after many decades of silvicultural research we still have greatly limited abilities to answer such fundamental questions as the response of an ecosystem process to varying densities of overstory trees.

Some replicated large-scale silvicultural experiments are necessary, as noted later in this chapter. Such experiments are very challenging logistically and financially so will be limited in number. There are already several excellent examples of such projects, including the DEMO (northwestern United States) and Warra (Tasmania) experiments mentioned in Chapter 8. Similar projects include harvest-cutting studies in the shortleaf pine (*Pinus echinata*)–hardwood forests of the Ouachita Mountains, Arkansas (United States) (Baker 1994), and two large-scale silvicultural experiments in mixed conifer forests in the northern Sierra Nevada in California (United States) (described in Franklin et al. 1999b).

An entire research network of experimental silvicultural sites, the Forest Ecosystem Research Network of Sites (FERNS) (Franklin et al. 1999b), has been established in Canada.

Inventory

A problem for forest managers in many jurisdictions is that the extent of the timber resources is poorly known. This contributes to overcommitment of resources with corresponding negative impacts on other forest values. Better, more accurate methods of forest inventory are needed to overcome this problem. There are many powerful new tools, including remote sensing techniques, that can be effectively used in broad-scale surveys. An entire collection of papers on such techniques and their application was published in the August 2001 issue of *Conservation Biology*. Lidar (laser altimetry) is an emerging technology that can directly measure the three-dimensional distribution of plant canopies and other structural attributes over large areas (Lefsky et al. 1999, 2002).

Landscape Heterogeneity and Contrasts between Natural and Human-Disturbed Landscapes

Responses of forest-dependent organisms to differences in landscape heterogeneity, including cut and uncut landscapes, is an important area for research (Carey et al. 1992; Cale and Hobbs 1994). Landscape heterogeneity is important for many taxa (Forman 1995; Bennett 1998). Results from some empirical investigations, which have used landscape indices to examine wildlife responses to landscape heterogeneity, have been equivocal (e.g., McGarigal and McComb 1995; Lindenmayer et al. 2001c; Tischendorf 2001). This is possibly because the metrics employed to characterize landscape heterogeneity (e.g., O'Neill et al. 1988; McGarigal and Marks 1994; Wegner 1994; Haines-Young and Chopping 1996) are not always particularly meaningful for assessing species response (Cale and Hobbs 1994). Other measures that are more closely related to animal movement (e.g., home range patterns) may be better predictors (Carey et al. 1992). In other cases, species respond to factors not captured by landscape metrics, such as time since fragmentation or structural features within habitat patches (Lindenmayer et al. 2001a). Most landscape metrics have not yet been rigorously tested (Noss 1999). For any or all of these reasons, present abilities to quantify spatial pattern have exceeded our capacity to interpret their relevance to biota (Levin 1992; Turner et al. 1995; Tickle et al. 1998).

Connectivity and Corridors

The effectiveness of wildlife corridors for conserving biodiversity is largely unknown. This is because designing field-based experiments to test knowledge gaps associated with corridor effectiveness is extremely difficult (Nicholls and Margules 1991; Inglis and Underwood 1992; Crome 1997). Although a number of important studies have begun (e.g., La Polla and Barrett 1993; Andreassen et al. 1996; Gilbert et al. 1998), other types of investigations must be conducted. Integrating studies of animal movement, dispersal, and matrix conditions would be useful in revealing new information. The value of these types of studies is in elucidating whether or not population models for heterogeneous (real) landscapes are prone to substantial error propagation (Karieva et al. 1997; Hanski 1999b).

Investigations of connectivity can be helped by intertwining demographic and genetic studies of effective interpatch dispersal with measures of matrix condition and landscape cover (such as the presence of corridors or retained vegetation). Such an approach using molecular genetic techniques (e.g., Wright 1969; Slatkin 1985; Slatkin and Barton 1989) has been employed in the Tumut Fragmentation Experiment for the bush rat and the greater glider (Hewittson 1997; Lindenmayer et al. 1999h; Lindenmayer and Peakall 2000; see Chapter 13).

Synergistic and Cumulative Effects

Timber harvesting in matrix lands may lead to synergistic and/or cumulative impacts on forest biota and ecosystem processes (see Chapter 5). These impacts can be difficult to quantify (Cocklin et al. 1992a), even with well-designed studies. For example, the long-term effects of changes in stand structure, such as gradual loss and depletion of biological legacies, may remain undetected within the time frames of most impact studies. Traditional forms of landscape analysis may not detect cumulative effects at a landscape scale. The potential for negative stand- and landscape-level cumulative effects and the current paucity of methods for detecting and assessing them (Cocklin et al. 1992b; Burris and Canter 1997) make studies of cumulative impacts a vital area for further work (Paine et al. 1998). Finally, studies of landscape heterogeneity in matrix lands must account for the combined impacts of human disturbances, such as logging, and natural disturbance regimes, such as wildfires, both on landscape patterns and forest biodiversity.

Metapopulation Modeling and Fragmentation Effects

In Chapter 2, we outlined some general themes associated with metapopulation dynamics and habitat fragmentation. New generations of metapopulation models and landscape models will need to include positive or negative conditions in the matrix to improve their predictive ability and take account of real landscape conditions. This may include redefining "patches" within these models to include surrounding areas in the matrix that undergo temporal and/or spatial changes in their suitability for organisms such as changes due to longer rotations. Sisk et al. (1997) provided an elegant example in which they modeled bird assemblages within California oak woodlands that were embedded in a matrix of either grassland or chaparral.

Other areas of work on fragmentation effects and matrix conditions are needed. A major knowledge gap at present is the lack of rigorous empirical studies on effects of threshold levels of habitat cover on species occurrence and species diversity (Andrén 1994; see the sections on fragmentation in Chapter 2). This work is urgently needed because of the fundamental importance of (1) habitat availability in species occurrence, and (2) separating effects of habitat loss from the effects of habitat fragmentation.

Effects of forest landscape modification on species response is a second, related area of research that is needed. Species occurrence has often been studied at the patch level rather than the landscape level (McGarigal and McComb 1999) and there is possibly a fundamental mismatch between the scale of landscape change and the scale at which biota are sampled. This problem is magnified when it is recognized that not all parts of a landscape are created equal and that typically the most productive areas are also the most perturbed (Chapters 4 and 5). These issues have been ignored by some workers who have compared perturbed high-productivity environments with unperturbed low productivity, found no differences in the target response (e.g., species abundance), and incorrectly concluded there have been no impacts of landscape change (Lindenmayer 1999b).

Reserve Effectiveness

Data on the relatively few species that can only occur in reserves are essential because they help set the bounds for the balance between reserve and matrix-based strategies (e.g., the relative areas of forest allocated to different uses, their spatial juxtaposition and the contribution that one form of land tenure makes to the other). These data are also important to further improve approaches to reserve design, such as selection algorithms (Margules et al. 1995).

Research is also needed to improve our understanding of the conservation value of midspatial-scale reserves (e.g., those 50–1,000 hectares in size; Zuidema et al. 1996). These types of protected areas are increasingly important for biodiversity conservation (often because of social, economic, and pragmatic reasons), even though they may not support viable populations of all taxa (Schwartz 1999; see Chapter 5). Establishment of such reserves can also be extremely controversial, particularly with regard to the marginal value of each size increment, as in the case of riparian reserve width. This is partially because such midspatial-scale (matrix-embedded) reserves are often proposed for private as well as government lands.

Methods and Approaches for Future Ecological Research

To enhance the conservation of biodiversity in production forests, management decision making must be better informed. Informed decision making is based on new and better information. The relevant knowledge will come from many different types of investigations.

Experiments

Designed field experiments are relatively rare in forest biodiversity research (although see Margules 1992; Schmiegelow and Hannon 1993; Phillips 1996). More are needed to address a wide range of questions spanning topics from the value of retained vegetation for enhancing the recolonization of logged areas (Gib-

bons and Lindenmayer 2002) to the effectiveness of wildlife corridors (Beier and Noss 1998).

While high-quality field experiments are invaluable, they also have limitations, including

- *Expense.* Some of the silvicultural experiments in the northwestern United States have a cumulative cost in excess of $US1 million per replication. They also may need to run for prolonged periods and so are vulnerable to budget cuts and termination before results are produced.
- *Lack of applicability to larger scales.* Experiments require the control of many potential sources of influence. As a result, findings from some experiments may have limited generality and capacity for extrapolation to other systems or problems at larger scales (McCarthy et al. 1997).
- *Unforeseen bias.* By carefully controlling some conditions (e.g., Tilman 1996), initially unforeseen biases can occur that confound the results and interpretation of effects (Huston 1997).
- *Limited Replication.* Experiments often have few replicates, which reduce their statistical power and necessitate caution when interpreting results.
- *Lack of Complexity.* Complex, interacting factors often cannot be readily examined in field experiments despite the fact that they can influence species persistence; examples include cumulative effects (sensu McComb et al. 1991; Burris and Canter 1997) and incremental long-term changes in landscape cover.

Conducting landscape-level experiments at spatial scales relevant to mobile groups such as birds and bats is a challenge (Wiens 1994). Some taxa will respond to the details of the landscape mosaic, but the effects of complexity created by interacting landscape components will be extremely difficult to test (Wiens 1999). Nevertheless, some important landscape-scale experiments have commenced (Schmiegelow and Hannon 1993; Schmiegelow et al. 1997), and many more are needed. Dooley and Bowers (1998) believed that large-scale experiments are "an important intermediary between the inherent abstraction of simulation modeling and what is observed in the real world." For example, stimulus from the material collated to write this book has resulted in a new large-scale experiment (the Nanangroe Experiment) in southern New South Wales (Australia), which was designed specifically to test the effects of changes in the landscape matrix on woodland vertebrates (Lindenmayer et al. 2001c; see Chapter 13).

Retrospective and Other Observational Studies

Adaptive·management studies will take time to generate results that are meaningful for applied forest management. Hence, other types of studies, including observational and retrospective studies (e.g., Recher et al. 1987), quasi-experiments (sensu Dunning et al. 1995), and analytical descriptions (Haila 1988) are useful.

Retrospective studies on sites logged or subject to natural disturbances in the past can provide useful information relatively quickly, although their value is dependent on the availability of good-quality data, especially for past-disturbance history. Analytical descriptions are investigations in which ecological phenomena are systematically described and related to key relevant ecological theory and underlying assumptions (Haila 1988). Quasi-experiments are studies in which a rigorous experimental design is employed to take advantage of existing landscape characteristics (Dunning et al. 1995). However, substituting space for time in ecological studies requires caution because results can be confounded by past histories, especially in systems where stochastic (disturbance) processes are important. In these cases, time-series studies may be essential to obtain unequivocal results (Pickett 1989).

Murphy and Noon's (1992) ecological principles approach to identifying conservation strategies for the northern spotted owl is another useful strategy that synthesizes retrospective information on a target species.

Computer Modeling

Better matrix management approaches can gain a great deal from computer modeling. For example, modeling approaches can help clarify (1) how many effective dispersal events are needed to maintain genetic variability under different landscape configurations (Lacy and Lindenmayer 1995), (2) the possible patterns and pathways of animal movement in heterogeneous forest landscapes (Gustafson and Gardner 1996), and (3) the contrasts in landscape heterogeneity arising from wildfires versus checkerboard harvesting regimes (McCarthy and Burgman 1995).

Modeling can be useful in exploring very large-scale processes that are often extremely difficult to test with experiments or observational studies (Bissonette 1997). The modeling work in the Biodiversity Pathways Concept (Carey et al. 1996) (see Chapter 8) is a good example.

Computer-based modeling can contribute to matrix management in other ways:

- It can act as a vehicle for synthesizing what is and what is not known about particular phenomena (e.g., dispersal patterns; Doak 1989) and where future studies are needed (Burgman et al. 1993).
- It can help identify hypotheses for future testing using field-based empirical studies (Pascual and Adkison 1994), such as in adaptive management programs (e.g., Walters 1986).
- It can provide a mechanism for comparing the relative effectiveness of management options (such as the risk of extinction of a given species under several conservation strategies; Possingham et al. 1993).
- It can assist managers in conceptualizing the possible long-term consequences of particular actions or strategies taken now (e.g., the cavity-tree abundance simulator developed by Ball et al. 1997, 1999).

Model-users should be clear about the limitations of models and modeling. Modeling is an approach to assist in understanding ecological systems and to facilitate problem analysis—but it cannot make precise predictions. It will be most useful if it is based upon good ecological data (Starfield and Bleloch 1992; Burgman et al. 1993); otherwise, models can generate outcomes of limited generic value and management applicability (e.g., Tilman et al. 1994; see critique by McCarthy et al. 1997). Indeed, empirical data provide the only true test of ecological theory (Franklin and MacMahon 2000).

In summary, there is no such thing as a perfect ecological study—all investigations will have limitations. Conversely, all scientific approaches have something to contribute; the key is to be aware of the strengths and limitations of each approach, identify the method or combinations of methods likely to give the best outcomes for a particular problem, and secure adequate funds to allow rigorous implementation and testing of the various approaches.

Other Areas for Research: The Adoption of New Knowledge

Many areas of ecological research are required to bridge knowledge gaps in matrix management. But the applied value of such research will only be realized if and when it is adopted by forest management agencies.

There are numerous instances where results of high-quality research are not adopted or only partially adopted by natural resource managers (Underwood 1995; Dovers and Lindenmayer 1997; Dargarvel 1998; Kirkpatrick 1998). Klenner and Vyse (1999) suggest that part of the problem could be the narrow focus of much research, which limits its applicability to forest management. Lautenschlager (1999) further speculates that research might have greater applied value if public concerns about natural resource management were included as part of the initial framing and design of studies. In cases where research is highly relevant to management, there can be a prolonged time lag before its adoption (such as in the Murray-Darling Basin in southeastern Australia; Ghassemi et al. 1995). It is important to determine how to promote information transfer to reduce the time lag in the uptake of new ecological ideas that are being generated so rapidly.

Work is needed on the research adoption process (Walker 1998) to determine (1) where it has been successful, where it has failed, and why; (2) the organizational and other structures that promote or impair capitalizing on new research (see Kirkpatrick 1998); and (3) the interplay among research, policy, and management.

Research should be undertaken to determine how to inject more science into policy—there are many cases where changes in forest management have occurred not because of ecological information, but as a result of public pressure, legal challenges, and media coverage (Lunney and Moon 1987; Franzreb 1993; Yaffee 1994). For example, Funtowicz and Ravetz (1991) describe situations typical of many recent forest debates where there have been a dearth of "hard" science, extreme conflict in resource and conservation values, and a need for urgency in decision making. They argue that "hard" policy decisions have often been driven by "soft" scientific input (Funtowicz and Ravetz 1991). Perhaps this is not surprising given the complexity of biodiversity conservation issues and that forest management debates have social, economic, and other dimensions. However, it is possible that research adoption might be facilitated by the development of better "scenario tools" that allow resource managers and policy makers to model "what if" questions (Walters 1986).

Several authors (e.g., Kanowski and Buchy 1999; Ison and Russell 2000) divide the process of adoption of research into two levels of research and development: first-order research and development, which has a strong disciplinary focus on technical natural resource management problems, and second-order research and development, which relates to problem solving via mutual learning and collaborative multidisciplinary partnerships between individuals and organizations with different expertise and responsibilities in natural resource management. Brand et al. (1993) highlighted the need for such second-order research and development in Canadian forests to promote ecologically sustainable forestry.

More interdisciplinary work is needed on the interfaces among the generation of new scientific knowledge, the stages of policy development, and institutional structures and organization (e.g., Clark and Kellert 1988; Clark 1993; Ison and Russell 2000). Such work is fundamental because the adaptive management process depends on feedback loops among research outcomes, policy development and change, and implementation of upgraded approaches on the ground (Chapter 16). Failure to communicate new research results (or failure to have results adopted as part of forest management) severs these links, leading to a breakdown of the adaptive management framework. Finally, future research should assess whether the adaptive management framework is itself an effective process or whether other frameworks are needed (Rogers 1997).

Development of comprehensive biodiversity conservation plans (and, as part of this, matrix management strategies) has powerful social and policy dimensions as noted earlier. These dimensions are key drivers in influencing research and its adoption, as well as being fundamentally important for approaches to natural resource use, such as adaptive management. We explore some of these dimensions very briefly in Chapter 18.

There is much evidence that, in many cases, credible science has ultimately been incorporated into management as the result of social factors—particularly laws and regulations—to which market pressures (for certification) may be added in the future. For example, science and scientific credibility were central in legal actions brought under the National Forest Management and Endangered Species Acts in the United States. Scientists and scientific syntheses were made central elements in development of plans, which were then judged for scientific credibility by courts.

Although litigation is probably not an optional approach to developing and evaluating ecosystem management plans, experience does suggest that "requirements" imposed by nonresource managers and organizations may be major factors driving development and application of science in the future. This circumstance is not without substantial risks and in fact may result in mounting public confusion as scientists increasingly contend with one another to present "scientific truth."

Social and Other Dimensions Associated with Matrix Management

Resource problems are, after all, human problems that are generated through economic and political systems which humans design.

—LUDWIG ET AL. (1993)

Allocating land to either reserves or commodity production has been the traditional approach to resolving disputes over management of forest resources. Continued dependence upon simplistic land allocation approaches will result in further losses of biodiversity as populations of many species are reduced or eliminated from matrix lands. Adopting conservation-based matrix management approaches, which blur the distinctions between forests in different land tenures, is the antithesis of the land allocation paradigm. The continuum of possibilities—from those based primarily on allocation to those based primarily on integrated management—provides national and global societies with important forest policy choices.

Integrated management of forest land to conserve biodiversity and ecosystem processes is philosophically and practically challenging for most stakeholders, including resource managers, conservationists, scientists, politicians, and landowners and their related institutions. Integrated approaches are more complex to apply and understand, require compromise and continued interactions among stakeholders, and can be more expensive than allocated solutions. Unfortunately, approaches based strictly on allocation of the forest-land base are not likely to achieve many societal goals, however appealing they may be in the short term. Integrated approaches will require adjustments by all stakeholders, the leveling of global playing fields, and incentives.

The balance between nonconsumptive and consumptive forest uses has been (and remains) a highly controversial issue throughout most of the world. It is a complex issue that is, nevertheless, often reduced to struggles between reservation and harvesting of native forests. Although the global focus often has been on tropical rainforests, some of the most intense controversies involve temperate forests. This has been the case for several decades in such diverse regions as North America (Thomas et al. 1990; Diem 1992; Forest Ecosystem Management Assessment Team 1993; Scientific Panel for Sustainable Forest Practices in Clayoquot Sound 1995; Sierra Nevada Ecosystem Project 1994; Yaffee 1994), Australia (Routley and Routley 1975; Lunney and Moon 1987; Kirkpatrick 1998), Scandinavia (Virkkala et al. 1994), and Chile (Armesto et al. 1998). Controversies over forest management are not confined to scientific issues (e.g., Duffy and Meier 1992 versus Johnson et al. 1992; Abbott and Christensen 1994 versus Calver et al. 1997). Rather, they have incorporated social, economic, and political issues (Resource Assessment Commission 1992; Yaffee 1994; Dargarvel 1995; Kirkpatrick 1998) as proponents of "exploitation" (the forest products industry) and "preservation" (conservationists) have debated expansion of wilderness areas (Dargarvel 1998; Lindenmayer and Recher 1998). Contributing to the intensity of the debates has been the intense focus of most forest management on commodity production, which is apparent from the rapid conversion of primeval forests to plantations, simplification of stands and landscapes, and limited consideration of nontimber forest values (Carey and Curtis 1996).

While substantial changes in forest policy and management have occurred in some regions, modified

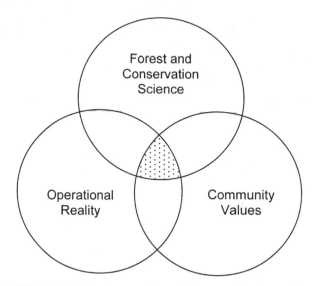

FIGURE 18.1. The intersection of forest and conservation science, community values, and the operational reality of implementing on-the-ground strategies for forest management. The shaded area in the center of the three core "spheres of influence" corresponds to actual on-the-ground forest management. Redrawn from a diagram presented by Linda Coady, Weyerhaeuser Company, Vancouver Island, British Columbia, Canada.

management of matrix lands for biodiversity and other ecological values has not received nearly as much attention as forest reservation. Policies recently adopted for some federal lands in the United States provide an important exception, although even in this case extensive reserves are typically a component of new regional plans. In Australia, formal efforts to incorporate matrix management, such as the Ecologically Sustainable Forest Management (ESFM) process, have emerged only recently; even these efforts can be compromised by intensified harvesting operations on matrix lands (Bauhaus 1999), which would transform some native forests into de facto plantations (e.g., Churton et al. 1996) (see Box 5.4 in Chapter 5).

Society ultimately will make the important decisions regarding the balance among forest uses and the strategies that will be used to achieve that balance. In this chapter, we explore some social, economic, and philosophical issues associated with matrix management. Adoption of such matrix-based approaches involves far more than forest and conservation science

since it intersects with society's valuation of various goods and services as well as the economics and operational reality of implementing forest operations (Figure 18.1).

Allocation Versus Integration as a Basis for Resolving Conflicts in Forest Use

Allocation and integration provide two contrasting approaches to resolving conflicts among forest uses, including conservation of biodiversity. At one end of the continuum of possibilities are policies in which the forest estate is allocated to intensively managed production forests ("fiber farms") and to reserves or wilderness areas reserved from commodity production and often any overt human manipulation. Near the opposite end of the continuum are policies that encourage integrated approaches that incorporate commodity and ecological goals on significant portions of the forest estate. Our emphasis in this book is on management of the matrix for a variety of forest values—in effect, for forest policies that incorporate integrated approaches. We feel that relying primarily on an allocation strategy will ultimately be less successful at achieving societal goals, including sustainable forestry and desired levels of biodiversity. However, many participants in forest debates have a different view and advocate—or at least accept—division of the forest estate into fiber farms and reserves as the preferred solution. Why is this?

Logically one might expect most human societies to favor integrated approaches that provide win-win solutions for a large number of stakeholders and variety of forest values—but this is not necessarily reflected in debates over forest policy. To the contrary, many important stakeholders on all sides of forest controversies advocate allocation as the solution. These include global wood products corporations, international conservation organizations, and intellectuals from diverse disciplines, including conservation biology, forestry, and social science. These stakeholders have a shared view of biodiversity conservation as being primarily a "set-aside" issue—a societal objective that is best or solely achievable by allocating lands to reserves, thereby freeing the remaining lands from significant conservation obligations. Surprisingly, such diverse stakeholders as timber corporations and large environmental organizations agree on this strategy and disagree only about how the forest pie is to be divided (i.e., which lands [location and total extent] are to be reserved.)

New Zealand adopted the allocation approach as national forest policy and is often used to exemplify the approach, which is sometimes referred to as the "New Zealand solution." In effect, the government reserved all native forests under its control from commodity management and concentrated wood production in plantations of exotic (nonnative) tree species on private and government lands. The issue was not whether some native forests could be managed to produce commodities and sustain biodiversity; indeed, a limited harvest program in native forests based on aerial removal of small patches of trees was proving both profitable and congruent with ecological values. Rather, the issue largely revolved around social issues, some of which are discussed below, and the solution was political rather than scientific. An accommodation between the timber industry and conservationists was apparently part of the solution, with most industrial players agreeing to forgo timber from native forests in return for a freer hand in intensively managing the plantations. Some limitations of the New Zealand solution are discussed below; there are serious questions about how well it has worked for either the native forests or for society at large, even in this small country.

Stakeholders tend to prefer allocated to integrated solutions for many reasons. Allocation is much simpler, technically and socially, than integrated approaches to resource management. Managing forest landscapes for both commodity and environmental values (including biodiversity) requires scientific information about many ecological attributes, complex analyses of trade-offs (optimization) among these values, and perpetual and exhausting social interactions among stakeholders. Managing forests to maximize a single value, such as production of timber, is much simpler, even if measures are required to mitigate environmental impacts.

Second, allocation solutions are politically attractive because they are simpler and, therefore, much

easier to understand and explain. They also appear to provide permanent solutions to the conflicts and certainty for all stakeholders, which are desirable outcomes politically.

Third, allocation solutions are philosophically satisfying—no compromises are required. Stakeholders differ profoundly in their valuation of forests, from the spiritual and aesthetic appreciations of natural forests held by many conservationists to the strong utilitarian values of many individual and corporate stakeholders. Allocated approaches tend to satisfy stakeholders' values, since once lands are allocated between commodity and conservation goals, few additional compromises are necessary. Overt human manipulation can be proscribed absolutely within the reserves and industry can pursue profit maximization on unreserved lands without biodiversity constraints. Furthermore, future interactions among stakeholders who have major philosophical differences can be limited; in contrast, integrated approaches require of stakeholders continuing—indeed perpetual—dialogue and willingness to compromise.

Fourth, allocated solutions appear to be cheaper and more efficient in achieving goals, at least with regard to production of commodities, such as wood. This is partially related to their simplicity (e.g., reduced information needs and lack of trade-offs) as well as to externalization of some environmental and long-term costs.

Fifth, allocated solutions typically serve institutional goals or imperatives better than integrated solutions do. Such approaches produce well-defined, apparently permanent outcomes (e.g., lines drawn on maps defining boundaries of reserves or wilderness areas) that provide satisfaction to institutional members and which can be used effectively in soliciting additional financial support. Furthermore, publicly considering matrix-based approaches can weaken the cases that institutions are making for specific goals. For example, agreeing that some species may be sustained on managed lands may be perceived as undercutting the case for reserves. (Indeed, there are some cases where there is no room for compromise on biological conservation [Woodwell 1989].) Presumably, allocated solutions also provide corporations with certainty regarding resource supplies and the regulatory environment that are attractive to investors and please corporate boards.

Sixth, evidence for the ability of integrated matrix-based approaches to achieve goals, such as successful demonstrations, pilot projects, experiments, and other scientific studies, is limited. Consequently, there are questions about the degree to which ecological conditions can be restored and biodiversity sustained in the matrix. As noted earlier, trading habitat that is known to be suitable (in existing native forests) for hypothetical habitat (in stands that are to be managed in new and enlightened ways) is viewed as risky to biodiversity goals. Of course, the long-term success of reserve-based approaches (allocation) in conserving biodiversity is equally unknown, particularly under realistic projections regarding the ultimate extent and geographic location of global forest reserves.

And, finally, the lack of trust and goodwill among stakeholders favors allocation approaches. Many conservationists doubt that the timber industry would ever adopt practices that reduce their return on investment; proposals for modified management by commodity producers are often viewed as public relations ploys rather than substantive proposals to accommodate biodiversity. Commodity groups often feel the need to clearly define the commodity landscape in order to resist the insatiable appetite of conservationists for additional reserves. Resource managers often lack credibility with stakeholders because of perceived biases toward commodity production and reluctance to innovate (Pittock 1994).

To summarize, land allocation is generally viewed as a much simpler means of resolving natural resource management issues than the more complex, multi-scaled strategies involved when large ecological reserves are integrated with matrix management, as outlined in this book. Integrated approaches are challenging for all stakeholders, including conservationists, politicians, resource managers, and the general public. Delineating production and conservation lands assists politicians and bureaucrats in moving forest debates off the political agenda (Dargarvel 1998), reduces pressures on resource managers to manage for multiple and often conflicting objectives, allows for relatively uncomplicated planning and management requirements on timberlands, and provides certainty that potentially negative effects of human activities are excluded from critical forest areas (at least in the short term).

There are many other factors that affect the willingness of stakeholders to work with one another and determine the acceptable balance point along the allocation-integration continuum. Competition in global markets is an important concern for corporations; how can a company hope to compete if they voluntarily adopt approaches that increase costs of production when others do not adopt such measures? Often debates in temperate and boreal regions are colored by attitudes or information acquired under the very different conditions found in tropical forests—concerns and circumstances that are not always relevant to forestry debates at higher latitudes. Related to this is the fact that most primary forests in temperate regions have already been incorporated into the matrix, particularly those found at lower elevations and on highly productive sites; hence, there is strong opposition to any additional loss.

Increasing the Viability of Integrated Approaches to Forest Landscapes

Our recurring theme is that allocating lands to reserves will not, by itself, achieve an acceptable level of biodiversity conservation. The establishment of large ecological reserves is necessary but not sufficient as a strategy (see Chapter 5) (Franklin 1993a; McNeely 1994a,b).

Allocation-based approaches, such as the one adopted in New Zealand, are demonstrably flawed and include many limitations (see also discussion of limitations of reserves in Chapter 5).

One such limitation is that many critical ecological values are so pervasive in the landscape that they can be neither captured nor mitigated by reserve systems. Stream and river systems, which provide habitat for aquatic biota and well-regulated, high-quality flows of water, are an excellent example. Riparian and aquatic habitats are a distributed system that must be conserved and protected wherever it occurs, including in commodity landscapes occupied by exotic plantations (as is increasingly understood in New Zealand). Hence, allocated approaches cannot free commodity landowners of all obligations to biodiversity and ecosystem processes.

A second limitation is that many (perhaps most) reserved forest landscapes require active management to achieve long-term goals. For example, management is often needed to restore natural conditions (remove roads, accelerate development of structurally complex stands, reduce unnatural fuel loadings) and protect ecosystems from unnatural disturbances, including exotic organisms. However, societies are typically unwilling to provide sufficient funding for restoration and management of areas that provide few direct economic benefits. For example, forests in New Zealand are beset by exotic organisms, including marsupials introduced from Australia, but limited funds are available for management or protection of the native forests. Even national parks in Canada, the United States, and Australia lack sufficient funding for adequate resource management programs. Lack of funding for management of reserves may become an even larger problem in the face of global environmental change.

A third limitation is that allocation approaches assume that the long-term productivity of matrix lands can be sustained under intensive management regimes that impact soil physical and chemical properties and reduce the diversity of fungi, invertebrates, and other small organisms that sustain key ecosystem processes. There is limited scientific research and empirical evidence regarding long-term productivity under very intense management regimes.

Last, allocation approaches often ignore the economic and social needs of local communities, including indigenous peoples. Powerful stakeholders (e.g., corporations and international conservation organizations) tend to achieve their goals but less-powerful stakeholders often do not.

One of the challenges of matrix-based management, then, is to create mechanisms that encourage the development, adoption, and implementation of integrated approaches to forest lands. In the remainder of this chapter we discuss

- Approaches that can be used to achieve this goal, including market-based incentives such as forest certification
- Problems that need to be addressed, such as adjustments in harvest levels
- Incentives for change among stakeholders

The Role of Globalization, Markets, and Certification

Globalization of the wood products industry has potential for both positive and negative impacts on sustainable management of forest landscapes (Daily and Walker 2000). The World Commission on Forests and Sustainable Development (1999) estimated that international trade in wood products exceeds $US100 billion annually, which is about 3 percent of world merchandise trade. Globalization can either reduce or exacerbate such market-based problems as inequities in environmental obligations among wood producers in different regions. Market-based incentives for ecologically sensitive management that would level the playing field for all producers would be an important positive development, for example. Global markets driven solely by price-based competition are a negative development.

Forest Certification

Forest certification represents a market-based mechanism that is potentially an important counter to ecological problems created by globalization of the wood products industry (Viana et al. 1996; Wallis et al. 1997). Forest certification provides standards of conduct of forest operations and periodic assessments of those operations—preferably by a third party—for conforming to these standards. Social and economic criteria are typically used along with environmental standards. As markets and purchasers require such certification processes, pressures increase for wood product firms to adopt environmentally sound practices in forest management. Certified forest products are typically labeled to identify that they have been produced in accordance with specified standards of sustainability, such as those defined by the Forest Stewardship Council (1993). Credible certification programs should ultimately result in enhanced matrix management on forest ownerships.

Certification arose partially from the 1992 Earth Summit in Rio de Janeiro. Attendees at this conference agreed that decisions regarding utilization of natural resources should link both producers and consumers and consider impacts on residents of other countries as well as future generations (Viana et al.

1996). International cooperation is important for species with distributions that span several countries (e.g., bird taxa in Scandinavia and Russia) or that migrate between continents (e.g., Neotropical songbirds).

The overall goal in certification is the adoption of standards that will ensure forest management is environmentally sensitive, socially aware, and economically viable (Upton and Bass 1996). Landowners (which can include governments) enter into an agreement with a certifying organization to conduct their forest management according to defined standards. Under systems where third-party certification is required, certifying organizations or organizations chartered by them periodically assess performance. Products derived from certified forests are typically labeled with a logo that allows consumers to identify these as "green" products from forests managed using environmentally responsible practices. Some "self-certification" programs have been created, such as the Sustainable Forestry Initiative (SFI) of the American Forest Products Association, but self-certification lacks the credibility of third-party assessments; recognizing this, some companies subscribing to the SFI standards are voluntarily obtaining third-party certification.

The leading third-party certification system is currently the program developed by the Forest Stewardship Council (FSC), which was established in 1993. A primary goal of this nongovernmental, nonprofit organization was to ensure that forestry practices in different nations were consistent with international conventions on ecological sustainability and the conservation of biodiversity (Wallis et al. 1997). FSC has adopted general principles for sustainable forestry practices; national or regional groups supplement these by developing detailed interpretations of those standards. This process occurred in Sweden in 1995–1996 (Angelstam and Pettersson 1997), and by 1999 approximately 25 percent of all forests in Sweden had been certified using detailed standards established by the Swedish Forest Stewardship Council (Hazell and Gustafsson 1999).

Actual certification of forest projects is conducted by private organizations, which are accredited by FSC to administer FSC standards; these may be either for profit or nonprofit organizations. As of mid-2000,

more than 18 million hectares of forest worldwide have been certified using FSC standards (Cauley et al. 2001). Certified ownerships vary widely in type and size, from privately owned forestry companies and tribal properties to government lands, such as the Pennsylvania state forests in the United States. Collins Pine Company, in northeastern California, was the first timber company to be certified by FSC standards; the fact that this is a privately held (family) business rather than a publicly held corporation is probably significant. Collins Pine Company committed itself to

- Training its employees in sustainability principles
- Improving energy efficiency and timber production
- Improving practices for water conservation (Daily and Walker 2000)

Certification and consumer environmental concerns have already impacted many existing and potential forestry projects (see the case study on the Rio Condor Project in Chapter 15). For example, large contracts to export timber products from Western Australia to Europe were lost because there was no credible evidence that these Australian forests were being managed according to ecologically sustainable principles. There is evidence that businesses can profit from improved environmental practices (Socolow et al. 1994; Daily and Walker 2000). Mönkkönen (1999) believes that certification could be a powerful tool to increase the economic benefits and values of ecologically sustainable forest management in Scandinavia.

Many problems need to be resolved before the potential benefits of certification for biodiversity conservation can be fully realized.

For instance, there is little interest in—and sometimes outright hostility toward—certification in some globally important forest regions, such as southeast Asia, and in some major markets for forest products, including countries such as Japan, China, and South Korea (Jenkins and Smith 1999). Some of these countries may for the foreseeable future continue to ignore the environmental impacts of their actions as producers and consumers of wood products. In many tropical countries, international certification agreements may have little impact since most harvested timber is consumed locally (Ghazoul 2001). Indeed, the total area of forest estimated to be certified is presently 90 million hectares or approximately 2 percent of the world's forested area (Food and Agriculture Organization of the United Nations 2001).

Another problem involves the competition that exists among alternative certification systems. Some certification systems are viewed as less scientific and less socially acceptable than others. As noted earlier, many stakeholders believe third-party certification to be more credible than self-certification. FSC is currently the leader in providing a global standard for certification, but it remains to be seen whether it will ultimately be accepted by both wood products corporations and environmental organizations. A proliferation of certification systems could confuse consumers and reduce the value of such approaches.

A third problem lies in defining sustainability and developing criteria that can be applied throughout the globe. This will be a difficult problem, although the two-tiered approach of FSC shows promise. Robust indicators for sustainability (e.g., those set out under the Montreal Process) have yet to be identified and rigorously tested (Tickle et al. 1998; Lindenmayer et al. 2001c).

Yet another problem lies in the fact that secondary impacts of logging on biodiversity are not adequately considered in existing criteria and indicators. For example, in many tropical countries logging roads lead to massive increases in hunting for game animals or bush meat (Redford 1992; Bennett 2000). In Africa alone, trade in bush meat exceeds 1 million metric tons annually (Bennett 2000), which has major negative impacts on wildlife populations and, ultimately, some ecosystem processes.

Despite these problems, certification could have a profound positive impact on forest practices and stimulate much wider adoption of ecologically sensitive matrix management (Putz and Romero 2001). This depends, of course, on broad adoption of credible and equitable certification systems with a robust set of protocols but which also incorporate flexibility to accommodate regional and national differences (Viana et al. 1996; Wallis et al. 1997). Even with the current status of certification, positive changes are already occurring in forest practices, not only in developed nations (Mönkönnen 1999) but also in less-developed

tropical nations (Nittler and Nash 1999; Putz et al. 2000; Ghazoul 2001). For example, the use of Reduced Impact Logging in Brazilian Amazonia by some companies has allowed certification of timber for export to Europe (De Marajò and Paragominas 2001).

Movement of Logs, Wood Chips, and Other Unprocessed Wood Products, and Live Woody Plants between Continents

A consequence of increased international trade in forest products during recent decades has been expanded shipments of raw logs, wood chips, and untreated wood products between continents. This trade represents a significant threat to native forest biodiversity because trade in unprocessed wood and wood products provides for movement of virulent forest pests and pathogens between continents and, consequently, into forests populated by species that have little or no resistance. Restoration of native forest biodiversity is possible following clearcutting or, conceivably, even following the potential disruptions of climate change but not if dominant tree species are no longer viable components of forest ecosystems.

The threat to native biodiversity is particularly severe where natural forests are dominated by a single species, because by decimating or extinguishing that species, a new pathogen or pest can dramatically alter the entire ecosystem. Obviously, the threat to the wood products industry is also immense where that economy depends upon extensive plantations of exotic species, such as radiata pine, which are potentially vulnerable to the introduction of pests or pathogens, often (but not always) from within their natural range. Forest regions that depend on one or two primary species as dominants in both natural forests and plantations—such as Douglas-fir in the case of northwestern North America—are most at risk.

The threat to forests from exotic insects and diseases as well as the role of live plants and unprocessed wood products as vectors for these organisms is evident from the historical record. North American forests have been profoundly impacted by introduced pests and pathogens most of which originated in Eurasia. Examples include chestnut blight (essentially extirpating American chestnut [*Castanea dentata*]), white pine blister rust (eliminating western white pine [*Pinus monticola*] as a commercial tree species and potentially extirpating whitebark pine [*Pinus albicaulis*], in much of its range), Dutch elm disease, woolly adelgid (decimating eastern hemlock [*Tsuga canadensis*]), gypsy moth (general forest defoliator in the eastern United States and a perpetual threat along the West Coast), and balsam woolly aphid (decimating true firs [*Abies* spp.] in both eastern and western North America). Pests are being introduced to other continents, however, as demonstrated by a virulent introduced nematode that has decimated native pines in Japan and diseases introduced into exotic plantations of radiata pine in South America.

Despite measures adopted to prevent such introductions, the escape of dangerous pests and pathogens continues today and is probably accelerating because of increases in intercontinental trade. It is simply impossible under current trade agreements to ensure that pests and pathogens will not slip through various treatment, inspection, and quarantine programs. As this book is written, a disease of unknown origin is spreading rapidly through hardwood forests in California, decimating species belonging to the oak family (*Quercus*). In 1999, Asian longhorned beetles were introduced to the Chicago and New York areas of the United States in shipping materials (crates, pallets) made of wood that was not kiln-dried or otherwise treated to destroy such pests. This was the most notable but not the only introduction of this aggressive wood-boring insect. In 2001 in the Seattle area in the northwestern United States, citrus longhorned beetles escaped from live trees imported from Korea, while these plants were still in quarantine. This Asiatic insect is a serious potential threat to native wild hardwood species and some conifers as well as to urban and orchard trees.

Global efforts to sustain native forest biodiversity must include a serious program to reduce the potential for intercontinental movements of forest pests and pathogens. This should be a part of a larger effort to prevent unwanted exotic organisms from becoming established and directly and indirectly impacting the composition, structure, and functioning of native ecosystems. Eliminating movement of logs, woodchips, and other unprocessed wood products between continents is logical from social as well as ecological perspectives. The primary manufacture of wood products should occur in originating countries or regions.

Importing logs to sustain a few manufacturing facilities, such as has occurred in the western United States, cannot be justified in view of high risks to the entire indigenous forest estate.

Revised Timber Yields and Greater Flexibility in Resource Allocation

Adopting matrix-based management without appropriately adjusting timber harvest levels can result in simply shifting ecological impacts of harvesting from one part of the landscape to another. As noted by many authors, wood production levels must reflect the productive capacity of the land base under a selected management regime and cannot be stipulated a priori.

Matrix-based management is going to result in reduction of timber yields, at least the short run, because

- Merchantable timber will be left behind on harvest units as retained structures.
- Growth of regenerated stands is reduced at least to some degree by competition from retained trees and understory plants (Incoll 1979; Rotherham 1983), although this may not always be an issue (Hansen et al. 1995b; Florence 1996).
- Aggregated retention removes a percentage of harvest areas entirely from production (Franklin et al. 1997).

Commodity impacts of various matrix management strategies, including modified silvicultural approaches, require adjustments in timber harvest levels on these lands. If adjustments are not made, then either the matrix-based strategy will fail or ecological impacts will be transferred to other lands through intensified timber harvesting. (Failure to adjust harvest levels in land tenures is, incidentally, a serious problem, with allocated resolutions of conflicts where timber harvest levels have not been adjusted to reflect reductions in the production land base associated with establishment of reserves.) In countries such as Australia, harvest adjustments are needed to meet stated objectives of policies and legislation such as the National Forest Policy Statement (Commonwealth of Australia 1992) and the National Biodiversity Strategy (Department of the Environment, Sports, and Territories 1995). Similarly, it was recognized that under the new forest practices code adopted by British Columbia, reductions in timber production would be 6 percent over ten years (Fenger 1996). On a smaller scale, shifting from clearcutting to the variable retention harvest system (see Chapter 8) in coastal British Columbia has been estimated to reduce timber yields by 12 to 15 percent (G. Dunsworth personal communication).

Long-term flexibility in calculated timber yields is also needed to accommodate a variety of new circumstances. Examples of such circumstances include additional scientific information (e.g., the need to protect habitat for newly discovered and/or rare species) and impacts of natural disturbances, such as wildfires, windstorms, and volcanic eruptions. Such flexibility is an essential part of adaptive management (see Chapter 16).

Reduced timber yields will often require adjustments in societal commitments to the wood products industry—an outcome that must be recognized by politicians, foresters, conservationists, and the general public. Policies and initiatives in some parts of the world have attempted to "lock in" guaranteed supplies of timber. Examples are the Australian Regional Forest Agreements (e.g., Commonwealth of Australia and Department of Natural Resources and Environment 1997) and Victorian Timber Industry Strategy (Government of Victoria 1986). Such commitments make it impossible to adjust harvest to levels consistent with ecologically sustainable matrix management strategies (Forsyth 1998). The success of matrix management strategies often hinges successful responses to the significant localized social costs that occur when timber harvest levels are adjusted.

Increased Costs and Other Tradeoffs

Increased costs and other trade-offs occur with adoption of matrix management for enhanced biodiversity values. Maintaining the spatial, structural, and floristic complexity of managed forests requires more complicated pre-logging survey and planning. Worker safety is another important factor regarding the use of modified silvicultural systems. These issues have arisen in studies of alternative cutting methods in Victorian and Tasmanian forests in southern Australia (Squire 1987;

Campbell 1997; Hickey and Neyland 2000) as well as in North American forests (Hope and McComb 1994).

Increased silvicultural and planning complexity can add to the costs of production in the short term, although long-term losses of forest productivity through the failure to maintain key ecosystem processes may have significant economic costs (Pimentel et al. 1992). Small-patch and shelterwood logging were found to be 10–38 percent more expensive than traditional clearcutting in the Montane Alternative Silvicultural Systems (MASS) project in western Canada (Phillips 1996). Elsewhere, strategies such as the retention of clumps of living trees were found to retard rates of regrowth on cutover sites (Incoll 1979; Rotherham 1983). Horne et al. (1991) estimated that modified harvesting methods would result in a substantial reduction in timber royalties during an eighty-year rotation across a 22,000-hectare forest management area in the Australian state of New South Wales. Different results have come from studies in other forests such as the Douglas-fir region of the United States. In that case, larger trees grown as a result of longer rotations and increased stand retention levels had increased relative values, which partially compensated for reduced growth rates and lower levels of extraction (Hansen et al. 1995b). Reduced impact logging (RIL) (sensu Putz et al. 2000) in tropical forests was actually significantly (more than 10 percent) cheaper than conventional logging methods, which did greater ecological damage (De Marajò and Paragomines 2001).

The trade-off between production costs and ecological benefits for biodiversity can be explored using integrated economic and ecological models (e.g., Hyde 1989; McKenney and Lindenmayer 1994; Hansen et al. 1995b; Carey et al. 1999b). Results from such models may reveal that conflicts between biodiversity conservation and commodity production objectives are less than initially anticipated, as in the case of second-growth conifer forests in Washington state (United States) (Carey et al. 1999b). In another example, Tarp et al. (2000) found superior ecological and economic outcomes from uneven-aged cutting practices in European beech (*Fagus sylvatica*) forests that mimicked natural disturbance regimes more successfully than clearcutting did.

In-depth economic assessments of matrix-based management are necessary. Analyses need to include the full cost to society of alternative management approaches, including effects on long-term site productivity, impacts on nontarget organisms, and long-term economic growth. Short-term job losses in the timber industry have most typically been the focus of economic analyses. However, matrix-based management may actually increase employment (Hueting 1996) and help reverse trends of declining employment associated with increasing mechanization in managed forests (Hammond 1991; Dargarvel 1995). Employment of forest workers in restoration and monitoring programs can make major short- and long-term contributions to local economies.

Trade-offs between Native Forest Management and Plantations

In a crude sense, matrix-based management is an approach to forests that is intermediate in intensity between preserves and plantations. Political, social, and economic objectives are ultimately translated to policies that are located on the continuum of possibilities (Figure 18.2).

In some cases, the best outcome for biodiversity conservation may be intensive plantation forestry in some areas traded off against enhanced matrix management (e.g., structural retention and longer rotations) or establishment of reserves in other areas. Hunter (1994) and Hunter and Calhoun (1996) discuss this issue using three broad classes of land use:

1. Plantation landscapes with very high levels of wood production
2. Multiple-use matrix lands that integrate some wood production with maintenance of biodiversity and other ecological values
3. Dedicated reserves with no wood production

Forests in the state of Maine (in the northeastern United States) are used to exemplify changes in land tenure that can arise from variations in the extent of plantation areas (Hunter 1994; Seymour and Hunter 1999) (see Figure 18.3). Since timber yields from plantation forests in Maine are three times greater than those from multiple-use matrix lands, wood production can be maintained while setting aside 3

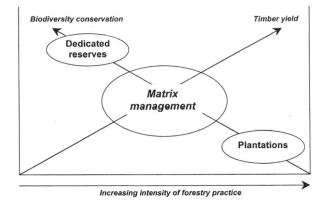

FIGURE 18.2. Simplified relationships between the intensity of forestry operations and timber yields, and biodiversity conservation.

hectares in dedicated reserves for each hectare allocated to intensive plantation forestry (Seymour and Hunter 1999). In Maine, 2 percent of the forest is currently in dedicated reserves, 90 percent in multiple-use forest management, and 6 percent in intensively managed plantations. A 2.5 percent increase in plantations could theoretically allow a 10 percent increase in dedicated reserves with no loss in wood production (Figure 18.3) (Hunter 1994). However, there are potential problems with such an approach (see below) and while it might be appropriate for Maine, it cannot be applied uncritically in all forest landscapes.

Simply allocating land to large ecological reserves and plantations is often complicated and controversial for several reasons. First, some conservationists may strongly oppose intensifying operations to the level of fiber farms because of potential losses of biodiversity (Routley and Routley 1975; Ray et al. 1983). A classic example is Tasmania, where extensive clearing of native forest and creation of exotic plantations will have major impacts on native forest diversity (see Box 5.5 in Chapter 5).

Second, areas committed to plantations, if not carefully selected, may foreclose any future options for conservation management, possibly impacting key elements of biodiversity (although ownership patterns often reduce opportunities to make such selections). Matrix management of native forests keeps open possible future options, while conversion of native forests to plantations may be difficult or impossible to reverse. Problems with attempting to rationally allocate

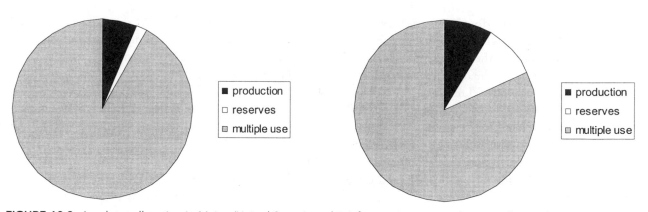

FIGURE 18.3. Land use allocation in Maine (United States) resulting from an increase in the area allocated to intensive plantation forestry. An increase in the area managed as plantations can enable a substantial increase in forest set aside as ecological reserves (from Hunter 1994). This approach has some advantages, but it also could have potentially negative impacts in some jurisdictions (see text).

land are further complicated by the fact that the flat and highly productive areas most suitable for plantations are also richest in biodiversity and sites of some key ecological processes (e.g., Braithwaite 1984; Harris 1984; Lindenmayer et al. 1997c; Spellerberg and Sawyer 1997).

Third, the long-term productivity of plantation forests depends on the presence of biota involved in nutrient and carbon cycling (see Chapter 1), but if intensive forestry operations do not provide for maintenance of such organisms, they ultimately jeopardize long-term forest health and productivity.

Fourth, intensively managed plantations often utilize significant chemical inputs in the form of fertilizers, herbicides, and pesticides. Such chemicals can have significant impacts on biota both within and outside plantations, including aquatic ecosystems. This could result in violations of criteria for sustainability (World Wide Fund for Nature 1996; Commonwealth of Australia 1997, 1998; Wallis et al. 1997) and thereby compromise the potential for forest certification.

Objectives of forest management will largely determine appropriate ecological, social, and economic trade-offs in land use allocations. For example, maintenance of timber yields from the state of Maine is the driver in the model described for Maine (Hunter 1994) (Figure 18.3). Assessments of alternatives, such as plantations versus integrated forestry approaches, need to include consideration of location and design

of plantations and impacts on biodiversity and ecological processes, as well as traditional cost-benefit analyses (McKenney and Common 1989; Hunter and Calhoun 1996). Also, conversion of native forests to plantations of exotic trees by widespread clearing of native forest is generally no longer viewed as ecologically acceptable.

Changing Roles and Challenges for Stakeholders in Forest Policy Development

The environment for development and implementation of forest policies in the twenty-first century is dramatically different than that of the previous century, regardless of what specific mix of policies are ultimately adopted. There are many reasons for the dramatic change, including emergence of the concept of the global village, with its attendant social as well as economic implications; the scale of many challenges, such as global environmental change; greatly expanded capabilities for mass and personal communication (related to this expansion is the greatly increased accessibility of information and the tools needed to interpret and manipulate that information); increased concerns for the environment, indigenous peoples, and social justice; and increased direct involvement of people in decisions related to the preceding factors.

Individual and institutional stakeholders in the management of forests and other natural resources are challenged to reconsider their roles, values, and attitudes by these changes in the development and implementation of forest policy. The challenges exist regardless of where regional, national, and global policies ultimately come down along the gradient of allocation and integration—although many changes are accentuated by the adoption of matrix-based approaches to conservation of biodiversity and ecosystem processes.

We begin this section by asserting that any stakeholder seriously committed to the goal of conserving forest biodiversity must recognize the importance of the matrix in achieving that goal and must work to incorporate matrix-based approaches into local, regional, national, and global conservation strategies. With that as a working premise, there will be challenges for everyone participating in the development of natural resource policies and projects. Decisions regarding the balance among various competing uses will be difficult and will involve trade-offs among environmental, economic, and social objectives, different ownerships, and even nations. Better communication, innovative institutional arrangements, very long term commitments, and some level of respect (if not trust) among stakeholders holding contrasting values are some of the challenges for all participants.

Challenges for Resource Professionals

Resource managers must assume new roles in the development and implementation of forest policies. The old resource managers (e.g., foresters and wildlife and fisheries managers) were often viewed by society and themselves as priesthoods or professional groups uniquely capable of comprehending complex scientific and technical matters and, hence, the appropriate sources of policy decisions. Today, the data sources (e.g., inventories and maps), management models, and tools (e.g., powerful computers equipped with geographic information systems) are available to many, if not most, stakeholders and their consultants, many of whom have scientific insights and technical capabilities equal to those of professional employees. Furthermore, these stakeholders often insist on being meaningfully involved in everything from policy decisions to on-the-ground project implementation. The

process of developing socially acceptable forest policies in infinitely varying natural and human landscapes is certainly extremely complex.

The roles of natural resource professionals are actually more demanding and exciting than they were in the previous century. They require comprehensive knowledge of the ecosystems and their constituent species, creativity in developing and analyzing management alternatives, and superior skills as communicators and as leaders in collaborations with stakeholders. Textbooks and manuals—professional orthodoxy—alone no longer suffice as guides.

Resource professionals have a primary role as the interface between ecosystems and society—perhaps not a new role but one that is rejuvenated and emphasized. Resource professionals are appropriately expected to be the individuals most knowledgeable regarding ecosystems under their stewardship and the limitations and potentials of these ecosystems under different policies and management regimes. Resource professionals must be able to communicate that knowledge objectively and comprehensively, particularly where public policies and public lands are involved. Selective use of information or abridgement of management alternatives, such as often occurred in the past, is not appropriate, but informing stakeholders of uncertainties and risks associated with various management alternatives is essential. Some implications for the education of resource professionals are obvious, including improved communication skills and expanded general knowledge of ecological processes (Jackson 1994) as well as intimate familiarity with the landscapes and ecosystems for which they have responsibility.

Challenges for Conservationists

Individuals and organizations involved with nature conservation need to evaluate their primary goals, particularly with regard to the relative importance of preserving "natural" ecosystems and landscapes versus the conservation of biodiversity. The highest priority for many conservationists and environmental organizations actually appears to be the reservation of areas lacking overt human activity—or, in other words, wild landscapes—and not threatened forest ecosystems or organisms (e.g., Armesto et al. 1998). Protecting all remaining primary forests is not congruent with the goal of sustaining the highest-possible level of forest biodiversity. Many policy decisions are zero-sum games in which objectives achieved in one quarter involve concessions in others; hence, short- and long-term trade-offs between various conservation objectives need more careful consideration than appears to have occurred in the past. Money, credibility, and political capital are not available in infinite supply. Hence, trade-offs need to be carefully considered in analyzing proposed programs and policies, particularly for potentially perverse policies that ultimately result in negative impacts on biodiversity conservation. For example, expansions in reserves that result in intensified management in the matrix can be a bad outcome for biodiversity conservation (Schwartz and van Mantgem 1997; see Chapter 5). Trading lands important for biodiversity for lands with high political appeal but that make little or no additional contribution to biodiversity conservation is another example of a perverse policy. Such circumstances do occur, as was the case recently in California. Well-meaning political initiatives to constrain management (e.g., to ban even-aged management) can also have perverse consequences if they create incentives to convert forestlands to other purposes, such as subdivisions.

Challenges for Scientists

Resource management in the twenty-first century provides many challenges to scientists. In our view, one challenge is to better understand that their professional role is to generate scientific information, not to make policy decisions. A part of that process involves recognizing the personal values and biases that they bring to the table and to separate those from the objective knowledge that they are expected to provide. Scientists need to improve their communication skills (Noble 1994). Increasing dialogue and collaboration among natural and social scientists is crucial, and natural scientists in particular need a much better understanding of policy-making processes.

Scientists need to extend their participation in policy processes beyond published reports and scientific papers. For example, the direct communication of scientific results to stakeholders, including resource managers and politicians, is important, particularly when the science has relevance to policy or practice (Wills and Hobbs 1998). Many scientists appear to operate under a "strategy of hope," in which they simply hope that their work will be useful for management professionals but do nothing to further that goal (Hamel and Prahalad 1989). Participation by scientists in the design and implementation of adaptive management activities is particularly important (see Chapter 16). Scientists can assist by contributing new concepts, particularly those that lie outside traditional approaches (i.e., paradigm shifts), such as the variable retention harvest system in northwestern North America (see Chapter 8).

Scientists need to use language that is understood by other stakeholders in defining problems and presenting solutions (Warren 1993). They have to work to make key concepts and themes understandable. As elegantly put by Naiman and Turner (2000), scientists not only need to "do good science but also [to do] useful science that is used in decision-making."

Resolution of disputes over interpretation of scientific studies is already becoming a challenging topic as face-offs between contending scientific consultants increase in legislative and judicial forums. Kirkpatrick (1998) has argued that such debates should be more open. Independent or third-party review has been proposed (and used) as a mechanism to ensure that environmental decisions and policy are based on the best-available scientific information (Meffe et al. 1998) and are open and transparent (Horowitz and Calver 1998). Calver et al. (1997) argued that scientific debates should include full publication of relevant datasets in the refereed scientific literature, careful evaluation and analysis (and reevaluation and reanalysis) of existing datasets, and targeted research programs to address knowledge deficiencies.

Finally, scientists need to be more aware that science is only part of the basis for policy decisions but not necessarily the primary basis. Social factors, including politics and economics, are also major and often dominant factors (Funtowicz and Ravetz 1991).

Challenges for Organizations and Institutions

If matrix management is to succeed, coordinated management will be required over different land tenures, from intensely managed forest landscapes to national parks and wilderness areas (Recher 1985; Norton and Lindenmayer 1991). This necessitates coordination and integration among organizations that have sometimes been antagonistic (Wright 1999) because of fundamentally different objectives and institutional cultures (Shea et al. 1997; Bissix and Rees 2001). In some cases there have been struggles over land jurisdiction, access to upper echelons of government, and legitimacy for their philosophy of land management. These antagonisms are added to other difficulties in collaboration, such as separate and incompatible databases. Another important constraint for corporate landowners in some countries is legal liabilities associated with collaborative planning for management of natural resources; such cooperative efforts can be viewed as illegal collusion among market competitors.

Organizations need to develop collaborative approaches and incentive structures to overcome technical barriers and past antagonisms (Bissix and Rees 2001). Models from other resource sectors, such as the water industry, may provide examples for forestry. For example, in Australia, the Murray-Darling Basin Commission has oversight of multisector interests across multiple states. In the United States, President Clinton directed all federal agencies to actively collaborate in the development and implementation of the Northwest Forest Plan, a goal that was effectively achieved and is reflected in the continued existence of the Regional Ecosystem Office staffed by individuals from all affected federal agencies (Pipkin 1998).

There are many other examples of coordinated forest research programs. The Canadian Forest Service maintains a section dedicated to technology transfer of research outcomes (D. McKenney personal communication). Klenner and Vyse (1999) even suggest the adoption of a reward system for researchers and managers to participate in multidisciplinary studies that address forest management problems in that country. Rogers (1997) has outlined a scientist/manager interface to improve natural resource management and discussed its value in conservation programs in Kruger National Park in South Africa.

Indeed, there is a very long history of collaboration among disparate forest organizations in the arena of forest protection. For example, cooperative fire suppression programs have existed in North America for nearly a century. Similar multi-owner, multi-agency programs have been successfully implemented with regard to other aspects of forest protection. There are certainly lessons that can be applied to collaborative efforts for conservation of biodiversity.

Changes in education and training also can contribute to better cooperation between organizations and lead potentially to integrated approaches, such as matrix management. Multidisciplinary training of natural resource professionals promotes an improved ability to manage across land tenures and provides the skills needed to tackle multifaceted issues.

Finally, organizations are challenged to devote adequate resources to management of databases, including the establishment of appropriate protocols (e.g., for documentation). High-quality, long-term, spatially explicit datasets on resource conditions, management activities (objectives, plans, and treatments), and monitoring are absolutely essential. Development of compatible approaches, at least among government agencies, is also an important goal.

Challenges to the Body Politic

Making the difficult choices between various alternative natural resource strategies is probably the greatest challenge for political bodies. As was very clear in the last decades of the twentieth century, decisions regarding environmental policy (including conservation of biodiversity) involve major trade-offs among environmental, social, and economic objectives. The debates and decision-making processes need to incorporate all relevant scientific and technical information, which is itself a challenge. Presenting alternatives and outcomes to political bodies in a clear and concise fashion can make explicit the costs, benefits, uncertainties, and risks to all parties (including decision makers) (i.e., it will often make clear that difficult choices must be made) (Johnson et al. 1991).

Providing and sustaining adequate financial support for conservation programs is an immense political challenge once policies and programs are adopted. Sustaining long-term financial support is particularly important in managing natural resources, including biodiversity, and a difficult challenge for governments, with their annual or biennial funding cycles and constant budgetary pressures (Yaffee 1994; Lugg 1998).

There are techniques to circumvent such problems, such as development of trusts or issuance of bonds. Ultimately, solutions to the problem of sustained, long-term support depend upon public acceptance of and commitment to these programs, which will require continuing education and involvement of citizenry.

There is no real alternative to public involvement and collaboration among all stakeholder groups in the development and implementation of policies for the conservation of forest biodiversity. We believe that this is the only way of generating policies that meet the final and absolute test: to be sustainable, policies must be socially acceptable.

Future Directions

. . . it is the history of all great industries . . . that the necessity for modification is not seen until the harm has been done and the results are felt.

—OVERTON PRICE [1902] IN GREGORY (1997)

More than 90 percent of the world's forests are in the matrix outside reserve systems. Strategies to manage matrix lands for the conservation of forest biodiversity are essential and are fundamental to any comprehensive strategy for biodiversity conservation. If the matrix is managed appropriately, the majority of forest biodiversity will actually persist there, augmenting the important but inadequate system of large ecological reserve systems (both current and potential).

Consequently, management strategies in the matrix will determine overall levels of forest biodiversity that are sustained, including (1) size, viability, and distribution of populations of most forest-dependent organisms, (2) connectivity, by facilitating or obstructing the movement of organisms, and (3) a level of buffering for sensitive areas, including reserves.

Management of the matrix to maintain ecosystem processes is also critical to sustained production of goods and services, such as wood, fiber, and well-regulated, high-quality water. Although we have not addressed this in detail, we strongly believe that societies must and will come to better comprehend and value the services provided by forests.

The matrix has been consistently overlooked or ignored in conservation biology despite its importance (Chapter 2). The focus has been on habitat (reserves) and nonhabitat (the matrix or the "rest" of the landscape). Reductionist theories and approaches have oversimplified the complexity of real landscapes. Many so-called general conservation principles have

not provided good service in attempts to conserve biodiversity. This has been elegantly noted by Harrison (1991):

> it may be both unrealistic and dangerous to promote general "principles of conservation biology," as is sometimes done on the grounds that non-academics must be presented with simple rules. The alternative is to accept that conservation biology is an essentially empirical science . . . and that in the practical arena, we may do better to explain than to hide the complexities and uncertainties involved.

We believe that explicit considerations of the matrix as a key landscape component will better inform ecological theory and advance efforts to conserve biodiversity. Greater emphasis should be placed on such topics as responses of species to matrix conditions, use and movement of biota through the matrix, and the influences of modified matrix management on edge effects. Relatively disparate disciplines within conservation biology and landscape ecology, such as conservation genetics, corridor design, and habitat fragmentation, might become better integrated if they focused more on the matrix, thereby increasing our understanding of the responses of biota to human-modified environments and management.

Maintenance of habitat across multiple spatial scales—from individual stands to landscapes to entire regions—must be the overarching objective of plans for biodiversity conservation if they are to be effective. Five principles encompass this important goal:

1. Maintenance of connectivity
2. Maintenance of landscape heterogeneity
3. Maintenance of stand structural complexity
4. Maintenance of aquatic system integrity
5. Adoption of "risk-spreading" whereby a range of management options is deployed to limit negative consequences if any one strategy proves ineffective or hostile (Chapter 3)

The use of knowledge and inferences from natural disturbance regimes to guide and modify human disturbances such as logging is a sixth guiding principle (Chapter 4) and can help achieve the goals discussed in Chapter 3.

Maintaining habitat across multiple spatial scales requires a comprehensive approach to conservation that includes large ecological reserves (Chapter 5), landscape-level approaches within the matrix (Chapters 6 and 7), and stand-level strategies (Chapter 8). Each of these broad components contributes to fulfilling the principles for enhanced biodiversity conservation listed in Chapter 3. These groupings illustrate that appropriate forest management requires careful management at a wide range of interconnected scales, including entire landscape mosaics, and not just ecological reserves or considerations of specific stand conditions such as primeval or old-growth forests.

The importance of the matrix for biodiversity conservation applies to private forest lands as well as government-owned production land (Gregory 1997). Indeed, private lands dominate the forest estate in some countries (e.g., Sweden, with more than 85 percent of forest land in private ownership; National Board of Forestry 1996a), and typically they occupy the most productive parts of the landscape (Scott et al. 2001a,b). Such high-productivity sites often are uniquely important for many elements of biodiversity (Braithwaite et al. 1993; Norton 1999) but have also suffered major human disturbances, such as conversion to agricultural (Armesto et al. 1998) and urban lands (Knight 1999).

Comprehensive plans for biodiversity conservation are underpinned by a hierarchical framework or checklist of strategies that includes not only large ecological reserves but also a suite of landscape and stand-level matrix management approaches. At the landscape level (see Chapter 6), these include

- Protection of specialized habitats, biodiversity hotspots, and sensitive areas
- Establishment of systems of retained habitat, such as wildlife corridors and riparian stream buffers
- Careful design, construction, and maintenance of road networks
- Long rotation periods

Approaches at the stand level include

- Retention of structural and floristic components from the original stand (e.g., large living trees, snags, logs, and intact thickets of understory vegetation)
- Active management of regenerated stands using such approaches as variable density thinning and decadence creation to accelerate development of key of forest structures
- Long rotation periods

It is important to consider each of the strategies provided in the checklist (presented in Chapters 6 and 8). Decisions as to which strategies are embraced vary according to the objectives of matrix management, the species targeted for conservation, and the natural history of the forest. Each forest landscape will have a unique solution, as is highlighted by the broad spectrum of outcomes in the case studies in Chapters 11–15.

Quantitative data to guide on-the-ground prescriptions are limited. This lack of knowledge is partly because few scientists and managers have focused on the role of the matrix in conserving biota. Adaptive management will facilitate the gathering of additional knowledge to guide effective matrix management; such information might include the types, numbers, and spatial patterns of retained vegetation needed to achieve specific goals on harvest units. Long-term monitoring is an essential component of the adaptive process, and ensuring adequate and sustained support for monitoring is a major challenge (Chapter 16). Implementation of well-designed and statistically valid monitoring programs (including adequate and sustained financial support) is a major challenge for stakeholders, managers, and decision makers worldwide. Without such programs there is no way to assess whether a forest is being managed in an ecologically sustainable way. Legal and market-based requirements for such assessments, such as the Montreal Criteria for Sustainability and forest certification agreements, should stimulate the adoption of more monitoring programs.

Comprehensive conservation plans that incorporate matrix management embrace a greater range of values than simple commodity production can. Enhanced forest stewardship for biodiversity conservation and the maintenance of ecosystem functions should enhance community support for forest management as a more socially acceptable use of native forests (Dargarvel 1995; Florence 1996). Managing for these additional forest values is seen by some parties as a constraint and an impediment (Thomas et al. 1988), but we choose to view such situations as new employment opportunities for stakeholders representing a broad range of interests (Chapter 18). Indeed, matrix management has the potential to increase employment in forest-related activities.

Past conflicts over use of forest resources has usually been resolved by land allocations and by dividing the forest estate into ecological reserves and commodity production areas because

- Land allocations create a perception of certainty in resource management and conservation outcomes.
- Land allocations are often the simplest option for policy makers and politicians seeking to resolve social and environmental conflict.
- Strong differences in values and high levels of distrust exist between conservationists and commodity producers, accentuating their mutual desires for segregation within the forest estate.

Reliance predominantly on the land allocation model, with its attempt at spatial segregation of the environmental and commodity values of the global forest estate, will ultimately have negative consequences for biodiversity (Chapter 5) and impair the productive capacity of forest ecosystems (Chapter 1).

If matrix management is to be more broadly adopted, stakeholders will have to put aside differences and collaborate in new working partnerships. Stakeholders must understand that although it is complex and adaptive (i.e., uncertain), matrix management

is essential to sustain both biodiversity and forest productivity (Armesto et al. 1998). There are encouraging examples from other countries (e.g., Sweden) that the necessary transitions can occur quickly.

Many countries and regions are in transition from traditional maximum sustained yield harvesting and "crop production" to "ecological forestry" (sensu Hunter 1999; Seymour and Hunter 1999), where biodiversity and ecological processes are considered to be important values of matrix forests along with commodity production (Chapter 18). Yet other regions (such as some parts of southeastern Australia) appear to be taking the opposite approach by instead intensifying commodity production in matrix lands (Bauhaus 1999). We argue in this book that intensification can have substantial *negative* implications for forest biodiversity conservation regardless of preservation strategies. Moreover, ecosystem integrity and forest productivity can also be severely compromised when land use practices are narrowly focused (Holling and Meffe 1996), repeating mistakes in forest management made repeatedly through human history (McNeely 1994b). Indeed, the history of natural resource use throughout the history of the world has been one of unwise use and overexploitation— whether it be fish stocks (Ludwig et al. 1993) or forests (Angelstam 1996).

Rapidly advancing technology and continued population growth mean that the potential impacts on the global forest estate and its biodiversity are unprecedented and may be permanent. Myers (1996) argued that "our present stewardship of the world's forests will affect the future of evolution for a period twenty times longer than humans have been humans."

Utilization of forests is inevitable due to the substantial human demands for commodities from forest landscapes. The critical importance of ecologically sustainable forest management in both the matrix and the reserves is now widely understood. We hope that present societies will increasingly appreciate the important conservation role the matrix plays and adopt appropriate matrix management strategies as part of comprehensive plans for biodiversity conservation, thereby increasing the potential for sustaining more of the world's biota.

LITERATURE CITED

Aapala, K., R. Heikkila, and T. Lindholm. 1996. Protecting the diversity of Finnish mires. Pp. 45–57 in *Peatlands in Finland*, ed. H. Vasander. Helsinki: Finnish Peatlands Society.

Aars, J., and R. A. Ims. 1999. The effect of habitat corridors on rates of transfer and interbreeding between vole demes. *Ecology* 80:1648–1655.

Abbe, T. B., and D. R. Montgomery. 1996. Large woody debris jams, channel hydraulics and habitat formation in large rivers. *Regulated Rivers Research and Management* 12:201–221.

Abbott, I., and P. Christensen. 1994. Applications of ecological and evolutionary principles to forest management in Western Australia. *Australian Forestry* 57:109–122.

Abensperg-Traun, M., and G. T. Smith. 2000. How small is too small for small animals? Four terrestrial arthropod species in different-sized remnants in agricultural Western Australia. *Biodiversity and Conservation* 8:709–726.

Aber, J., N. Christensen, I. Fernandez, J. Franklin, L. Hidinger, M. Hunter, J. MacMahon, D. Mladenoff, J. Pastor, D. Perry, R. Slagen, and H. van Miegroet. 2000. Applying ecological principles to management of the U.S. national forests. *Issues in Ecology* 6:1–20.

Åberg, J., J. E. Swenson, and P. Angelstam. 1995. The effect of matrix on the occurrence of hazel grouse (*Bonasa bonasia*) in isolated habitat fragments. *Oecologia* 103:265–269.

Acker, S. A., T. E. Sabin, L. M. Ganio, and W. A. McKee. 1998. Development of old-growth structure and timber volume growth trends in maturing Douglas-fir stands. *Forest Ecology and Management* 104:265–280.

Adams, E. M., and M. L. Morrison. 1993. Effects of forest stand structure and composition on red-breasted nuthatches and brown creepers. *Journal of Wildlife Management* 57:616–629.

Adams, L. W., and A. D. Geis. 1983. Effects of roads on small mammals. *Journal of Applied Ecology* 20:403–415.

Adams, M. H., and P. M. Attiwill. 1984. The role of *Acacia* spp. in nutrient balance and cycling in regenerating *Eucalyptus regnans* F. Muell. forests. 1. Temporal changes in biomass and nutrient content. *Australian Journal of Botany* 32:205–215.

Adler, G. H., and R. Levins. 1994. The island syndrome in rodent populations. *Quarterly Review of Biology* 69:473–490.

Agee, J. K. 1993. *Fire ecology of the Pacific Northwest forests*. Washington, D.C.: Island Press.

———. 1999. Fire effects on landscape fragmentation in interior west forests. Pp. 43–60 in *Forest fragmentation: Wildlife management implications*, ed. J. A. Rochelle, L. A. Lehmann, and J. Wisniewski. Leiden, Germany: Brill.

Agee, J. K., B. Bahro, M. A. Finney, P. N. Omi, D. B. Sapsis, C. N. Skinner, J. W. van Wagtendonk, and C. P. Weatherspoon. 2000. The use of shaded fuelbreaks in landscape fire management. *Forest Ecology and Management* 127:55–66.

Agee, J. K., and M. H. Huff. 1987. Fuel succession in a western hemlock/Douglas fir forest. *Canadian Journal of Forest Research* 17:697–704.

Akçakaya, H. R., and S. Ferson. 1990. *RAMAS/space user manual: Spatially structured population models for conservation biology*. Setauket, N.Y.: Exeter Software.

Alaback, P. B. 1982. Dynamics of understory biomass in Sitka spruce–western hemlock forests of southeast Alaska. *Ecology* 63:1932–1948.

———. 1984. *Plant succession following logging in the Sitka spruce–western hemlock forests of southeast Alaska: Implications for management*. USDA Forest Service General Technical Report PNW-173.

Alban, D. H. 1969. The influence of western hemlock and western redcedar on soil properties. *Soil Science Society of America Proceedings* 33:453–459.

Aldrich, P. R., and J. L. Hamrick. 1998. Reproductive dominance of pasture trees in a fragmented tropical forest mosaic. *Science* 281:103–105.

Alexander, R. R. 1964. Minimizing windfall around clear cuttings in spruce-fir forests. *Forest Science* 10:130–142.

Allee, W. C. 1931. *Animal aggregations: A study in general sociology*. Chicago: University of Chicago Press.

Allee, W. C., A. E. Emerson, O. Park, T. Park, and K. P. Schmidt. 1949. *Principles of animal ecology*. Philadelphia: Saunders.

Allen, A. S., J. A. Andrews, A. C. Finzi, R. Matamala, D. D. Richter, and W. H. Schlesinger. 2000. Effects of free-air CO_2 enrichment (FACE) on belowground processes in a *Pinus taeda* forest. *Ecological Applications* 10:437–448.

Allen, D. H. 1991. *An insert technique for constructing artificial red-cockaded woodpecker cavities*. USDA Forest Service General Technical Report SE-73.

Allen, R., K. Platt, and S. Wiser. 1995. Biodiversity in New Zealand plantations. *New Zealand Forestry* 39:26–29.

Allen, T. F. H., and T. W. Hoekstra. 1992. *Toward a unified ecology*. New York: Columbia University Press.

Allen, T. F. H., and T. B. Starr. 1988. *Hierarchy: Prespectives for ecological complexity*. Chicago: University of Chicago Press.

Alverson, W. S., W. Kuhlmann, and D. M. Waller. 1994. *Wild forests conservation biology and public policy*. Washington, D.C.: Island Press.

Amaranthus, M. P. 1997. *Forest sustainability: An approach to definition and assessment at the landcape level*. December. General Technical Report PNW-GTR-416. Pacific Northwest Research Station, USDA Forest Service.

Amaranthus, M. P., and D. A. Perry. 1994. The functioning of ectomycorrzhial fungi in the field: Linkages in space and time. *Plant and Soil* 159:133–140.

Amaranthus, M., J. M. Trappe, L. Bednar, and D. Arthur. 1994. Hypogeous fungal production in mature Douglas-fir forest fragments and surrounding plantations and its relation to coarse woody debris and animal mycophagy. *Canadian Journal of Forest Research* 24:2157–2165.

Ambrose, G. J. 1982. An ecological and behavioural study of vertebrates using hollows in eucalypt branches. Ph.D. diss., La Trobe University, Melbourne, Australia.

Ambrose, J. P., and S. P. Bratton. 1990. Trends in landscape heterogeneity along the borders of the Great Smoky Mountains National Park. *Conservation Biology* 4:135–143.

Ambuel, B., and S. Temple. 1983. Area-dependent changes in the bird communities and vegetation of southern Wisconsin forests. *Ecology* 64:1057–1068.

Andelman, S. J., and W. F. Fagan. 2000. Umbrellas and flagships: Efficient conservation surrogates or expensive mistakes? *Proceedings of the National Academy of Sciences* 97:5954–5959.

Anderson, H. W., M. D. Hoover, and K. G. Reinhart. 1976. *Forests and water: Effects of forest management on floods, sedimentation and water supply*. USDA Forest Service, Berkeley, Calif.

Anderson, R. C., J. A. Fralisch, and J. M. Baskin. 1999. *Savannas, barrens, and rock outcrop communities of North America*. Cambridge: Cambridge University Press.

Andreassen, H. P., S. Halle, and R. Ims. 1996. Optimal width of movement corridors for root voles: Not too narrow and not too wide. *Journal of Applied Ecology* 33:63–70.

Andrén, H. 1992. Corvid density and nest predation in relation to forest fragmentation: A landscape perspective. *Ecology* 73:794–804.

———. 1994. Effects of habitat fragmentation on birds and mammals in landscapes with different proportions of suitable habitat: A review. *Oikos* 71:355–366.

———. 1996. Population responses to habitat fragmentation: Statistical power and the random sample hypothesis. *Oikos* 76:235–242.

———. 1997. Habitat fragmentation and changes in biodiversity. *Ecological Bulletins* 46:171–181.

———. 1999. Habitat fragmentation, the random sample hypothesis and critical thresholds. *Oikos* 84:306–308.

Andrén, H., and P. Angelstam. 1988. Elevated predation rates as an edge effect in habitat islands: Experimental evidence. *Ecology* 6:544–547.

Andrén, H., and A. Delin. 1994. Habitat selection in the Eurasian red squirrel, *Sciurus vulgaris*, in relation to forest fragmentation. *Oikos* 70:43–48.

Andrew, N., L. Rodgerson, and A. York. 2000. Frequent fuel-reduction burning: The role of logs and associated leaf litter in the conservation of ant biodiversity. *Austral Ecology* 25:99–107.

Angelstam, P. 1992. Conservation of communities—the importance of edges, surroundings and landscape mosaic structure. Pp. 9–70 in *Ecological principles of nature conservation*, ed. L. Hansson. Barking, U.K.: Elsevier.

———. 1996. The ghost of forest past—natural disturbance regimes as a basis for reconstruction for biologically diverse forests in Europe. Pp. 287–337 in *Conservation of faunal diversity in forested landscapes*, ed. R. M. DeGraaf and R. I. Miller. London: Chapman and Hall.

———. 1997. Landscape analysis as a tool for the scientific management of biodiversity. *Ecological Bulletins* 46:140–170.

Angelstam, P., P. Majewski, and S. Bondrup-Neilsen. 1995. West-east cooperation in Europe for sustainable boreal forests. *Water, Air and Soil Pollution* 82:3–11.

Angelstam, P., and B. Pettersson. 1997. Principles of present Swedish forestry biodiversity management. *Ecological Bulletins* 46:191–203.

Angermeier, P. I. 1995. Ecological attributes of extinction-prone species: Loss of freshwater fishes of Virginia. *Conservation Biology* 9:143–158.

Aplet, G. H., and W. S. Keeton. 1999. Application of historic range of variability concepts to biodiversity conservation. Pp. 71–86 in *Practical approaches to the conservation of biological diversity*, ed. Richard K. Baydack, Henry Campa III, and Jonathan B. Haufler. Washington, D.C.: Island Press.

Anonymous. 1992. *Guidelines for selecting reserve trees*. Washington Department of Natural Resources; Olympia, Wash.

———. 1995. *Biodiversity guidebook. Forest Practices Code of British Columbia*. Ministry of Forests, British Columbia and B.C. Environment, Vancouver, British Columbia.

———. 1996. *National forest conservation reserves*. Commonwealth proposed criteria. Commonwealth of Australia. July. Canberra.

Arborvitae. 1995. *Arborvitae*. IUCN/WWF Forest Conservation Newsletter, 1 (September): 5.

Armbruster, P., and R. Lande. 1993. A population viability analysis for African elephant (*Luxodonta africana*): How big should reserves be? *Conservation Biology* 7:602–610.

Armesto, J. J., R. Rozzi, C. Smith-Ramirez, and M. T. Arroyo. 1998. Conservation targets in South American temperate forests. *Science* 282:1271–1272.

Arnold, G. W. 1983. The influence of ditch and hedgerow structure, length of hedgerows, and area of woodland and garden on bird numbers on farmland. *Journal of Applied Ecology* 20:731–750.

Arnold, G. W., D. E. Steven, and J. R. Weeldenburg. 1993. Influences of remnant size, spacing pattern and connectivity on population boundaries and demography in Euros *Macropus robustus* living in a fragmented landscape. *Biological Conservation* 64:219–230.

Arnold, G. W., and J. R. Weeldenburg. 1998. The effects of isolation, habitat fragmentation and degradation by livestock grazing on the use by birds of patches of Gimlet *Eucalyptus salubris* woodland in the wheatbelt of Western Australia. *Pacific Conservation Biology* 4:155–163.

Arnott, J. T., W. J. Beese, A. K. Mitchell, and J. Peterson. 1995. *Montane alternative silvicultural systems (MASS)*. FRDA Report no. 238. Forest Resource Development Agreement. Victoria, British Columbia, Canada.

Arrhenius, O. 1921. Species and area. *Journal of Ecology* 9:95–99.

Arroyo, M. T. K., C. Donoso, R. E. Murua, E. E. Pisano, R. P. Schlatter, and I. A. Serey. 1996. *Toward an ecologically sustainable forestry project: Concepts, analysis and recommendations. Protecting biodiversity and ecosystem processes in the Rio Condor Project, Tierra del Fuego*. Departamento de Investigacion y Desarrollo, Universidad de Chile, Santiago, Chile.

Ås, S. 1999. Invasion of matrix species in small habitat patches. *Conservation Ecology* (online) 3(1). http://ns2.resalliance.org/pub/www/Journal/vol3/iss1/art1/index.html.

Ashton, D. H. 1976. The development of even-aged stands of *Eucalyptus regnans* F. Muell. in central Victoria. *Australian Journal of Botany* 24:397–414.

———. 1986. Ecology of bryophytic communities in mature *Eucalyptus regnans* F. Muell. forest at Wallaby Creek, Victoria. *Australian Journal of Botany* 34:107–129.

Asia-Pacific Forestry Commission. 1997. *Code of practice for forest harvesting in Asia-Pacific*. November. Asia-Pacific Forestry Commission.

Askins, R. A., and M. J. Philbrick. 1987. Effects of changes in regional forest abundance on the decline and recovery of a forest bird community. *Wilson Bulletin* 99:7–21.

Askins, R. A., M. J. Philbrick, and D. S. Sugeno. 1987. Relationships between the regional abundance of forest and the composition of bird communities. *Biological Conservation* 39:129–152.

Attiwill, P. M. 1994. Ecological disturbance and the conservative management of eucalypt forests in Australia. *Forest Ecology and Management* 63:301–346.

Aust, W. M., and R. Lea. 1992. Comparative effects of aerial and ground logging on soil properties in a tupelo-cypress wetland. *Forest Ecology and Management* 50:57–73.

Austin, M. P. 1999. A silent clash of paradigms: Some inconsistencies in community ecology. *Oikos* 86:170–178.

Austin, M. P., E. M. Cawsey, B. L. Baker, M. M. Yialeloglou, D. J. Grice, and S. V. Briggs. 2000. *Predicted vegetation cover in the Central Lachlan Region. Report and Appendices*. Project conducted under the Bushcare program of the Natural Heritage Trust. September. CSIRO. NSW National Parks and Wildlife Service. Natural Heritage Trust.

Austin, M. P., and C. R. Margules. 1986. Assessing representativeness. Pp. 45–67 in *Wildlife conservation evaluation*, ed. M. B. Usher. London: Chapman and Hall.

Austin, M. P., A. O. Nicholls, and C. R. Margules. 1990. Measurement of the realized qualitative niche: Environmental niches of five *Eucalyptus* species. *Ecological Monographs* 60:161–177.

Awimbo, J. A., D. A. Norton, and F. B. Overmars. 1996. An evaluation of representativeness for nature conservation, Hokitika Ecological District, New Zealand. *Biological Conservation* 75:177–186.

Backhouse, F., and J. D. Lousier. 1991. *Silviculture systems research: Wildlife tree problem analysis*. Report for the Ministry of Forests,

Ministry of Environment and B.C. Wildlife Tree Committee. July. Conducted by W.F.S. Enterprises, Nanaimo, British Columbia, Canada.

Backhouse, F., and T. Manning. 1996. *Predator habitat enhancement as a strategy for managing small mammal damage in B.C. forests: A problem analysis.* Guidelines prepared for Ministry of Environment, B.C. Forest Service, British Columbia, Canada.

Bader, P., S. Jansson, and B. Jonsson. 1995. Wood-inhabiting fungi and substratum decline in selectively logged boreal forest. *Biological Conservation* 72:355–362.

Baillie, S. R., W. J. Sutherland, S. N. Freeman, R. D. Gregory, and E. Paradis. 2000. Consequences of large-scale processes for the conservation of birds. *Journal of Applied Ecology* 37 (Supplement 1):88–102.

Baker, J. B., comp. 1994. *Proceedings of the symposium on ecosystem management research in the Ouchita Mountains: Pretreatment conditions and preliminary findings.* USDA Forest Service General Technical Report SO-112.

Baker, W. L. 1992. The landscape ecology of large disturbances in the design and management of nature reserves. *Landscape Ecology* 7:181–194.

———. 1995. Long-term response of disturbance landscapes to human intervention and global change. *Landscape Ecology* 10:143–159.

Ball, I., D. B. Lindenmayer, and H. P. Possingham. 1997. A tree hollow simulation model for forest managers: The dynamics of the absence of wood in trees. MODSIM. *Proceedings of the Simulation Society Conference, Hobart:* 1580–1585.

———. 1999. HOLSIM: A model for simulating hollow availability in managed forest stands. *Forest Ecology and Management* 123:179–194.

Banks, J. C. 1993. *Tree-ring analysis of two mountain ash trees* Eucalyptus regnans *F. Muell from the Watts and O'Shannassy Catchments, Central Highlands, Victoria.* A Report to the Central Highlands Old Growth Forest Project. August. Department of Conservation and Natural Resources, Melbourne, Australia.

Barbour, M. G., and W. D. Billings, eds. 2000. *North American terrestrial vegetation.* 2nd ed. Cambridge: Cambridge University Press.

Barbour, M. G., and J. Major, eds. 1977. *Terrestrial vegetation of California.* New York: John Wiley and Sons.

Barbour, M. S., and J. A. Litvaitis. 1993. Niche dimensions of New England cottontails in relation to habitat patch size. *Oecologia* 95:321–327.

Barclay, S. D., D. M. Rowell, and J. E. Ash. 1999. Phermonally mediated colonization patterns in Onychophora. *Journal of Zoology* 250:437–446.

Barker, P., and J. B. Kirkpatrick. 1994. *Phyllocladus aspleniifolius:* Variability in the population structure, the regeneration niche and dispersion patterns in Tasmanian forest. *Australian Journal of Botany* 42:163–190.

Barker, P. C., T. J. Wardlaw, and M. J. Brown. 1996. Selection and design of *Phytophthora* management areas for the conservation of threatened flora in Tasmania. *Biological Conservation* 95:187–193.

Barling, R. D., and I. D. Moore. 1994. The role of buffer strips in the management of waterway pollution: A review. *Environmental Management* 18:543–558.

Barnett, J. L., R. A. How, and W. F. Humphreys. 1978. The use of habitat components by small mammals in eastern Australia. *Australian Journal of Ecology* 3:277–285.

Barrett, G. W., H. A. Ford, and H. F. Recher. 1994. Conservation of woodland birds in a fragmented rural landscape. *Pacific Conservation Biology* 1:245–256.

Bart, J., and E. D. Forsman. 1992. Dependence on northern spotted owl *Strix occidentalis caurina* on old growth forests in western USA. *Biological Conservation* 62:95–100.

Barton, D. R., W. D. Taylor, and R. M. Biette. 1985. Dimensions of riparian buffer strips required to maintain ba habitat in southern Ontario streams. *North American Journal of Fisheries Management* 5:364–378.

Barton, J. L., and P. E. Davies. 1993. Buffer strips and streamwater contamination by atrazine and pyrethenoids aerially applied to *Eucalyptus nitens* plantations. *Australian Forestry* 56:201–210.

Baudry, J. 1984. Effects of landscape structure on biological communities: The case of hedgerow network landscapes. Pp. 55–65 in *Proceedings of the First International Seminar on Methodology in Landscape Ecological Research and Planning,* vol. 1, ed. J. Brandt and P. Agger. Roskilde, Denmark: Roskilde Universitetsforlag Geo.

Bauhaus, J. 1999. Silvicultural practices in Australian native forests—an introduction. *Australian Forestry* 62:217–222.

Baumgartner, L. L. 1939. Fox squirrel dens. *Journal of Mammalogy* 20:456–465.

Baur, A., and B. Baur. 1990. Are roads barriers to dispersal in the land snail *Arianta arbustorum? Canadian Journal of Zoology* 68:613–617.

———. 1992. Effect of corridor width on animal dispersal: A simulation study. *Global Ecology and Biogeography Letters* 2:52–56.

Baxter, C. V., C. A. Frissell, and F. R. Hauer. 1999. Geomorphology, logging roads, and the distribution of bull trout (*Salvelinus confluentus*) spawning in a forested river basin: Implications for management and conservation. *Transactions of the North American Fisheries Society* 128:854–867.

Bayley, P. B. 1995. Understanding large river-floodplain ecosystems. *BioScience* 45:153–158.

Bayne, E. M., and K. A. Hobson. 1997. Comparing the effects of landscape fragmentation by forestry and agriculture on predation of artificial nests. *Conservation Biology* 11:1418–1429.

———. 1998. The effects of habitat fragmentation by forestry and agriculture on the abundance of small mammals in the southern boreal mixedwood forest. *Canadian Journal of Zoology* 76:62–69.

BC Environment. 1993. *Guidelines for maintaining biodiversity during juvenile spacing.* Canada–British Columbia Forest Resource Development Agreement. B.C. Environment and B.C. Ministry of Forestry. Victoria, British Columbia, Canada.

———. 1995. *A bibliography of selected literature on wildlife trees with annotations and abstracts.* Wildlife Working Report no. 66. Wildlife Tree Committee of British Columbia, B.C. Ministry of Environment, Land and Parks. British Columbia, Canada.

Beaty, R. M., and A. H. Taylor. 2001. Spatial and temporal variation of fire regimes in a mixed conifer forest landscape, Southern Cascades, California, U.S.A. *Journal of Biogeography* 28:955–966.

Bedward, M., R. L. Pressey, and D. A. Keith. 1992. A new approach for selecting fully representative reserve networks: Addressing efficiency, reserve design and land suitability with iterative analysis. *Biological Conservation* 62:115–125.

Beese, W. J., and A. A. Bryant. 1999. Effect of alternative silvicultural systems on vegetation and bird communities in coastal montane forests of British Columbia, Canada. *Forest Ecology and Management* 115:231–242.

Behan, R. W. 1990. Multi resource forest management: A paradigmatic challenge to professional forestry. *Journal of Forestry* 15:12–18.

Beier, P. 1993. Determining minimum habitat areas and habitat corridors for cougars. *Conservation Biology* 7:94–108.

Beier, P., and R. Noss. 1998. Do habitat corridors provide connectivity? *Conservation Biology* 12:1241–1252.

Bellamy, P. E., S. A. Hinsley, and I. Newton. 1996. Factors influencing bird species numbers in small woods in south-east England. *Journal of Applied Ecology* 33:249–262.

Bellinger, R. G., F. W. Ravlin, and M. L. McManus. 1989. Forest edge effects and their influence on the gypsy moth (Lepidoptera: Lymantriidae) egg mass distribution. *Environmental Entomology* 18:840–843.

Belthoff, J. R., and G. Ritchison. 1989. Natal dispersal of eastern screech-owls. *Condor* 91:254–265.

Bender, D. J., T. A. Contreras, and L. Fahrig. 1998. Habitat loss and population decline: A meta-analysis of the patch size effect. *Ecology* 79:517–533.

Benkman, C. W. 1993. Logging, conifers, and the conservation of crossbills. *Conservation Biology* 5:115–119.

Bennett, A. F. 1990a. *Habitat corridors: Their role in wildlife management and conservation*. Department of Conservation and Environment, Melbourne, Australia.

———. 1990b. Land use, forest fragmentation and the mammalian fauna at Naringal, south-western Victoria. *Australian Wildlife Research* 17:325–347.

———. 1991. Roads, roadsides and wildlife conservation: A review. Pp. 99–117 in *Nature conservation 2: The role of corridors*, ed. D. A. Saunders and R. J. Hobbs. Sydney: Surrey Beatty and Sons.

———. 1998. *Linkages in the landscape: The role of corridors and connectivity in wildlife conservation*. IUCN, Gland, Switzerland.

Bennett, A. F., and L. A. Ford. 1997. Land use, habitat change and the conservation of birds in fragmented rural environments: A landscape perspective from the Northern Plains, Victoria, Australia. *Pacifiic Conservation Biology* 3:244–261.

Bennett, A. F., K. Henein, and G. Merriam. 1994. Determinants of corridor quality: Chipmunks and fencerows in a farmland mosaic. *Biological Conservation* 68:155–165.

Bennett, E. L. 2000. Timber certification: Where is the voice of the biologist? *Conservation Biology* 14:921–923.

Benninger-Traux, M., J. L. Vankat, and R. L. Schaefer. 1992. Trail corridors as habitat and conduits for movement of plant species in Rocky Mountain National Park, Colorado, USA. *Landscape Ecology* 6:269–278.

Berg, A., S. G. Nilsson, and U. Bostrom. 1992. Predation on artificial wader nests on large and small bogs along a south-north gradient. *Ornis Scandinavia* 23:13–16.

Berg, A., B. Ehnstrom, L. Gustaffson, T. Hallingback, M. Jonsell, and J. Weslien. 1994. Threatened plant, animal and fungus species in Swedish forests: Distribution and habitat associations. *Conservation Biology* 8:718–731.

Berg, D. R., W. A. McKee, and M. Maki. In press. Floodplain restoration: Management of a dynamic system. In *Restoration of Puget Sound rivers*, ed. D. R. Montgomery, S. Bolton, and D. Booth. Seattle: University of Washington Press.

Berger, J. 1990. Persistence of different-sized populations: An empirical assessment of rapid extinctions in bighorn sheep. *Conservation Biology* 4:91–98.

Bergeron, Y., R. Bradshaw, and O. Engelmark. 1993. Disturbance dynamics in boreal forest. Uppsala, Sweden: Opulus Press.

Bergeron, Y., B. Harvey, A. Leduc, and S. Gauthier. 1999. Forest management guidelines based on natural disturbance dynamics: Stand- and forest-level considerations. *Forestry Chronicle* 75:49–54.

Bergeron, Y., P. J. Richard, C. Carcailler, S. Gauthier, M. Flannigan, and Y. T. Prarie. 1998. Variability in fire frequency and forest composition in Canada's southeastern boreal forest: A challenge for sustainable forest management. *Conservation Ecology* (online) 8. http://www.consecol.org/vol2/iss2/art8.

Beschta, R. L., R. E. Bilby, G. W. Brown, L. B. Holtby, and T. D. Hofstra. 1987. Stream temperature and aquatic habitat: Fisheries and forestry interactions. Pp. 191–232 in *Streamside management: Forestry and fisheries interactions*, ed. E. O. Salo and T. W. Cundy. Institute of Forest Resources Contribution no. 59. University of Washington, Seattle.

Beschta, R. L., and W. S. Platts. 1986. Morphological features of small streams: Significance and function. *Water Resources Bulletin* 22:369–379.

Bierregaard, R. O., and T. E. Lovejoy. 1989. Effects of forest fragmentation on Amazonian understorey bird communities. *Acta Amazonica* 19:215–241.

Bierregaard, R. O., T. E. Lovejoy, V. Kapos, A. Santos, and R. W. Hutchings. 1992. The biological dynamics of tropical rainforest fragments. *BioScience* 42:859–866.

Bierregaard, R. O., and P. C. Stouffer. 1997. Understorey birds and dynamic habitat mosaics in Amazonian rainforests. Pp. 138–153 in *Tropical forest remnants: Ecology, management and conservation of fragmented communities*, ed. W. F. Laurance and R. O. Bierregaard. Chicago: University of Chicago Press.

Bilby, R. E., and P. A. Bisson. 1998. Function and distribution of large woody debris. Pp. 324–346 in *River ecology and management: Lessons from the Pacific Coastal Ecoregion*, ed. R. J. Naiman and R. E. Bilby. New York: Springer-Verlag.

Billings, W. D., and A. F. Mark. 1957. Factors involved in the persistence of montane treeless balds. *Ecology* 38:140–142.

Billington, H. L. 1991. Effect of population size on genetic variation in a dioecious conifer. *Conservation Biology* 5:115–119.

BirdLife International. 2000. *Threatened birds of the world*. London: Lynx Editions.

Bissix, G., and J. A. Rees. 2001. Can strategic ecosystem management succeed in multiagency environments? *Ecological Applications* 11:570–583.

Bissonette, J., ed. 1997. *Wildlife and landscape ecology: Effects of pattern and scale*. New York: Springer-Verlag.

Björse, G., and R. Bradshaw. 1998. Two thousand years of forest dynamics in southern Sweden: Suggestions for forest management. *Forest Ecology and Management* 104:15–26.

Blake, J. G. 1983. The trophic structure of bird communities in forest patches in east-central Illinois. *Wilson Bulletin* 95:416–430.

Block, W. M., and L. A. Brennan. 1993. The habitat concept in ornithology: Theory and applications. Pp. 35–91 in *Current ornithology 11*, ed. D. M. Power. New York: Plenum Press.

Boecklen, W. J. 1986. Optimal reserve design: Consequences of genetic drift. *Biological Conservation* 38:323–338.

Boland, D. J., M. I. Brooker, G. M. Chippendale, N. Hall, B. P. Hyland, R. D. Johnston, D. A. Kleinig, and J. D. Turner. 1984. *Forest trees of Australia*. Melbourne: CSRIO Publishing.

Bolger, D. T., A. V. Suarez, K. R. Crooks, S. A. Morrison, and T. J. Case. 2000. Arthropods in urban habitat fragments in southern California: Area, age, and edge effects. *Ecological Applications* 10:1230–1248.

Boone, R. B., and M. L. Hunter. 1996. Using diffusion models to simulate the effects of land use on grizzly bear dispersal in the Rocky Mountains. *Landscape Ecology* 11:51–64.

Boose, E., K. Chamberlain, and D. Foster. 2001. Landscape and regional impacts of hurricanes in New England. *Ecological Monographs* 71:27–48.

Borg, H., P. D. King, and I. C. Loh. 1997. *Stream and groundwater response to logging and subsequent regeneration in the southern forest of Western Australia: Interim results of paired catchment studies*. Water Authority of Western Australia. WH34, Perth, Western Australia.

Bormann, F. H., and G. Likens. 1979. *Patterns and process in a forested ecosystem*. New York: Springer-Verlag.

Botkin, D. B., and L. M. Talbot. 1992. Biological diversity and forests. Pp. 47–74 in *Contemporary issues in forest management: Policy implications*, ed. N. Sharma. Washington, D.C.: The World Bank.

Boubée, J., I. Jowlett, S. Nichols, and E. Williams. 1999. *Fish passage at culverts: A review, with possible solutions for New Zealand indigenous species*. Department of Conservation, Wellington, New Zealand.

Boulinier, T., J. D. Nichols, J. E. Hines, J. R. Sauer, C. H. Flather, and K. H. Pollock. 2001. Forest fragmentation and bird community dynamics: Inference at regional scales. *Ecology* 82:1159–1169.

Bowman, D. M. 1998. Tansley Review no. 101. The impact of aboriginal landscape burning on the Australian biota. *New Phytologist* 140:385–410.

———. 1999. *Australian rainforests: Islands of green in a land of fire.* Melbourne, Australia: Cambridge University Press.

Bowne, D. R., J. D. Peles, and G. W. Barrett. 1999. Effects of landscape spatial structure on movement patterns of the hispid cotton rat (*Sigmodon hispidus*). *Landscape Ecology* 14:53–65.

Boyce, S. G. 1995. *Landscape forestry.* New York: John Wiley and Sons.

Boyce, S. G., and N. D. Cost. 1978. *Forest diversity—new concepts and applications.* USDA Forest Service Research Paper SE-194.

Braithwaite, L. W. 1984. The identification of conservation areas for possums and gliders in the Eden woodpulp concession district. Pp. 501–508 in *Possums and gliders*, ed. A. P. Smith and I. D. Hume. Sydney, Australia: Surrey Beatty and Sons.

Braithwaite, L. W., L. Belbin, J. Ive, and M. P. Austin. 1993. Land use allocation and biological conservation in the Batemans Bay forests of New South Wales. *Australian Forestry* 56:4–21.

Brand, D. G., R. W. Roberts, and R. Kemp. 1993. International initiatives to achieve sustainable management of forests: Canada's model forests, the Commonwealth Forestry Initiative, and the development assistance community. *Commonwealth Forestry Review* 72:297–302.

Bren, L. J. 1997. Effects of increasing riparian buffer widths on timber resource availability: A case study. *Australian Forestry* 60:260–263.

Brereton, R. 1997. *Management prescriptions for the swift parrot in production forests.* Report to the Tasmanian Regional Forest Agreement and Heritage Technical Committee. June. Tasmanian Parks and Wildlife Service.

Brereton, R., S. Bennett, and I. Mansergh. 1995. Enhanced greenhouse climate change and its potential effects on selected fauna of south-eastern Australia: A trend analysis. *Biological Conservation* 72:203–214.

Bright, P. W. 1998. Behaviour of specialist species in habitat corridors: Arboreal doormice avoid corridor gaps. *Animal Behaviour* 56:1485–1490.

Brinson, M. M., and J. Verhoeven. 1999. Riparian forests. Pp. 265–299 in *Maintaining biodiversity in forest ecosystems*, ed. M. Hunter Jr. Cambridge: Cambridge University Press.

British Columbia Ministry of Forests. 1995. *Forest practices code of British Columbia: Biodiversity guidebook.* Victoria, British Columbia: Queens Printer.

Brokaw, N. V., and R. A. Lent. 1999. Vertical structure. Pp. 373–399 in *Maintaining biodiversity in forest ecosystems*, ed. M. Hunter Jr. Cambridge: Cambridge University Press.

Brooks, T. M., S. L. Pimm, and J. O. Oyugi. 1999. Time lag between deforestation and bird extinction in tropical forest fragments. *Conservation Biology* 13:1140–1150.

Brosofske, K. D., J. Chen, R. J. Naiman, and J. F. Franklin. 1997. Harvesting effects on microclimatic gradients from small streams to uplands in western Washington. *Ecological Applications* 7:1188–2000.

Brothers, T. S., and A. Spingarn. 1992. Forest fragmentation and alien plant invasion of central Indiana old-growth forests. *Conservation Biology* 6:91–100.

Brown, G. W., J. L. Nelson, and K. A. Cherry. 1997. The influence of habitat structure on insectivorous bat activity in montane ash forests of the Central Highlands of Victoria. *Australian Forestry* 60:138–146.

Brown, J. H., and A. Kodric-Brown. 1977. Turnover rates in insular biogeography: Effect of immigration on extinction. *Ecology* 58:445–449.

Brown, K., and R. W. Hutchings. 1997. Disturbance, fragmentation and the dynamics of diversity in Amazonian forest butterflies. Pp. 138–153 in *Tropical forest remnants: Ecology, management and conservation of fragmented communities*, ed. W. F. Laurance and R. O. Bierregaard. Chicago: University of Chicago Press.

Brown, M. J. 1985. Benign neglect and active management in Tasmania's forests: A dynamic balance or ecological collapse? *Forest Ecology and Management* 85:279–289.

———. 1988. *Distribution and conservation of King Billy pine.* Forestry Commission of Tasmania, Hobart, Tasmania, Australia.

Brown, M. J., and J. Hickey. 1990. Tasmanian forest—genes or wilderness. *Search* 21:86–87.

Brown, M. J., G. Kantvilas, and S. J. Jarman. 1994. Conservation of non-vascular plants in Tasmania, with particular reference to lichens. *Biodiversity and Conservation* 3:263–278.

Brown, S., A. J. Gillespie, and A. E. Lugo. 1989. Biomass estimation methods for tropical forests with applications to forest inventory data. *Forest Science* 35:895.

Brown, T. G., and T. McMahon. 1988. Winter ecology of the juvenile coho salmon in Carnation Creek: Summary of findings and management implications. Pp. 108–117 in *Proceedings of the workshop: Applying fifteen years of Carnation Creek results*, ed. T. W. Chamberlain. Carnation Creek Steering Committee, Pacific Biological Station, Nanaimo, British Columbia, Canada.

Bruce-Jones, K., and B. R. Riddle. 1996. Regional scale monitoring of biodiversity. Pp. 193–252 in *Biodiversity in managed landscapes: Theory and practice*, ed. R. C. Szaro and D. W. Johnston. New York: Oxford University Press.

Bruns, H. 1960. The economic importance of birds in forests. *Bird Study* 7:193–208.

Budd, W. W., P. L. Cohen, P. R. Saunders, and F. R. Steiner. 1987. Stream corridor management in the Pacific Northwest. I. Determination of stream-corridor widths. *Environmental Management* 11:587–597.

Bull, E. L., and A. D. Partridge. 1986. Methods of killing trees for use by cavity nesters. *Wildlife Society Bulletin* 14:142–146.

Bull, E. L., A. D. Partridge, and W. G. Williams. 1981. *Creating snags with explosives.* USDA Forest Service PNW 393. Pacific Northwest Forest and Range Station, Portland, Ore.

Bunnell, F. 1995. Forest-dwelling fauna and natural fire regimes in British Columbia: Patterns and implications for conservation. *Conservation Biology* 9:636–644.

———. 1999a. Let's kill a panchreston. Giving fragmentation a meaning. Foreword to (pp. vii–xiii) *Forest wildlife and fragmentation: Management implications*, ed. J. Rochelle, L. A. Lehmann, and J. Wisniewski. Leiden, Germany: Brill.

———. 1999b. What habitat is an island? Pp. 1–31 in *Forest wildlife and fragmentation: Management implications*, ed. J. Rochelle, L. A. Lehmann, and J. Wisniewski. Leiden, Germany: Brill.

Bunnell, F., and I. Kremsater. 1990. Sustaining wildlife in managed forests. *Northwest Environmental Journal* 6:243–269.

Bunnell, F. L., L. L. Kremsater, and E. Wind. 1999. Managing to sustain vertebrate richness in forests of the Pacific Northwest: Relationships within stands. *Environmental Review* 7:97–146.

Burbidge, A. A., K. A. Johnson, P. Fuller, and R. I. Southgate. 1988. Aboriginal knowledge of the mammals of the central deserts of Australia. *Australian Wildlife Research* 15:9–39.

Burbidge, A. A., and N. L. Mckenzie. 1989. Patterns in the modern decline of Western Australia's vertebrate fauna: Causes and conservation implications. *Biological Conservation* 50:143–198.

Burdon J. J., and G. A. Chilvers. 1994. Demographic changes and the development of competition in a native eucalypt forest invaded by exotic pines. *Oecologia* 97:419–423.

Burgman, M. A. 1996. Characterization and delineation of the eucalypt old-growth forest estate in Australia: A review. *Forest Ecology and Management* 83:149–161.

Burgman, M., S. Ferson, and H. R. Akçakaya. 1993. *Risk assessment in conservation biology*. New York: Chapman and Hall.

Burgman, M. A., H. R. Akçakaya, and S. S. Loew. 1988. The use of extinction models for species conservation. *Biological Conservation* 43:9–25.

Burgman, M. A., and I. S. Ferguson. 1995. *Rainforest in Victoria: A review of the scientific basis of current and proposed protection measures*. Report to Victorian Department of Conservation and Natural Resources. Forest Services Technical Reports 95-4.

Burgman, M. A., and D. B. Lindenmayer. 1998. *Conservation biology for the Australian environment*. Chipping Norton, Australia: Surrey Beatty and Sons.

Burke, D. M., and E. Nol. 1998. Influence of food abundance, nest-site habitat, and forest fragmentation on breeding ovenbirds. *Auk* 115:96–104.

Burke, V. J., ed. 2000. Gap analysis for landscape conservation. Special issue. *Landscape Ecology* 15.

Burkey, T. V. 1989. Extinction in nature reserves: The effect of fragmentation and the importance of migration between reserve fragments. *Oikos* 55:75–81.

———. 1993. Edge effects in seed and egg predation at two neotropical rainforest sites. *Biological Conservation* 66:139–143.

Burnett, S. 1992. Effects of a rainforest road on movements of small mammals: Mechanisms and implications. *Wildlife Research* 19:95–104.

Burris, R. K., and L. W. Canter. 1997. Cumulative impacts are not properly addressed in environmental assessments. *Environmental Impact Assessment Review* 67:5–18.

Burrows, N., L. McCaw, and G. Friend. 1989. *Fire management on nature conservation lands*. Department of Conservation and Land Management. Occasional paper 1/89. Perth, Western Australia.

Byre, D. 1991. Aboriginal archaeology in forests—circles around the past. Pp. 385–392 in *Conservation of Australia's forest fauna*, ed. D. Lunney. Royal Zoological Society of NSW, Sydney, Australia.

Cabeza, M., and A. Moilanen. In press. Reserve design and the persistence of biodiversity. *Trends in Evolution and Ecology* (in press).

Cale, P. 1990. The value of road reserves for the avifauna of the central wheatbelt of Western Australia. Pp. 359–367 in *Australian ecosystems: Two hundred years of utilisation, degradation and reconstruction*, ed. D. A. Saunders, A. J. Hopkins, and R. A. How. Chipping Norton, Australia: Surrey Beatty and Sons.

———. 1999. The spatial dynamics of the white-browed babbler in a fragmented agricultural landscape. Ph.D. diss., University of New England, Armidale, New South Wales, Australia.

Cale, P. G., and R. J. Hobbs. 1994. Landscape heterogeneity indices: Problems of scale and applicability, with particular reference to animal habitat description. *Pacific Conservation Biology* 1:183–193.

Calhoun, A. 1999. Forested wetlands. Pp. 300–331 in *Maintaining biodiversity in forest ecosystems*, ed. M. Hunter Jr. Cambridge: Cambridge University Press.

Calver, M. C., R. J. Hobbs, P. Horwitz, and A. R. Main. 1996. Science, principles and forest management: A response to Abbott and Christensen. *Australian Forestry* 59:1–6.

Calver, M. C., C. R. Dickman, M. C. Feller, R. J. Hobbs, P. Horwitz, H. F. Recher, and G. Wardell-Johnson. 1997. Towards resolving conflict between forestry and conservation in Western Australia. *Australian Forestry* 61:258–266.

Campbell, I. C., K. R. James, B. T. Hart, and A. Devereaux. 1992. Allochthonous coarse particulate organic material in forest and pasture reaches of two south-eastern Australian streams. I. Litter accession. *Freshwater Biology* 27:341–352.

Campbell, I. C., and T. J. Doeg. 1989. Impact of timber harvesting and production on streams: A review. *Australian Journal of Marine and Freshwater Research* 40:519–539.

Campbell, R. G. 1984. The eucalypt forests. Pp. 1–12 in *Silvicultural and environmental aspects of harvesting some major commercial eucalypt forests in Victoria: A review*, ed. R. G. Campbell, E. A. Chesterfield, F. G. Craig, P. C. Fagg, P. W. Farrell, G. R. Featherstone, D. W. Flinn, P. Hopmans, J. D. Kellas, C. J. Leitch, R. H. Loyn, M. A. Macfarlane, L. A. Pederick, R. O. Squire, H. T. Stewart, and G. C. Suckling. Forests Commission Victoria, Division of Education and Research. Forests Commission Victoria, Melbourne, Australia.

———. comp. 1997. Evaluation and development of sustainable silvicultural systems for mountain ash forests. Discussion paper. *Value Adding and Silvicultural Systems Report*. VSP Technical Report no. 28. July. Forests Service, Department of Natural Resources and Environment, Melbourne.

Cannon, C. H., D. R. Peart, and M. Leighton. 1998. Tree species diversity in commercially logged Bornean rainforest. *Science* 281:1366–1368.

Canters, K., ed. 1997. *Habitat fragmentation and infrastructure*. Ministry of Transport, Public Works and Water Management, Delft, The Netherlands.

Carey, A. B. 1993. *The forest ecosystem study: Experimental manipulation of managed stands to provide habitat for spotted owls and to enhance plant and animal diversity*. A summary and background for the interagency experiment at Fort Lewis, Wash. Forestry Sciences Laboratory, Olympia, Wash.

———. 1994. *Forest ecosystem study: Summary, background, and rationale*. Olympia Forest Science Laboratory. Northwest Research Station. USDA, Olympia, Wash.

———. 1995. Sciurids in Pacific Northwest managed and old-growth forests. *Ecological Applications* 5:648–661.

———. 2000. Effects of new forest management strategies on squirrel populations. *Ecological Applications* 10:248–257.

Carey, A. B., and R. O. Curtis. 1996. Conservation of biodiversity: A useful paradigm for forest ecosystem management. *Wildlife Society Bulletin* 24:610–620.

Carey, A. B., C. Elliott, B. R. Lippke, J. Session, C. J. Chambers, C. D. Oliver, J. F. Franklin, and M. G. Raphael. 1996. *Washington Forest Landscape Management Project—a pragmatic, ecological approach to small-landscape management*. Washington Forest Landscape Management Project Report no. 2. Department of Natural Resources, Olympia, Wash.

Carey, A. B., and J. D. Gill. 1983. Direct habitat improvement—some recent advances. Pp. 80–87 in *Snag habitat management: Proceedings of the symposium*. U.S. Forest Service General Technical Report RM-99.

Carey, A. B., S. P. Horton, and B. L. Biswell. 1992. Northern spotted owls: Influence of prey base and landscape character. *Ecological Monographs* 62:223–250.

Carey, A. B., and M. L. Johnson. 1995. Small mammals in managed, naturally young, and old-growth forests. *Ecological Applications* 5:336–352.

Carey, A. B., J. Kershner, B. Biswell, and L. Dominguez de Toledo. 1999a. Ecological scale and forest development: Squirrels, dietary fungi, and vascular plants in managed and unmanaged forests. *Wildlife Monographs* 142:1–71.

Carey, A. B., B. R. Lippke, and J. Sessions. 1999b. Intentional systems management: Managing forests for biodiversity. *Journal of Sustainable Forestry* 9(3/4):83–125.

Carey, A. B., and H. R. Sanderson. 1981. Routine to accelerate tree cavity formation. *Wildlife Society Bulletin* 9:14–21.

Carlson, A., and P. Edenhamn. 2000. Extinction dynamics and the regional persistence of a tree frog metapopulation. *Proceedings of the Royal Society of London Series B* 267:1311–1313.

Caro, T. M. 2001. Species richness and abundance of small mammals inside and outside an African national park. *Biological Conservation* 98:251–257.

Carroll, C., R. F. Noss, and P. C. Paquet. 2001. Carnivores as focal species for conservation planning in the Rocky Mountains. *Ecological Applications* 11:961–980.

Carthew, S. M., and R. L. Goldingay. 1997. Non-flying mammals as pollinators. *Trends in Evolution and Ecology* 12:104–108.

Cascade Center for Ecosystem Management. 1993. *Young managed stands*. Cascade Center for Ecosystem Management Communique, Oregon State University, Corvallis.

———. 1995. *Residual trees as biological legacies*. Cascade Center for Ecosystem Management Communique no. 2. Oregon State University, Corvallis.

Catling, P. C., and R. J. Burt. 1995. Studies of the ground-dwelling mammals of the eucalypt forests in south-eastern New South Wales: The effect of habitat variables on distribution and abundance. *Wildlife Research* 22:271–288.

Catterall, C. P., R. J. Green, and D. N. Jones. 1991. Habitat use by birds across a forest-suburb interface in Brisbane: Implications for corridors. Pp. 247–258 in *Nature conservation 2: The role of corridors*, ed. D. A. Saunders and R. J. Hobbs. Chipping Norton, Australia: Surrey Beatty and Sons.

Catterall, C. P., S. D. Piper, S. E. Bunn, and J. M. Arthur. 2001. Flora and fauna assemblages vary with local topography in a subtropical eucalypt forest. *Austral Ecology* 26:56–69.

Caughley, G. 1978. *The analysis of vertebrate populations*. London: John Wiley and Sons.

Caughley, G. C., and A. Gunn. 1995. *Conservation biology in theory and practice*. Cambridge, Mass.: Blackwell Science.

Cauley, H. A., C. M. Peters, R. Z. Donovan, and J. M. O'Connor. 2001. Forest Stewardship Council Forest Certification. *Conservation Biology* 15:311–312.

Chambers, C. L., T. Carrigan, T. Sabin, J. Tappeiner, and W. C. McComb. 1997. Use of artificially created Douglas-fir snags by cavity-nesting birds. *Western Journal of Applied Forestry* 12:93–97.

Chambers, C. L., W. C. McComb, and J. C. Tappeiner. 1999. Breeding bird responses to three silvicultural treatments in the Oregon Coast Range. *Ecological Applications* 9:171–185.

Chapin, F. S., B. Walker, R. J. Hobbs, D. U. Hooper, J. H. Lawton, O. E. Sala, and D. Tilman. 1997. Biotic control over the functioning of ecosystems. *Science* 277:500–504.

Chapin, F. S., and G. Whiteman. 1998. Sustainable development of the boreal forest: Interaction of ecological, social and business feedbacks. *Conservation Ecology* (online) 8. http://www.consecol.org/vol2/iss2/art8.

Chapin, T. G., D. J. Harrison, and D. D. Katnik. 1998. Influence of landscape pattern on habitat use by American marten in an industrial forest. *Conservation Biology* 12:1327–1337.

Chen, J. 1991. Edge effects: Microclimatic pattern and biological responses in old-growth Douglas-fir forests. Ph.D. diss., University of Washington, Seattle.

Chen, J., J. F. Franklin, and T. A. Spies. 1990. Microclimatic pattern and basic biological responses at the clearcut edges of old-growth Douglas-fir stands. *Northwest Environmental Journal* 6:424–425.

———. 1992. Vegetation responses to edge environments in old-growth Douglas-fir forests. *Ecological Applications* 2:387–396.

Chesser, R. K. 1983. Isolation by distance: Relationship to the management of genetic resources. Pp. 66–77 in *Genetics and conservation: A reference for managing wild animal and plant populations*, ed. C. M. Schonewald-Cox, S. M. Chambers, B. MacBryde, and W. L. Thomas. Menlo Park, Calif.: Benjamin/Cummings.

Chesterfield, E. A., J. McCormick, and G. Hepworth. 1991. The effect of low root temperatures on the growth of mountain forest eucalypts in relation to the ecology of *Eucalyptus nitens*. *Proceedings of the Royal Society of Victoria* 103:67–76.

Chindarsi, K. A. 1997. The logging of Australian native forests: A critique. *Australian Quarterly* 69:86–104.

Chou, Y. H., R. A. Minnich, and R. J. Dezzani. 1993. Do fire sizes differ between southern California and Baja California? *Forest Science* 39:835–844.

Christensen, N. L. 2000. Vegetation of the southeastern coastal plain. Pp. 397–448 in *North American terrestrial vegetation*, ed. M. T. Barbour and W. D. Billings. 2nd ed. Cambridge: Cambridge University Press.

Christensen, N., A. M. Bartuska, J. H. Brown, S. Carpenter, C. D'Antonio, R. Francis, J. F. Franklin, J. A. MacMahon, R. F. Noss, D. J. Parsons, C. H. Peterson, M. G. Turner, and R. G. Woodmansee. 1996. The report of the Ecological Society of America on the scientific basis for ecosystem management. *Ecological Applications* 6:665–691.

Churchill, S. 1998. *Australian bats*. Sydney, Australia: Reed New Holland.

Churton, N. L., L. J. Bren, and C. M. Kerruish. 1996. Efficiency of mechanism mountain ash thinning in the Central Highlands of Victoria. *Australian Forestry* 62:72–78.

Ciancio, O., and S. Nocentini. 2000. Forest management from Positivism to the Culture of Complexity. Pp. 47–58 in *Methods and approaches in forest history*, ed. M. Agnoletti and S. Anderson. Tampere, Finland: CABI Publishing in association with The Internation Union of Forestry Research Organisations (IUFRO).

Cissel, J. H., J. S. Frederick, G. E. Grant, D. H. Olson, S. V. Gregory, S. L. Garman, L. R. Ashkenas, M. G. Hunter, J. A. Kertis, J. H. Mayo, M. D. McSwain, S. G. Swetland, K. A. Swindle, and D. O. Wallin. 1998. *A landscape plan based on historical fire regimes for a managed forest ecosystem: The Augusta Creek study*. USDA Forest Service General Technical Report PNW-GTR-422.

Cissel, J. H., F. J. Swanson, and P. J. Weisberg. 1999. Landscape management using historical fire regimes: Blue River, Oregon. *Ecological Applications* 9:1217–1231.

Claridge, A. W. 1993. Hypogeal fungi as a food resource for wildlife in the managed forests of south-eastern Australia. Ph.D. diss., The Australian National University, Canberra.

Claridge, A. W., and D. B. Lindenmayer. 1993. The mountain brushtail possum, *Trichosurus caninus* Ogilby, as a disseminator of fungi in the mountain ash forests of the Central Highlands of Victoria. *Victorian Naturalist* 110:91–95.

———. 1994. The need for a more sophisticated approach toward wildlife corridor design in the multiple-use forests of southeastern Australia: The case for mammals. *Pacific Conservation Biology* 1:301–307.

———. 1998. Temporal and spatial variation in the consumption of hypogeal fungi by the mountain brushtail possum, *Trichosurus caninus*. *Mycological Research* 102:269–272.

Clark, T. W. 1993. Creating and using knowledge for species and ecosystem conservation: Science, organizations, and policy. *Perspectives in Biology and Medicine* 36:497–525.

Clark, T. W., and S. R. Kellert. 1988. Toward a policy paradigm of the wildlife sciences. *Renewable Resources Journal* 6:7–16.

Clark, T. W., and S. C. Minta. 1994. Greater Yellowstone's future: Prospects for ecosystem science, management and policy. Moose, Wyo.: Homestead Publishing.

Clark, T. W., R. M. Warneke, and G. G. George. 1990. Management and conservation of small populations. Pp. 1–18 in *Management and conservation of small populations*, ed. T. W. Clark and J. H. Seebeck. Proceedings of the conference on the Management and Conservation of Small Populations. Melbourne, Australia, 26–27 September 1989.

Clark, T. W., and D. Zaunbrecher. 1987. The Greater Yellowstone Ecosystem: The ecosystem concept in natural resource policy and management. *Renewable Resources Journal* 5:8–16.

Clement, J. P., and D. C. Shaw. 1999. Crown structure and the distribution of epiphyte functional group mass in old-growth *Pseudotsuga menziesii* trees. *Ecoscience* 6:243–254.

Clements, F. E. 1916. *Plant succession: An analysis of the development of vegetation.* Carnegie Institute Publication 242. Washington, D.C.

Clevenger, A. P., and N. Waltho. 2000. Factors influencing the effectiveness of wildlife underpasses in Banff National Park, Alberta, Canada. *Conservation Biology* 14:47–56.

Clinnick, P. F. 1985. Buffer strip management in forest operations. *Australian Forestry* 48:34–45.

Clout, M. N. 1984. Improving exotic forests for native birds. *New Zealand Journal of Forestry* 29:193–200.

Clout, M. N., and P. D. Gaze. 1984. Effects of plantation forestry on birds in New Zealand. *Journal of Applied Ecology* 21:795–815.

Coates, K. D., and P. J. Burton. 1997. A gap-based approach for development of silvicultural systems to address ecosystem management objectives. *Forest Ecology and Management* 99:337–354.

Cocklin, C., S. Parker, and J. Hay. 1992a. Notes on cumulative environmental change I: Concepts and issues. *Journal of Environmental Management* 35:31–49.

———. 1992b. Notes on cumulative environmental change II: A contribution to methodology. *Journal of Environmental Management* 35:51–67.

Cocks, K. D., and I. A. Baird. 1991. The role of geographic information systems in the collection, extrapolation and use of survey data. Pp. 74–80 in *Nature conservation: Cost effective biological surveys and data analysis*, ed. C. R. Margules and M. P. Austin. Australia: CSIRO Publishing.

Cogbill, C. V. 1996. Black growth and fiddlebutts: The nature of old-growth red spruce. Pp. 113–125 in *Eastern old-growth forests: Prospects for rediscovery and recovery*, ed. M. B. Davies. Washington, D.C.: Island Press.

Cogger, H. 1995. *Reptiles and amphibians of Australia.* 5th ed. Sydney, Australia: Reed.

Colbourne, R., and R. Kleinpaste. 1983. A banding study of the North Island brown kiwi in an exotic forest. *Notornis* 30:109–124.

Cole, E. K., M. D. Pope, and R. G. Anthony. 1997. Effects of road management on movement and survival of Roosevelt elk. *Journal of Wildlife Management* 61:1115–1126.

Collinge, S. K., and R. T. T. Forman. 1998. A conceptual model of land conversion processes: Predictions and evidence from a microlandscape experiment with grassland insects. *Oikos* 82:66–84.

Collingham, Y. C., and B. Huntley. 2000. Impacts of habitat fragmentation and patch size on migration routes. *Ecological Applications* 10:131–144.

Commonwealth of Australia. 1992. *National forest policy statement.* Perth: Advance Press.

———. 1997. *Australia's first approximation report for the Montreal Process.* June. Commonwealth of Australia, Canberra.

———. 1998. *A framework of regional (sub-national) level criteria and indicators of sustainable forest management in Australia.* August. Commonwealth of Australia, Canberra.

———. 1999. *International forest conservation: Protected areas and beyond.* A discussion paper for the intergovernmental forum on forests. March. Commonwealth of Australia, Canberra.

Commonwealth of Australia and Department of Natural Resources and Environment. 1997. *Comprehensive regional assessment—biodiversity. Central Highlands of Victoria.* Commonwealth of Australia, and Department of Natural Resources and Environment, Canberra.

Connell, M. J., R. J. Raison, and A. G. Brown, eds. 1999. *Intensive management of regrowth forest for wood production in Australia.* Proceedings of the National Workshop, 18–20 May. Orbost, Victoria, Australia. Melbourne, Australia: CSIRO Publishing.

Conner, R. N. 1988. Wildlife populations: Minimally viable or ecologically functional? *Wildlife Society Bulletin* 16:80–84.

Conner, R. N., J. G. Dickson, and B. A. Locke. 1981. Herbicide killed trees infected by fungi: Potential cavity sites for woodpeckers. *Wildlife Society Bulletin* 9:308–310.

Conner, R. N., and D. C. Rudolph. 1989. *Red-cockaded woodpecker colony status and trends on the Angelina, Davy Crockett and Sabine national forests.* Research Paper SO-250. USDA Forest Service Southern Forest Experimental Station, New Orleans, La.

Connor, E. F., and E. D. McCoy. 1979. The statistics and biology of the species-area relationship. *American Naturalist* 113:791–833.

Conradt, L., E. J. Bodsworth, T. J. Roper, and C. D. Thomas. 2000. Non-random dispersal in the butterfly *Manioal jurtina*: Implications for metapopulation models. *Proceedings of the Royal Society of London. Series B* 267:1505–1510.

Cooper-Ellis, S., D. A. Foster, G. Carlton, and A. Lezberg. 1999. Forest response to catastrophic wind: Results from an experimental hurricane. *Ecology* 80:2683–2696.

Copeyon, C. K. 1990. A technique for constructing cavities for the red-cockaded woodpecker. *Wildlife Society Bulletin* 18:303–311.

Cork, S. 1997. The contribution of science to resolving ecological issues in temperate Australian forests. Pp. 52–93 in *Saving our natural heritage: The role of science in managing Australian ecosystems*, ed. C. Copeland and D. Lewis. Sydney, Australia: Halstead Press.

Costa, R., and R. E. F. Escano. 1989. *Red-cockaded woodpecker status and management in the southern region in 1986.* USDA Forest Service Southern Region Technical Publication R8-TP 12.

Costanza, R., R. d'Arge, R. de Groot, S. Farber, M. Grasso, B. Hannon, K. Limburg, S. Nacem, R. V. O'Neill, J. Paruelo, R. G. Raskin, P. Sutton, and C. van den Belt. 1997. The value of the world's ecosystem services and natural capital. *Nature* 387:253–260.

Cowardin, L. M., V. Carter, F. C. Golet, and E. T. LaRoe. 1979. *Classification of wetlands and deepwater habitats of the United States.* USDI Fish and Wildlife Service FWS/OBS-79/31.

Craighead, J. J., and F. C. Craighead Jr. 1969. *Hawks, owls, and wildlife.* New York: Dover Publishing.

Cranston, P. S. 1990. Biomonitoring and invertebrate taxonomy. *Environmental Monitoring and Assessment* 14:265–273.

Cremer, K. W. 1962. The effects of fire on eucalypts reserved for seeding. *Australian Forestry* 26:129–154.

Croizat, L. C. 1960. *Principia botanica: Or, beginnings of botany.* Codicator, Hitchin, England: Weldon and Wesley.

Crome, F. H. 1985. Problems of wildlife management in relation to forestry practices in the southern hemisphere. Pp. 15–21 in *Wildlife management in the forests and the forestry-controlled lands in the tropics and the southern hemisphere*, ed. J. Kikkawa. Proceedings of IUFRO Workshop. Brisbane, Queensland, Australia. July 1984.

———. 1994. Tropical forest fragmentation: Some conceptual and methodological issues. Pp. 61–76 in *Conservation biology in Australia and Oceania*, ed. C. Moritz and J. Kikkawa. Chipping Norton, Australia: Surrey Beatty and Sons.

———. 1997. Research on tropical fragmentation: Shall we keep on doing what we are doing? Pp. 485–501 in *Tropical forest remnants: Ecology, management, and conservation of fragmented communities*, ed. W. F. Laurance and R. O. Bierregaard. Chicago: University of Chicago Press.

Crome, F. H., J. Isaacs, and L. Moore. 1994. The utility to birds and mammals of remnant riparian vegetation and associated windbreaks in the tropical Queensland uplands. *Pacific Conservation Biology* 1:328–343.

Crow, T. R., and E. J. Gustafson. 1997. Ecosystem management: Managing natural resources in time and space. Pp. 215–228 in *Creating a forestry for the twenty-first century: The science of ecosystem management*, ed. K. A. Kohm and J. F. Franklin. Washington, D.C.: Island Press.

Crowe, M. P., J. Paxton, and G. Tyers. 1984. Felling dead trees with explosives. *Australian Forestry* 47:84–87.

Crumpacker, D. W., S. W. Hodge, D. F. Friedley, and W. P. Gregg. 1988. A preliminary assessment of the status of major terrestrial and wetland ecosystems on federal and Indian lands in the United States. *Conservation Biology* 2:103–115.

Cubbage, F. W., W. S. Dvorak, R. C. Abt, and G. Pacheco. 1996. World timber supply and prospects: Models, projections, plantations and implications. Paper presented at Central America and Mexico Coniferous (CAMCORE) Annual Meeting. Bali, Indonesia.

Culver, D. C., L. S. Master, M. C. Christman, and H. H. Hobbs. 2000. Obligate cave fauna of the forty-eight contiguous United States. *Conservation Biology* 14:386–401.

Cunningham, M., and C. Moritz. 1998. Genetic effects of forest fragmentation on a rainforest restricted lizard (Scincidae, *Gnypetoscincus queenslandiae*). *Biological Conservation* 83:19–30.

Cunningham, S. A. 2000. Depressed pollination in habitat fragments causes low fruit set. *Proceedings of the Royal Society of London. Series B* 267:1149–1152.

Cunningham, T. M. 1960. The natural regeneration of *Eucalyptus regnans*. *Bulletin of the School of Forestry*, University of Melbourne 1:1–158.

Curry, G. N. 1991. The influence of proximity to plantation of diversity and abundance of bird species in an exotic pine plantation in north-eastern New South Wales. *Wildlife Research* 18:299–314.

Curtis, R. O. 1994. *Some simulation estimates of mean annual increment of Douglas Fir: Results, limitations, implications for management.* U.S. Forest Service Research Paper PNW-RP-485.

_____. 1995. *Extended rotations and culmination ages of coast Douglas-fir: Old studies speak to current issues.* USDA Forest Service Research Paper PNW-RP-489.

_____. 1997. The role of extended rotations. Pp. 165–170 in *Creating a forestry for the twenty-first century: The science of ecosystem management*, ed. K. A. Kohm and J. F. Franklin. Washington, D.C.: Island Press.

_____. 1998. "Selective cutting" in Douglas fir: History revisited. *Journal of Forestry* 96:40–46.

Curtis, R. O., and A. Carey. 1996. Timber supply in the Pacific Northwest: Managing economic and ecologic values. *Journal of Forestry* 94:4–7, 35–37.

Curtis, R. O., and D. D. Marshall. 1993. Douglas-fir rotations—time for reappraisal. *Western Journal of Applied Forestry* 8:81–85.

Cutler, A. 1991. Nested faunas and extinction in fragmented habitats. *Conservation Biology* 5:496–505.

Daily, G. C., and B. H. Walker. 2000. Seeking the great transition. *Nature* 403:243–245.

Dale, V. H., S. M. Pearson, J. L. Offerman, and R. V. O'Neill. 1994. Relating patterns of land-use change to faunal diversity in the central Amazon. *Conservation Biology* 8:1027–1036.

Daniels, R. B., and J. W. Gilliam. 1996. Sediment and chemical load reduction by grass and riparian filters. *Soil Science Society of America Journal* 60:246–251.

Danielson, B. J., and M. W. Hubbard. 2000. The influence of corridors on the movement behavior of individual *Peromyscus polionotus* in experimental landscapes. *Landscape Ecology* 15:323–331.

Dargarvel, J. 1995. *Fashioning Australia's forests.* Melbourne, Australia: Oxford University Press.

_____. 1998. Politics, policy and process in the forests. *Australian Journal of Environmental Management* 5:25–30.

Darveau, M., P. Beauchesne, L. Belanger, J. Hout, and P. Larue. 1995. Riparian forest strips as habitat for breeding birds in boreal forest. *Journal of Wildlife Management* 59:67–78.

Date, E. M., H. F. Recher, H. A. Ford, and D. A. Stewart. 1996. The conservation and ecology of rainforest pigeons in northern New South Wales. *Pacific Conservation Biology* 2:299–308.

Daubenmire, R. 1970. *Steppe vegetation of Washington.* Washington State University Agricultural Experiment Station Technical Bulletin 62. Pullman, Wash.

Davey, S. M. 1989. Thoughts towards a forest wildlife management strategy. *Australian Forestry* 52:56–67.

Davie, J. 1997. Is biodiversity really the link between conservation and ecologically sustainable management? A reflection on paradigm and practice. *Pacific Conservation Biology* 3:83–90.

Davies, K. F., C. Gascon, and C. R. Margules. 2001. Habitat fragmentation: Consequences, management, and future directions. Pp. 81–97 in *Conservation Biology: Research priorities for the next decade*, ed. M. E. Soulé and G. H. Orians. Washington, D.C.: Island Press.

Davies, K. F., and C. R. Margules. 1998. Effects of habitat fragmentation on carabid beetles: Experimental evidence. *Journal of Animal Ecology* 67:460–471.

Davies, K. F., C. R. Margules, and J. F. Lawrence. 2000. Which traits of species predict population declines in experimental forest fragments? *Ecology* 81:1450–1461.

Davis, G. E., and W. L. Halvorson. 1988. *Inventory and monitoring of natural resources in Channel Islands National Park, California.* USDI National Park Service. Ventura, Calif.

Davis, L. S., K. N. Johnson, P. S. Bettinger, and T. E. Howard. 2001. *Forest management to sustain ecological, economic, and social values.* 4th ed. New York: McGraw-Hill.

Davis, P. E., and M. Nelson. 1994. Relationships between riparian buffer widths and the effects of logging on stream habitat, invertebrate community composition and fish abundance. *Australian Journal of Marine and Freshwater Research* 45:1289–1305.

Deacon, J. N. and R. Mac Nally. 1998. Local extinction and nestedness of small mammal faunas in fragmented forest of central Victoria, Australia. *Pacific Conservation Biology* 4:122–131.

DeBell, D. S., R. O. Curtis, C. A. Harrington, and J. C. Tappeiner. 1997. Shaping stand development through silvicultural practices. Pp. 141–149 in *Creating a forestry for the twenty-first century: The science of ecosystem management*, ed. K. A. Kohm and J. F. Franklin. Washington, D.C.: Island Press.

Debinski, D. M., and R. D. Holt. 2000. A survey and overview of habitat fragmentation experiments. *Conservation Biology* 14:342–355.

Debinski, D. M., C. Ray, and E. H. Saveraid. 2001. Species diversity and the scale of the landscape mosaic: Do scales of movement and patch size affect diversity? *Biological Conservation* 98:179–190.

DeGraaf, R., and I. Miller, eds. 1996. *Conservation of faunal diversity in forested landscapes.* London: Chapman and Hall.

DeGraaf, R., and A. L. Shigo. 1985. *Managing cavity trees for wildlife in the northeast.* USDA Forest Service General Technical Report GTR-NE-101. Upper Darby, Penn.

Delcourt, H. R., and P. A. Delcourt. 2000. Eastern deciduous forests. Pp. 357–395 in *North American terrestrial vegetation*, ed. M. G. Barbour and W. D. Billings. 2nd ed. Cambridge: Cambridge University Press.

Delin, A. E., and H. Andrén. 1999. Effects of habitat fragmentation on Eurasian red squirrel (*Sciurus vulgaris*) in a forest landscape. *Landscape Ecology* 14:67–72.

Dellasala, D. A., J. C. Hagar, K. A. Engel, W. C. McComb, R. L. Fairbanks, and E. G. Campbell. 1996. Effects of silvicultural modifications of temperate rainforest on breeding and wintering bird communities, Prince of Wales Island, southeast Alaska. *Condor* 98:706–721.

Delong, S. C., and W. B. Kessler. 2000. Ecological characteristics of mature forest remnants left by wildfire. *Forest Ecology and Management* 131:93–106.

deMaynadier, P., and M. Hunter. 1995. The relationship between forest management and amphibian ecology: A review of the North American literature. *Environment Review* 3:230–261.

———. 2000. Road effects on amphibian movements in a forested landscape. *Natural Areas Journal* 20:56–65.

De Marajò, I., and C. Paragominas. 2001. Conservation in Brazil: Managing the rainforests. *Economist*, 12–18 May, 87–89.

Dempster, K. 1962. Internal memorandum to Director (of Fisheries and Wildlife Division). 2 October. Fisheries and Wildlife Division, Melbourne, Australia.

Denton, S. J. 1976. Status of prairie falcon breeding in Oregon. Master's thesis, Oregon State University, Corvallis.

Department of Conservation, Forests and Lands. 1989. *Code of practice. Code of forest practices for timber production.* Revision no. 1. May. Department of Conservation, Forests and Lands, Melbourne, Australia.

Department of Natural Resources and Environment. 1996. *Code of practice. Code of forest practices for timber production.* Revision no. 2. November. Department of Natural Resources and Environment, Melbourne, Australia.

Department of Primary Industries and Energy. 1997. *Plantations for Australia: The 2020 vision.* Joint Report from Ministerial Council on Forestry, Fisheries and Aquaculture, Standing Committee on Forestry, Plantations Australia, Australian Forest Growers, National Association of Forest Industries. Department of Primary Industries and Energy, Canberra.

Department of the Environment, Sports, and Territories. 1995. *National strategy for the conservation of Australia's biological diversity.* Canberra: Australian Government Publishing Service.

Desrochers, A., and S. J. Hannon. 1997. Gap crossing decisions by forest songbirds during the post-fledging period. *Conservation Biology* 11:1204–1210.

Diamond, J. M. 1973. Distributional ecology of New Guinea birds. *Science* 179:759–769.

———. 1975. The island dilemma: Lessons of modern biogeographic studies for the design of natural preserves. *Biological Conservation* 7:129–146.

———. 1976. Island biogeography and conservation: Strategy and limitations. *Science* 193:1027–1029.

Diamond, J. M., K. D. Bishop, and S. van Balen. 1987. Bird survival in an isolated Javan woodlot: Island or mirror? *Conservation Biology* 2:132–142.

Diamond, J. M., and R. M. May. 1976. Island biogeography and the design of nature reserves. Pp. 163–186 in *Theoretical ecology: Principles and applications*, ed. R. M. May. Philadelphia: W. B. Saunders.

Díaz, J. A., R. Carbonell, E. Virgós, T. Santos, and J. L. Tellería. 2000. Effects of forest fragmentation on the distribution of the lizard *Psammodromus algirus*. *Animal Conservation* 3:235–240.

Dickson, J. G., and J. C. Huntley. 1987. Riparian zones and wildlife in southern forests: The problem and squirrel relationships. Pp. 37–39 in *Managing southern forests for wildlife and fish*. USDA, Southern Forest and Experiment Station General Technical Report SO-65.

Dickson, R., T. Aldred, and I. Baird. 1997. National forest conservation reserves: Recent developments. Pp. 359–368 in *National parks and protected areas: Selection, delimitation and management*, ed. J. J. Pigram and R. Sundell. Centre for Water Policy Research, University of New England, Armidale, Australia.

Didham, R. K. 1997. The influence of edge effects and forest fragmentation on leaf litter invertebrates in central Amazonia. Pp. 55–70 in *Tropical forest remnants: Ecology, mangement and conservation*, ed. W. F. Laurance and R. O. Bierregaard. Chicago: University of Chicago Press.

Diem, A. 1992. Clearcutting British Columbia. *Ecologist* 22:261–270.

Doak, D. 1989. Spotted owls and old growth logging in the Pacific Northwest. *Conservation Biology* 3:389–396.

Doak, D., and L. S. Mills. 1994. A useful role for theory in conservation. *Ecology* 75:615–626.

Dobson, A. P., J. P. Rodriguez, and W. M. Roberts. 2001. Synoptic tinkering: Integrating strategies for large-scale conservation. *Ecological Applications* 11:1019–1026.

Doeg, T. J., and J. D. Koehn. 1990. *A review of Australian studies on the effects of forestry practices on aquatic values.* Silvicultural Systems Project Technical Report no. 5. Fisheries Division, Department of Conservation and Environment, Melbourne, Australia.

Dood, A. R., R. D. Brannon, and R. D. Mace. 1985. Management of grizzly bears in the northern continental divide ecosystems, Montana. *Transactions of the Fifty-first North American Wildlife and Natural Resources Conference* 51:162–177.

Dooley, J. L., and M. A. Bowers. 1998. Demographic responses to habitat fragmentation: Experimental tests at the landscape and patch scale. *Ecology* 79:969–980.

Doughty, R. W. 2001. *The eucalyptus.* Baltimore: Johns Hopkins University Press.

Douglas, M. 1997. Forests of East Gippsland before Europeans. Pp. 231–246 in *Australia's ever-changing forests III: Proceedings of the Third National Conference on Australian Forest History*, ed. J. Dargarvel. Australian Forest History Society, the Australian National University, Canberra, Australia.

Dovers, S., and D. B. Lindenmayer. 1997. Managing the environment: Rhetoric, policy and reality. *Australian Journal of Public Administration* 56:65–80.

Dovers, S. R., and T. W. Norton. 1994. Towards an ecological approach to sustainability: Considerations for ecosystem management. *Pacific Conservation Biology* 1:283–293.

Dramstad, W. E., J. D. Olson, and R. T. Forman. 1996. *Landscape ecology principles in landscape architecture and land-use planning.* Sponsored by the Harvard University Graduate School of Design. Washington, D.C.: Island Press.

Duffy, D. C., and A. J. Meier. 1992. Do Appalachian herbaceous understories ever recover from clearcutting? *Conservation Biology* 6:196–201.

Duhig, N., S. Munks, and M. Wapstra. 2000. Designing better wildlife habitat clumps. *Forest Practices News* 2:11.

Duncan, B. D., and G. Isaac. 1986. *Ferns and allied plants of Victoria, Tasmania and South Australia.* Melbourne, Australia: Melbourne University Press.

Dunning, J. B., R. Borgella, K. Clements, and G. K. Meffe. 1995. Patch isolation, corridor effects, and colonisation by a resident sparrow in a managed pine woodland. *Conservation Biology* 9:542–550.

Dunning, J. B., B. J. Danielson, and H. R. Pulliman. 1992. Ecological processes that affect populations in complex landscapes. *Oikos* 65:169–175.

Dunstan, C. E., and B. J. Fox. 1996. The effects of fragmentation and disturbance of rainforest on ground-dwelling mammals on the Robertson Plateau, New South Wales, Australia. *Journal of Biogeography* 23:187–201.

Dunsworth, G., and B. Beese. 2000. New approaches in managing temperate rainforests. Pp. 24–25 in *Mountain forests and sustainable development*, prepared for The Commission on Sustainable Development (CSD) and its 2000 spring session by Mountain Agenda. Swiss Agency for Development and Cooperation, Berne, Switzerland.

Dwyer, P. D. 1983. Little bent-wing bat. Pp. 338–339 in *Complete book of Australian mammals*, ed. R. Strahan. Sydney, Australia: Angus and Robertson.

Dyck, W. J. 2000. Nature conservation in New Zealand plantation forestry. Pp. 35–43 in *Nature Conservation 5: Nature conservation in production environments: Managing the matrix*, ed. J. L. Craig, N. Mitchell, and D. A. Saunders. Chipping Norton, Australia: Surrey Beatty and Sons.

Dyrness, C. T., L. A. Viereck, and K. Van Cleve. 1986. Fire in Taiga communities of interior Alaska. Pp. 74–86 in *Forest ecosystems in the Alaskan Taiga: A synthesis of structure and function*, ed. K. Van

Cleve, F. S. Chapin, P. W. Flanagan, L. A. Viereck, and C. T. Dyrness. New York: Springer-Verlag.

East, R. 1981. Species-area curves and populations of large mammals in African savannah reserves. *Biological Conservation* 21:111–126.

Eberhart, K. E., and P. M. Woodard. 1987. Distribution of residual vegetation associated with large fires in Alberta. *Canadian Journal of Forest Research* 117:1207–1212.

Edwards, R. T. 1998. The hyporheic zone. Pp. 399–429 in *River ecology and management: Lessons from the Pacific coastal ecoregion*, ed. R. J. Naiman and R. E. Bilby. New York: Springer-Verlag.

Egler, F. E. 1954. Vegetation science concepts. I. Initial floristic composition. A factor in old field vegetation development. *Vegetatio* 4:412–417.

Ehmann, H., and H. Cogger. 1985. Australia's endangered herpetofauna: A review of criteria and policies. Pp. 435–447 in *Biology and Australasian frogs and reptiles*, ed. G. Grigg, R. Shine, and H. Ehmann. Sydney, Australia: Surrey Beatty and Sons.

Ehrlich, P. R. 1997. *A world of wounds: Ecologists and the human dilemma*. Ecology Institute, Oldendorf, Germany.

Eldridge, J. 1971. Some observations on the dispersion of small mammals in hedgerows. *Journal of Zoology* 165:530–534.

Eldridge, K., J. Davidson, C. Harwood, and G. van Wyk. 1994. *Eucalypt domestication and breeding*. Oxford: Clarendon Press.

Elkie, P. C., and R. S. Rempel. 2001. Detecting scales of pattern in boreal forest landscapes. *Forest Ecology and Management* 147:253–261.

Elliott, H. J., R. Bashford, A. Greener, and S. G. Candy. 1992. Integrated pest management in the Tasmanian *Eucalyptus* leaf beetle, *Chrysophtharta bimaculata* (Oliver) (Coleoptera: Chrysomelidae). *Forest Ecology and Management* 53:29–38.

Elton, C. S. 1927. *Animal ecology*. London: Methuen.

Engstrom, R. T., L. A. Brennan, W. L. Neel, R. M. Farrar, S. T. Lindenman, W. K. Moser, and S. M. Hermann. 1996. Silvicultural practices and red-cockaded woodpecker management: A reply to Rudolph and Conner. *Wildlife Society Bulletin* 24:334–338.

Enoksson, B., P. Angelstam, and K. Larsson. 1995. Deciduous forest and resident birds: The problem of fragmentation within a coniferous forest landscape. *Landscape Ecology* 10:267–275.

Erwin, T. L. 1982. Tropical forests: Their richness in coleoptera and other species. *Coleopterists Bulletin* 36:74–75.

Esseen, P. 1994. Tree mortality patterns after experimental fragmentation of an old-growth conifer forest. *Biological Conservation* 68:19–28.

Esseen, P., B. Ehnström, L. Ericson, and K. Sjöberg. 1997. Boreal forests. *Ecological Bulletins* 46:16–47.

Esseen, P., B. Ehnström, and K. Sjöberg. 1992. Boreal forests—the focal habitats of Fennoscandia. Pp. 252–325 in *Ecological principles of nature conservation*, ed. L. Hansson. London: Elsevier.

Esseen, P., K. Renhorn, and R. B. Pettersson. 1996. Epiphytic lichen biomass in managed and old-growth boreal forests: Effect of branch quality. *Ecological Applications* 6:228–238.

Estades, C. F. 2001. The effect of breeding-habitat patch size on bird population density. *Landscape Ecology* 16:161–173.

Estades, C. F., and S. A. Temple. 1999. Deciduous-forest bird communities in a fragmented landscape dominated by exotic pine plantations. *Ecological Applications* 9:573–585.

Evans, J., and B. G. Hibberd. 1990. Managing to diversify forests. *Arboriculture Journal* 14:373–378.

Evans, P. 1993. Assessing heritage values of sawmills and tramways in Central Victoria. Pp. 163–186 in *Australia's ever-changing forests II. Proceedings of the Second National Conference on Australian Forest History*, ed. J. Dargavel and S. Feary. Australian Forest History Society, The Australian National University, Canberra, Australia.

Evink, G. L., P. Garrett, D. Zeigler, and J. Berry, eds. 1996. *Trends in addressing transportation related wildlife mortality*. Report FL-ER-58-96. Florida Department of Transportation, Tallahassee, Fla.

Fahrig, L. 1992. Relative importance of spatial and temporal scales in a patchy environment. *Theoretical Population Biology* 41:300–314.

———. 1997. Relative effects of habitat fragmentation and habitat loss on population extinction. *Journal of Wildlife Management* 61:603–610.

———. 1998. When does fragmentation of breeding habitat affect population survival? *Ecological Modelling* 105: 273–292.

———. 1999. Forest loss and fragmentation: Which has the greater effect on persistence of forest-dwelling animals? Pp. 87–95 in *Forest fragmentation: Wildlife management implications*, ed. J. A. Rochelle, L. A. Lehmann, and J. Wisniewski, Leiden, Germany: Brill.

Fahrig, L., and G. Merriam. 1994. Conservation of fragmented populations. *Conservation Biology* 8:50–59.

Fahrig, L., and J. Paloheimo. 1988. Effect of spatial arrangement of habitat patches on local population size. *Ecology* 69:468–475.

Fairweather, P. G. 1991. Statistical power and design requirements for environmental monitoring. *Australian Journal of Marine and Freshwater Research* 42:555–568.

Fenger, M. 1996. Implementing biodiversity conservation through the British Columbia forest practices code. *Forest Ecology and Management* 85:67–77.

Fensham, R. J. 1996. Land clearance and conservation of inland dry rainforest in north Queensland, Australia. *Biological Conservation* 75:289–298.

Fetherston, K. L., R. J. Naiman, and R. E. Bilby. 1995. Large woody debris, physical process, and riparian forest development in montaine river networks for the Pacific Northwest. *Geomorphology* 13:133–144.

Fiedler, C. E., S. F. Arno, C. E. Keegan, and K. A. Blatner. 2001. Overcoming America's wood deficit: An overlooked option. *BioScience* 51:53–58.

Fischer, J., and D. B. Lindenmayer. 2000. A review of relocation as a conservation management tool. *Biological Conservation* 96:1–11.

———. 2002. The conservation value of paddock trees for birds in a variegated landscape in southern New South Wales. I. Species composition and site occupancy patterns. *Biodiversity and Conservation* (in press).

Fischer, W. C., and B. R. McClelland. 1983. *A cavity-nesting bird bibliography including related titles on forest snags, fire, insects, diseases and decay*. March. General Technical Report INT-140. Intermountain Forest and Range Experiment Station, Ogden, Utah.

Fisher, A. M., and D. C. Goldney. 1997. Use by birds of riparian vegetation in an extensively fragmented landscape. *Pacific Conservation Biology* 3:275–288.

———. 1998. Native forest fragments as critical bird habitat in a softwood landscape. *Australian Forestry* 61:287–295.

Fitzgibbon, C. D. 1997. Small mammals in farm woodlands: The effects of habitat, isolation and surrounding land-use patterns. *Journal of Applied Ecology* 34:530–539.

Flannery, T. F. 1994. *The future eaters*. Sydney, Australia: Reed Books.

Flather, C. H., and J. H. Sauer. 1996. Using landscape ecology to test hypotheses about large-scale abundance patterns in migratory birds. *Ecology* 77:28–35.

Flood, J. 1980. *The moth hunters: Aboriginal prehistory of the Australian Alps*. Australian Institute of Aboriginal Studies, Canberra, Australia.

Florence, R. 1996. *Ecology and silviculture of eucalypts*. Melbourne, Australia: CSIRO Publishing.

Food and Agriculture Organization of the United Nations. 2001. *State of the World's forests*. Rome, Italy: FAO.

Forbes, S. H., and D. K. Boyd. 1997. Genetic structure and migration in native and reintroduced Rocky Mountain wolf populations. *Conservation Biology* 11:1126–1234.

Ford, E. D. 1999. Using long-term investigations to develop silvicultural theory for new forestry. *Forestry Chronicle* 75:379–383.

Foreman, D., J. Davis, D. Johns, R. Noss, and M. Soulé. 1992. The wildlands project mission statement. *Wild Earth* (Special Issue):3–4.

Forest Ecosystem Management Assessment Team. 1993. *Forest ecosystem management: An ecological, economic, and social assessment*. USDA Forest Service, Portland, Ore.

Forest Practices Board. 1998. *Threatened fauna manual for production forests in Tasmania*. Forest Practices Board, Hobart, Tasmania, Australia.

———. 1999a. *Fauna Technical Note 9. Guidelines for the design and maintenance of stream crossings—culverts*. Forest Practices Board, Hobart, Tasmania, Australia.

———. 1999b. *Fauna Technical Note 7. Wildlife habitat clumps*. Forest Practices Board, Hobart, Tasmania, Australia.

Forestry Commission of Tasmania. 1990. *Geomorphology manual*. Forestry Commission of Tasmania, Hobart, Tasmania, Australia.

———. 1993. *Forestry practices code*. Forestry Commission of Tasmania, Hobart, Tasmania, Australia.

Forestry Tasmania. 1999. *Annual Report*. Forestry Tasmania, Hobart, Tasmania, Australia.

Forest Stewardship Council. 1993. *FSC principles and criteria for natural forest management*. Forest Stewardship Council, Oaxaca, Mexico.

———. 1996. *FSC Document no. 1.2. Principles and criteria for forest management*. Revised. March. Forest Stewardship Council, Oaxaca, Mexico.

Forman, R. T. 1964. Growth under controlled conditions to explain the hierarchical distributions of a moss, *Tetraphis pellucida*. *Ecological Monographs* 34:1–25.

———. 1995. *Land mosaics: The ecology of landscapes and regions*. New York: Cambridge University Press.

———. 1998. Roads and their major ecological effects. *Annual Review of Ecology and Systematics* 29:207–231.

———. 2000. Estimate of the area affected ecologically by the road system in the United States. *Conservation Biology* 14:31–35.

Forman, R. T., and R. D. Deblinger. 2000. The ecological road-effect zone of a Massachusetts (USA) suburban highway. *Conservation Biology* 14:36–46.

Forman, R. T., and M. Godron. 1986. *Landscape ecology*. New York: John Wiley and Sons.

Forsman, E. D., E. C. Meslow, and H. Wight. 1984. Distribution and ecology of the spotted owl in Oregon. *Wildlife Monographs* 87:1–64.

Forsyth, J. 1998. Anarchy in the forests: A plethora of rules, an absence of enforceability. *Environmental and Planning Law Journal* 15:338–349.

Foster, D. R. 1983. The history and pattern of fire in the boreal forest of south-eastern Labrador. *Canadian Journal of Botany* 61:2459–2471.

Foster, D. R., J. B. Aber, J. M. Melillo, R. D. Bowden, and F. A. Bazzaz. 1997. Forest response to disturbance and anthropogenic stress. *BioScience* 47:437–445.

Foster, D. R., and E. R. Boose. 1992. Patterns of forest damage resulting from catastrophic wind in central New England, USA. *Journal of Ecology* 80:79–98.

Foster, D. R., D. H. Knight, and J. F. Franklin. 1998. Landscape patterns and legacies resulting from large infrequent forest disturbances. *Ecosystems* 1:497–510.

Foster, R. B. 1980. Heterogeneity and disturbance in tropical vegetation. Pp. 75–92 in *Conservation biology: An evolutionary-ecological perspective*, ed. M. E. Soulé and B. A. Wilcox. Sunderland, Mass.: Sinauer Associates.

Fox, B. J., and M. D. Fox. 2000. Factors determining mammal species richness on habitat islands and isolates: Habitat diversity, disturbance, species interactions and guild assembly rules. *Global Ecology and Biogeography* 9:19–37.

Frankham, R. 1996. Relationship of genetic variation to population size in wildlife. *Conservation Biology* 10:1500–1508.

Franklin, A. B., D. R. Anderson, R. J. Gutierrez, and K. P. Burnham. 2000a. Climate, habitat quality, and fitness in northern spotted owl populations in northwestern California. *Ecological Monographs* 70:539–590.

Franklin, J. F. 1988. Structural and functional diversity in temperate forests. Pp. 166–175 in *Biodiversity*, ed. E. O. Wilson. Washington, D.C.: National Academy Press.

———. 1990. Biological legacies: A critical management concept from Mount St. Helens. *Transactions of the Fifty-fifth North American Wildlife and Natural Resource Conference* 216–219.

———. 1992. Scientific basis for new perspectives in forests and streams. Pp. 25–72 in *Watershed management: Balancing sustainability and environmental change*, ed. R. J. Naiman. New York: Springer-Verlag.

———. 1993a. Preserving biodiversity: Species, ecosystems or landscapes? *Ecological Applications* 3:202–205.

———. 1993b. Lessons from old-growth. *Journal of Forestry* (December):11–13.

Franklin, J. F., D. E. Berg, D. A. Thornburgh, and J. C. Tappeiner. 1997. Alternative silvicultural approaches to timber harvesting: Variable retention harvest systems. Pp. 111–139 in *Creating a forestry for the twenty-first century: The science of ecosystem management*, ed. K. A. Kohm and J. F. Franklin. Washington, D.C.: Island Press.

Franklin, J. F., K. Cromack, W. Denison, A. McKee, C. Maser, J. Sedell, F. Swanson, and G. Juday. 1981. *Ecological attributes of old-growth Douglas-fir forests*. USDA Forest Service General Technical Report PNW-118. Pacific Northwest Forest and Range Experimental Station, Portland, Ore.

Franklin, J. F., and C. T. Dyrness. 1988. *Natural vegetation of Oregon and Washington*. Corvallis: Oregon State University Press.

Franklin, J. F., and J. A. Fites-Kaufmann. 1996. Assessment of late-successional forests of the Sierra Nevada. Pp. 627–656 in *Sierra Nevada Ecosystem Project: Final Report to Congress*. Vol. 2. *Assessment and scientific basis for management options*. Center for Water and Wildland Resources, University of California, Davis.

Franklin, J. F., and R. T. Forman. 1987. Creating landscape patterns by forest cutting: Ecological consequences and principles. *Landscape Ecology* 1:5–18.

Franklin, J. F., and M. A. Hemstrom. 1981. Aspects of succession in the coniferous forests of the Pacific Northwest. Pp. 212–229 in *Forest succession: Concepts and application*, ed. D. C. West, H. H. Shugart, and D. B. Botkin. New York: Springer-Verlag.

Franklin, J. F., M. E. Harmon, and F. J. Swanson. 1999a. Complementary roles of research and monitoring: Lessons from the U.S. LTER Program and Tierra del Fuego. Paper presented at the symposium Toward a Unified Framework for Inventorying and Monitoring Forest Ecosystem Resources, November 1998, Guadalajara, Mexico.

Franklin, J. F., D. B. Lindenmayer, J. A. MacMahon, A. McKee, J. Magnusson, D. A. Perry, R. Waide, and D. R. Foster. 2000b. Threads of continuity: Ecosystem disturbances, biological legacies and ecosystem recovery. *Conservation Biology in Practice* 1:8–16.

Franklin, J. F., and J. A. MacMahon. 2000. Messages from a mountain. *Science* 288:1183–1185.

Franklin, J. F., J. A. MacMahon, F. J. Swanson, and J. R. Sedell. 1985. Ecosystem responses to the eruption of Mount St. Helens. *National Geographic Research* (Spring):198–216.

Franklin, J. F., L. A. Norris, D. R. Berg, and G. R. Smith. 1999b. The history of DEMO: An experiment in regeneration harvest of northwestern forest ecosystems. *North West Science* 73 (Special Issue):3–11.

Franklin, J. F., H. H. Shugart, and M. E. Harmon. 1987. Tree death as an ecological process. *BioScience* 37:550–556.

Franklin, J. F., T. A. Spies, R. Van Pelt, A. B. Carey, D. A. Thornburgh, D. R. Berg, D. B. Lindenmayer, M. E. Harmon, W. S. Keeton, D. C. Shaw, K. Bible, and J. Chen. 2002. Disturbances and structural development of natural forest ecosystems with silvicultural implications, using Douglas-fir forests as an example. *Forest Ecology and Management* 155:399–423.

Franzreb, K. E. 1993. Perspectives on the landmark decision designating the northern spotted owl (*Strix occidentalis caurina*) as a threatened subspecies. *Environmental Management* 17:445–452.

Freemark, K., and B. Collins. 1992. Landscape ecology of birds breeding in temperate forest fragments. Pp. 443–454 in *Conservation of neotropical migrants*, ed. J. Hagan and D. J. Johnston. Washington, D.C.: Smithsonian Institution.

Friend, G. R. 1982. Mammal populations in exotic pine plantations and indigenous eucalypt forests in Gippsland, Victoria. *Australian Forestry* 45:3–18.

Fries, C., M. Carlsson, B. Dahlin, T. Lamas, and O. Sallnas. 1998. A review of conceptual landscape models for multi-objective forestry in Sweden. *Canadian Journal of Forest Research* 8:159–167.

Fries, C., O. Johansson, B. Pettersson, and P. Simonsson. 1997. Silvicultural models to maintain and restore natural stand structures in Swedish boreal forests. *Forest Ecology and Management* 94:89–103.

Fritz, R. 1979. Consequences of insular population structure: Distribution and extinction of spruce grouse populations. *Oecologia* 42:57–65.

Frumhoff, P. C. 1995. Conserving wildlife in tropical forests managed for timber. *BioScience* 45:456–464.

Fulé, P. Z., and W. W. Covington. 1999. Fire regime changes in la Michilía Biosphere Reserve, Durango, Mexico. *Conservation Biology* 13:640–652.

Funtowicz, S. O., and J. R. Ravetz. 1991. A new scientific methodology for global environmental issues. Pp. 137–152 in *Ecological economics: The science and management of sustainability*, ed. R. Costanza. New York: Columbia University Press.

Gardner, R. H., B. T. Milne, M. G. Turner, and R. V. O'Neill. 1987. Neutral models for analysis of broad-scale landscape patterns. *Landscape Ecology* 1:19–28.

Garman, S. L., F. J. Swansson, and T. A. Spies. 1999. Past, present and future landscape patterns in the Douglas-fir region of the Pacific Northwest. Pp. 61–68 in *Forest wildlife and fragmentation: Management implications*, ed. J. A. Rochelle, L. A. Lehmann, and J. Wisnewski. Leiden, Germany: Brill.

Garrett, M. G., and W. L. Franklin. 1988. Behavioral ecology of dispersal in the black-tailed prairie dog. *Journal of Mammalogy* 69:236–250.

Gascon, C. 1993. Breeding-habitat use by five Amazonian frogs at forest edge. *Biodiversity and Conservation* 2:438–444.

Gascon, C., T. Lovejoy, R. O. Bierregaard, J. R. Malcolm, P. C. Stouffer, H. L. Vasconcelos, W. F. Laurance, B. Zimmerman, M. Tocher, and S. Borges. 1999. Matrix habitat and species richness in tropical forest remnants. *Biological Conservation* 91:223–229.

Gascon, C., and T. E. Lovejoy. 1998. Ecological impacts of forest fragmentation in central Amazonia. *Zoology* 101:273–280.

Gaston, K. J. 1994. *Rarity*. London: Chapman and Hall.

Gaston, K. J., and T. M. Blackburn. 1995. Birds, body size and the threat of extinction. *Philosophical Transactions of the Royal Society of London Series B: Biological Sciences* 347:205–212.

Gates, J. E., and L. W. Gysel. 1978. Avian nest dispersion and fledging success in field-forest ecotones. *Ecology* 59:871–883.

Gayer, K. 1886. Der gemischte Wald—seine Begründung and Pflege, insbesondere durch Horst- und Gruppenwirtschaft. Verlag Paul Parey, Berlin.

Gentry, A. H., and J. Lopez-Parodi. 1980. Deforestation and increased flooding of the upper Amazon. *Science* 210:1354–1355.

Gerrand, A. M. 1997. Management decision classification: A system of zoning land managed by Forestry Tasmania. Pp. 480–487 in *Conservation outside nature reserves*, ed. P. Hale and D. Lamb. Centre for Conservation Biology, Brisbane, Australia.

Ghassemi, F., A. J. Jakeman, and H. A. Nix. 1995. Salinisation of land and water resources. Sydney, Australia: University of New South Wales Press.

Ghazoul, J. 2001. Barriers to biodiversity conservation in forest certification. *Conservation Biology* 15:315–317.

Gibbons, P. 1999. Habitat tree retention in wood production forests. Ph.D. diss., the Australian National University, Canberra.

Gibbons, P., and D. B. Lindenmayer. 1996. A review of issues associated with the retention of trees with hollows in wood production forests. *Forest Ecology and Management* 83:245–279.

———. 1997. *Conserving hollow-dependent fauna in timber-production forests New South Wales*. National Parks and Wildlife Service, Environmental Heritage Monograph Series 3:1–110.

———. 2002. *Hollows and wildlife conservation in Australia*. Melbourne, Australia: CSIRO Publishing.

Gibbons, P. G., D. B. Lindenmayer, and M. Tanton. 2000. The effects of slash burning on the mortality and collapse of trees retained on logged sites in south-eastern Australia. *Forest Ecology and Management* 139:51–61.

Gibson, N., M. J. Brown, K. Williams, and A. V. Brown. 1992. Flora and vegetation of ultramafic areas in Tasmania. *Australian Journal of Ecology* 17:297–303.

Gilbert, F., A. Gonzalez, and I. Evens-Freke. 1998. Corridors maintain species richness in the fragmented landscapes of a microsystem. *Proceedings of the Royal Society of London. Series B* 265:577–582.

Gilbert, J. M. 1959. Forest succession in the Florentine Valley, Tasmania. *Papers and Proceedings Royal Society Tasmania* 93:129–151.

Gilbert, L. E. 1980. The equilibrium theory of island biogeography: Fact or fiction? *Journal of Biogeography* 7:209–235.

Gill, A. M., and M. A. McCarthy. 1998. Intervals between prescribed fires in Australia: What intrinsic variation should apply? *Biological Conservation* 85:161–169.

Gill, A. M., J. C. Woinarski, and A. York. 1999. *Australia's biodiversity—responses to fire. Plants, birds and invertebrates*. Environment Australia Biodiversity Technical Paper 1:1–266.

Gill, F. B. 1995. *Ornithology*. 2nd ed. New York: W. H. Freeman and Company.

Gilmore, A. M. 1990. Plantation forestry: Conservation impacts on terrestrial vertebrate fauna. Pp. 377–388 in *Prospects for Australian plantations*, ed. J. Dargavel and N. Semple. Centre for Resource and Environmental Studies, The Australian National University, Canberra.

Gilpin, M. E. 1987. Spatial structure and population vulnerability. Pp. 126–139 in *Viable populations for conservation*, ed. M. E. Soulé. New York: Cambridge University Press.

Gippel, C. J., B. L. Finlayson, and I. C. O'Neill. 1996. Disturbance and hydraulic significance of large woody debris in a lowland Australian river. *Hydrobiologica* 318:179–194.

Gleason, H. A. 1926. The individualistic concept of the plant association. *Bulletin of the Torrey Botanical Club* 53:7–26.

Goldingay, R. L. 2000a. Small dasyurid marsupials—are they effective pollinators? *Australian Journal of Zoology* 48:597–606.

———. 2000b. Use of sap trees by the yellow-bellied glider in the Shoalhaven region of New South Wales. *Wildlife Research* 27:217–222.

Goldingay, R. L., S. M. Carthew, and R. J. Whelan. 1991. The importance of non-flying mammals in pollination. *Oikos* 61:79–87.

Goldingay, R. G., and R. P. Kavanagh. 1991. The yellow-bellied glider: A review of its ecology and management considerations. Pp. 365–375 in *Conservation of Australia's forest fauna*, ed. D. Lunney. Royal Zoological Society of New South Wales, Sydney, Australia.

Goldsmith, B., ed. 1991. *Monitoring for conservation and ecology*. London: Chapman and Hall.

Gooday, P., P. Whish-Wilson, and L. Weston. 1997. Regional Forest Agreements. Central Highlands of Victoria. Pp. 1–11 in *Australian forest products statistics*. Australian Bureau of Agricultural and Resource Economics, Canberra, Australia. September Quarter.

Goosem, M. 1997. Internal fragmentation: The effects of roads, highways and powerline clearings on movements and mortality on rainforest vertebrates. Pp. 241–255 in *Tropical forest remnants: Ecology, management and conservation of fragmented communities*, ed. W. F. Laurance and R. O. Bierregaard. Chicago: University of Chicago Press.

———. 2001. Effects of tropical rainforest roads on small mammals: Inhibition of crossing movements. *Wildlife Research* 28:351–364.

Gordon, G., A. S. Brown, and T. Pulsford. 1988. A Koala (*Phascolartos cinereus* Goldfuss) population crash during drought and heatwave conditions in southwestern Queensland. *Australian Journal of Ecology* 13:451–461.

Government of Victoria. 1986. *Timber Industry Strategy*. Government Statement no. 9. Melbourne, Australia: Government Printer.

———. 1988. *Flora and Fauna Guarantee Act. no. 47 of 1988*. Melbourne, Australia: Government Printer.

———. 1992. *Flora and fauna guarantee strategy: Conservation of Victoria's biodiversity*. Department of Conservation and Environment, Melbourne, Australia.

Graham, R. L. 1982. Biomass dynamics of dead Douglas-fir and western hemlock boles in mid-elevation forests of the Cascade range. Ph.D. diss., Oregon State University, Corvallis.

Graham, R. T., A. E. Harvey, M. F. Jurgensen, T. B. Jain, M. J. R. Tonn, and D. S. Page-Dumroese. 1994. *Managing coarse woody debris in forests of the Rocky Mountains*. USDA Forest Service General Technical Report INT-RP-477. Intermountain Research Station, Ogden, Utah.

Graham, R. T., and T. B. Jain. 1998. Silviculture's role in managing boreal forests. *Conservation Ecology* (online) 8. http://www.consecol.org/vol2/iss2/art8.

Gratkowski, H. J. 1956. Windthrow around staggered settings in old-growth Douglas-fir. *Forest Science* 2:60–74.

Graynoth, E. 1989. Effects of logging on stream environments and faunas in Nelson. *New Zealand Journal of Marine and Freshwater Research* 13:79–109.

Grayson, R. B., B. L Finlayson, M. D. Jayasuriya, and P. O'Shaughnessy. 1992. *The effects of a clearfelling and harvesting operation on the quality of streamflow from a mountain ash forest*. Land and Water Resources Research and Development Corporation Project UME 1. Land and Water Resources Research and Development Corporation, Canberra, Australian Capital Territory.

Gregory, S. V. 1997. Riparian management in the twenty-first century. Pp. 69–85 in *Creating a forestry for the twenty-first century: The science of ecosystem management*, ed. K. A. Kohm and J. F. Franklin. Washington, D.C.: Island Press.

Gregory, S. V., F. J. Swanson, and W. A. McKee. 1991. An ecosystem perspective of riparian zones. *BioScience* 40:540–551.

Grief, G. E., and O. W. Archibold. 2000. Standing dead-tree component of the boreal forest in central Saskatchewan. *Forest Ecology and Management* 131:37–46.

Grodinski, C., and M. Stüwe. 1987. With lots of help alpine ibex return to their mountains. *Smithsonian* 18:68–77.

Groombridge, B., ed. 1992. *IUCN red list of threatened animals*. IUCN, Gland, Switzerland.

Grove, S. J. 2001. Extent and decomposition of dead wood in Australian lowland tropical rainforest with different management histories. *Forest Ecology and Management* 154:35–53.

Grover, D. R., and P. J. Slater. 1994. Conservation value to birds of remnants of *Melaleuca* forest in suburban Brisbane. *Wildlife Research* 21:433–444.

Grumbine, R. E. 1990. Viable populations, reserve design, and federal lands management: A critique. *Conservation Biology* 4:127–134.

Gucinski, H., M. J. Furniss, R. R. Ziemer, and M. H. Brookes, eds. 2000. *Forest roads: A synthesis of scientific information*. USDA Forest Service, Washington, D.C.

Gullan, P. J., and P. S. Cranston. 1994. *The insects. An outline of entomology*. Melbourne, Australia: Chapman and Hall.

Gullan, P. K., and A. C. Robinson. 1980. Vegetation and small mammals of a Victorian forest. *Australian Mammalogy* 3:87–96.

Gundersen, V., and J. Rolstad. 1998. Nøkkelbiotoper i skog. En vurdering av nøkkelbiotoper som forvaltningstiltak for bevaring av biologisk mangfold i skog. NISK, Ås.

Gurd, D. B., and T. D. Nudds. 1999. Insular biogeography of mammals in Canadian national parks: A re-analysis. *Journal of Biogeography* 26:973–982.

Gustafson, E. J., and R. H. Gardner. 1996. The effect of landscape heterogeneity on the probability of patch colonisation. *Ecology* 77:94–107.

Gustafson, E. J., and G. R. Parker. 1992. Relationships between landcover proportion and indices of landscape spatial pattern. *Landscape Ecology* 7:101–110.

Gustafsson, L. 2000. Red-listed species and indicators: Vascular plants in woodland key habitats and surrounding production forests in Sweden. *Biological Conservation* 92:35–43.

Gustafsson, L., and I. Ahlen. 1996. *Geography of plants and animals: National atlas of Sweden*. SNA (National Atlas of Sweden), Italy.

Gustafsson, L., J. de Jong, and M. Noren. 1999. Evaluation of Swedish woodland key habitats using red-listed bryophytes and lichens. *Biodiversity and Conservation* 8:1101–1114.

Gustafsson, L., and L. Hansson. 1997. Corridors as a conservation tool. *Ecological Bulletins* 46:182–190.

Gutzwiller, K. J., and S. H. Anderson. 1987. Multiscale associations between cavity-nesting birds and features of Wyoming streamside woodlands. *Condor* 89:534–548.

Haila, Y., 1988. The multiple faces of ecological data and theory. *Oikos* 53:408–411.

———. 1999. Islands. Pp. 234–264 in *Maintaining biodiversity in forest ecosystems*, ed. M. L. Hunter. Cambridge: Cambridge University Press.

Haila, Y., I. K. Hanski, J. Niemelä, P. Puntilla, S. Raivio, and H. Tukia. 1993. Forestry and the boreal fauna: Matching management with natural forest dynamics. *Annales Zoologica Fennici* 31:187–202.

Haines-Young, R., and M. Chopping. 1996. Quantifying landscape structure: A review of landscape indices and their application to forested environments. *Progress in Physical Geography* 20:418–445.

Hale, P., and D. Lamb, eds. 1997. *Conservation outside reserves*. Centre for Conservation Biology, University of Queensland, Brisbane, Australia.

Hall, C. M. 1988. The "worthless lands hypothesis" and Australia's national parks and reserves. Pp. 441–459 in *Australia's ever-*

changing forests, ed. K. J. Frawley and N. M. Semple. Australian Defense Force Academy, Canberra.

Halley, J. M., C. F. Thomas, and P. C. Jepson. 1996. A model for the spatial dynamics of linyphiid spiders in farmland. *Journal of Applied Ecology* 33:471–492.

Halpern, C. B. 1988. Early successional pathways and the resistance and resilience of forest communities. *Ecology* 69:1703–1715.

Halpern, C. B., S. A. Evands, C. R. Nelson, D. McKenzie, D. A. Liguori, D. E. Hibbs, and M. G. Halaj. 1999. Responses of forest vegetation to varying levels and patterns of green-tree retention: An overview of a long-term experiment. *Northwest Science* 73 (Special Issue):27–44.

Halpern, C. B., and J. F. Franklin. 1990. Physiognomic development of *Pseudotsuga* forests in relation to initial structure and disturbance intensity. *Journal of Vegetation Science* 1:475–482.

Halpern, C. B., and M. G. Raphael, eds. 1999. Special issue on retention harvests in northwestern forest ecosystems. Demonstration of Ecosystem Management Options (DEMO) study. *Northwest Science* 73 (Special Issue):1–125.

Halpern, C. B., and T. A. Spies. 1995. Plant species diversity in natural and managed forests of the Pacific Northwest. *Ecological Applications* 5:913–934.

Halme, E., and J. Niemelä. 1993. Carabid beetles in fragments of coniferous forest. *Annales Zoologica Fennici* 30:17–30.

Hamel, G., and C. K. Prahalad. 1989. Strategic intent. *Harvard Business Review* 89:63–76.

Hammond, H. 1991. *Seeing the forest among the trees*. Vancouver, British Columbia: Polestar Press.

Hanley, T. A. 1993. Balancing economic development, biological conservation, and human culture: The Sitka black-tailed deer *Odocoileus hemionus sitkensis* as an ecological indicator. *Biological Conservation* 66:61–67.

Hanley, T. A., C. T. Robbins, and D. E. Spalinger. 1989. *Forest habitats and the nutritional ecology of Sitka black-tailed deer: A research synthesis with implications for forest management*. USDA Forest Service General Technical Report GTR-PNW-230.

Hanley, T. A., and C. L. Rose. 1987. *Influence of overstory on snow depth and density in hemlock-spruce stands: Implications for management of deer habitat in southeastern Alaska*. USDA Forest Service Research Note PNW-RN-459.

Hannah, L., J. L. Carr, and A. Lankerani. 1995. Human disturbance and natural habitat: A biome level analysis of a global data set. *Biodiversity and Conservation* 4:128–155.

Hannon, S. J., and S. E. Cotterill. 1998. Nest predation in Aspen woodlots in an agricultural area in Alberta: The enemy from within. *Auk* 115:16–25.

Hanowski, J. M., G. J. Niemi, and D. C. Christian. 1997. Influence of within-plantation heterogeneity and surrounding landscape composition on avian communities in hybrid popular plantations. *Conservation Biology* 11:936–944.

Hansen, A., and J. Rotella. 1999. Abiotic factors. Pp. 161–209 in *Maintaining biodiversity in forest ecosystems*, ed. M. Hunter Jr. Cambridge: Cambridge University Press.

Hansen, A. J., S. L. Garman, B. Marks, and D. L. Urban. 1993. An approach to managing vertebrate diversity across multiple-use landscapes. *Ecological Applications* 3:481–496.

Hansen, A. J., S. L. Garman, J. F. Weigand, D. L. Urban, W. C. McComb, and M. G. Raphael. 1995b. Alternative silvicultural regimes in the Pacific Northwest: Simulations of ecological and economic effects. *Ecological Applications* 5:535–554.

Hansen, A. J., and P. Hounihan. 1996. Canopy tree retention and avian diversity in the Oregon Cascades. Pp. 402–421 in *Biodiversity in managed landscapes: Theory and practice*, ed. R. C. Szaro and D. W. Johnston. New York: Oxford University Press.

Hansen, A. J., W. C. McComb, R. Vega, M. G. Raphael, and M. Hunter. 1995a. Bird habitat relationships in natural and man-aged forests in the west Cascades of Oregon. *Ecological Applications* 5:555–569.

Hansen, A. J., T. A. Spies, F. J. Swanson, and J. L. Ohmann. 1991. Conserving biodiversity in managed forests. *BioScience* 41: 382–392.

Hanski, I. 1994a. Patch occupancy dynamics in fragmented landscapes. *Trends in Evolution and Ecology* 9:131–134.

———. 1994b. A practical model of metapopulation dynamics. *Journal of Animal Ecology* 63:151–162.

———. 1998. Metapopulation dynamics. *Nature* 396:41–49.

———. 1999a. *Metapopulation ecology*. Oxford: Oxford University Press.

———. 1999b. Habitat connectivity, habitat continuity, and metapopulations in dynamic landscapes. *Oikos* 87:209–219.

Hanski, I., and M. Gilpin. 1991. Metapopulation dynamics: Brief history and conceptual domain. *Biological Journal of the Linnean Society* 42:3–16.

Hanski, I., and M. Gyllenberg. 1993. Two general metapopulation models and the core-satellite hypothesis. *American Naturalist* 142:17–41.

Hanski, I., and P. Hammond. 1995. Biodiversity in boreal forests. *Trends in Evolution and Ecology* 10:5–6.

Hanski, I., T. Pakkala, M. Kuussaari, and G. Lei. 1995. Metapopulation structure of an endangered butterfly in a fragmented landscape. *Oikos* 72:21–28.

Hanski, I., and D. Simberloff. 1997. The metapopulation approach, its history, conceptual domain, and application to conservation. Pp. 5–26 in *Metapopulation biology: Ecology, genetics and evolution*, ed. I. Hanski and M. E. Gilpin. San Diego: Academic Press.

Hanski, I., and C. D. Thomas. 1994. Metapopulation dynamics and conservation: A spatially explicit model applied to butterflies. *Biological Conservation* 68:167–180.

Hanski, I. K., T. J. Fenske, and G. J. Niemi. 1996. Lack of edge effect in nesting success of breeding birds in managed forest landscapes. *Auk* 113:578–585.

Hansson, L. 1983. Bird numbers across edges between mature conifer and clearcuts in central Sweden. *Ornis Scandinavia* 14:97–103.

———., ed. 1992. *Ecological principles of nature conservation*. London: Elsevier.

———. 1998. Nestedness as a conservation tool: Plants and birds of oak-hazel woodland in Sweden. *Ecology Letters* 1:142–145.

Hansson, L., and P. Angelstam. 1990. Landscape ecology as a theoretical basis for nature conservation. *Landscape Ecology* 5:191–201.

Haramis, G. M., and D. Q. Thompson. 1985. Density-production characteristics of box-nesting wood ducks in a northern greentree impoundment. *Journal of Wildlife Management* 49:429–436.

Hargis, C. D., J. A. Bissonette, and J. L. David. 1998. The behavior of landscape metrics commonly used in the study of habitat fragmentation. *Landscape Ecology* 13:167–178.

Harmon, M. E. 2001. Carbon sequestration in forests: Addressing the scale question. *Journal of Forestry* 99:24–29.

Harmon, M. E., and J. F. Franklin. 1989. Tree seedlings on logs in *Picea-Tsuga* forests of Oregon and Washington. *Ecology* 70:48–59.

Harmon, M., J. F. Franklin, F. Swanson, P. Sollins, S. V. Gregory, J. D. Lattin, N. H. Anderson, S. P. Cline, N. G. Aumen, J. R. Sedell, G. W. Lienkaemper, K. Cromack, and K. Cummins. 1986. Ecology of coarse woody debris in temperate ecosystems. *Advances in Ecological Research* 15:133–302.

Harmon, M. E., W. K. Ferrell, and J. F. Franklin. 1990. Effects of carbon storage on conversion of old-growth forest to young forests. *Science* 9:699.

Harper, K. A., and S. E. Macdonald. 2001. Structure and composition of riparian boreal forest: New methods for analyzing edge influence. *Ecology* 82:649–659.

Harr, R. D. 1986. Effects of clearcutting on rain-on-snow runoff in western Oregon: A new look at old studies. *Water Resources Bulletin* 22:1095–1100.

Harris, A. S., and W. A. Farr. 1974. *The forest ecosystem of southeast Alaska. 7: Forest ecology and timber management.* General Technical Report PNW-25. USDA, Forest Service, Pacific Northwest Forest and Range Experiment Station, Portland, Ore.

Harris, L. D. 1984. *The fragmented forest.* Chicago: University of Chicago Press.

Harris, L. D., and J. Scheck. 1991. From implications to applications: The dispersal corridor principle applied to the conservation of biology diversity. Pp. 189–220 in *Nature Conservation 2: The role of corridors*, ed. D. A. Saunders and R. J. Hobbs. Chipping Norton, Australia: Surrey Beatty and Sons.

Harrison, R. L. 1992. Toward a theory of inter-refuge corridor design. *Conservation Biology* 6:293–295.

Harrison, S. 1991. Metapopulations and conservation. Pp. 111–128 in *Large-scale ecology and conservation biology*, ed. P. J. Edwards, R. M. May, and N. R. Webb. Oxford: Blackwell Science.

Harrison, S., and E. Bruna. 2000. Habitat fragmentation and large-scale conservation: What do we know for sure? *Ecography* 22:225–232.

Harrison, S., J. Maron, and G. Huxel. 2000. Regional turnover and fluctuation of five plants confined to serpentine seeps. *Conservation Biology* 14:769–779.

Harrison, S., D. D. Murphy, and P. Ehrlich. 1988. Distribution of the bay checkerspot butterfly, *Euphydras editha bayensi*: Evidence for a metapopulation model. *American Naturalist* 132:360–382.

Harrison, S., and A. D. Taylor. 1997. Empirical evidence for metapopulation dynamics. Pp. 27–42 in *Metapopulation biology: Ecology, genetics and evolution*, ed. I. Hanski and M. E. Gilpin. San Diego: Academic Press.

Harvey, A. E., M. F. Jurgensen, M. J. Larsen, and R. T. Graham. 1987. *Decaying organic materials and soil quality in the inland Northwest: A management opportunity.* USDA Forest Service General Technical Report INT-225. Intermountain Research Station, Ogden, Utah.

Haesler, M., and R. Taylor. 1993. Use of tree hollows by birds in sclerophyll forests in north-eastern Tasmania. *Tasforests* (October):51–56.

Haskell, D. G. 2000. Effects of forest roads on macroinvertebrate soil fauna of the southern Appalachian Moutains. *Conservation Biology* 14:57–63.

Hastings, A. 1993. Complex interactions between dispersal and dynamics: Lessons from coupled logistic equations. *Ecology* 74:1362–1372.

Hastings, A., and S. Harrison. 1994. Metapopulation dynamics and genetics. *Annual Review of Ecology and Systematics* 25:167–188.

Haycock, N. E., T. P. Burt, K. W. Goulding, and G. Pinay, eds. 1997. *Buffer zones: Their processes and potential in water protection.* Quest Environmental, Hertfordshire, U.K.

Haynes, R. W., and G. P. Perez, eds. 2001. *Northwest Forest Plan research synthesis.* USDA Forest Service General Technical Report PNW-GTR-498.

Hazell, P., and L. Gustafsson. 1999. Retention of trees at final harvest—evaluation of a conservation technique using ephiphytic bryophyte and lichen transplants. *Biological Conservation* 90:133–142.

Hellawell, J. M. 1991. Development of a rationale for monitoring. Pp. 1–14 in *Monitoring for conservation and ecology*, ed. F. B. Goldsmith. London: Chapman and Hall.

Helms, J. A., ed. 1998. *The dictionary of forestry.* Bethesda, Md.: Society of American Foresters.

Hemstrom, M. A., and J. F. Franklin. 1982. Fire and other disturbances of the forest in Mount Rainier National Park. *Quaternary Research* 18:32–51.

Hemstrom, M., T. Spies, C. Palmer, R. Kiester, J. Teply, P. McDonald, and R. Warbington. 1998. *Late-successional and old-growth forest effectiveness monitoring plan for the Northwest Forest Plan.* USDA Forest Service General Technical Report GTR-PNW-438.

Henderson, J. A. 1994. The ecological consequences of long-rotation forestry. Pp. 4–26 in *High quality forestry workshop: The idea of long rotations*, ed. J. F. Weigand, R. W. Haynes, and J. L. Mikowski. University of Washington Center for International Trade in Forest Products, Seattle, Wash.

Hennon, P. E., and E. M. Loopstra. 1991. *Persistence of western hemlock and western redcedar trees 38 years after girdling at Cat Island in southeast Alaska.* USDA Forest Service Research Note PNW-RN-507.

Hewittson, H. 1997. The genetic consequences of habitat fragmentation on the bush rat (*Rattus fuscipes*) in a pine plantation near Tumut, NSW. Honors thesis, Division of Botany and Zoology, The Australian National University, Canberra.

Heywood, V. H., and S. N. Stuart. 1992. Species extinction in tropical forests. Pp. 91–117 in *Tropical deforestation and species extinction*, ed. T. C. Whitmore and J. A. Sayer. London: IUCN and Chapman and Hall.

Hickey, J. E., and M. G. Neyland. 2001. Testing silvicultural options for mixed forest (*Eucalyptus-Nothofagus*) regeneration in Tasmania. *Sustainable Management of Indigenous Forest Symposium*, Southern Connection Congress, Christchurch, New Zealand.

Hickey, J. E., M. G. Neyland, L. G. Edwards, and J. K. Dingle. 1999. Testing alternative silvicultural systems for wet eucalypt forests in Tasmania. Pp. 136–141 in *Practising Forestry Today: 18th Biennial Conference of the Institute of Foresters of Australia*, ed. R. C. Ellis and P. J. Smethurst. 3–8 October. Hobart, Tasmania.

Hicks, B. J., J. D. Hall, P. A. Bisson, and J. R. Sedell. 1991. Responses of salmonids to habitat changes. Pp. 438–517 in *Influences of forest and rangeland management on salmonid fishes and their habitats*, ed. W. R. Meehan. American Fisheries Society Special Publication no. 19. Bethesda, Md.

Highley, T. L., and T. Kent Kirk. 1979. Mechanisms of wood decay and unique features of heartrots. *Phytopathology* 69:1151–1157.

Hilborn, R. 1992. Can fisheries agencies learn from experience? *Fisheries* 17:6–14.

Hill, C. J. 1995. Linear strips of rain forest vegetation as potential dispersal corridors for rain forest insects. *Conservation Biology* 9:1559–1566.

Hill, C. J., A. Gillison, and R. E. Jones. 1992. The spatial distribution of rainforest butterflies at three sites in North Queensland. *Australian Journal of Tropical Ecology* 8:37–46.

Hill, J. K., C. D. Thomas, and O. T. Lewis. 1996. Effects of habitat patch size and isolation on dispersal by *Hesperia comma* butterflies: Implications for metapopulation structure. *Journal of Animal Ecology* 65:725–735.

——. 1999. Flight morphology in fragmented populations of a rare British butterfly, *Hesperia comma*. *Biological Conservation* 87:277–283.

Hilty, J., and A. Merenlender. 2000. Faunal indicator taxa selection for monitoring ecosystem health. *Biological Conservation* 92:185–197.

Hinds, W. T. 1984. Towards monitoring of long-term trends in terrestrial ecosystems. *Environmental Conservation* 11:11–18.

Hinsley, S. A., P. Rothery, and P. E. Bellamy. 1999. Influence of woodland area on breeding success in great tits *Parus major* and blue tits *Parus caeruleus*. *Journal of Avian Biology* 30:271–281.

Hobbs, R. J. 1992. The role of corridors in conservation: Solution or bandwagon? *Trends in Ecology and Evolution* 7:389–392.

——. 1997. The right path or the roads to nowhere? *Nature Australia* (Autumn):57–63.

Hobbs, R., and L. F. Huenneke. 1992. Disturbance, diversity, and invasion: Implications for conservation. *Conservation Biology* 6:324–337.

Hobbs, R., D. A. Saunders, and G. W. Arnold. 1993. Integrated landscape ecology: A Western Australian perspective. *Biological Conservation* 64:231–238.

Hobson, K. A., and J. Schieck. 1999. Changes in bird communities in boreal mixedwood forest: Harvest and wildfire effects over thirty years. *Ecological Applications* 9:849–863.

Hokit, D. G., B. M. Stith, and L. C. Branch. 1999. Effects of landscape structure in Florida scrub: A population perspective. *Ecological Applications* 9:1124–1134.

Holdsworth, A. R., and C. Uhl. 1997. Fire in Amazonian selectively logged rain forest and the potential for fire reduction. *Ecological Applications* 7:713–725.

Holland, R., and S. Jain. 1977. Vernal pools. Pp. 515–533 in *Terrestrial vegetation of California*, ed. M. G. Barbour and J. Major. New York: John Wiley and Sons.

Holling, C. S. 1973. Resilience and stability of ecological systems. *Annual Review of Ecology and Systematics* 4:1–23.

———. 1992a. Cross-scale morphology, geometry, and dynamics of ecosystems. *Ecological Monographs* 62:447–502.

———. 1992b. The role of forest insects in structuring the boreal landscape. Pp. 170–195 in *A systems analysis of the global boreal forest*, ed. H. H. Shugart, R. Leemans, and G. B. Bonan. Cambridge: Cambridge University Press.

———, ed. 1978. *Adaptive environmental assessment and management.* International Series on Applied Systems Analysis 3, International Institute for Applied Systems Analysis. Toronto: John Wiley and Sons.

Holling, C. S., and M. Meffe. 1996. Command and control and the pathology of natural resource management. *Conservation Biology* 10:328–337.

Hollstedt, C., and A. Vyse, eds. 1997. *Sicamous Creek silvicultural systems project: Workshop proceedings, 24–25 April 1996. Kamloops, British Columbia.* Working Paper 24/1997. Research Branch, B.C. Ministry of Forests, Victoria, British Columbia.

Hooper, D. U., D. E. Bignell, V. K. Brown, L. Brussaard, J. M. Dangerfield, D. H. Wall, D. H. Wardle, D. C. Coleman, K. E. Giller, P. Lavelle, W. H. Van der Putten, P. C. De Ruiter, J. Rusek, W. L. Silver, J. M. Tiedje, and V. Wolters. 2000. Interactions between aboveground and belowground biodiversity in terrestrial ecosystems: Patterns, mechanisms, and feedbacks. *BioScience* 50:1049–1061.

Hooper, M. D. 1971. The size and surroundings of nature reserves. Pp. 555–561 in *The scientific management of animal and plant communities for conservation*, ed. E. Duffey and A. S. Watt. Oxford: Blackwell Scientific Publications.

Hooper, R. G. 1988. Longleaf pines used for cavities by red-cockaded woodpeckers. *Journal of Wildlife Management* 52:392–398.

Hooper, R. G., M. R. Lennartz, and H. D. Muse. 1991. Heart rot and cavity tree selection by red-cockaded woodpeckers. *Journal of Wildlife Management* 55:323–327.

Hooper, R. G., C. J. Watson, and R.E.F. Escano. 1990. Hurricane Hugo's initial effects on red-cockaded woodpeckers in the Francis Marion National Forest. Pp. 220–224 in *Transactions of the 55th North American Wildlife and Natural Resource Conference.*

Hope, S., and W. C. McComb. 1994. Perceptions of implementing and monitoring wildlife tree prescriptions on national forests in western Washington and Oregon. *Wildlife Society Bulletin* 22:383–392.

Hopper, S. D. 1996. The use of genetic information in establishing reserves for nature conservation, Pp. 253–260 in *Biodiversity in managed landscapes: Theory and practice*, ed. R. C. Szaro and D. W. Johnston. New York: Oxford University Press.

Horne, R., G. Watts, and G. Robinson. 1991. Current forms and extent of retention areas within selectively logged blackbutt forest in NSW: A case study. *Australian Forestry* 54:148–153.

Horowitz, P., and M. Calver. 1998. Credible science? Evaluating the regional forest agreement process in Western Australia. *Australian Journal of Environmental Management* 5:213–225.

Horton, S. P., and R. W. Mannan. 1988. Effects of prescribed fire on snags and cavity-nesting birds in southeastern Arizona. *Wildlife Society Bulletin* 16:37–44.

Howard, T. M., 1973. Studies in the ecology of *Nothofagus cunninghamii* Oerst. I. Natural regeneration on the Mt. Donna Buang massif, Victoria. *Australian Journal of Botany* 21:67–78.

Howard, T. M., and D. H. Ashton. 1973. The distribution of *Nothofagus cunninghamii* rainforest. *Proceedings of the Royal Society of Victoria* 86:47–75.

Howe, R. G., G. J. Davies, and V. Mosca. 1991. The demographic significance of "sink" populations. *Biological Conservation* 57:239–255.

Howe, R. W. 1984. Local dynamics of bird assemblages in small forest habitat islands in Australia and North America. *Ecology* 65:1585–1601.

Hudson, W. E., ed. 1991. *Landscape linkages and biodiversity.* Washington, D.C.: Island Press.

Hueting, R. 1996. Three persistent myths in the environment debate. *Ecological Economics* 18:81–88.

Huggard, O. J. 1993. Prey selectivity of wolves in Banff National Park. II. Age, sex and condition of elk. *Canadian Journal of Zoology* 71:140–147.

Huijser, M. P., and P. J. Bergers. 2000. The effect of roads and traffic on hedgehog (*Erinaceus europaeus*) populations. *Biological Conservation* 95:111–116.

Humphrey, J. W., A. C. Newton, A. J. Pearce, and E. Holden. 2000. The importance of conifer plantations in northern Britain as a habitat for fungi. *Biological Conservation* 96:241–252.

Hunter, M. G. 2001. *Management in young forests.* Cascades Center for Ecosystem Management Communique. July. Cascades Center for Ecosystem Management, Department of Forest Science, Oregon State University, Corvallis, Ore.

Hunter, M. L., Jr. 1990. *Wildlife, forests and forestry: Principles for managing forests for biological diversity.* Englewood Cliffs: Prentice Hall.

———. 1991. Coping with ignorance: The coarse filter strategy for maintaining biodiversity. Pp. 266–281 in *Balancing on the edge of extinction*, ed. K. Kohm. Washington, D.C.: Island Press.

———. 1993. Natural fire regimes as spatial models for managing boreal forests. *Biological Conservation* 65:115–120.

———. 1994. *Fundamentals of conservation biology.* Cambridge, Mass.: Blackwell Science.

———. 1996. Benchmarking for managing natural ecosystems: Are human activities natural? *Conservation Biology* 10:695–697.

———., ed. 1999. *Maintaining biodiversity in forest ecosystems.* Cambridge: Cambridge University Press.

———. 2002. *Fundamentals of conservation biology.* 2nd ed. Abingdon, England: Blackwell Science.

Hunter, M. L., and A. Calhoun. 1996. A triad approach to land use allocation. Pp. 446–491 in *Biodiversity in managed landscapes: Theory and practice*, ed. R. C. Szaro and D. W. Johnson. New York: Oxford University Press.

Hunter, M. L., and P. Yonzon. 1993. Altitudinal distributions of birds, mammals, people, forests, and parks in Nepal. *Conservation Biology* 7:420–423.

Huston, M. A. 1997. Hidden treatments in ecological experiments: Re-evaluating the ecosystem function of biodiversity. *Oecologia* 110:449–460.

Hutchings, H. R. W. 1991. Dinâmica de três comunidades de Papilionoidea (Insecta: Lepidoptera) en fragmentos de floresta na

Amazonia central. MSc thesis, INPA/Fundação Universidade de Amazonas, Manaus, Brazil.

Hutha, E., J. Jokimaki, and P. Rahko. 1998. Distribution and reproductive success of the pied flycatcher *Fidecula hypoleuca* in relation to forest patch size and vegetation characteristics: The effects of scale. *Ibis* 140:214–222.

Hutto, R. L. 1995. Composition of bird communities following stand replacing fires in northern Rocky Mountain (USA) conifer forests. *Conservation Biology* 10:1041–1058.

Hyatt, T. L., and R. J. Naiman. 2001. The residence time of large woody debris in the Queets River, Washington, USA. *Ecological Applications* 11:191–202.

Hyde, W. F. 1989. Marginal costs of managing endangered species: The case of the red-cockaded woodpecker. *Journal of Agricultural Economic Research* 41:12–19.

Imbeau, I., M. Monkokonnen, and A. Desrochers. 2001. Long-term effects of forestry on birds of the eastern Canadian boreal forests: A comparison with Fennoscandia. *Conservation Biology* 15:1151–1162.

Ims, R. A., J. Rolstad, and P. Wegge. 1993. Predicting space use responses to habitat fragmentation: Can voles *Microtus oeconomus* serve as an experimental model system (EMS) for capercaillie grouse *Tetrao urogallus* in boreal forest? *Biological Conservation* 63:261–268.

Incoll, R. D. 1995. Landscape ecology of glider populations in old-growth forest patches. Honors thesis, Department of Zoology, University of Melbourne, Australia.

Incoll, R. D., R. H. Loyn, S. J. Ward, R. B. Cunningham, and C. F. Donnelly. 2000. The occurrence of gliding possums in old-growth patches of mountain ash (*Eucalyptus regnans*) in the central highlands of Victoria. *Biological Conservation* 98:77–88.

Incoll, W. D. 1979. Effects of overwood trees on growth of young stands of *Eucalyptus sieberi*. *Australian Forestry* 42:110–116.

Ingham, D. S., and M. J. Samways. 1996. Application of fragmentation and variegation models to epigaeic invertebrates in South Africa. *Conservation Biology* 10:1353–1358.

Inglis, G., and A. J. Underwood. 1992. Comments on some designs proposed for experiments on the biological importance of corridors. *Conservation Biology* 6:581–586.

Inions, G., M. T. Tanton, and S. M. Davey. 1989. Effects of fire on the availability of hollows in trees used by the common brushtail possum, *Trichosurus vulpecula* Kerr 1792, and the ringtail possum, *Pseudocheirus peregrinus* Boddaerts 1785. *Australian Wildlife Research* 16:449–458.

Innes, J., R. Hay, I. Flux, P. Bradfield, H. Speed, and P. Jansen. 1999. Successful recovery of North Island kokako *Callaeus cinerea wilsoni* populations by adaptive management. *Biological Conservation* 87:201–214.

Isaac, L. A. 1943. *Reproductive habits of Douglas-fir*. Charles Lathrop Pack Forestry Foundation, Washington, D.C.

Ison, R. L., and D. B. Russell, eds. 2000. *Agricultural extension and rural development: Breaking out of traditions*. Cambridge: Cambridge University Press.

IUCN (International Union for the Conservation of Nature and Natural Resources). 1980. *World Conservation Strategy*. Gland, Switzerland.

Jackson, J. A. 1978. Red-cockaded woodpeckers and pine red heart disease. *Auk* 94:160–163.

———. 1994. The red-cockaded woodpecker recovery program: Professional obstacles to cooperation. Pp. 157–181 in *Endangered species recovery: Finding the lessons, improving the process*, ed. T. W. Clark, R. P. Reading, and A. I. Clarke. Washington, D.C.: Island Press.

Jäggi, C., and B. Baur. 1999. Overgrown forest as a possible cause for the local extinction of *Vipera aspis* in the northern Swiss Jura mountains. *Amphibia-Reptilia* 20:25–34.

Jansson, G., and P. Angelstam. 1999. Threshold levels of habitat composition for the presence of the long-tailed tit (*Aegithalos caudatus*) in a boreal landscape. *Landscape Ecology* 14:283–290.

Janzen, D. H. 1983. No park is an island: Increase in interference from outside as park size decreases. *Oikos* 41:402–410.

Jaquet, N. 1996. How spatial and temporal scales influence understanding of sperm whale distribution: A review. *Mammal Review* 26:51–65.

Jarvinen, O., K. Kuusela, and R. A. Vaisanen. 1977. Effects of modern forestry on the numbers of breeding birds in Finald 1945–1975 (in Finnish with English summary). *Silva Fennica* 11:284–294.

Jenkins, M. B., and E. T. Smith. 1999. *The business of sustainable forestry: Strategies for an industry in transition*. Washington, D.C.: Island Press.

Jensen, W. R., T. K. Fuller, and W. L. Robinson. 1986. Wolf, *Canis lupus*, distribution on the Ontario-Michigan border near Sault Ste Marie. *Canadian Field Naturalist* 100:363–366.

Johns, A. G. 1996. Bird population persistence in Sabahan logging concessions. *Biological Conservation* 75:3–10.

Johnson, A. S., W. M. Ford, and P. E. Hale. 1993. The effects of clearcutting on herbaceous understories are still not fully known. *Conservation Biology* 7:433–439.

Johnson, A. S., P. E. Hale, W. M. Ford, J. M. Wentworth, J. R. French, O. F. Anderson, and G. B. Pullen. 1995. White-tailed deer foraging in relation to successional stage, overstory type and management of southern Appalachian forests. *American Midland Naturalist* 133:18–35.

Johnson, D. H., and T. A. O'Neil, eds. 2001. *Wildlife-habitat relationships in Oregon and Washington*. Corvallis: Oregon State University Press.

Johnson, K. N., J. F. Franklin, J. W. Thomas, and J. Gordon. 1991. *Alternatives for management of late-successional forests of the Pacific Northwest*. A report to the Agricultural Committee and the Merchant Marine Committee of the U.S. House of Representatives. College of Forestry, Oregon State University, Corvallis.

Johnson, K. N., J. Sessions, J. F. Franklin, and J. Gabriel. 1998. Integrating wildfire into strategic planning for Sierra Nevada forests. *Journal of Forestry* (January):42–49.

Johnson, P. 1994. Environmental ambassadors or global canaries? *Park Watch* (March):4–7.

Johnson, R. A. 1981. Application of the guild concept to environmental impact analysis of terrestrial vegetation. *Journal of Environmental Management* 13:205–222.

Johnson, S. 1992. Protected areas. Pp. 447–478 in *Global diversity: Status of the earth's living resources*. World Conservation Monitoring Centre. London: Chapman and Hall.

Johnson, S. L., F. J. Swanson, G. E. Grant, and S. M. Wondzell. 2000. Riparian forest disturbances by a mountain flood—the influence of floated wood. *Hydrological Processes* 14:3031–3050.

Jokimaki, J., and E. Huhta. 1996. Effects of landscape matrix and habitat structure on a bird community in northern Finland: A multiscale approach. *Ornis Fennica* 73:97–113.

Jones, J. A. 2000a. Hydrologic processes and peak discharge response to forest removal, regrowth, and roads in 10 small experimental basins, western Cascades, Oregon. *Water Resources Research* 36:2621–2642.

Jones, J. A., and G. E. Grant. 1996. Peak flow responses to clearcutting and roads in small and large basins, western Cascades, Oregon. *Water Resources Research* 32:959–974.

Jones, J. A., F. J. Swanson, B. C. Wemple, and K. U. Snyder. 2000. Effects of roads on hydrology, geomorphology, and disturbance patches in stream networks. *Conservation Biology* 14:76–85.

Jones, M. 2000b. Road upgrade, road mortality and remedial measures: Impacts on a population of eastern quolls and Tasmanian devils. *Wildlife Research* 27:289–296.

Jonsson, B. G., and M. Jonsell. 1999. Exploring potential biodiversity indicators in boreal forests. *Biodiversity and Conservation* 8:1417–1433.

Kaila, L., P. Martikainen, and P. Puntilla. 1997. Dead trees left in clearcuts benefit Coleoptera adapted to natural disturbances in boreal forests. *Biodiversity and Conservation* 6:1–8.

Kanowski, J. 2001. Effects of elevated CO_2 on the foliar chemistry of seedlings of two rainforest trees from north-east Australia: Implications for folivorous marsupials. *Austral Ecology* 26:165–172.

Kanowski, P., and M. Buchy. 1999. Pp. 78–84 in *Intensive management of regrowth forest for wood production in Australia*, ed. M. J. Connell, R. J. Raison, and A. G. Brown. Proceedings of the National Workshop. 18–20 May, Orbost, Victoria, Australia. Melbourne, Australia: CSIRO Publishing.

Kapos, V. 1989. Effects of isolation on the water status of forest patches in Brazilian Amazonia. *Journal of Tropical Ecology* 5:173–185.

Karieva, P. 1990. Population dynamics in a spatially complex environment: Theory and data. *Philosophical Transactions of the Royal Society of London Series B. Biological Sciences* 330:165–190.

Karieva, P., D. Skelly, and M. Ruckleshaus. 1997. Reevaluation of the use of models to predict the consequences of habitat loss and fragmentation. Pp. 156–166 in *The ecological basis for conservation*, ed. S. T. A. Pickett, R. S. Ostfeld, M. Shachak, and G. Likens. New York: Chapman and Hall.

Karr, J. R. 1982. Population variability and extinction in the avifauna of a tropical land bridge island. *Ecology* 63:1975–1978.

Kauffman, J. B., and C. Uhl. 1991. Interactions of anthropegenic activities, fire, and rainforests in the Amazon Basin. Pp. 117–134 in *Fire in the tropical biota*, ed. J. G. Goldammer. New York: Springer-Verlag.

Kavanagh, R. P., and R. J. Turner. 1994. Birds in eucalypt plantations: The likely role of retained habitat trees. *Australian Birds* 28:32–41.

Kavanagh, R. P., and G. A. Webb. 1998. Effects of variable intensity logging on mammals, reptiles and amphibians at Waratah Creek, south-eastern New South Wales. *Pacific Conservation Biology* 4:326–347.

Keals, N., and J. D. Majer. 1991. The conservation of ant communities along the Wubin-Perenjori corridor. Pp. 387–393 in *Nature Conservation 2: The role of corridors*, ed. D. A. Saunders and R. J. Hobbs. Chipping Norton, Australia: Surrey Beatty and Sons.

Keast, A. 1968. Seasonal movements in the Australian honeyeaters (Meliphagidae) and their ecological significance. *Emu* 67:159–210.

———, ed. 1981. *Ecological biogeography of Australia*. The Hague: Junk.

Keeton, W. S. 2000. Occurrence and reproductive role of remnant old-growth trees in mature Douglas-fir forests, southern Washington Cascade Range. Ph.D. diss., University of Washington, Seattle.

Keller, C.M.E., C. S. Robbins, and J. S. Hatfield. 1993. Avian communities in riparian forests of different widths in Maryland and Delaware. *Wetlands* 13:137–144.

Keller, E. M., and F. J. Swanson. 1979. Effects of large organic material on channel form and fluvial processes. *Earth Surface Processes* 4:361–380.

Keller, V., and H. P. Pfister. 1997. Wildlife passages as a means of mitigating effects of habitat fragmentation by roads and railway lines. Pp. 70–80 in *Habitat fragmentation and infrastructure*, ed. K. Canters. Ministry of Transportation, Public Works and Water Management, Delft, The Netherlands.

Kelly, P. A., and J. T. Rotenberry. 1993. Buffer zones and ecological reserves in California: Replacing guesswork with science. Pp. 85–92 in *Interface between ecology and land development in Califor-*nia, ed. J. E. Keeley. Southern California Academy of Sciences, Los Angeles.

Khan, M. L., S. Menon, and K. S. Bawa. 1997. Effectiveness of the protected area network in biodiversity conservation: A case study of Meghalaya state. *Biodiversity and Conservation* 6:853–868.

Kick, D. D. 1990. *Draft environmental impact statement Sunken Camp area, Washburn Ranger District Chequamegon National Forest, Bayfield County, Wisconsin*. USDA Forest Service, Park Falls, Wisc.

Kiilsgaard, C. W., S. E. Greene, and S. G. Stafford. 1987. Nutrient concentrations in litterfall from some western conifers with special reference to calcium. *Plant and Soil* 102:223–227.

King, A. P. 1963. T*he influences of colonization on the forests and the prevalence of bushfires in Australia*. CSIRO, Division Physical Chemistry, Melbourne, Australia.

Kinnear, J. E., M. L. Onus, and R. N. Bromilow. 1988. Fox control and rock-wallaby population dynamics. *Australian Wildlife Research* 15:435–450.

Kirkpatrick, J. 1994. *A continent transformed: Human impact on the natural vegetation of Australia*. Melbourne, Australia: Oxford University Press.

Kirkpatrick, J., and M. Brown. 1994. A comparison of direct and environmental domain approaches to planning reservation of forest higher plant communities and species in Tasmania. *Conservation Biology* 8:217–223.

Kirkpatrick, J., and L. Gilfedder. 1995. Maintaining integrity compared with maintaining rare and threatened taxa in remnant bushland in subhumid Tasmania. *Biological Conservation* 74:1–8.

Kirkpatrick, J. B. 1983. An iterative method for establishing priorities for the selection of nature reserves: An example from Tasmania. *Biological Conservation* 25:127–134.

———. 1998. Nature conservation and the regional forest agreement process. *Australian Journal of Environmental Management* 5:31–37.

Klein, B. C. 1989. Effects of forest fragmentation on dung and carrion beetle communities in central Amazonia. *Ecology* 70:1715–1725.

Klenner, W., and D. W. Kroeker. 1990. Denning behaviour of black bears, *Ursus americanus*, in western Manitoba. *Canadian Field Naturalist* 104:540–544.

Klenner, W., and A. Vyse. 1999. Interdisciplinary research approaches to address complex forest management issues. *Forestry Chronicle* 75:473–476.

Klinka, K., V. J. Krajina, A. Ceska, and A. M. Scagel. 1989. *Indicator plants of coastal British Columbia*. Vancouver, British Columbia: UBC Press.

Knight, F. R. 1997. Spatial patterns and large tree densities of old-growth mixed-conifer and pure-pine forests in the Sierra Nevada and Pacific Northwest. Master's thesis, University of Washington, Seattle.

Knight, R. L. 1999. Private lands: The neglected geography. *Conservation Biology* 13:223–224.

Knight, R. R., B. M. Blanchard, and L. L. Eberhardt. 1988. Mortality patterns and population sinks for Yellowstone grizzly bears, 1973–1985. *Wildlife Society Bulletin* 16:121–125.

Knowles, L., and E. T. Witkowski. 2000. Conservation biology of the succulent shrub, *Euphorbia barnardii*, a serpentine endemic of the Northern Province, South Africa. *Austral Ecology* 25:241–252.

Koehn, J. 1993. Fish need trees. *Victorian Naturalist* 110:255–257.

Koenig, W. D. 1998. Spatial autocorrelation in California land birds. *Conservation Biology* 12:612–619.

Kohm, K. A., and J. F. Franklin, eds. 1997. *Creating a forestry for the twenty-first century: The science of ecosystem management*. Washington, D.C.: Island Press.

Kondolff, M. G., R. Kattelmann, M. Embury, and D. C. Erman. 1996. Status of riparian habitat. Pp. 1009–1030 in *Sierra Nevada*

Ecosystem Project. Final report to Congress [on] Status of the Sierra Nevada. Vol. 2. *Assessments and scientific basis for management options.* Report 37. Wildland Resources Center, University of California, Davis.

Kotliar, N. B., and J. A. Wiens. 1990. Multiple scales of patchiness and patch structure: A hierarchical framework for the study of heterogeneity. *Oikos* 59:253–260.

Krear, H. R. 1965. An ecological and ethological study of the pika (*Ochotona princeps saxatilis* Bangs) in the Front Range of Colorado. Ph.D thesis, University of Colorado, Boulder.

Krebs, C. J. 1978. *The experimental analysis of distribution and abundance.* New York: Harper and Row.

Krebs, C. J., B. L. Keller, and R. H. Tamarin. 1969. *Microtus* population biology: Demographic changes in fluctuating populations of *M. ochrogaster* and *M. pennyslvanicus. Ecology* 50:587–607.

Krebs, E. A. 1998. Breeding biology of crimson rosellas (*Platycercus elegans*) on Black Mountain, Australian Capital Territory. *Australian Journal of Zoology* 46:119–136.

Kremen, C., V. Razafimahatratra, R. P. Guillery, J. Rakotomalala, A. Weiss, and J. Ratsisompatrarivo. 1999. Designing the Masoala National Park in Madagascar based on biological and socioeconomic data. *Conservation Biology* 13:1055–1068.

Kremsater, L., and F. L. Bunnell. 1999. Edge effects: Theory, evidence and implications to management of western North American forests. Pp. 117–153 in *Forest wildlife and fragmentation: Management implications,* ed. J. Rochelle, L. A. Lehmann, and J. Wisniewski. Leiden, Germany: Brill.

Kruckeberg, A. R. 1967. Ecotypic response to ultramafic soils by some plant species of northwestern United States. *Brittonia* 19:133–151.

Krusic, R. A., M. Yamasaki, C. D. Neefus, and P. J. Pekeins. 1996. Bat habitat use in White Mountain National Forest. *Journal of Wildlife Management* 60:625–631.

Kuchling, G., J. P. Dejose, A. A. Burbidge, and S. D. Bradshaw. 1992. Beyond captive breeding: The western swamp tortoise *Pseudemydura umbrina* recovery programme. *International Zoo Year Book* 31:37–41.

Lacy, R. C. 1987. Loss of genetic diversity from managed populations: Interacting effects of drift, mutation, immigration, selection, and population subdivision. *Conservation Biology* 1:143–158.

_____. 1993a. Impacts of inbreeding in natural and captive populations of vertebrates: Implications for conservation. *Perspectives in Biology and Medicine* 36:480–496.

_____. 1993b. VORTEX—a model for use in population viability analysis. *Wildlife Research* 20:45–65.

Lacy, R. C., and D. B. Lindenmayer. 1995. Using population viability analysis (PVA) to explore the impacts of population sub-division on the mountain brushtail possum, *Trichosurus caninus* Ogilby (Phalangeridae: Marsupialia) in south-eastern Australia. II. Changes in genetic variability in sub-divided populations. *Biological Conservation* 73:131–142.

LaHaye, W. S., R. J. Gutierrez, and H. R. Akçakaya. 1994. Spotted owl metapopulation dynamics in southern California. *Journal of Animal Ecology* 63:775–785.

Lahti, D. C. 2001. The "edge effect on nest predation" hypothesis after twenty years. *Biological Conservation* 99:365–374.

Lamb, D. 1991. Variability and change in nutrient cycling in Australian rainforests. Pp. 83–92 in *The rainforest legacy,* ed. G. Werren and P. Kershaw. Special Australian Heritage Publication Series no. 7 (2). Canberra: Australian Government Publishing Service.

Lamb, D., A. Smith, G. Wilkinson, and R. Loyn. 1998. *Managing habitat trees in Queensland forests.* A report by the Habitat Tree Technical Advisory Group to the Queensland Department of Natural Resources. Queensland Department of Primary Industries, Brisbane, Australia.

Lamb, F. B. 1966. *Mahogany in tropical America.* Ann Arbor: University of Michigan Press.

Lambeck, R. J. 1997. Focal species: A multi-species umbrella for nature conservation. *Conservation Biology* 11:849–856.

_____. 1999. *Landscape planning for biodiversity conservation in agricultural regions: A case study from the wheatbelt of Western Australia.* Biodiversity Technical Paper no. 2, 1–96. Environment Australia, Canberra.

Lamberson, R. H., R. McKelvey, B. R. Noon, and C. Voss. 1992. A dynamic analysis of northern spotted owl viability in a fragmented forest landscape. *Conservation Biology* 6:505–512.

Lamberson, R. H., B. R. Noon, C. Voss, and R. McKelvey. 1994. Reserve design for territorial species: The effects of patch size and spacing on the viability of the northern spotted owl. *Conservation Biology* 8:185–195.

Land, D., W. R. Marion, and T. E. O'Meara. 1989. Snag availability and cavity nesting birds in slash pine plantations. *Journal of Wildlife Management* 53:1165–1171.

Land Conservation Council. 1994. *Final recommendations.* Melbourne Area. District 2 Review. Land Conservation Council, Melbourne, Australia.

Landers, J. L., R. J. Hamilton, A. S. Johnson, and R. L. Marchington. 1979. Food and habitat of black bears in southeastern Carolina. *Journal of Wildlife Management* 43:143–153.

Landres, P. B., J. Verner, and J. W. Thomas. 1988. Ecological use of vertebrate indicator species: A critique. *Conservation Biology* 2:316–328.

Langford, K. J., R. J. Moran, and P. J. O'Shaughnessy. 1982. The Coranderk Experiment—the effects of roading and timber harvesting in mature mountain ash forest on streamflow and quality. Pp. 92–102 in *The first national symposium on forest hydrology,* ed. E. M. O'Loughlin and L. J. Bren. Institution of Engineers, Canberra, Australia.

La Marche, J. L., and D. P. Lettenmaier. 2001. Effects of forest roads on flood flows in the Deschutes River, Washington. *Earth Surface Processes and Landforms* 26:115–134.

La Polla, V. N., and G. W. Barrett. 1993. Effects of corridor width and presence on the population dynamics of the meadow vole (*Microtus pennsylvanicus*). *Landscape Ecology* 8:25–37.

Larson, D. W., U. Matthes, and P. E. Kelly. 2000. *Cliff ecology: Patterns and processes in cliff ecosystems.* Cambridge: Cambridge University Press.

Laurance, S. G., and W. F. Laurance. 1999. Tropical wildlife corridors: Use of linear rainforest remnants by arboreal marsupials. *Biological Conservation* 91:231–239.

Laurance, W. F. 1990. Comparative responses of five arboreal marsupials to tropical forest fragmentation. *Journal of Mammalogy* 71:641–653.

_____. 1991a. Ecological correlates of extinction proneness in Australian tropical rain forest mammals. *Conservation Biology* 5:79–89.

_____. 1991b. Edge effects in tropical forest fragments: Application of a model for the design of nature reserves. *Biological Conservation* 57:205–219.

_____. 1997a. Responses of mammals to rainforest fragmentation in tropical Queensland: A review and synthesis. *Wildlife Research* 24:603–612.

_____. 1997b. Hyper-disturbed parks: Edge effects and the ecology of isolated reserves in isolated rainforest reserves in tropical Australia. Pp. 71–784 in *Tropical forest remnants: Ecology, management and conservation of fragmented communities,* ed. W. F. Laurance and R. O. Bierregaard. Chicago: University of Chicago Press.

_____. 2000. Rainforest fragmentation kills big trees. *Nature* 404:836.

Laurance, W. F., and R. O. Bierregaard. 1997. *Tropical forest remnants: Ecology, management and conservation of fragmented communities.* Chicago: University of Chicago Press.

Laurance, W. F., R. O. Bierregaard, C. Gascon, R. K. Didham, A. P. Smith, A. J. Lynam, V. M. Viana, T. E. Lovejoy, K. E. Sieving, J. W. Sites, M. Andersen, M. D. Tocher, E. A. Kramer, C. Restrepo, and C. Moritz. 1997. Tropical forest fragmentation: Synthesis of a diverse and dynamic discipline. Pp. 502–525 in *Tropical forest remnants: Ecology, management and conservation of fragmented communities,* ed. W. F. Laurance and R. O. Bierregaard. Chicago: University of Chicago Press.

Laurance, W. F., L. V. Ferreira, J. M. Rankin-de Mérona, S. G. Laurance, R. Hutchings, and T. E. Lovejoy. 1998. Effects of forest fragmentation on recruitment patterns in Amazonian tree communities. *Conservation Biology* 12:460–464.

Laurance, W. F., D. Pérez-Slicrup, P. Delamônica, P. M. Fearnside, S. D'Angelo, A. Jerozolinski, L. Pohl, and T. E. Lovejoy. 2001. Rain forest fragmentation and the structure of Amazonian liana communities. *Ecology* 82:105–116.

Laurance, W. F., H. L. Vasconcelos, and T. E. Lovejoy. 2000. Forest loss and fragmentation in the Amazon: Implications for wildlife conservation. *Oryx* 34:39–45.

Laurance, W. F., and E. Yensen. 1991. Predicting the impacts of edge effects in fragmented habitats. *Biological Conservation* 55:77–92.

Lautenschlager, R. A. 1999. Improving long-term forest ecology research for the twenty-first century. *Forestry Chronicle* 75:477–480.

Lavorel, S., M. Stafford Smith, and N. Reid. 1999. Spread of mistletoes (*Amyema preissii*) in fragmented Australian woodlands: A simulation study. *Landscape Ecology* 14:147–160.

Law, B., and C. R. Dickman. 1998. The use of habitat mosaics by terrestrial vertebrate fauna: Implications for conservation and management. *Biodiversity and Conservation* 7:323–333.

Law, B. S., J. Anderson, and M. Chidel. 1999. Bat communities in a fragmented forest landscape on the south-west slopes of New South Wales, Australia. *Biological Conservation* 88:333–345.

Lawler, S. P. 1993. Species richness, species composition and population dynamics of protists in experimental microcosms. *Journal of Animal Ecology* 62:711–719.

Lawton, J. 1997. The science and non-science of conservation biology. *Oikos* 79:3–5.

Lee, A. K., and R. Martin. 1988. *The koala.* Australian Natural History Series. Sydney: University of New South Wales Press.

Lefkovitch, L. P., and L. Fahrig. 1985. Spatial characteristics of habitat patches and population survival. *Ecological Modelling* 30:297–308.

Lefroy, T., and R. Hobbs, 2000. Functional mimics of natural ecosystems as a basis for sustainable agriculture. Pp. 179–187 in *Nature Conservation 5: Nature conservation in production environments: Managing the matrix,* ed. J. L. Craig, N. Mitchell, and D. A. Saunders. Chipping Norton, Australia: Surrey Beatty and Sons.

Lefsky, M. A., W. B. Cohen, S. A. Acker, G. G. Parker, T. A. Spies, and D. Harding. 1999. Lidar remote sensing of the canopy structure and biophysical properties of Douglas-fir western hemlock forests. *Remote Sensing of Environment* 70:339–361.

Lefsky, M. A., W. B. Cohen, G. G. Parker, and D. J. Harding. 2002. Lidar remote sensing for ecosystem studies. *BioScience* 52:19–30.

Lehmkuhl, J. F., S. D. West, C. L. Chambers, W. C. McComb, D. A. Manuwai, K. A. Aubry, J. L. Erickson, R. A. Gitzen, and M. Lau. 1999. An experiment for assessing vertebrate response to varying levels and patterns of green-tree retention. *Northwest Science* 73 (Special Issue):45–63.

Lemckert, F., and T. Brassil. 2000. Movements and habitat use of the endangered giant barred river frog (*Mixophyes iteratus*) and the implications for its conservation in timber production forests. *Biological Conservation* 96:177–184.

Leopold, A. 1933. *Game management.* New York: Charles Scribners.

Lertzman, K. P., G. Sutherland, A. Inselberg, and S. Saunders. 1996. Canopy gaps and the landscape mosaic in a temperate rainforest. *Ecology* 77:1254–1270.

Lesica, P. 1996. Using fire history models to estimate proportions of old growth forest in northwest Montana, USA. *Biological Conservation* 77:33–39.

Lesslie, R., and M. Maslen. 1995. *National wilderness inventory handbook.* 2nd ed. Australian Heritage Commission. Canberra: Australian Government Publishing Service.

Leung, K. P., C. R. Dickman, and L. A. Moore. 1993. Genetic variation in fragmented populations of an Australian rainforest rodent, *Melomys cervinipes. Pacific Conservation Biology* 1:58–65.

Levin, S. 1992. The problem of pattern and scale in ecology. *Ecology* 73:1943–1967.

Levins, R. A. 1970. Extinction. *Lecture Notes in Mathematics and Life Sciences* 2:75–107.

Lewis, A., J. L. Stein, J. A. Stein, H. A. Nix, B. G. Mackey, and J. K. Bowyer, 1991. *An assessment of regional conservation adequacy: Tasmania.* Consultancy Report to the Resource Assessment Commission (Forest and Timber Inquiry). Canberra, Australia.

Li, H., J. F. Franklin, F. J. Swanson, and T. A. Spies. 1993. Developing alternative forest cutting patterns: A simulation approach. *Landscape Ecology* 8:63–75.

Lidicker, W. Z. 1999. Responses of mammals to habitat edges: An overview. *Landscape Ecology* 14:333–343.

Likens, G. E., ed. 1985. *An ecosystem approach to aquatic ecology: Mirror Lake and its environment.* New York: Springer-Verlag.

Lindenmayer, D. B. 1989. The ecology and habitat requirements of Leadbeater's possum. Ph.D diss., The Australian National University, Canberra.

———. 1992. *The ecology and habitat requirements of arboreal marsupials in the montane ash forests of the Central Highlands of Victoria: A summary of studies.* Value Adding and Silvicultural Systems Program, no. 6. Native Forest Research. Department of Conservation and Environment, Melbourne, Australia.

———. 1994a. Timber harvesting impacts on wildlife: Implications for ecologically sustainable forest use. *Australian Journal of Environmental Management* 1:56–68.

———. 1994b. Wildlife corridors and the mitigation of logging impacts on forest fauna in south-eastern Australia. *Wildlife Research* 21:323–340.

———. 1995. Disturbance, forest wildlife conservation and a conservative basis for forest management in the mountain ash forests of Victoria. *Forest Ecology and Management* 74:223–231.

———. 1996. *Wildlife and woodchips: Leadbeater's possum as a test case of sustainable forestry.* Sydney, Australia: University of New South Wales Press.

———. 1998. *The design of wildlife corridors in wood production forests.* New South Wales National Parks and Wildlife Service, Occasional Paper Series. Forest Issues Paper, no. 4. Pp. 1–41.

———. 1999a. Using environmental history and ecological evidence to appraise management regimes in forests: A case study from the mountain ash forests of Victoria. Pp. 74–96 in *Environmental history,* ed. S. Dovers. Melbourne, Australia: Oxford University Press.

———. 1999b. Future directions for biodiversity conservation in managed forests: Indicator species, impact studies and monitoring programs. *Forest Ecology and Management* 115:277–287.

———. 2000. Factors at multiple scales affecting distribution patterns and its implications for animal conservation—Leadbeater's possum as a case study. *Biodiversity and Conservation* 9:15–35.

Lindenmayer, D. B., I. Ball, H. P. Possingham, M. A. McCarthy, and M. L. Pope. 2001e. A landscape-scale test of the predictive ability of a meta-population model in an Australian fragmented forest ecosystem. *Journal of Applied Ecology* 38:36–48.

Lindenmayer, D. B., and R. B. Cunningham. 1996. A habitat-based microscale forest classification system for zoning wood production areas to conserve a rare species threatened by logging operations in south-eastern Australia. *Environmental Monitoring and Assessment* 39:543–557.

———. 1997. Patterns of co-occurrence among arboreal marsupials in the forests of central Victoria, south-eastern Australia. *Australian Journal of Ecology* 22:340–346.

Lindenmayer, D. B., R. B. Cunningham, and C. F. Donnelly. 1993a. The conservation of arboreal marsupials in the montane ash forests of the Central Highlands of Victoria, south-east Australia. IV. The distribution and abundance of arboreal marsupials in retained linear strips (wildlife corridors) in timber production forests. *Biological Conservation* 66:207–221.

———. 1994a. The conservation of arboreal marsupials in the montane ash forests of the Central Highlands of Victoria, south-east Australia. VI. Tests of the performance of models of nest tree and habitat requirements of arboreal marsupials. *Biological Conservation* 70:143–147.

———. 1997a. Tree decline and collapse in Australian forests: Implications for arboreal marsupials. *Ecological Applications* 7:625–641.

Lindenmayer, D. B., R. B. Cunningham, C. F. Donnelly, and J. F. Franklin. 2000c. Structural features of Australian old growth montane ash forests. *Forest Ecology and Management* 134: 189–204.

Lindenmayer, D. B., R. B. Cunningham, C. F. Donnelly, and H. A. Nix. 2001a. The distribution of birds in a fragmented forest landscape. *Ecological Monographs* 78:1–18.

Lindenmayer, D. B., R. B. Cunningham, C. F. Donnelly, M. T. Tanton, and H. A. Nix. 1993b. The abundance and development of cavities in montane ash-type eucalypt trees in the montane forests of the Central Highlands of Victoria, south-eastern Australia. *Forest Ecology and Management* 60:77–104.

Lindenmayer, D. B., R. B. Cunningham, C. F. Donnelly, B. J. Triggs, and M. Belvedere. 1994b. The conservation of arboreal marsupials in the montane ash forests of the central highlands of Victoria, south-east Australia. V. Patterns of use and the micro-habitat requirements of the mountain brushtail possum, *Trichosurus caninus* Ogilby in retained linear strips (wildlife corridors). *Biological Conservation* 68:43–51.

———. 1994c. The diversity, abundance and microhabitat requirements of terrestrial mammals in contiguous forests and retained linear strips in the montane ash forests of the central highlands of Victoria. *Forest Ecology and Management* 67:113–133.

Lindenmayer, D. B., R. B. Cunningham, R. Lesslie, and C. F. Donnelly. 2001c. Landscape indices and vertebrate species response in two Australian forest landscapes. *Forest Ecology and Management* (in press).

Lindenmayer, D. B., R. B. Cunningham, C. MacGregor, C. R. Tribolet, and C. F. Donnelly. 2001d. The Nannangroe landscape experiment—baseline data for mammals, reptiles and nocturnal birds. *Biological Conservation* 101:157–169.

Lindenmayer, D. B., R. B. Cunningham, and M. A. McCarthy. 1999a. The conservation of arboreal marsupials in the montane ash forests of the Central Highlands of Victoria, south-eastern Australia. VIII. Landscape analysis of the occurrence of arboreal marsupials in the montane ash forests. *Biological Conservation* 89:83–92.

Lindenmayer, D. B., R. B. Cunningham, H. A. Nix, M. T. Tanton, and A. P. Smith. 1991b. Predicting the abundance of hollow-bearing trees in montane ash forests of south-eastern Australia. *Australian Journal of Ecology* 16:91–98.

Lindenmayer, D. B., R. B. Cunningham, M. Pope, and C. F. Donnelly. 1999b. The Tumut fragmentation experiment in south-

eastern Australia: The effects of landscape context and fragmentation of arboreal marsupials. *Ecological Applications* 9:594–611.

———. 1999c. A field-based experiment to examine the response of mammals to landscape context and habitat fragmentation. *Biological Conservation* 88:387–403.

Lindenmayer, D. B., R. B. Cunningham, M. Pope, C. F. Donnelly, H. A. Nix, and R. D. Incoll. 1997c. *The Tumut fragmentation experiment in south-eastern Australia: The effects of landscape context and fragmentation on arboreal marsupials.* CRES Working Paper, 1997/4.

Lindenmayer, D. B., R. B. Cunningham, M. T. Tanton, H. A. Nix, and A. P. Smith. 1991a. The conservation of arboreal marsupials in the montane ash forests of the central highlands of Victoria, south-east Australia. III. The habitat requirements of Leadbeater's possum, *Gymnobelideus leadbeateri* McCoy and models of the diversity and abundance of arboreal marsupials. *Biological Conservation* 56:295–315.

Lindenmayer, D. B., R. B. Cunningham, M. T. Tanton, and A. P. Smith. 1990a. The conservation of arboreal marsupials in the montane ash forests of the central highlands of Victoria, south-east Australia. II. The loss of trees with hollows and its implications for the conservation of Leadbeaters possum *Gymnobelideus leadbeateri* McCoy (Marsupialia: Petauridae). *Biological Conservation* 54:133–145.

———. 1990b. The conservation of arboreal marsupials in the montane ash forests of the central highlands of Victoria, south-east Australia, I. Factors affecting the occupancy of trees with hollows. *Biological Conservation* 54: 111–131.

Lindenmayer, D. B., R. B. Cunningham, M. T. Tanton, A. P. Smith, and H. A. Nix. 1991c. Characteristics of hollow-bearing trees occupied by arboreal marsupials in the montane ash forests of the Central Highlands of Victoria, south-eastern Australia. *Forest Ecology and Management* 40:289–308.

Lindenmayer, D. B., and J. F. Franklin. 1997a. Forest structure and sustainable temperate forestry: A case study from Australia. *Conservation Biology* 11:1053–1068.

———. 1997b. Re-inventing forestry as a discipline—a forest ecology perspective. *Australian Forestry* 60:53–55.

———. 2000. Managing unreserved forest for biodiversity conservation: The importance of matrix. Pp. 13–25 in *Nature Conservation 5: Nature conservation in production environments: Managing the matrix,* ed. J. Craig, N. Mitchell, and D. A. Saunders. Sydney, Australia: Surrey Beatty and Sons.

Lindenmayer, D. B., and R. D. Incoll. 1998. Community-based monitoring of vertebrates in Victorian forests. *On the Brink* 11:11–12.

Lindenmayer, D. B., R. D. Incoll, R. B. Cunningham, and C. F. Donnelly. 1999g. Attributes of logs in the mountain ash forests in south-eastern Australia. *Forest Ecology and Management* 123:195–203.

Lindenmayer, D. B., and R. C. Lacy. 1995a. Metapopulation viability of arboreal marsupials in fragmented old growth forests: A comparison among species. *Ecological Applications* 5:183–199.

———. 1995b. Using population viability analysis (PVA) to explore the impacts of population sub-division on the mountain brushtail possum, *Trichosurus caninus* Ogilby (Phalangeridae: Marsupialia) in south-eastern Australia. I. Demographic stability and population persistence. *Biological Conservation* 73:119–129.

Lindenmayer, D. B., R. C. Lacy, and M. L. Pope. 2000a. Testing a population viability analysis model. *Ecological Applications* 10:580–597.

Lindenmayer, D. B., R. C. Lacy, H. Tyndale-Biscoe, A. Taylor, K. Viggers, and M. L. Pope. 1999h. Integrating demographic and genetic studies of the greater glider (*Petauroides volans*) at Tumut, south-eastern Australia: Setting hypotheses for future testing. *Pacific Conservation Biology* 5:2–8.

Lindenmayer, D. B., B. G. Mackey, R. B. Cunningham, C. F. Donnelly, I. Mullen, M. A. McCarthy, and A. M. Gill. 2000d. Statistical and environmental modelling of myrtle beech (*Nothofagus cunninghamii*) in southern Australia. *Journal of Biogeography* 27:1001–1009.

Lindenmayer, D. B., B. Mackey, I. Mullins, M. A. McCarthy, A. M. Gill, R. B. Cunningham, and C. F. Donnelly. 1999f. Stand structure within forest types—are there environmental determinants? *Forest Ecology and Management* 123:55–63.

Lindenmayer, D. B., C. R. Margules, and D. Botkin. 2000b. Indicators of forest sustainability biodiversity: The selection of forest indicator species. *Conservation Biology* 14:941–950.

Lindenmayer, D., and M. A. McCarthy. 2002a. Congruence between natural and human forest disturbance: A case study from Australian montane ash forests. *Forest Ecology and Management* 155:319–335.

_____. 2002b. The spatial anatomy of two weed invasions. *Biological Conservation* 102:77–87.

Lindenmayer, D. B., M. A. McCarthy, and M. L. Pope. 1999d. A test of Hanski's simple model for metapopulation model. *Oikos* 84:99–109.

Lindenmayer, D. B., M. A. McCarthy, H. P. Possingham, and S. Legge. 2001b. Testing PVA models—kingfishers, habitat patches and landscape fragmentation. *Oikos* 92:445–458.

Lindenmayer, D. B., and R. A. Meggs. 1996. Use of den trees by Leadbeater's Possum (*Gymnobelideus leadbeateri*). *Australian Journal of Zoology* 44:625–638.

Lindenmayer, D. B., and H. A. Nix. 1993. Ecological principles for the design of wildlife corridors. *Conservation Biology* 7:627–630.

Lindenmayer, D. B., H. A. Nix, J. P. MacMahon, M. F. Hutchinson, and M. T. Tanton. 1991d. The conservation of Leadbeater's possum, *Gymnobelideus leadbeateri* McCoy, a case study of the use of bioclimatic modelling. *Journal of Biogeography* 18:371–383.

Lindenmayer, D. B., and T. W. Norton. 1993. The conservation of Leadbeater's possum in south-eastern Australia and the spotted owl in the Pacific Northwest of the USA: Management issues, strategies and lessons. *Pacific Conservation Biology* 1:13–18.

Lindenmayer, D. B., T. W. Norton, and M. T. Tanton. 1991e. Differences between the effects of wildfire and clearfelling in montane ash forests of Victoria and its implications for fauna dependent on tree hollows. *Australian Forestry* 53:61–68.

Lindenmayer, D. B., and R. H. Peakall. 2000. The Tumut experiment—integrating demographic and genetic studies to unravel fragmentation effects: A case study of the native bush rat. Pp. 173–201 in *Genetics, demography and viability of fragmented populations*, ed. A. Young and G. Clarke. Cambridge: Cambridge University Press.

Lindenmayer, D. B., and M. L. Pope. 2000. The design of exotic softwood plantations to enhance wildlife conservation: Preliminary lessons from the Tumut fragmentation experiment. Pp. 44–49 in *Nature conservation 5: Nature conservation in production environments: Managing the matrix*, ed. J. Craig, N. Mitchell, and D. A. Saunders. Sydney: Surrey Beatty and Sons.

Lindenmayer, D. B., M. L. Pope, and R. B. Cunningham. 1999e. Roads and nest predation: An experimental study in a modified forest ecosystem. *Emu* 99:148–152.

Lindenmayer, D. B., and H. P. Possingham. 1994. The risk of extinction: Ranking management options for Leadbeater's possum. Centre for Resource and Environmental Studies, The Australian National University and The Australian Nature Conservation Agency, Canberra.

_____. 1995a. Modeling the impacts of wildfire on metapopulation behaviour of the Australian arboreal marsupial, Leadbeater's possum, *Gymnobelideus leadbeateri*. *Forest Ecology and Management* 74:197–222.

_____. 1995b. The conservation of arboreal marsupials in the montane ash forests of the central highlands of Victoria, south-

east Australia. VII. The contribution of changed wood production practices to metapopulation persistence. *Biological Conservation* 73:239–257.

_____. 1996. Modeling the relationships between habitat connectivity, corridor design and wildlife conservation within intensively logged wood production forests of south-eastern Australia. *Landscape Ecology* 11:79–105.

Lindenmayer, D. B., and H. F. Recher. 1998. Aspects of ecologically sustainable forestry in temperate eucalypt forests—beyond an expanded reserve system. *Pacific Conservation Biology* 4:4–10.

Lindenmayer, D. B., A. P. Smith, S. A. Craig, and L. F. Lumsden. 1990. A survey of the distribution of Leadbeater's possum, *Gymnobelideus leadbeateri* McCoy, in the Central Highlands of Victoria. (Appendix). *Victorian Naturalist* 107:136–137.

Lindenmayer, D. B., A. Welsh, and C. F. Donnelly. 1997b. The use of nest trees by the mountain brushtail possum (*Trichosurus caninus*) (Phalangeridae: Marsupialia). III. Spatial configuration and co-occupancy of nest trees. *Wildlife Research* 24:661–677.

Lindenmayer, D. B., A. Welsh, C. F. Donnelly, and R. Meggs. 1996. The use of nest trees by the mountain brushtail possum (*Trichosurus caninus*) (Phalangeridae: Marsupialia). I. Number of occupied trees and frequency of tree use. *Wildlife Research* 23:343–361.

Linder, P., and L. Östlund. 1998. Structural changes in three mid-boreal Swedish forest landscapes, 1885–1996. *Biological Conservation* 85:9–19.

Lint, J., B. Noon, R. Anthony, E. Forsman, M. Raphael, M. Collopy, and E. Starkey. 1999. *Northern spotted owl effectiveness monitoring plan for the Northwest Forest Plan*. USDA Forest Service General Technical Report PNW-GTR-440.

Loman, J., and T. von Schantz. 1991. Birds in a farmland—more species in small than in large habitat islands. *Conservation Biology* 5:176–188.

Long, C. J., C. Whitlock, P. J. Bartlein, and S. H. Millspaugh. 1998. A 9000-year fire history from the Oregon Coast Range, based on a high-resolution charcoal study. *Canadian Journal of Forest Research* 28:774–787.

Lord, J. M., and D. A. Norton. 1990. Scale and the spatial concept of fragmentation. *Conservation Biology* 4: 197–202.

Lorimer, C. G., and L. E. Frelich. 1994. Natural disturbance regimes in old-growth northern hardwoods. *Journal of Forestry* (January):33–38.

Lovejoy, T. E. 1980. Discontinuous wilderness: Minimum areas for conservation. *Parks* 5:13–15.

Lovejoy, T. E., R. O. Bierregaard, A. B. Rylands, J. R. Malcolm, C. E. Quintela, L. H. Harper, K. S. Brown, A. H. Powell, H. O. Schubert, and M. B. Hays. 1986. Edge and other effects of isolation on Amazon forest fragments. Pp. 257–285 in *Conservation biology: The science of scarcity and diversity*, ed. M. E. Soulé. Sunderland, Mass.: Sinauer Associates.

Lovejoy, T. E., and D. C. Oren. 1981. The minimum critical size of ecosystems. Pp. 7–12 in *Forest island dynamics in man-dominated landscapes*, ed. R. L. Burgess and D. M. Sharpe. New York: Springer-Verlag.

Lovejoy, T. E., J. M. Rankin, R. O. Bierregaard, K. S. Brown, L. H. Emmons, and M. E. van der Voort. 1984. Extinctions. Pp. 295–325 in *Ecosystem decay of Amazon remnants*, ed. M. H. Niteki. Chicago: University of Chicago Press.

Lovett, S., and P. Price, eds. 1999. *Riparian land management technical guidelines*. Vol. 1. *Principles of sound management*. Land and Water Resources Research and Development Corporation, Canberra, Australia.

Loyn, R. H. 1985a. Bird populations in successional forests of mountain ash *Eucalyptus regnans* in Central Victoria. *Emu* 85:213–230.

_____. 1985b. Strategies for conserving wildlife in commercially productive eucalypt forest. *Australian Forestry* 48:95–101.

———. 1987. Effects of patch area and habitat on bird abundances, species numbers and tree health in fragmented Victorian forests. Pp. 65–77 in *Nature conservation: The role of remnants of native vegetation*, ed. D. A. Saunders, G. Arnold, A. Burbidge, and A. Hopkins. Chipping Norton, Australia: Surrey Beatty and Sons.

———. 1998. Birds in patches of old-growth ash forest, in a matrix of younger forest. *Pacific Conservation Biology* 4:111–121.

Loyn, R. H., E. A. Chesterfield, and M. A. Macfarlane. 1980. *Forest utilization and the flora and fauna in Boola Boola State Forest in south-eastern Victoria*. Forest Commission of Victoria Bulletin 28. Melbourne: Victoria.

Loyn, R. H., E. G. McNabb, L. Volodina, and R. Willig. 2001. Modelling landscape distributions of large forest owls as applied to managing forests in north-east Victoria, Australia. *Biological Conservation* 97:361–376.

Lucas, A. E., and J. A. Synden. 1970. The concept of multiple purpose land use—myth or management system? *Australian Forestry* 34:73–83.

Luck, G. W., H. P. Possingham, and D. C. Paton. 1999. Bird responses at inherent and induced edges in the Murray Mallee, South Australia. I. Differences in abundance and diversity. *Emu* 99:157–169.

Ludwig, D. 1999. Is it meaningful to estimate a probability of extinction? *Ecology* 80:298–310.

Ludwig, D., R. Hilborn, and C. Walters. 1993. Uncertainty, resource exploitation and conservation: Lessons from history. *Science* 260:17–36.

Lugg, A. 1998. Social impacts of a forest policy on a dependent rural community: Bombshell or blessing? *Australian Forestry* 61:173–184.

Luke, R. H., and A. G. McArthur. 1978. *Bushfires in Australia*. Canberra: Australian Government Publishing Service.

Lumsden, L. F., J.S.A. Alexander, F.A.R. Hill, S. P. Krasna, and C. E. Silveira. 1991. *The vertebrate fauna of the Land Conservation Council Melbourne-2 study area*. Arthur Rylah Institute for Environmental Research. Technical Report Series no. 115. Department of Conservation and Environment, Melbourne, Australia.

Lumsden, L. F., A. F. Bennett, J. Silins, and S. Krasna. 1994. *Fauna in a remnant vegetation-farmland mosaic, movement, roosts and foraging ecology of bats, a report to the Australian Nature Conservation Agency "Save the Bush Program."* Flora and Fauna Branch, Department of Conservation and Natural Resources, Melbourne, Australia.

Lunney, D., ed. 1991. *The conservation of Australia's forest fauna*. Royal Zoological Society of New South Wales, Sydney, Australia.

Lunney, D., and C. Moon. 1987. The Eden woodchip debate (1969–86). *Search* 18:15–20.

Luoma, J. R. 1999. *The hidden forest: The biography of an ecosystem*. New York: Henry Holt.

Lutze, M. T., R. G. Campbell, and P. C. Fagg. 1999. Development of silviculture in the native state forests of Victoria. *Australian Forestry* 62:236–244.

Lynch, J., W. F. Carmen, D. A. Saunders, and P. Cale. 1995. Short-term use of vegetated road verges and habitat by four bird species in the central wheatbelt of Western Australia. Pp. 34–42 in *Nature conservation 4: The role of networks in conservation*, ed. D. A. Saunders, J. L. Craig, and E. M. Mattiske. Chipping Norton, Australia: Surrey Beatty and Sons.

Lyon, L. J. 1983. Road density models describing habitat effectiveness for elk. *Journal of Forestry* 81:592–595.

Lyons, M. T., R. R. Brooks, and D. C. Craig. 1974. The influence of soil composition of the Coolac Serpentinite belt in New South Wales. *Journal and Proceedings, Royal Society of New South Wales* 107:67–75.

Macarthur, R. H. 1972. *Geographical ecology: Patterns in the distribution of species*. Princeton, N.J.: Princeton University Press.

Macarthur, R. H., and E. O. Wilson. 1963. An equilibrium theory of insular zoogeography. *Evolution* 17:373–387.

———. 1967. *The theory of island biogeography*. Princeton, N.J.: Princeton University Press.

Macdonald, L. H., and A. Smart. 1993. Beyond the guidelines: Practical lessons for monitoring. *Environmental Monitoring and Assessment* 26:203–218.

Mace, R. D., J. S. Walker, T. D. Manley, L. J. Lyon, and H. Zuuring. 1996. Relationships among grizzly bears, roads and habitat in the Swan Mountains, Montana. *Journal of Applied Ecology* 33:1395–1404.

Mace, R. D., J. S. Walker, T. D. Manley, L. J. Lyon, and H. Zuuring. 1996. Relationships among grizzly bears, roads, and habitat in the Swan Mountains. *Journal of Applied Ecology* 33:1395–1404.

Macfarlane, M. A. 1988. Mammal populations in mountain ash (*Eucalyptus regnans*) forests of various ages in the Central Highlands of Victoria. *Australian Forestry* 51:14–27.

Macfarlane, M. A., and J. H. Seebeck. 1991. *Draft management strategies for the conservation of Leadbeater's possum*, Gymnobelideus leadbeateri, *in Victoria*. Arthur Rylah Institute Technical Report Series 111. Department of Conservation and Environment, Melbourne, Australia.

Macfarlane, M. A., J. Smith, and K. Lowe. 1998. *Leadbeater's possum recovery plan. 1998–2002*. Department of Natural Resources and Environment, Government of Victoria, Australia.

Machtans, C. S., M. A. Villard, and S. J. Hannon. 1996. Use of riparian buffer strips as movement corridors for birds. *Conservation Biology* 10:1366–1379.

Mackey, B. G. 1993. A spatial analysis of the environmental relations of rainforest structural types. *Journal of Biogeography* 20:303–336.

———. 1994. Predicting the potential distribution of rainforest structural types. *Journal of Vegetation Science* 5:43–54.

Mackey, B. G., R. G. Lesslie, D. B. Lindenmayer, R. D. Incoll, and H. A. Nix. 1999. *The role of wilderness and wild rivers in nature conservation*. Major report to Environment Australia. http://www.environment.gov.au/heritage/wwr/ anlr_0999/code/pub.html.

Mackey, B. G., R. G. Lesslie, D. B. Lindenmayer, and H. A. Nix. 1998. The role of wilderness and wild rivers in nature conservation. *Pacific Conservation Biology* 4:182–185.

Mackey, B. G., D. B. Lindenmayer, A. M. Gill, and M. A. McCarthy. 2002. *The central highlands ecosystem study—integrating ecological and statistical modelling for identifying conservation refugia*. Department of Geography and The Centre for Resource and Environmental Studies, The Australian National University, Canberra.

Mackey, B. G., D. W. McKenney, Y. Yang, J. P. MacMahon, and M. F. Hutchinson. 1996. Site regions revisited: A climatic analysis of Hills' site regions for the province of Ontario using a parametric method. *Canadian Journal of Forest Research* 26:333–354.

Mackey, B. G., H. A. Nix, M. F. Hutchinson, J. P. MacMahon, and P. M. Fleming. 1988. Assessing representativeness of places for conservation reserves and heritage listing. *Environmental Management* 12:501–504.

Mackowski, C. M. 1984. The ontogeny of hollows in blackbutt, *Eucalyptus pilularis* and its relevance to the management of forests for possums, gliders and timber. Pp. 517–525 in *Possums and gliders*, ed. A. P. Smith and I. D. Hume. Chipping Norton, Australia: Surrey Beatty and Sons.

———. 1987. Wildlife hollows and timber management in blackbutt forest. Master's thesis, University of New England, Armidale, Australia.

MacLaren, J. P. 1996. *Environmental effects of planted forests*. New Zealand Forest Research Institute. Brebner Print, New Zealand.

MacMillan Blodel Limited. 1999. *Summary of the first year critique workshop on the MacMillan Blodel BC Coastal Forest Project, July*

14–16 1999. Prepared by Dovetail Consulting for MacMillan Blodel, Vancouver, British Columbia.

Mac Nally, R. 1994. Habitat-specific guild structure of forest birds in south-eastern Australia: A regional scale perspective. *Journal of Animal Ecology* 63:988–1001.

———. 1997. Monitoring forest bird communities for impact assessment: The influence of sampling intensity and spatial scale. *Biological Conservation* 82:355–367.

Mac Nally, R., A. F. Bennett, and G. Horrocks. 2000. Forecasting the impacts of habitat fragmentation. Evaluation of species-specific predictions of the impact of habitat fragmentation on birds in the box-ironbark forests of central Victoria, Australia. *Biological Conservation* 95:7–29.

Mac Nally, R., and G. Horrocks. 2000. Landscape-scale conservation of an endangered migrant: The swift parrot (*Lathamus discolor*) in its winter range. *Biological Conservation* 92:335–343.

Madej, M. A. 2001. Erosion and sediment delivery following removal of forest roads. *Earth Surface Processes and Landforms* 26:175–190.

Mader, H. J. 1984. Animal isolation by roads and agricultural fields. *Biological Conservation* 29:81–96.

Mader, H. J., C. Scell, and P. Kornacher. 1990. Linear barriers to arthropod movements in the landscape. *Biological Conservation* 54:209–222.

Madsen, S., D. Evans, T. Hamer, P. Henson, S. Miller, S. K. Nelson, D. Roby, and M. Stapanian. 1999a. *Marbled murrelet effectiveness monitoring plan for the Northwest Forest Plan.* USDA Forest Service General Technical Report PNW-GTR-439.

Madsen, T., R. Shine, M. Olsson, and H. Wittzell. 1999b. Restoration of an inbred adder population. *Nature* 402:34–35.

Majer, J. D., H. F. Recher, and A. C. Postle. 1994. Comparison of arthropod species richness in eastern and Western Australian canopies: A contribution to the species number debate. Memoirs of the Queensland Museum 36:121–131.

Malcolm, J. R. 1997. Biomass and diversity of small mammals in Amazonian forest fragments. Pp. 207–221 in *Tropical Forest Remnants: Ecology, management and conservation of fragmented communities*, ed. W. F. Laurance and R. O. Bierregaard. Chicago: University of Chicago Press.

Malcolm, J. R., and J. C. Ray. 2000. Influence of timber extraction routes on central African small-mammal communities, forest structure, and tree diversity. *Conservation Biology* 14:1623–1638.

Malcolm, S. B., and M. P. Zalucki. 1993. *Biology and conservation of the monarch butterfly.* Natural History Museum of Los Angeles County, Los Angeles, Calif.

Malmqvist, B., and A. Eriksson. 1995. Benthic insects in Swedish lake-outlet streams—patterns in species richness and assemblage structure. *Freshwater Biology* 34:285–296.

Manion, P. D., and R. A. Zabel. 1979. Stem decay perspectives—an introduction to the mechanisms of tree defense and decay patterns. *Phytopathology* 69:1136–1138.

Manley, P., G. E. Brogan, C. Cook, M. E. Flores, D. G. Fullmer, S. Husari, T. M. Jimerson, L. M. Lux, M. E. McCain, J. A. Rose, G. Schmitt, J. C. Schuyler, and M. J. Skinner. 1995. *Sustaining ecosystems: A conceptual framework.* USDA Forest Service Pacific Southwest Region Publication R5-EM-TP-001.

Mansergh, I., and S. Bennett. 1989. "Greenhouse" and wildlife management in Victoria. *Victorian Naturalist* 106:243–251.

Mapstone, B. D. 1995. Scalable decision rules for environmental impact studies: Effect size, Type I and Type II errors. *Ecological Applications* 5:401–410.

Marcot, B. G. 1997. Biodiversity of old forests of the west: A lesson from our elders. Pp. 87–105 in *Creating a forestry for the twenty-first century: The science of ecosytem management*, ed. K. A. Kohm and J. F. Franklin. Washington, D.C.: Island Press.

Margules, C. R. 1992. The Wog Wog habitat fragmentation experiment. *Environmental Conservation* 19:316–325.

Margules, C. R., A. J. Higgs, and R. W. Rafe. 1982. Modern biogeographic theory: Are there any lessons for nature reserve design? *Biological Conservation* 24:115–128.

Margules, C. R., and D. B. Lindenmayer. 1996. Landscape level concepts and indicators for the conservation of forest biodiversity and sustainable forest management. Pp. 65–83 in *Proceedings of the UBC-UPM Conference on the Ecological, Social and Political Issues of the Certification of Forest Management.*

Margules, C. R., G. A. Milkovits, and G. T. Smith. 1994b. Contrasting effects of habitat fragmentation on the scorpion *Cercophonius squama* and an amphipod. *Ecology* 75:2033–2042.

Margules, C. R., A. O. Nicholls, and R. L. Pressey. 1988. Selecting networks of reserves to maximise biological diversity. *Biological Conservation* 43:663–676.

Margules, C. R., A. O. Nicholls, and M. B. Usher. 1994a. Apparent species turnover, probability of extinction and the selection of nature reserves: A case study of the Ingleborough limestone pavements. *Conservation Biology* 8:398–409.

Margules, C. R., and R. L. Pressey. 2000. Systematic conservation planning. *Nature* 405:243–253.

Margules, C. R., T. D. Redhead, M. F. Hutchinson, and D. P. Faith. 1995. *BioRap: Guidelines for using the BioRap methodology and tools.* Canberra, Australia: CSIRO Publishing.

Martikainen, P., J. Sitonen, L. Kaila, and P. Puntilla. 1996. Intensity of forest management and bark beetles in non-epidemic conditions: A comparison between Finnish and Russian Karelia. *Journal of Applied Entomology* 120:257–264.

Martin, T. E., and J. R. Karr. 1986. Patch utilisation by migrating birds: Resource orientated? *Ornis Scandinavia* 17:165–174.

Marzluff, J. M., and M. Restani. 1999. The effect of forest fragmentation on avian nest predation. Pp. 155–169 in *Forest wildlife and fragmentation: Management implications*, ed. J. Rochelle, L. A. Lehmann, and J. Wisniewski. Leiden, Germany: Brill.

Maser, C., R. F. Tarrant, J. M. Trappe, and J. F. Franklin. 1988. *From the forest to the sea: A story of fallen trees.* USDA Forest Service General Technical Report PNW-GTR-229.

Maser, C., and J. M. Trappe. 1984. *The seen and unseen world of the fallen tree.* USDA Forest Service General Technical Report PNW-GTR-164.

Maser, C., J. M. Trappe, and R. A. Nussbaum. 1978. Fungi–small mammal interrelationships with emphasis on Oregon coniferous forests. *Ecology* 59:799–809.

Maser, C., J. M. Trappe, and D. C. Ure. 1977. Implications of small mammal mycophagy to the management of western coniferous forests. Pp. 78–88 in *Transactions of the North American Wildlife and Natural Resources Conference.*

Matlack, G. R. 1993. Microenvironment variation within and among forest edge sites in the eastern United States. *Biological Conservation* 66:185–194.

Matlack, G. R., and J. A. Litvaitis. 1999. Forest edges. Pp. 210–233 in *Managing biodiversity in forest ecosystems*, ed. M. Hunter Jr. Cambridge: Cambridge University Press.

May, R. 1999. Unanswered questions in ecology. *Philosophical Transactions Royal Society of London Series B* 354:1951–1959.

May, S. A. 2001. Aspects of the ecology of the cat, dog and fox in the south-east forests of New South Wales. Ph.D. diss., The Australian National University, Canberra.

May, S. A., and T. W. Norton. 1996. Influence of fragmentation and disturbance on the potential impact of feral predators on native fauna in Australian forest ecosystems. *Wildlife Research* 23:387–400.

McAlpine, C. A., J. J. Mott, G. C. Grigg, and P. Sharma. 1999. The influence of landscape structure on kangaroo abundance in disturbed semi-arid woodland. *Rangeland Journal* 21:104–134.

McCarthy, M. A., and M. A. Burgman. 1995. Coping with uncertainty in forest wildlife planning. *Forest Ecology and Management* 74:23–36.

McCarthy, M. A., D. C. Franklin, and M. A. Burgman. 1994. The importance of demographic uncertainty: An example from the helmeted honeyeater. *Biological Conservation* 67:135–142.

McCarthy, M. A., A. M. Gill, and D. B. Lindenmayer. 1999. Fire regimes in mountain ash forest: Evidence from forest age structure, extinction models and wildlife habitat. *Forest Ecology and Management* 124:193–203.

McCarthy, M. A., and D. B. Lindenmayer. 1998. Development of multi-aged mountain ash (*Eucalyptus regnans*) forest under natural disturbance regimes: Implications for wildlife conservation and logging practices. *Forest Ecology and Management* 104:43–56.

_____. 1999a. Incorporating metapopulation dynamics of greater gliders into reserve design in disturbed landscapes. *Ecology* 80:651–667.

_____. 1999b. Spatially correlated extinction in metapopulation models. *Biodiversity and Conservation* 9:47–63.

_____. 2000. Spatially correlated extinction in a metapopulation model of Leadbeater's possum. *Biodiversity and Conservation* 9:47–63.

McCarthy, M., D. B. Lindenmayer, and M. Dreschler. 1997. Extinction debts and the risks faced by abundant species. *Conservation Biology* 11:221–226.

McCarthy, M. A., D. B. Lindenmayer, and H. P. Possingham. 2000. Australian treecreepers and landscape fragmentation: A test of a spatially explicit PVA model. *Ecological Applications* 10:1722–1731.

_____. 2001. Assessing spatial PVA models of arboreal marsupials using significance tests and Bayesian statistics. *Biological Conservation* 98:191–200.

McCarty, J. P. 2001. Ecological consequences of recent climate change. *Conservation Biology* 15:320–331.

McCay, T. S. 2000. Use of woody debris by cotton mice (*Peromyscus gossypinus*) in a southeastern pine forest. *Journal of Mammalogy* 81:527–535.

McClellan, B. N., and D. M. Shackleton. 1988. Grizzly bears and resource extraction industries: Effects of roads on behaviour, habitat use and demography. *Journal of Applied Ecology* 25:451–460.

McClellan, M. H., D. N. Swanston, P. E. Hennon, R. L. Deal, T. L. De Santo, and M. S. Wipfli. 2000. *Alternatives to clearcutting in the old-growth forests of southeast Alaska: Study plan and establishment report.* USDA Forest Service General Technical Report PNW-GTR-494.

McClelland, B. R., and C. Frissell. 1975. Identifying forest snags useful for hole-nesting birds. *Journal of Forestry* (July):414–417.

McCollin, D. 1998. Forest edges and habitat selection in birds: A functional approach. *Ecography* 21:247–260.

McComb, W. C., and D. B. Lindenmayer. 1999. Dying, dead and down trees. Pp. 335–372 in *Maintaining biodiversity in forest ecosystems*, ed. M. Hunter Jr. Cambridge: Cambridge University Press.

McComb, W. C., K. McGarigal, J. D. Fraser, and W. H. Davis. 1991. Planning for basin-level cumulative effects in the Appalachian coal fields. Pp. 138–151 in *Wildlife habitats in managed landscapes*, ed. J. E. Rodiek and E. C. Bolen. Washington, D.C.: Island Press.

McComb, W. C., T. A. Spies, and W. H. Emmingham. 1993. Douglas-fir forests: Managing for timber and mature-forest habitat. *Journal of Forestry* (December):31–42.

McConnell, A. 1993. How many ships sail in the forest? Inventory of historic places in Tasmania's forests. Pp. 261–272 in *Australia's ever-changing forests II*, ed. J. Dargavel and S. Feary. Proceedings of the Second National Conference on Australian forest history, Canberra.

_____. 1995. *Archeological potential zoning: A strategy for the protection of Aboriginal archaeological sites in Tasmanian state forests.* Vol. 1. *The archeological potential zoning project: Background, methods,*

use and discussion. Report to Forestry Tasmania. Hobart, Tasmania: Forestry Tasmania.

McCoy, E. D., and H. R. Mushinsky. 1999. Habitat fragmentation and the abundances of vertebrates in the Florida scrub. *Ecology* 80:2526–2538.

McCullough, D., ed. 1997. *Metapopulations and wildlife conservation.* Washington, D.C.: Island Press.

McCune, B. 1993. Gradients in epiphyte biomass in three *Pseudotsuga-Tsuga* forests of different ages in western Oregon and Washington. *Bryologist* 96:405–411.

McGarigal, K., and B. J. Marks. 1994. *FRAGSTATS: Spatial analysis program for quantifying landscape structure.* Version 2.0. Forest Science Department, Oregon State University, Corvallis.

McGarigal, K., and W. C. McComb. 1992. Streamside versus upslope breeding bird communities in the central Oregon Coast Range. *Journal of Wildlife Management* 56:10–23.

_____. 1995. Relationships between landscape structure and breeding birds in the Oregon Coast Range. *Ecological Monographs* 65:235–260.

_____. 1999. Forest fragmentation effects on breeding bird communities in the Oregon Coast Range. Pp. 1–31 in *Forest wildlife and fragmentation: Management implications*, ed. J. Rochelle, L. A. Lehmann, and J. Wisniewski. Leiden, Germany: Brill.

McGarigal, K., W. H. Romme, M. Crist, and E. Roworth. 2001. Cumulative effects of roads and logging on landscape structure in the San Juan Mountains, Colorado (USA). *Landscape Ecology* 16:327–349.

McGrady-Steed, J., P. M. Harris, and P. J. Morin. 1997. Biodiversity regulates ecosystem predictability. *Nature* 390:162–165.

McIntosh, B. A., J. R. Sedell, R. F. Thurlow, S. E. Clarke, and G. L. Chandler. 2000. Historical changes in pool habitats in the Columbia River Basin. *Ecological Applications* 10:1478–1496.

McIntyre, S. 1994. Integrating agriculture and land use and management for conservation of a native grassland flora in a variegated landscape. *Pacific Conservation Biology* 1:236–244.

McIntyre, S., and G. W. Barrett. 1992. Habitat variegation, an alternative to fragmentation. *Conservation Biology* 6:146–147.

McIntyre, S., G. W. Barrett, and H. A. Ford. 1996. Communities and ecosystems. Pp. 154–170 in *Conservation Biology*, ed. I. F. Spellerberg. Harlow, England: Longman.

McIntyre, S., and R. Hobbs. 1999. A framework for conceptualizing human effects on landscapes and its relevance to management and research models. *Conservation Biology* 13:1282–1292.

McIntyre, S., J. G. McIvor, and N. D. MacLeod. 2000. Principles for sustainable grazing in eucalypt woodlands: Landscape-scale indicators and the search for thresholds. Pp. 92–100 in *Management for sustainable ecosystems*, ed. P. Hale, D. Moloney, and P. Sattler. Centre for Conservation Biology, The University of Queensland, Brisbane, Australia.

McKenney, D. W., and M. S. Common. 1989. *The economic analysis of public sector forest management using linear programming: An Australian application.* Centre for Resource and Environmental Studies Working Paper 1989/4. Canberra, Australia.

McKenney, D. W., and D. B. Lindenmayer. 1994. An economic assessment of a nest box strategy for the conservation of an endangered species. *Canadian Journal of Forest Research* 24:2012–2019.

McNeely, J. A. 1994a. Protected areas for the twenty-first century: Working to provide benefits to society. *Biodiversity and Conservation* 3:390–405.

_____. 1994b. Lessons from the past: Forests and biodiversity. *Biodiversity and Conservation* 3:3–20.

McShea, W. J., and J. H. Rappole. 2000. Managing the abundance and diversity of breeding bird populations through manipulation of deer populations. *Conservation Biology* 14:1161–1170.

Mech, L. D. 1970. *The wolf.* New York: Natural History Press.

_____. 1973. *Wolf numbers in the Superior National Forest of Minnesota.* USDA Forest Service Research Publication NC-97.

Mech, L. D., S. H. Fritts, G. L. Radde, and W. L. Paul. 1988. Wolf distribution and road density in Minnesota. *Wildlife Society Bulletin* 16:85–87.

Mech, S. G., and J. G. Hallett. 2001. Evaluating the effectiveness of corridors: A genetic approach. *Conservation Biology* 15:467–474.

Meffe, G. K., D. Boersma, D. D. Murphy, B. R. Noon, H. R. Pulliam, M. E. Soulé, and D. M. Waller. 1998. Independent scientific review in natural resource management. *Conservation Biology* 12:268–270.

Meffe, G. K., and C. R. Carroll. 1995. *Principles of conservation biology.* Sunderland, Mass.: Sinauer Associates.

Meggs, J. M. 1997. *Simsons stag beetle,* Hoplogonus simonsi, *in north-east Tasmania: Distribution, habitat characteristics and conservation requirements.* A report to the Forest Practices Board of Tasmania. November. Hobart, Tasmania.

Meredith, C. W. 1984. Possums or poles? The effects of silvicultural management on the possums of Chiltern State Park, north-east Victoria. Pp. 517–526 in *Possums and gliders,* ed. A. P. Smith and I. D. Hume. Chipping Norton, Australia: Surrey Beatty and Sons.

Merriam, G., K. Henein, and K. Stuart-Smith. 1991. Landscape dynamics models. Pp. 399–412 in *Quantitative methods in landscape ecology,* ed. M. G. Turner and R. H. Gardner. New York: Springer-Verlag.

Merriam, G., and A. Lanoue. 1990. Corridor use by small mammals: Field measurements for three experimental types of *Peromyscus leucopus. Landscape Ecology* 4:123–131.

Mesquita, R. C., P. Delamonica, and W. F. Laurance. 1999. Effect of surrounding vegetation on edge-related tree mortality in Amazonian forest fragments. *Biological Conservation* 91:129–134.

Metzeling, L., T. Doeg, and W. O'Conner. 1995. The impact of salinisation and sedimentation on aquatic biota. Pp. 126–136 in *Conserving biodiversity: Threats and solutions,* ed. R. A. Bradstock, T. A. Auld, D. A. Keith, R. T. Kingsford, D. Lunney, and D. P. Sivertsen. Chipping Norton, Australia: Surrey Beatty and Sons.

Metzger, J. P. 1997. Relationships between landscape structure and tree species diversity in tropical forests of south-east Brazil. *Landscape and Urban Planning* 37:29–35.

Michaelis, F. B. 1984. Possible effects of forestry on inland waters of Tasmania: A review. *Environmental Conservation* 11:331–343.

Michener, W. K., and J. W. Brunt. 2000. Ecological data: Design, management, and processing. Oxford, United Kingdom: Blackwell Science.

Milledge, D. R., C. L. Palmer, and J. L. Nelson. 1991. Barometers of change: The distribution of large owls and gliders in mountain ash forests of the Victorian Central Highlands and their potential as management indicators. Pp. 55–65 in *Conservation of Australia's forest fauna,* ed. D. Lunney. Royal Zoological Society of New South Wales, Sydney, Australia.

Miller, C., and D. L. Urban. 2000. Modeling the effects of fire management alternatives on Sierra Nevada mixed-conifer forests. *Ecological Applications* 10:85–94.

Miller, D. R., J. D. Lin, and Z. N. Lu. 1991. Some effects of surrounding forest canopy architecture on the wind field in small clearings. *Forest Ecology and Management* 45:79–91.

Miller, J. R., L. A. Joyce, R. L. Knight, and R. M. King. 1996. Forest roads and landscape structure in the southern Rocky Mountains. *Landscape Ecology* 11:115–127.

Miller, S. G., S. P. Bratton, and J. Hadidian. 1992. Impacts of white-tailed deer on endangered and threatened vascular plants. *Natural Areas Journal* 12:67–74.

Miller, W. 1970. Factors influencing the status of eastern and mountain bluebirds in southwestern Manitoba. *Blue Jay* 28:38–46.

Mills, L. S. 1995. Edge effects and isolation: Red-backed voles on forest remnants. *Conservation Biology* 9:395–403.

Mills, L. S., and F. W. Allendorf. 1996. The one-migrant-per-generation rule in conservation and management. *Conservation Biology* 10:1509–1518.

Mills, L. S., M. E. Soulé, and D. F. Doak. 1993. The keystone species concept in ecology and conservation. *BioScience* 43:219–224.

Milne, B. T., M. G. Turner, J. A. Wiens, and A. R. Johnson. 1992. Interactions between fractal geometry of landscapes and allometric herbivory. *Theoretical Population Biology* 41:337–353.

Mitchell, R. J., W. L. Neel, J. K. Hiers, F. T. Cole, and J. B. Atkinson Jr. 2000. *A model management plan for conservation easements in longleaf pine-dominated landscapes.* Joseph E. Jones Ecological Research Center, Newton, Georgia.

Mitchell, S. 1999. *Longleaf pine-wiregrass community. The natural communities of Georgia.* Joseph E. Jones Ecological Research Center, Newton, Georgia.

Mladenoff, D. J., J. White, T. R. Crow, and J. Pastor. 1994. Applying principles of landscape design and management to integrate old-growth forest enhancement and commodity use. *Conservation Biology* 8:752–762.

Mladenoff, D. J., J. White, J. Pastor, and T. R. Crow. 1993. Comparing spatial pattern in unaltered old-growth and disturbed forest landscapes. *Ecological Applications* 3:294–306.

Moilanen, A., and M. Cabeza. In press. Single-species dynamic site selection. *Ecological Applications.*

Mönkkönen, M. 1999. Managing Nordic boreal forest landscapes for biodiversity: Ecological and economic perspectives. *Biodiversity and Conservation* 8:85–99.

Mönkkönen, M., and P. Reunanen. 1999. On critical thresholds in landscape connectivity: A management perspective. *Oikos* 84:302–305.

Montgomery, D. R. 1994. Road surface drainage, channel initiation, and slope stability. *Water Resources Research* 30:1925–1939.

Montréal Process Liaison Office. 2000. *The Montréal Process: Progress and innovation in implementing criteria and indicators for the conservation and sustainable management of temperate and boreal forests.* Year 2000 Progress Report, Canadian Forest Service.

Moore, M. K. 1977. Factors contributing to blowdown in streamside leave strips on Vancouver Island. *Land Management Report* 3:1–34. British Columbia Forest Service, Vancouver.

Moore, S. E., and H. L. Allen. 1999. Plantation forestry. Pp. 400–433 in *Maintaining biodiversity in forest ecosystems,* ed. M. L. Hunter Jr. Cambridge: Cambridge University Press.

Moriarty, J. J., and W. C. McComb. 1983. The long-term effect of timber stand improvement on snag and cavity densities in the central Appalachians. Pp. 40–44 in *Snag habitat management,* ed. J. W. Davis, G. A. Goodwin, and R. A. Ockenfels. Proceedings of the symposium. 7–9 June 1983, Northern Arizona University, Flagstaff.

Morrison, M. L., and B. G. Marcot. 1995. An evaluation of resource inventory and monitoring programs used in national forest planning. *Environmental Management* 19:147–156.

Morrison, M. L., B. G. Marcot, and R. W. Mannan. 1992. *Wildlife habitat relationships: Concepts and applications.* Madison: University of Wisconsin Press.

Morrison, P. H., and F. J. Swanson. 1990. *Fire history and pattern in a Cascade Mountain landscape.* USDA Forest Service General Technical Report PNW-GTR-254. Portland, Ore.

Morton, S. 1996. Looking after our land. A future for Australia's biodiversity. *Search* 27:124–126.

Moyle, P. B., and G. M. Sato. 1991. On the design of preserves to protect native fishes. Pp. 155–169 in *Battle against extinction: Native fish management in the American West,* ed. W. L. Minckley and J. E. Deacon. Tucson: University of Arizona Press.

Mueck, S. G. 1990. *The floristic composition of mountain ash and alpine ash forests in Victoria.* SSP Technical Report no. 4. Department of Conservation and Environment, Victoria, Australia.

Mueck, S. G., K. Ough, and J. C. Banks. 1996. How old are wet forest understories? *Australian Journal of Ecology* 21:345–348.

Mueck, S. G., and R. Peacock. 1992. *Impacts of intensive timber harvesting on the forests of East Gippsland, Victoria.* VSP Technical Report no. 15. Department of Conservation and Environment, Melbourne, Australia.

Mulder, B. S., B. R. Noon, T. A. Spies, M. G. Raphael, C. J. Palmer, A. R. Olsen, G. H. Reeves, and H. H. Welsh. 1999. *The strategy and design of the effectiveness monitoring program for the Northwest Forest Plan.* USDA Forest Service General Technical Report PNW-GTR-437.

Murcia, C. 1995. Edge effects on fragmented forests: Implications for conservation. *Trends in Ecology and Evolution* 10:58–62.

Murphy, D. D. 1989. Conservation and confusion: Wrong species, wrong scale, wrong conclusion. *Conservation Biology* 3:82–84.

Murphy, D. D., K. E. Freas, and S. T. Weiss. 1990. An environment-metapopulation approach to population viability for a threatened invertebrate. *Conservation Biology* 4:41–51.

Murphy, D. D., and B. R. Noon. 1992. Integrating scientific methods with habitat conservation planning: Reserve design for northern spotted owls. *Ecological Applications* 2:3–17.

Myers, N. 1996. The world's forests: Problems and potentials. *Environmental Conservation* 23:156–168.

Naeem, S. 1998. Species redundancy and ecosystem reliability. *Conservation Biology* 12:39–45.

Naiman, R. J., ed. 1992. *Watershed management: Balancing sustainability and environmental change.* New York: Springer-Verlag.

Naiman, R. J., and R. E. Bilby, eds. 1998. *River ecology and management: Lessons from the Pacific coastal ecoregion.* New York: Springer-Verlag.

Naiman, R. J., R. E. Bilby, and P. A. Bisson. 2000. Riparian ecology and management in the Pacific coastal rain forest. *BioScience* 50:996–1011.

Naiman, R. J., H. Decamps, and H. Pollock. 1993. The role of riparian corridors in maintaining regional biodiversity. *Ecological Applications* 3:209–212.

Naiman, R. J., and M. G. Turner. 2000. A future perspective on North America's freshwater ecosystems. *Ecological Applications* 10:958–970.

National Association of Forest Industries. 1989. *Wood production and the environment: Working in harmony with nature.* National Association of Forest Industries, Canberra, Australia.

National Board of Forestry [Sweden]. 1995. *A richer forest.* National Board of Forestry [Sweden], Jonkoping, Sweden.

———. 1996a. *The Swedish forest: A compilation of facts on forestry and the forest industries in Sweden.* National Board of Forestry, Jonkoping, Sweden.

———. 1996b. *Action plan for biological diversity and sustainable forestry.* National Board of Forestry, Jonkoping, Sweden.

National Research Council. 1990. *Forestry research: A mandate for change.* Washington, D.C.: National Academy Press.

Naughton, G. P., C. B. Henderson, K. R. Foresman, and R. L. McGraw II. 2000. Long-toed salamanders in harvested and intact Douglas-fir forests of western Montana. *Ecological Applications* 10:1681–1689.

Naugle, D. E., K. F. Higgins, S. M. Nusser, and W. C. Johnson. 1999. Scale-dependent habitat use in three species of prairie wetland birds. *Landscape Ecology* 14:267–276.

Neitlich, P. N., and B. McCune. 1997. Hotspots of epiphytic lichen diversity in two young managed forests. *Conservation Biology* 11:172–182.

Nelson, J. 1991. Beyond parks and protected areas: From public and private stewardship to landscape planning and management. *Environments* 21:23–34.

Nelson, J. L., and B. J. Morris. 1994. Nesting requirements of the yellow-tailed black cockatoo *Calyptorynchus funereus* in mountain ash forest (*Eucalyptus regnans*) and implications for forest management. *Wildlife Research* 21:267–278.

Nelson, M. E. 1993. Natal dispersal and gene flow in white-tailed deer in northeastern Minnesota. *Journal of Mammalogy* 74:316–322.

Nève, G., B. R. Barascud, J. Hughes, H. Aubert, P. Descimon, P. Lebrun, and M. Baguette. 1996. Dispersal, colonization power and metapopulation structure in the vulnerable butterfly *Proclossiana eunomia* (Lepidoptera: Nymphalidae). *Journal of Applied Ecology* 33:14–22.

New, T. R. 1995. Onychophora in invertebrate conservation: Priorities, practice and prospects. In *Onychophora: Past and present*, ed. M. H. Walker and D. B. Norman. *Zoological Journal of the Linnean Society* 14:77–89.

———. 2000. *Conservation biology: An introduction for southern Australia.* Melbourne, Australia: Oxford University Press.

Newmark, W. D. 1985. Legal and biotic boundaries of western North American national parks: A problem of congruence. *Biological Conservation* 33:197–208.

———. 1987. A land-bridge island perspective on mammalian extinctions in western North American parks. *Nature* 325:430–432.

Newton, I. 1994. The role of nest sites in limiting the numbers of hole-nesting birds—a review. *Biological Conservation* 70:265–276.

———. 1998. *Population limitation in birds.* London: Academic Press.

Nicholls, A. O., and C. R. Margules. 1991. The design of studies to demonstrate the biological importance of corridors. Pp. 49–61 in *Nature Conservation 2: The role of corridors*, ed. D. A. Saunders and R. J. Hobbs. Chipping Norton, Australia: Surrey Beatty and Sons.

———. 1993. An upgraded reserve selection algorithm. *Biological Conservation* 64:165–169.

Niemelä, J. 1997. Invertebrates and boreal forest management. *Conservation Biology* 11:601–610.

Niemelä, J., Y. Hiala, and P. Puntitila. 1996. The importance of small-scale heterogeneity in boreal forests: Variation in diversity in forest-floor invertebrates across the succession gradient. *Ecography* 19:352–368.

Niemelä, J., D. Langor, and J. R. Spence. 1993. Effects of clear-cut harvesting on boreal ground-beetle assemblages (Coleoptera: Carabidae) in western Canada. *Conservation Biology* 7:551–561.

Niemelä, J., P. Renvall, and R. Penttila. 1995. Interactions of fungi at late stages of wood decomposition. *Annual Botanicol Fennici* 32:141–152.

Niemien, M., and I. Hanski. 1998. Metapopulations of moths on islands: A test of two contrasting models. *Journal of Animal Ecology* 67:149–160.

Nittler, J. B., and D. W. Nash. 1999. The certification model for forestry in Bolivia. *Journal of Forestry* 97:32–36.

Nix, H. A. 1978. Determinants of environmental tolerance limits in plants. Pp. 195–206 in *Biology and quarternary environments*, ed. D. Walker and J. C. Guppy. Canberra: Australian Academy of Science.

———. 1994. The brigalow. Pp. 198–233 in *Australian Environmental History*, ed. S. Dovers. Melbourne, Australia: Oxford University Press.

———. 1997. Management of parks and reserves for the conservation of biological diversity. Pp. 11–36 in *National parks and protected areas: Selection, delimitation and management*, ed. J. J. Pigram and R. C. Sundell. Centre for Water Policy Research, University of New England, Armidale, Australia.

Nix, H. A., and M. A. Switzer. 1991. Rainforest animals. Atlas of vertebrates endemic to Australia's wet tropics. *Kowari* 1:1–112.

Noble, I. R. 1994. Science, bureaucracy, politics and ecologically sustainable development. Pp. 117–125 in *Ecology and sustainability of southern temperate ecosystems*, ed. T. W. Norton and S. R. Dovers. Melbourne, Australia: CSIRO Publishing.

Noble, I. R., and R. O. Slatyer. 1980. The use of vital attributes to predict successional changes in plant communities subject to recurrent disturbances. *Vegetatio* 43:5–21.

Noel, J. M., W. J. Platt, and E. B. Moser. 1998. Structural characteristics of old- and second-growth stands of longleaf pine

(*Pinus palustris*) in the Gulf coastal region of the USA. *Conservation Biology* 12:533–548.

Norse, E., K. Rosenbaum, D. Wilcove, B. Wilcox, W. Romme, D. Johnston, and M. Stout. 1986. *Conserving biological diversity in our national forests*. Washington, D.C.: The Wilderness Society.

North, M. 1993. Stand structure and truffle abundance associated with northern spotted owl habitat. Ph.D thesis, University of Washington, Seattle.

Norton, D. A. 1996. Monitoring biodiversity in New Zealand's terrestrial ecosystems. Pp. 19–41 in *Papers from a seminar series on biodiversity*, comp. B. McFadgen and P. Simpson. Department of Conservation, Wellington, New Zealand.

———. 1999. Forest reserves. Pp. 525–555 in *Maintaining biodiversity in forest ecosystems*, ed. M. Hunter Jr. Cambridge: Cambridge University Press.

Norton, T. W., and D. B. Lindenmayer. 1991. Integrated management of forest wildlife: Towards a coherent strategy across state borders and land tenures. Pp. 237–244 in *Conservation of Australia's forest fauna*, ed. D. Lunney. Chipping Norton, Australia: Surrey Beatty and Sons.

Noss, R. F. 1983. A regional landscape approach to maintain diversity. *BioScience* 33:700–706.

———. 1987. Corridors in real landscapes: A reply to Simberloff and Cox. *Conservation Biology* 1:159–164.

———. 1991. Landscape connectivity: Different functions at different scales. Pp. 27–39 in *Landscape Linkages and Biodiversity*, ed. W. E. Hudson. Washington, D.C.: Island Press.

———. 1999. Assessing and monitoring forest biodiversity: A suggested framework for indicators. *Forest Ecology and Management* 115:135–146.

Noss, R. F., and A. Y. Cooperrider. 1994. *Saving nature's legacy: Protecting and restoring biodiversity*. Washington, D.C.: Island Press.

Noss, R. F., and L. D. Harris. 1986. Nodes, networks, and MUMs: Preserving diversity at all scales. *Environmental Management* 10:299–309.

Novacek, M. J., and E. E. Cleland. 2001. The current biodiversity extinction event: Scenarios for mitigation and recovery. *Proceedings of the National Academy of Science* 98:5466–5470.

Nowacki, G. J., and M. D. Abrams. 1997. Radial-growth averaging criteria for reconstructing disturbance histories from presettlement-origin oaks. *Ecological Monographs* 67:225–249.

NSW National Parks and Wildlife Service and State Forests of NSW. 1996. *Conservation protocols for timber harvesting on state forests for the duration of the IFA decision*. 29 November. NSW National Parks and Wildlife Service and State Forests of NSW.

O'Brien, G. P. 1983. Power pole damage by acorn woodpeckers in southeastern Arizona. Pp. 14–18 in *Snag habitat management*, ed. J. W. Davis, G. A. Goodwin, and R. A. Ockenfels. Proceedings of the symposium. 7–9 June. Northern Arizona University, Flagstaff.

Ogden, J., G. H. Stewart, and R. B. Allen. 1996. Ecology of New Zealand *Nothofagus* forests. Pp. 25–82 in *The ecology and biogeography of* Nothofagus *forests*, ed. T. T. Veblen, R. S. Hill, and J. Read. New Haven, Conn.: Yale University Press.

Ohmann, J. L., W. C. McComb, and A. A. Zumrawi. 1994. Snag abundance for primary cavity-nesting birds on nonfederal forest lands in Oregon and Washington. *Wildlife Society Bulletin* 22:607–620.

Ohmann, J. L., and T. A. Spies. 1998. Regional gradient analysis and spatial patterns of woody plant communities of Oregon forests. *Ecological Monographs* 68:151–182.

Økland, B. 1996. Unlogged forests: Important sites for preserving the diversity of mycetophilids (Diptera: Sciaroidea). *Biological Conservation* 76:297–310.

Oliver, C. D. 1981. Forest development in North America following major disturbances. *Forest Ecology and Management* 3:153–168.

Oliver, C. D., D. Adams, T. Bonnicksen, J. Bowyer, F. Cubbage, N. Sampson, S. Schlarblum, R. Whaley, and H. Wiant. 1997. *Report on forest health of the United States*. Chartered by Charles Taylor, member, United States Congress, 11th District, North Carolina.

Oliver, C. D., and B. C. Larson. 1996. *Forest stand dynamics*. New York: John Wiley and Sons.

Olsen, P. 1995. *Australian birds of prey*. Sydney, Australia: University of New South Wales Press.

O'Neill, G., and P. Attiwill. 1997. Getting ecological paradigms into the political debate: Or will the messenger be shot? Pp. 351–357 in *The ecological basis of conservation: Heterogeneity, ecosystems and biodiversity*, ed. S.T.A. Pickett, R. S. Ostfeld, M. Shackak, and G. E. Likens. New York: Chapman and Hall.

O'Neill, R. V., J. R. Krummel, R. H. Gardner, G. Sugihara, B. Jackson, D. L. DeAngelis, B. T. Milne, M. G. Turner, B. Zygmunt, S. W. Christensen, V. H. Dale, and R. L. Graham. 1988. Indices of landscape pattern. *Landscape Ecology* 1:153–162.

Orians, G. H. 1986. The place of science in environmental problem-solving. *Environment* 28:12–17, 38–41.

Osborne, P. 1984. Bird numbers and habitat characteristics in farmland hedgerows. *Journal of Applied Ecology* 21:63–82.

O'Shaughnessy, P., and J. Jayasuriya. 1991. Managing the ash-type forest for water production in Victoria. Pp. 341–363 in *Forest management in Australia*, ed. F. H. McKinnell, E. R. Hopkins, and J.E.D. Fox. Chipping Norton, Australia: Surrey Beatty and Sons.

Ough, K., and A. Murphy. 1996. The effect of clearfell logging on tree-ferns in Victorian wet forest. *Australian Forestry* 59:178–188.

———. 1998. *Understorey islands: A method of protecting understorey flora during clearfelling operations*. Department of Natural Resources and Environment. Internal VSP Report no. 29. Department of Natural Resources and Environment, Melbourne, Australia.

Ough, K., and J. Ross. 1992. *Floristics, fire and clearfelling in wet forests of the Central Highlands of Victoria*. Silvicultural Systems Project, Technical Report no. 11. Department of Conservation and Environment, Melbourne, Australia.

Ozanne, C.M.P., M. R. Speight, C. Hambler, and H. F. Evans. 2000. Isolated trees and forest patches: Patterns in canopy arthropod abundance and diversity in *Pinus sylvestris* (scots pine). *Forest Ecology and Management* 137:53–63.

Paine, R. T., M. J. Tegner, and E. A. Johnson. 1998. Compounded perturbations yield ecological surprises. *Ecosystems* 1:535–545.

Palmer, C., and J. C. Woinarski. 1999. Seasonal roosts and foraging movements of the black flying fox (*Pteropus alecto*) in the Northern Territory: Resource tracking in a landscape mosaic. *Wildlife Research* 26:823–838.

Palmer, M. A., A. P. Covich, S. Lake, P. Biro, J. J. Brooks, J. Cole, C. Dahm, J. Gibert, W. Goedkoop, L. Martens, J. Verhoeven, and W. J. Van de Bund. 2000. Linkages between aquatic sediment biota and life among sediments as potential drivers of biodiversity and ecological processes. *BioScience* 50:1062–1075.

Palmer, M. E. 1987. A critical look at rare plant monitoring in the United States. *Biological Conservation* 39:113–127.

Parker, G. G. 1997. Canopy structure and light environment of an old-growth Douglas-fir/western hemlock forest. *Northwest Science* 71:261–270.

Parker, G. G., and M. J. Brown. 2000. Forest canopy stratification—is it useful? *American Naturalist* 155:473–484.

Parker, L. 1997. Restaging an evolutionary drama: Thinking big on the Chequamegon and Nicolet National Forests. Pp. 218–219 in *Creating a forestry for the twenty-first century: The science of ecosystem management*, ed. K. A. Kohm and J. F. Franklin. Washington, D.C.: Island Press.

Parks, C. G., E. M. Bull, and G. M. Filip. 1995. Using artificial in-oculated decay fungi to create wildlife habitat. Pp. 175–177 in *Partnerships for sustainable forest ecosystem management*, ed. C. Aguirre-Bravo, L. Eskew, A. B. Vilal-Salas, and C. E. Gonzalez-Vicente. USDA Forest Service General Technical Report RM-GTR-266.

Parmesan, C. 1996. Climate and species' range. *Nature* 383:765–766.

Parry, B. B. 1997. Abiotic edge effects in wet sclerophyll forest in the central highlands of Victoria. Master's thesis, School of Botany, University of Melbourne, Australia.

Parsons, P. 1991. Biodiversity conservation under global climatic change: The insect Drosophila as a biological indicator? *Global Ecology and Biogeography Letters* 1:77–83.

Pärt, T., and B. Söderstrom, 1999. Conservation value of semi-nat-ural pastures in Sweden: Contrasting botanical and avian meas-ures. *Conservation Biology* 13:755–765.

Parviainen, J., W. Bücking, K. Vandekerkhove, A. Schuck, and R. Päivinen. 2000. Strict forest reserves in Europe: Efforts to en-hance biodiversity and research on forests left for free develop-ment in Europe (EU-COST-Action E4). *Forestry* 73:107–118.

Pascual, M. A., and M. D. Adkison. 1994. The decline of the stellar sea lion in the northeast Pacific: Demography, harvest or envi-ronment? *Ecological Applications* 4:393–403.

Paton, P. W. 1994. The effect of edge on avian nest success: How strong is the evidence? *Conservation Biology* 8:17–26.

Pattemore, V., and J. Kikkawa. 1975. Comparison of bird popula-tions in logged and unlogged rainforest at Wiangarie State For-est, NSW. *Australian Forestry* 37:188–198.

Patterson, B. D. 1987. The principle of nested subsets and its impli-cations for biological conservation. *Conservation Biology* 1:247–293.

Patterson, B. D., and W. Atmar. 1986. Nested subsets and the struc-ture of insular mammalian faunas and archipelagos. *Biological Journal of the Linnean Society* 28:65–82.

Patton, D. R. 1974. Patting increases deer and edges of a pine for-est in Arizona. *Journal of Forestry* (December):764–766.

———. 1992. *Wildlife habitat relationships in forested ecosystems*. Port-land, Ore.: Timber Press.

Pauw, A. 1998. Pollen transfer on birds' tongues. *Nature* 394:731–734.

Payette, S., C. Morneau, L. Sirois, and M. Desponts. 1989. Recent fire history of the northern Quebec biomes. *Ecology* 70:656–673.

Pearce, J. L., M. A. Burgman, and D. C. Franklin. 1994. Habitat se-lection by helmeted honeyeaters. *Wildlife Research* 21:53–63.

Pearson, S. F., and D. A. Manuwal. 2001. Breeding bird response to riparian buffer width in managed Pacific Northwest Douglas-fir forests. *Ecological Applications* 11:840–853.

Pearson, S. M. 1993. The spatial extent and relative influence of landscape-level factors on wintering bird populations. *Landscape Ecology* 8:3–18.

Pearson, S. M., M. G. Turner, R. H. Gardner, and R. V. O'Neill. 1996. An organism-based perspective of habitat fragmentation. Pp. 77–95 in *Biodiversity in managed landscapes: Theory and prac-tice*, ed. R. C. Szaro and D. W. Johnston. New York: Oxford University Press.

Peck, J. E., and B. McCune. 1997. Remnant trees and canopy lichen communities in western Oregon: A retrospective approach. *Eco-logical Applications* 7:1181–1187.

Pengelly, W. L. 1972. Clearcutting: Detrimental aspects for wildlife resources. *Journal of Soil and Water Conservation* (November/De-cember):255–258.

Perry, D. A. 1994. *Forest ecosystems*. Baltimore: Johns Hopkins Uni-versity Press.

Perry, D. A., and M. P. Amaranthus. 1997. Disturbance, recovery, and stability. Pp. 31–56 in *Creating a forestry for the twenty-first century: The science of ecosystem management*, ed. K. A. Kohm and J. F. Franklin. Washington, D.C.: Island Press.

Perry, D. H., M. Lenz, and J. A. Watson. 1985. Relationships be-tween fire, fungal rots and termite damage in Australian forest trees. *Australian Forestry* 48:46–53.

Peterken, G. F. 1996. *Natural woodland: Ecology and conservation in northern temperate regions*. Cambridge: Cambridge University Press.

———. 1999. Applying natural forestry concepts in an intensively managed landscape. *Global Ecology and Biogeography* 8:321–328.

Peterken, G. F., D. Ausherman, M. Buchenau, and R. T. Forman. 1992. Old-growth conservation within British upland conifer plantations. *Forestry* 65:127–144.

Peterken, G. F., and J. L. Francis. 1999. Open spaces as habitats for vascular ground flora species in the woods of central Lin-colnshire, U.K. *Biological Conservation* 91:55–72.

Peters, R. L., and J. D. Darling. 1985. The greenhouse effect and nature reserves. *BioScience* 35:707–717.

Peters, R. L., and T. E. Lovejoy, eds. 1992. *Global warming and bio-logical diversity*. New Haven, Conn.: Yale University Press.

Peterson, C. J., and S. T. Pickett. 1995. Forest reorganization: A case study in an old-growth forest catastrophic blowdown. *Ecol-ogy* 76:763–774.

Petty, S. J., G. Shaw, and D.I.K. Anderson. 1994. Value of nest boxes for population studies and conservation of owls in conifer-ous forests in Britain. *Journal of Raptor Research* 28:134–142.

Pharo, E., and A. J. Beattie. 2001. Management types as a surrogate for vascular plant, bryophyte, and lichen diversity. *Australian Journal of Botany* 49:23–30.

Phillips, E. J. 1996. *Comparing silvicultural systems in a coastal mon-tane forest: Productivity and cost of harvesting operations*. FERIC Special Report, no. SR-109. Canadian Forest Service and BC Ministry of Forests: Victoria, British Columbia.

Pickett, S. T. 1996. Sustainable forestry in Chilean Tierra del Fuego. *Trends in Evolution and Ecology* 11:450–451.

Pickett, S. T., and J. H. Thompson. 1978. Patch dynamics and the design of nature reserves. *Biological Conservation* 13:27–37.

Pickett, S.T.A. 1989. Space-for-time substitution as an alternative to long-term studies. Pp. 110–135 in *Longterm studies in ecology: Approaches and alternatives*, ed. G. E. Likens. New York: Springer-Verlag.

Pierson, E. D., T. Elmqvist, W. E. Rainey, and P. A. Cox. 1996. Ef-fects of tropical cyclonic storms on flying fox populations in the South Pacific Islands of Samoa. *Conservation Biology* 10:438–451.

Pigram, J. J., and R. C. Sundell, eds. 1997. *National parks and pro-tected areas: Selection, delimitation and management*. Centre for Water Policy Research, University of New England, Armidale, Australia.

Pimentel, D., U. Stachow, D. A. Takacs, H. W. Brubaker, A. R. Dumas, J. J. Meaney, J. A. O'Neill, D. E. Onsi, and D. B. Corzilius. 1992. Conserving biological diversity in agricul-tural/forestry systems. *BioScience* 42:354–362.

Pimentel, D., C. Wilson, C. McCullum, R. Huang, P. Dwen, J. Flack, Q. Tran, T. Saltman, and B. Cliff. 1997. Economic and environmental benefits of biodiversity. *BioScience* 47:747–757.

Pimm, S. L., H. L. Jones, and J. Diamond. 1988. On the risk of ex-tinction. *American Naturalist* 132:757–785.

Pinard, M. A., and F. E. Putz. 1997. Monitoring carbon sequestrian benefits associated with a reduced logging impact project in Malaysia. Mitigation and adaptation strategies for global change. *Biotropica* 2:203–215.

Pipkin, J. 1998. *The Northwest Forest Plan revisited*. Washington, D.C.: U.S. Department of Interior.

Pither, J., and P. D. Taylor. 1998. An experimental assessment of landscape connectivity. *Oikos* 83:166–174.

Pittock, J. 1994. Logging abuse. *Parkwatch* 31:6–8.

Poff, N. L. 1997. Landscape filters and species traits: Towards mechanistic understanding and prediction in stream ecology. *Journal of the North American Benthic Society* 16:391–409.

Pope, S. E., L. Fahrig, and H. G. Merriam. 2000. Landscape complementation and metapopulation effects on leopard frog populations. *Ecology* 81:2498–2508.

Possingham, H. P., and I. Davies. 1995. ALEX: A model for the viability analysis of spatially structured populations. *Biological Conservation* 73:143–150.

Possingham, H. P., D. B. Lindenmayer, and T. W. Norton. 1993. A framework for improved threatened species management using population viability analysis. *Pacific Conservation Biology* 1:39–45.

Possingham, H. P., D. B. Lindenmayer, T. W. Norton, and I. Davies. 1994. Metapopulation viability of the Greater Glider in a wood production forest in south-eastern Australia. *Biological Conservation* 70:265–276.

Potter, M. A. 1990. Movement of North Island brown kiwi (*Apteryx australis mantelli*) between forest remnants. *New Zealand Journal of Ecology* 14:17–24.

Pouliquen-Young, O. 1997. Evolution of the system of protected areas in Western Australia. *Environmental Conservation* 24:168–181.

Powell, A. H., and G. V. Powell. 1987. Population dynamics of male euglossine bees in Amazonian forest fragments. *Biotropica* 19:176–179.

Powell, G. V., and R. Björk. 1995. Implications of intratropical migration on reserve design: A case study using *Pharomachrus mocinno. Conservation Biology* 9:354–362.

Prance, G. T. 1991. Rates of loss of biodiversity: A global view. Pp. 27–44 in *The scientific management of temperate communities for conservation*, ed. I. F. Spellerberg, F. B. Goldsmith, and M. G. Morris. Oxford: Blackwell Scientific Publications.

Prendergast, J. R., and B. C. Eversham. 1997. Species richness covariance in higher taxa: Empirical tests of the biodiversity indicator concept. *Ecography* 20:210–216.

Prendergast, J. R., R. M. Quinn, and J. H. Lawton. 1999. The gaps between theory and practice in selecting nature reserves. *Conservation Biology* 13:484–492.

Prendergast, J. R., R. M. Quinn, J. H. Lawton, B. C. Eversham, and D. W. Gibbons. 1993. Rare species, the coincidence of diversity hotspots and conservation strategies. *Nature* 365:335–336.

Prentice, I. C., W. Cramwe, S. P. Harrison, R. Leemans, R. A. Monserud, and A. M. Solomon. 1992. A global biome model based on plant physiology and dominance, soil properties and climate. *Journal of Biogeography* 19:117–134.

Pressey, R., and R. M. Cowling. 2001. Reserve selection algorithms and the real world. *Conservation Biology* 15:275–277.

Pressey, R. L. 1994a. Ad hoc reservations: Forward or backward steps in developing representative reserve systems. *Conservation Biology* 8:662–668.

———. 1994b. Land classifications are necessary for conservation planning but what do they tell us about fauna? Pp. 31–41 in *Future of the fauna of western New South Wales*, ed. D. Lunney, S. Hand, P. Redd, and D. Butcher. Royal Society of New South Wales, Mosman, Sydney.

———. 1995. Conservation reserves in New South Wales. Crown jewels or leftovers? *Search* 26:47–51.

———. 1997. Algorithms, politics and timber: An example of the role of science in a public, political negotation process over new conservation areas in production forests. Pp. 73–87 in *Ecology for everyone: Communicating ecology to scientists, the public and the politicans*, ed. R. T. Wills, R. J. Hobbs, and M. D. Fox. Chipping Norton, Australia: Surrey Beatty and Sons.

Pressey, R. L., S. Ferrier, T. C. Hager, C. A. Woods, S. L. Tully, and K. M. Weinman. 1996. How well protected are the forests of north-eastern New South Wales? Analyses of forest environments in relation to formal protection measures, land tenure,

and vulnerability to clearing. *Forest Ecology and Management* 85:311–333.

Pressey, R. L., C. J. Humphries, C. R. Margules, R. I. Vane-Wright, and P. H. Williams. 1993. Beyond opportunism: Key principles for systematic reserve selection. *Trends in Evolution and Ecology* 8:124–128.

Pressey, R. L., and V. S. Logan. 1994. Level of geographical subdivision and its effects on assessments of reserve coverage: A review of regional studies. *Conservation Biology* 8:1037–1046.

Pressey, R. L., H. P. Possingham, and J. R. Day. 1997. Effectiveness of alternative heuristic algorithms for identifying minimum requirements for conservation reserves. *Biological Conservation* 80:207–219.

Pressey, R. L., and S. L. Tully. 1994. The cost of ad hoc reservation: A case study in western New South Wales. *Australian Journal of Ecology* 19:375–384.

Preston, F. W. 1962. The canonical distribution of commonness and rarity. *Ecology* 43:185–215, 410–432.

Price, K., and G. Hochachka. 2001. Epiphytic lichen abundance: Effects of stand age and composition in coastal British Columbia. *Ecological Applications* 11:904–913.

Price, O. 1999. Conservation of frugivorous birds and monsoon rainforest patches in the Northern Territory. Ph.D. diss., The Australian National University, Canberra.

Price, O., J. C. Woinarski, D. L. Liddle, and J. Russell-Smith. 1995. Patterns of species composition and reserve design for a fragmented estate: Monsoon rainforests in the Northern Territory. *Biological Conservation* 74:9–19.

Price, O. F., J.C.Z. Woinarski, and D. Robinson. 1999. Very large requirements for frugivorous birds in monsoon rainforests of the Northern Territory, Australia. *Biological Conservation* 91:169–180.

Primack, R. B. 1993. *Essentials of conservation biology.* Sunderland, Mass.: Sinauer Associates.

Pringle, C. M. 2001. Hydrologic connectivity and the management of biological reserves: A global perspective. *Ecological Applications* 11:981–998.

Prober, S., and M. A. Austin. 1990. Habitat peculiarity as a cause of rarity in *Eucalyptus paliformis. Australian Journal of Ecology* 16:189–205.

Prober, S., and K. R. Thiele. 1995. Conservation of the grassy white box woodlands: Relative contributions of size and disturbance to floristic composition and diversity of remnants. *Australian Journal of Botany* 43:349–366.

Proctor, J., and R. J. Woodell. 1975. The ecology of ultramafic soils. *Advances in Ecological Research* 9:255–366.

Pulliam, H. R., J. B. Dunning, and J. Liu. 1992. Population dynamics in complex landscapes: A case study. *Ecological Applications* 2:165–177.

Putz, F. E., K. H. Redford, J. G. Robinson, R. Fimbel, and G. M. Bate. 2000. *Biodiversity conservation in the context of tropical forest management.* World Bank Environment Department Papers. Paper no. 75. Biodiversity Series—Impact studies 1. The World Bank, Washington, D.C.

Putz, F. E., and C. Romero. 2001. Biologists and timber certification. *Conservation Biology* 15:313–314.

Quinn, J. F., and A. Hastings. 1987. Extinction in sub-divided habitats. *Conservation Biology* 1:198–208.

Quinn, J. F., C. L. Wolin, and M. L. Judge. 1989. An experimental analysis of patch size, habitat subdivision, and extinction in a marine intertidal snail. *Conservation Biology* 3:242–251.

Rab, M. A. 1998. Rehabilitation of snig tracks and landings following logging of *Eucalyptus regnans* forests in the Victorian Central Highlands—a review. *Australian Forestry* 61:103–113.

Raedeke, K. J. 1988. Introduction. Pp. xiii–xv in *Streamside management: Riparian wildlife and forestry interactions*, ed. K. J. Raedeke.

Contribution no. 59. College of Forest Resources, University of Washington, Seattle.

Ralph, C. J., G. L. Hunt, M. G. Raphael, and J. F. Platt, eds. 1995. *Ecology and conservation of the marbled murrelet*. USDA Forest Service General Technical Report PSW-152. Albany, Calif.

Rambo, T. R. 2001. Decaying logs and habitat heterogeneity: Implications for bryophyte diversity in western Oregon forests. *Northwest Science* 75:270–277.

Rankin-De Merona, J. M., G. T. Prance, R. W. Hutchings, F. M. Silva, W. A. Rodrigues, and M. E. Euhling. 1992. Preliminary results of large-scale tree inventory of upland rainforest. *Acta Amazonica* 22:494–534.

Raphael, M. G., and M. White. 1984. Use of snags by cavity-nesting birds in the Sierra Nevada. *Wildlife Monographs* 86:1–66.

Rawlinson, P. A., and P. R. Brown. 1980. Forestry practices threaten wildlife—The case for leadbeater's possum. Pp. 57–61 in *What state is the garden in?*, ed. P. Westbrook and J. Farhall. Victoria, Melbourne, Australia: Conservation Council.

Ray, S., D. Robinson, and G. Werren. 1983. *Pines versus native forests: An annotated bibliography*. Native Forests Action Council, Melbourne, Australia.

Read, J., and M. Brown. 1996. Ecology of Australian *Nothofagus* forests. Pp. 131–181 in *The ecology and biogeography of* Nothofagus *forests*, ed. T. T. Veblen, R. S. Hill, and J. Read. New Haven, Conn.: Yale University Press.

Rebertus, A. J., T. Kitzberger, T. T. Veblen, and L. M. Roovers. 1997. Blowdown history and landscape patterns in the Andes of Tierra del Fuego, Argentina. *Ecology* 78:678–692.

Recher, H. F. 1985. A diminishing resource: Mature forest and its role in forest management. Pp. 28–33 in *Wildlife management in the forests and forest-controlled lands in the tropics and the southern hemisphere*, ed. J. Kikkawa. IUFRO and University of Queensland, Brisbane, Australia.

———. 1996. Conservation and management of eucalypt forest vertebrates. Pp. 339–388 in *Conservation of faunal diversity in forested landscapes*, ed. R. DeGraff and I. Miller. London: Chapman and Hall.

Recher, H. F., J. D. Majer, and S. Ganesh. 1996. Eucalypts, arthropods and birds: On the relation between foliar nutrients and species richness. *Forest Ecology and Management* 85:177–196.

Recher, H. F., W. Rohan-Jones, and P. Smith. 1980. *Effects of the Eden woodchip industry on terrestrial vertebrates with recommendations for management*. Research Note no. 42. Forestry Commission of New South Wales.

Recher, H. F., J. Shields, R. Kavanagh, and G. Webb. 1987. Retaining remnant mature forest for nature conservation at New South Wales. Pp. 177–194 in *Nature Conservation: The role of remnants of native vegetation*, ed. D. A. Saunders, G. W. Arnold, A. A. Burbridge, and A. J. Hopkins. Chipping Norton: Surrey Beatty and Sons.

Redford, K., and G. da Fonseca. 1986. The role of gallery forests in the zoogeography of the Cerrado's non-volant mammalian fauna. *Biotropica* 18:126–135.

Redford, K. H. 1992. The empty forest. *BioScience* 42:412–422.

Reed, R. A., J. Johnson-Barnard, and W. L. Baker. 1996. Contribution of roads to forest fragmentation in the Rocky Mountains. *Conservation Biology* 10:1098–1106.

Reeves, G. H., and J. R. Sedell. 1992. An ecosystem approach to the conservation and management of freshwater habitat for anadromous salmonids in the Pacific Northwest. *Transactions of the North American Wildlife and Natural Resources Conference* 57:408–415.

Reeves, G. H., L. E. Benda, K. M. Burnett, P. A. Bisson, and J. R. Sedell. 1995. A disturbance-based ecosystem approach to maintaining and restoring freshwater habitats of evolutionary significant units of anadromous salmonids in the Pacific Northwest. American Fisheries Society. *Symposium* 17:334–349.

Reh, W., and W. Seitz. 1990. The influence of land use on the genetic structure of populations of the common frog *Rana temporaria*. *Biological Conservation* 54:239–249.

Reiner, R., and T. Griggs. 1989. The Nature Conservancy undertakes riparian restoration projects in California. *Restoration and Management Notes* 7:3–8.

Renjifo, L. M. 2001. Effect of natural and anthropogenic landscape matrices on the abundance of sub-Andean bird species. *Ecological Applications* 11:14–31.

Resource Assessment Commission. 1992. *Forest and timber inquiry*. Final Report. Vol. 1. Canberra: Australian Government Printing Service.

Reville, B. J., J. D. Tranter, and H. D. Yorkston. 1990. Impact of forest clearing on the endangered seabird *Sula abbotti*. *Biological Conservation* 51:23–38.

Rich, A. C., D. S. Dobkin, and L. J. Niles. 1994. Defining forest fragmentation by corridor width: The influence of narrow forest-dividing corridors on forest-nesting birds in southern New Jersey. *Conservation Biology* 8:1109–1121.

Richards, B. N., R. G. Bridges, R. A. Curtin, H. A. Nix, K. R. Shepherd, and J. Turner. 1990. *Biological conservation of the south-east forests*. Report of the Joint Scientific Committee. Canberra: Australian Government Publishing Service.

Richardson, C. J. 2000. Freshwater wetlands. Pp. 449–499 in *North American terrestrial vegetation*, ed. M. G. Barbour and W. D. Billings. 2nd ed. Cambridge: Cambridge University Press.

Richardson, D. M., P. A. Williams, and R. J. Hobbs. 1994. Pine invasions in the southern hemisphere: Determinants of spread and invadability. *Journal of Biogeography* 21:511–527.

Ricketts, T. H. 2001. The matrix matters: Effective isolation in fragmented landscapes. *American Naturalist* 158:87–99.

Ricketts, T., W. Dinerstein, D. Olson, C. Loudes, W. Eichbaum, K. Kavanagh, P. Hedao, P. Hurley, K. Carney, R. Abell, and S. Walters. 1999. *A conservation assessment of the terrestrial ecosystems of North America*. Vol. 1. *The United States and Canada*. Washington, D.C.: Island Press.

Rieman, B. E., D. C. Lee, and R. F. Furrow. 1997. Distribution, status, and likely future trends of bull trout within the Columbia River and Klamath River basins. *North American Journal of Fisheries Management* 17:1111–1125.

Ringold, P. L., J. Alegria, R. L. Czaplewksi, B. S. Mulder, T. Tolle, and K. Burnett. 1996. Adaptive monitoring design for ecosystem management. *Ecological Applications* 6:745–747.

Ripple, W. J., G. A. Bradshaw, and T. A. Spies. 1991. Measuring forest landscape patterns in the Cascade range of Oregon, USA. *Biological Conservation* 57:73–88.

Ripple, W. J., K. T. Hershey, and R. Anthony. 2000. Historical forest patterns of Oregon's central Coast Range. *Biological Conservation* 93:127–133.

Risser, P. G. 1988. General concepts for measuring cumulative impacts on wetland ecosystems. *Environmental Management* 12:585–589.

Roberts, K. A. 1991. Field monitoring: Confessions of an addict. Pp. 179–212 in *Monitoring for conservation and ecology*, ed. F. B. Goldsmith. London: Chapman and Hall.

Robertshaw, J. D., and R. H. Harden. 1989. Predation on macropodoidea: A review. Pp. 735–753 in *Kangaroos, wallabies and rat kangaroos*, ed. G. Grigg, P. Jarman, and I. Hume. Chipping Norton, Australia: Surrey Beatty and Sons.

Robertson, A. W., D. Kelly, J. J. Ladley, and A. D. Sparrow. 1999. Effects of pollinator loss on endemic New Zealand mistletoes (Loranthaceae). *Conservation Biology* 13:499–508.

Robinson, G. R., R. D. Holt, M. S. Gaines, S. P. Hamburg, M. L. Johnson, H. S. Fitch, and E. A. Martinko. 1992. Diverse and contrasting effects of habitat fragmentation. *Science* 257:524–526.

Robinson, G. R., and J. F. Quinn. 1988. Extinction, turnover and species diversity in an experimentally fragmented California annual grassland. *Ecology* 76:71–82.

Robinson, S. K. 1998. Another threat posed by forest fragmentation: Reduced food supply. *Auk* 115:1–3.

Robinson, S. K., E. R. Thompson, T. M. Donovan, D. R. Whitehead, and J. Faaborg. 1995. Regional forest fragmentation and the nesting success of migratory birds. *Science* 267:1987–1990.

Robinson, W. D. 1999. Long-term changes in the avi-fauna of Barro Colorado Island, Panama, a tropical forest isolate. *Conservation Biology* 13:85–97.

Rochelle, J. A., L. A. Lehmann, and J. Wisniewski, eds. 1999. *Forest Fragmentation. Wildlife management implications*. Leiden, Germany: Brill.

Rodenhouse, N. L., and L. B. Best. 1994. Foraging patterns of vesper sparrows (*Pooecetes gramineus*) breeding in cropland. *American Midland Naturalist* 131:196–206.

Rodrigues, A., and K. Gaston. 2001. How large do reserve networks need to be? *Ecology Letters* 4:602–609.

Rodrigues, A. S., R. D. Gregory, and K. J. Gaston. 2000. Robustness of reserve selection procedures under temporal species turnover. *Proceedings of the Royal Society of London. Series B* 267:49–55.

Rogers, K. 1997. Operationalizing ecology under a new paradigm: An African perspective. Pp. 60–77 in *The ecological basis for conservation: Heterogeneity, ecosystems, and biodiversity*, ed. S. T. Pickett, R. S. Ostfeld, M. Shachak, and G. E. Likens. New York: Chapman and Hall.

Rolstad, J., I. Gjerde, K. Storaunet, and E. Rolstad. 2001. Epiphytic lichens in Norwegian coastal spruce forest: Historic logging and present forest structure. *Ecological Applications* 11:421–436.

Rolstad, J., and P. Wegge. 1987. Distribution and size of capercaillie leks in relation to old forest fragmentation. *Oecologia* 72:389–394.

Romme, W. H., and D. G. Despain. 1989. Historical perspectives on the Yellowstone fires of 1988. *BioScience* 39:695–699.

Rose, C. L., B. G. Marcot, T. K. Mellen, J. L. Ohmann, K. L. Waddell, D. L. Lindley, and B. Schreiber. 2001. Decaying wood in Pacific Northwest forests: Concept and tools for habitat management. Pp. 580–623 in *Wildlife-habitat relationships in Oregon and Washington*, eds. D. H. Johnson and T. A. O'Neill. Corvallis: Oregon State University Press.

Rose, C. R., and P. S. Muir. 1997. Green tree retention: Consequences for timber production in forests of the Western Cascades, Oregon. *Ecological Applications* 7:209–217.

Rosenberg, D. K., B. R. Noon, and E. C. Meslow. 1997. Biological corridors: Form, function and efficacy. *BioScience* 47:677–687.

Rosenzweig, M. L. 1995. *Species diversity in space and time*. Cambridge: Cambridge University Press.

Rotherham, I. 1983. Suppression of surrounding vegetation by veteran trees in karri (*Eucalyptus diversicolor*). *Australian Forestry* 46:8–13.

Routley, R., and V. Routley. 1975. *The fight for the forests: The takeover of Australian forests for pines, woodchips and intensive forestry*. Research School of Social Sciences, The Australian National University, Camberra, Australia.

Rowell, D. M., A. V. Higgins, D. A. Briscoe, and N. N. Tait. 1995. The use of chromosomal data in the systematics of the viviparous onychophorans from Australia (Onychophora: Peripatopsidae). In *Onychophora: Past and present*, ed. M. H. Walker and D. B. Norman. *Zoological Journal of the Linnean Society* 14:139–153.

Rowley, I., and G. Chapman. 1991. The breeding biology, food, social organisation, demography and conservation of the major mitchell or pink cockatoo, *Cacatua leadbeateri*, on the margin of the Western Australian wheatbelt. *Australian Journal of Zoology* 39:211–261.

Rudnicky, T. C., and M. L. Hunter. 1993. Avian nest predation in clearcuts, forests, and edges in a forest-dominated landscape. *Journal of Wildlife Management* 57:358–364.

Rudolph, D. C., and R. N. Conner. 1996. Red-cockaded woodpeckers and silvicultural practice: Is uneven-aged silviculture preferable to even-aged? *Wildlife Society Bulletin* 24:330–333.

Ruggerio, L. F., K. B. Aubry, A. B. Carey, and M. H. Huff (technical coordinators). 1991. *Wildlife and vegetation of unmanaged Douglas-fir forests*. General Technical Report PNW-GTR-285. May. USDA Forest Service, Pacific Northwest Research Station, Portland, Ore.

Rülcker, C., P. Angelstam, and P. Rosenberg. 1994. Natural forest-fire dynamics can guide conservation and silviculture in boreal forests. *SkogForsk* 2:1–4.

Ruth, R. H., and A. S. Harris. 1979. *Management of western hemlock–Sitka spruce forests for timber production*. USDA Forest Service General Technical Report PNW-88. Portland, Ore.

Rutherford, I., K. White, N. Marsh, and K. Jerie. 2000. Some observations on the amount and distribution of large woody debris in Australian streams. *RipRap* 16:10–16.

Saab, V. 1999. Importance of spatial scale to habitat use by breeding birds in riparian forests: A hierarchical approach. *Ecological Applications* 9:135–151.

Saari, L., J. Aberg, and J. E. Swenson. 1998. Factors influencing the dynamics of occurrence of the hazel grouse in a fine-grained managed landscape. *Conservation Biology* 12:586–592.

Saccheri, I., M. Kuussaari, M. Kankare, P. Vikman, W. Fortelius, and I. Hanski. 1998. Inbreeding and extinction in a butterfly metapopulation. *Nature* 392:491–494.

Samuelsson, J., L. Gustafsson, and T. Ingelög. 1994. *Dying and dead trees—a review of their importance for biodiversity*. Swedish Threatened Species Unit, Uppsala, Sweden.

Sanderson, H. R. 1975. Den-tree management for gray squirrels. *Wildlife Society Bulletin* 3:125–131.

Santiago Declaration. 1995. *The Montreal process: Criteria and indicators for the conservation and sustainable management of temperate and boreal forests*. Canadian Forest Service, Hull, Quebec.

Sargent, R. A., J. C. Kilgo, B. R. Chapman, and K. V. Miller. 1998. Predation of artificial nests in hardwood fragments enclosed by pine and agricultural habitats. *Journal of Wildlife Management* 62:1438–1442.

Sarre, S. 1995. Mitochondrial DNA variation among populations of *Oedura reticulata* (Gekkonidae) in remnant vegetation: Implications for metapopulation structure and population decline. *Molecular Ecology* 4:395–405.

Sarre, S., G. T. Smith, and J. A. Meyers. 1995. Persistence of two species of gecko (*Oedura reticulata* and *Gehyra variegata*) in remnant habitat. *Biological Conservation* 71:25–33.

Saunders, D. A. 1979. The availability of tree hollows for use as nest sites by white-tailed cockatoos. *Australian Wildlife Research* 6:205–216.

Saunders, D. A., G. W. Arnold, A. A. Burbridge, and A. J. Hopkins, eds. 1987. *Nature conservation: The role of remnants of native vegetation*. Chipping Norton, Australia: Surrey Beatty and Sons.

Saunders, D. A., and R. J. Hobbs, eds. 1991. *Nature conservation 2: The role of corridors*. Chipping Norton, Australia: Surrey Beatty and Sons.

Saunders, D. A., R. J. Hobbs, and C. R. Margules. 1991. Biological consequences of ecosystem fragmentation: A review. *Conservation Biology* 5:18–32.

Saunders, D. A., and J. Ingram. 1995. *Birds of southwestern Australia*. Chipping Norton, Australia: Surrey Beatty and Sons.

Saunders, D. A., G. T. Smith, and I. Rowley. 1982. The availability and dimensions of tree hollows that provide nest sites for cockatoos (Psittaciformes) in Western Australia. *Australian Wildlife Research* 9:541–546.

Saurez, A. V., D. T. Bolger, and T. J. Case. 1998. Effects of fragmentation and invasion on native ant communities in coastal southern California. *Ecology* 79:2041–2056.

Saveneh, A. G., and P. Dignan. 1998. The use of shelterwood in *Eucalyptus regnans* forest: The effect of overwood removal at three years on regeneration stocking and health. *Australian Forestry* 4:251–259.

Savidge, J. A. 1987. Extinction of an island forest avifauna by an introduced snake. *Ecology* 68:660–668.

Savill, P. S. 1983. Silviculture in windy climates. *Commonwealth Forestry Bureau* 44:473–488.

Sax, J. L. 1980. *Mountains without handrails: Reflections on the national parks.* Ann Arbor: University of Michigan Press.

Schelnas, J., and R. Breenberg, eds. 1996. *Forest patches in tropical landscapes.* Washington, D.C.: Island Press.

Schieck, J., K. Lertzman, N. Nyberg, and R. Page. 1995. Effects of patch size in old growth montane forests. *Conservation Biology* 9:1072–1084.

Schiegg, K. 2001. Saproxylic insect diversity of beech: Limbs are richer than trunks. *Forest Ecology and Management* 149:295–330.

Schmidt, W. 1989. Plant dispersal by motor cars. *Vegetatio* 80:147–152.

Schmiegelow, F. K., and S. J. Hannon. 1993. Adaptive management, adaptive science and the effects of forest fragmentation on boreal birds in northern Alberta. *Transactions of the North American Wildlife and Natural Resources Conference* 58:584–598.

Schmiegelow, F. K., C. S. Machtans, and S. J. Hannon. 1997. Are boreal birds resilient to forest fragmentation? An experimental study of short-term community responses. *Ecology* 78:1914–1932.

Schneider, D. C. 1994. *Quantitative ecology: Temporal and spatial scaling.* San Diego: Academic Press.

Schnitzler, A., and F. Borlea. 1998. Lessons from natural forests as key for sustainable management and improvement of naturalness in managed broadleaved forests. *Forest Ecology and Management* 109:293–303.

Schonewald-Cox, C. M. 1988. Boundaries in the protection of nature reserves. *BioScience* 38:480–486.

Schonewald-Cox, C. M., and M. Buechner. 1990. Park protection and public roads. Pp. 373–395 in *Conservation biology: The theory and practice of nature conservation, preservation and management,* ed. P. L. Fiedler and S. K. Jain. New York: Chapman and Hall.

Schowalter, T. D. 1989. Canopy arthropod community structure and herbivory in old-growth and regenerating forests in western Oregon. *Canadian Journal of Forest Research* 19:318–322.

Schultz, C. B. 1998. Dispersal behavior and its implications for reserve design in a rare Oregon butterfly. *Conservation Biology* 12:284–292.

Schütz, J. 2001. Opportunities and strategies of transforming regular forests to irregular forests. *Forest Ecology and Management* 151:87–94.

Schwartz, M. W., ed. 1997. *Conservation in highly fragmented landscapes.* New York: Chapman and Hall.

Schwartz, M. W. 1999. Choosing the appropriate scale of reserves for conservation. *Annual Review of Ecology and Systematics* 30:83–108.

Schwartz, M. W., and P. J. van Mantgem. 1997. The value of small preserves in chronically fragmented landscapes. Pp. 379–394 in *Conservation in highly fragmented landscapes,* ed. M. W. Schwartz. New York: Chapman and Hall.

Scientific Advisory Group. 1995. *National Forest Conservation Reserves, Commonwealth proposed criteria: A position paper.* Canberra: Australian Government Publishing Service.

Scientific Panel for Sustainable Forest Practices in Clayoquot Sound. 1995. *Report 5. Sustainable ecosystem management in Clayoquot Sound: Planning and practices.* Cortex Consultants, Victoria, British Columbia.

Scientific Panel on Ecosystem Based Forest Management. 2000. *Simplified forest management to achieve watershed and forest health: A critique.* Report to the National Wildlife Federation, Seattle, Wash.

Scott, J. M. 1999. Vulnerability of forested ecosystems in the Pacific Northwest to loss of area. Pp. 33–42 in *Forest wildlife and fragmentation: Management implications,* ed. J. Rochelle, L. A. Lehmann, and J. Wisniewski. Leiden, Germany: Brill.

Scott, J. M., R. J. F. Abbitt, and C. R. Groves. 2001a. What are we protecting? *Conservation Biology in Practice* 2:18–19.

Scott, J. M., F. Davis, B. Csuti, R. Noss, B. Butterfield, C. Groves, H. Anderson, S. Caicco, F. D'Erchia, T. C. Edwards Jr., J. Ulliman, and R. G. Wright. 1993. Gap analysis: A geographic approach to protection of biological diversity. *Wildlife Monographs* 123:1–41.

Scott, J. M., F. W. Davis, R. G. McGhie, R. G. Wright, C. Groves, and J. Estes. 2001b. Nature reserves: Do they capture the full range of America's biological diversity? *Ecological Applications* 11:999–1007.

Scotts, D. J. 1991. Old-growth forests: Their ecological characteristics and value to forest-dependent vertebrate fauna of south-east Australia. Pp. 147–159 in *Conservation of Australia's forest fauna,* ed. D. Lunney. Royal Zoological Society of New South Wales, Sydney, Australia.

———. 1994. Sustaining sensitive wildlife within temperate forest landscapes: Regional systems of retained habitat as a planning framework. Pp. 85–106 in *Ecology and sustainability of southern temperate ecosystems,* ed. T. Norton and S. Dovers. Melbourne, Australia: CSIRO Publishing.

Scougall, S. A., J. D. Majer, and R. J. Hobbs. 1993. Edge effects in grazed and ungrazed Western Australia wheatbelt remnants in relation to ecosystem reconstruction. Pp. 163–178 in *Nature conservation 3: Reconstruction of fragmented ecosystems,* ed. D. A. Saunders, R. J. Hobbs, and P. R. Ehrlich. Chipping Norton, Australia: Surrey Beatty and Sons.

Seabrook, W. A., and E. B. Dettmann. 1996. Roads as activity corridors for cane toads in Australia. *Journal of Wildlife Management* 60:363–368.

Seagle, S. W., and H. H. Shugart. 1985. Faunal richness and turnover on dynamics landscapes: A simulation study. *Journal of Biogeography* 15:759–774.

Sedell, J. R., P. A. Bisson, F. J. Swanson, and S. V. Gregory. 1988. What we know about large trees that fall into streams and rivers. Pp. 47–81 in *From the forest to the sea: A story of fallen trees,* ed. C. Maser, R. F. Tarrant, J. M. Trappe, and J. F. Franklin. USDA Forest Service General Technical Report PNW-GTR-229.

Seebeck, J. H., R. M. Warneke, and B. J. Baxter. 1984. Diet of the bobuck, *Trichosurus caninus* (Ogilby) (Marsupialia: Phalangeridae) in a mountain forest in Victoria. Pp. 145–154 in *Possums and gliders,* ed. A. P. Smith and I. D. Hume. Chipping Norton, Australia: Surrey Beatty and Sons.

Semlitsch, R. D., and J. R. Bodie. 1998. Are small, isolated wetlands expendable? *Conservation Biology* 12:1129–1133.

Serena, M., ed. 1995. *Reintroduction biology of Australasian fauna.* Chipping Norton, Australia: Surrey Beatty and Sons.

Seymour, R. S., and M. L. Hunter. 1999. Principles of ecological forestry. Pp. 22–61 in *Maintaining biodiversity in forest ecosystems,* ed. M. L. Hunter. Cambridge: Cambridge University Press.

Shafer, C. L. 1990. *Nature reserves: Island theory and conservation practice.* Washington, D.C.: Smithsonian Institute Press.

Shaffer, M. L. 1981. Minimum population sizes for species conservation. *BioScience* 31:131–134.

Shaffer, M. L., and F. B. Samson. 1985. Population size and extinction: A note on determining critical population size. *American Naturalist* 125:144–152.

Sharitz, R. R., L. R. Boring, D. H. van Lear, and J. E. Pinder. 1992. Integrating ecological concepts with natural resource management of southern forests. *Ecological Applications* 2:226–237.

Shaw, D., J. Greenleaf, and D. Berg. 1993. Monitoring new forestry. *Environmental Monitoring and Assessment* 26:187–193.

Shea, S. R., I. Abbott, J. A. Armstrong, and K. J. McNamara. 1997. Sustainable conservation: A new integrated approach to nature conservation in Australia. Pp. 39–48 in *Conservation outside nature reserves*, ed. P. Hale and D. Lamb. University of Queensland, Brisbane, Australia.

Shepherd, N., and G. C. Caughley. 1987. Options for managment of kangaroos. Pp. 188–219 in *Kangaroos and management in the sheep rangelands of Australia*, ed. G. C. Caughley, N. Shepherd, and J. Short. Sydney, Australia: Cambridge University Press.

Shepherd, T. G., M. J. Saxon, D. B. Lindenmayer, T. W. Norton, and H. P. Possingham. 1992. *A proposed management strategy for the Nalbaugh Special Prescription Area based on guiding ecological principles*. South East Forest Series no. 2. Threatened Species Research. NSW National Parks and Wildlife Service, Sydney, Australia.

Shine, R., and M. Fitzgerald. 1989. Conservation and reproduction of an endangered species: The broad-headed snake, *Hoplocephalus bungaroides* (Elapidae). *Australian Zoologist* 25:65–66.

Shinneman, D. J., and W. L. Baker. 1997. Nonequilibrium dynamics between catastrophic disturbances and old growth forests in ponderosa pine landscapes of the Black Hills. *Conservation Biology* 11:1276–1288.

Sierra Nevada Ecosystem Project. 1994. *Progress report*. Wildland Resources Center, University of California, Davis.

———. 1996. *Final report to Congress on status of the Sierra Nevada*. Vol. 1. *Assessment summaries and management strategies*. Report 36. Wildland Resources Center, University of California, Davis.

———. 1997. *Final report to Congress on the status of the Sierra Nevada. Addendum*. Report 40. Wildland Resources Center, University of California, Davis.

Sieving, K. E., M. F. Willson, and T. L. de Santo. 2000. Defining corridor functions for endemic birds in fragmented south-temperate rainforest. *Conservation Biology* 14:1120–1132.

Siitonen, J., and P. Martikainen. 1994. Occurrence of rare and threatened insects living on decayed *Populus tremula*: A comparison between Finnish and Russian Karelia. *Scandinavian Journal of Forest Research* 9:185–191.

Siitonen, J., P. Martikainen, P. Punttila, and J. Rauh. 2000. Coarse woody debris and stand characteristics in mature managed and old-growth boreal mesic forests in southern Finland. *Forest Ecology and Management* 128:211–225.

Sillett, S. C., B. McCune, J. E. Peck, T. R. Rambo, and A. Ruchty. 2000. Dispersal limitations of epiphytic lichens result in species dependent on old-growth forests. *Ecological Applications* 10:789–799.

Silsbee, D. G., and G. L. Larson. 1983. A comparison of streams in logged and unlogged areas of the Great Smoky Mountains National Park. *Hydrobiology* 102:99–111.

Silsbee, D. G., and D. L. Peterson. 1993. Planning for implementation of long-term resource monitoring programs. *Environmental Monitoring and Assessment* 26:177–185.

Simard, S. W., D. A. Perry, M. D. Jones, D. D. Myrold, D. M. Durall, and R. Molina. 1997. Net transfer of carbon between ectomycorrhizal tree species in the field. *Nature* 188:579–582.

Simberloff, D. A. 1988. The contribution of population and community biology to conservation science. *Annual Review of Ecology and Systematics* 19:473–511.

———. 1998. Flagships, umbrellas, and keystones: Is single-species management passe in the landscape era? *Biological Conservation* 83:247–257.

Simberloff, D., and L. G. Abele. 1982. Refuge design and island geographic theory: Effects of fragmentation. *American Naturalist* 120:41–45.

Simberloff, D., and T. Dayan. 1991. The guild concept and the structure of ecological communities. *Annual Review of Ecology and Systematics* 22:115–143.

Simberloff, D. A., J. A. Farr, J. Cox, and D. W. Mehlman. 1992. Movement corridors: Conservation bargains or poor investments? *Conservation Biology* 6:493–504.

Simberloff, D. S., and J. Cox. 1987. Consequences and costs of conservation corridors. *Conservation Biology* 1:63–71.

Sims, R. A., I. G. W. Corns, and K. Klinka, eds. 1995. *Global to local: Ecological land classification*. London: Kluwer Academic Publishers.

Sinton, D. S., J. A. Jones, J. L. Ohmann, and F. J. Swanson. 2000. Windthrow disturbance, forest composition, and structure in the Bull Run Basin, Oregon. *Ecology* 81:2539–2556.

Sisk, T., N. M. Haddad, and P. R. Ehrlich. 1997. Bird assemblages in patchy woodlands: Modeling the effects of edge and matrix habitats. *Ecological Applications* 7:1170–1180.

Sizer, N., and E. V. Tanner. 1999. Responses of woody plant seedlings to edge formation in a lowland tropical rainforest, Amazonia. *Biological Conservation* 91:135–142.

Sjörberg, K., and L. Ericson. 1992. Forested and open wetlands. Pp. 326–351 in *Ecological principles of nature conservation*, ed. L. Hansson. London: Elsevier.

Skonhoft, A., N. G. Yoccoz, N. Stenseth, J. Gilliard, and A. Loison. In press. Optimal management of a wildlife species moving between a protected core area and a surrounding hunting area: The Chamois (*Rupicapra rupicapra*) in the French Alps as an example. *Ecological Applications* (in press).

Slatkin, M. 1985. Rare alleles as indicators of gene flow. *Evolution* 39:53–65.

Slatkin, M., and N. H. Barton. 1989. A comparison of three indirect methods for estimating average levels of gene flow. *Evolution* 43:1349–1368.

Smith, A. P. 1984. Diet of Leadbeater's possum *Gymnobelideus leadbeateri* (Marsupialia). *Australian Wildlife Research* 11:265–273.

Smith, A. P., N. Horning, and D. Moore. 1997. Regional biodiversity planning and lemur conservation with GIS in western Madagascar. *Conservation Biology* 11:498–512.

Smith, A. P., and D. B. Lindenmayer. 1992. Forest succession, timber production and conservation of Leadbeater's possum (*Gymnobelideus leadbeateri* Marsupialia: Petauridae). *Forest Ecology and Management* 49:311–332.

Smith, A. P., D. M. Moore, and S. P. Andrew. 1992. *Proposed forestry operations in the Glen Innes Forest Management Area: Fauna impact statement*. Austeco Pty. Ltd. and the Forestry Commission of New South Wales, Armidale, New South Wales, Australia.

Smith, A. P., G. S. Wellham, and S. W. Green. 1989. Seasonal foraging activity and microhabitat selection by echidnas (*Tachyglossus aculeatus*) on the New England tablelands. *Australian Journal of Ecology* 14:457–466.

Smith, A. T. 1980. Temporal changes in insular populations of the pika (*Ochotona principes*). *Ecology* 61:8–13.

Smith, D. M., B. C. Larson, M. J. Kelty, P. Mark, and S. Ashton. 1999. *The practice of silviculture: Applied forest ecology*. 9th ed. New York: John Wiley and Sons.

Smith, D. S., and P. C. Hellmund. 1993. *Ecology of greenways: Design and function of linear conservation areas*. Minneapolis: University of Minnesota Press.

Smith, P. 1985. Effects of intensive logging on birds in eucalypt forest near Bega, New South Wales. *Emu* 85:15–21.

Smith, P. A. 1994. Autocorrelation in logistic regression modeling of species' distribution. *Global Ecology and Biogeography Letters* 4:47–61.

Smith P. L, B. R. Wilson, C. N. Nadolny, and D. Lang. 2000. *The ecological role of native vegetation in New South Wales*. Background Paper No 2. The Native Vegetation Advisory Council of New South Wales, Sydney, Australia.

Smith, R. B., and P. Woodgate 1985. Appraisal of fire damage for timber salvage by remote sensing in mountain ash forests. *Australian Forestry* 48:252–263.

Smith, R. J. 2000. An investigation into the relationships between anthropogenic forest disturbance patterns, population viability and landscape indices. Master's thesis, Institute of Land and Food Resources, University of Melbourne, Australia.

Society of American Foresters. 1984. *Scheduling the harvest of old growth*. Society of American Foresters, Bethesda, Md.

Socolow, R., C. Andrews, F. Berkhout, and V. Thomas, eds. 1994. *Industrial ecology and global change*. Cambridge: Cambridge University Press.

Soderquist, T. R., and R. Mac Nally. 2000. The conservation value of mesic gullies in dry forest landscapes: Mammal populations in the box-ironbark ecosystem of southern Australia. *Biological Conservation* 93:281–291.

Soulé, M. E., ed. 1987. *Viable populations for conservation*. New York: Cambridge University Press.

Soulé, M. E., and M. A. Sanjayan. 1998. Conservation targets: Do they help. *Science* 279:2060–2061.

Soulé, M. E., and D. Simberloff. 1986. What do genetics and ecology tell us about the design of nature reserves? *Biological Conservation* 35:19–40.

Spackman, S. C., and J. W. Hughes. 1995. Assessment of minimum stream corridor width for biological conservation: Species richness and distribution along mid-order strems in Vermont, USA. *Biological Conservation* 71:325–332.

Sparrow, H. R., T. D. Sisk, P. R. Ehrlich, and D. D. Murphy. 1994. Techniques and guidelines for monitoring neotropical butterflies. *Conservation Biology* 8:800–809.

Specht, R. L., A. Specht, M. B. Whelan, and E. E. Hegarty. 1995. *Conservation atlas of plant communities in Australia*. Lismore, New South Wales: Centre for Coastal Management and Southern Cross University Press.

Spellerberg, I. F. 1994. *Monitoring ecological change*. 2nd ed. Cambridge: Cambridge University Press.

———. 1998. Ecological effects of roads and traffic: A literature review. *Global Ecology and Biogeography Letters* 7:317–333.

Spellerberg, I. F., and J. Sawyer. 1996. Standards for biodiversity: A proposal based on biodiversity standards for forest plantations. *Biodiversity and Conservation* 5:447–459.

———. 1997. Biological diversity in plantation forests. Pp. 517–522 in *Conservation outside nature reserves*, ed. P. Hale and D. Lamb. Centre for Conservation Biology, The University of Queensland, Brisbane, Australia.

Spence, J. R., D. W. Langor, J. Niemelä, H. E. Cárcamo, and C. R. Currie. 1996. Northern forestry and carabids: The case for concern about old-growth species. *Annales Zoologici Fennici* 33:173–184.

Spies, T. A., and B. V. Barnes. 1985. A multi-factor ecological classification of the northern hardwood and conifer ecosystems of the Sylvania Recreation Area, Upper Peninsula, Michigan. *Canadian Journal of Forest Research* 15:949–960.

Spies, T. A., and J. F. Franklin. 1988. Old growth and forest dynamics in the Douglas-fir region of western Oregon and Washington. *Natural Areas Journal* 8:190–201.

Spies, T. A., W. J. Ripple, and G. A. Bradshaw. 1994. Dynamics and pattern of a managed coniferous forest landscape in Oregon. *Ecological Applications* 4:555–568.

Spies, T. A., and M. G. Turner. 1999. Dynamic forest mosaics. Pp. 95–160 in *Maintaining biodiversity in forest ecosystems*, ed. M. Hunter Jr. Cambridge: Cambridge University Press.

Squire, R. O. 1987. *Revised treatments, design and implementation strategy for the Silvicultural Systems Project*. August. Lands and Forests Division, Department of Conservation, Forests and Lands, Melbourne, Australia.

———. 1990. *Report on the progress of the Silvicultural Systems Project, July 1986–June 1989*. Department of Conservation and Environment, Melbourne, Australia.

———. 1993. The professional challenge of balancing sustained wood production and ecosystem conservation in the native forests of south-eastern Australia. *Australian Forestry* 56:237–248.

Squire, R. O., R. G. Campbell, K. J. Wareing, and G. R. Featherston. 1991. The mountain ash forests of Victoria: Ecology, silviculture and management for wood production. Pp. 38–57 in *Forest management in Australia*, ed. F. H. McKinnell, E. R. Hopkins, and J. E. D. Fox. Chipping Norton, Australia: Surrey Beatty and Sons.

Squire, R. O., B. Dexter, R. Smith, A. Manderson, and D. Flinn. 1987. *Evaluation of alternative systems for Victoria's commercially important mountain forests*. Project brief. Lands and Forests Division, Department of Conservation, Forests and Lands, Melbourne, Australia.

Srivastava, D. S., and J. H. Lawton. 1998. Why more productive sites have more species: An experimental test of theory using tree-hole communities. *American Naturalist* 152:510–529.

Stacey, P. B., and M. Taper. 1992. Environmental variation and the persistence of small populations. *Ecological Applications* 2:18–29.

Stamps, J. A., M. Buechner, and V. V. Krishnan. 1987a. The effects of edge permeability and habitat geometry on emigration from patches of habitat. *American Naturalist* 129:533–552.

———. 1987b. The effects of habitat geometry on territorial defense costs: Intruder pressure in bounded habitats. *American Zoologist* 27:307–325.

Stanford, J. A., and J. V. Ward. 1993. An ecosystem perspective of alluvial rivers: Connectivity and the hyporheic corridor. *Journal of the North American Benthological Society* 12:48–60.

Stanford, J. A., J. V. Ward, and B. K. Ellis. 1994. Ecology of the alluvial aquifers of the Flathead River, Montana. Pp. 367–390 in *Groundwater ecology*, ed. J. Gilbert, D. Danielopol, and J. A. Stanford. San Diego: Academic Press.

Starfield, A. M., and A. L. Bleloch. 1992. *Building models for conservation and wildlife management*. Burgess International Group, Edina, Minn.

Stauffer, D. F., and L. B. Best. 1980. Habitat selection by birds of riparian communities. *Journal of Wildlife Management* 44:1–15.

Steeger, C., M. Machmer, and E. Walters. Undated. *Ecology and management of woodpeckers and wildlife trees in British Columbia*. Report for the Fraser River action plan, Environment Canada, conducted by Pandion Ecological Research Ltd. Vancouver, British Columbia.

Steinblums, I. J., H. A. Froehlich, and J. K. Lyons. 1984. Designing stable buffer strips for stream protection. *Journal of Forestry* 82:49–52.

Stelfox, J. G. 1971. Bighorn sheep in the Canadian Rockies: A history 1800–1970. *Canadian Field Naturalist* 85:101–122.

Stenseth, N., and W. Lidicker. 1992. *Animal dispersal*. London: Chapman and Hall.

Stohlgren, T. J. 1992. Resilience of a heavily logged grove of giant sequoia (*Sequoiadendron giganteum*) in Kings Canyon National Park, California. *Forest Ecology and Management* 54:115–140.

Stokes, B. J., and A. Schilling. 1997. Improved harvesting methods for wet sites. *Forest Ecology and Management* 90:155–160.

Stoms, D. M., F. W. Davis, K. L. Driese, K. M. Cassidy, and M. P. Murray. 1998. Gap analysis of the vegetation of the intermountain semi-desert ecoregion. *Great Basin Naturalist* 58:199–216.

Storch, I. 1997. The importance of scale in habitat conservation for an endangered species: The capercaillie in central Europe. Pp.

310–330 in *Wildlife and landscape ecology: Effects of pattern and scale*, ed. J. Bissonette. New York: Springer-Verlag.

Stouffer, P. C., and R. O. Bierregaard. 1995. Use of Amazonian forest fragments by understorey insectivorous birds. *Ecology* 76:2429–2445.

Strahler, A. N. 1957. Quantitative analysis of watershed geomorphology. *Transactions of American Geophysical Union* 38:913–920.

Strayer, D. L. 1999. Statistical power of presence-absence data to detect population deciles. *Conservation Biology* 13:1034–1038.

Styskel, E. W. 1983. Problems in snag management implementation—a case study. Pp. 24–27 in *Snag habitat management*, ed. J. W. Davis, G. A. Goodwin, and R. A. Ockenfels. Proceedings of the symposium. 7–9 June. Northern Arizona University, Flagstaff.

Suckling, G. C. 1982. Value of reserved habitat for mammal conservation in plantations. *Australian Forestry* 45:19–27.

Suckling, G. C., E. Backen, A. Heislers, and F. G. Neumann. 1976. *The flora and fauna of radiata pine plantations in north-eastern Victoria*. Forest Commission of Victoria Bulletin no. 24. Forest Commission of Victoria, Melbourne, Australia.

Sugal, C. 1997. Most forests have no protection. *World Watch* 10:9.

Sverdrup-Thygeson, A. 2000. Forest management and conservation: Woodland key habitats, indicator species and tree retention. Ph.D. diss., Faculty of Mathematics and Natural Sciences, University of Oslo, Norway.

Sverdrup-Thygeson, A., and D. B. Lindenmayer. 1999. Indikatorarter I skogforvaltningen. Fauna. Norwegian Zoological Society. *Fauna* 51:150–159.

———. 2002. Ecological continuity and assumed indicator fungi in boreal forest: The importance of the landscape matrix. *Forest Ecology and Management* (in press).

Swanson, F. J., S. V. Gregory, J. R. Sedell, and A. G. Campbell. 1982. Land-water interactions: The riparian zone. Pp. 267–291 in *Analysis of coniferous forest ecosystems in the western United States*, ed. R. L. Edmonds. US/IBP Synthesis Series 14. Stroudsburg, Penn.: Hutchinson Ross Publishing.

Swanson, F. J., S. L. Johnson, S. V. Gregory, and S. A. Acker. 1998. Flood disturbance in a forested mountain landscape. *BioScience* 48:681–689.

Swanson, F. J., T. J. Kratz, N. Caine, and R. G. Woodmansee. 1988. Landform effects on ecosystem patterns and processes. *BioScience* 38:92–98.

Swanson, F., J. Jones, B. Wemple, and K. Snyder. 2000. Roads in forest watersheds—assessing effects from a landscape perspective. In *Proceedings of the Seventh Biennial Watershed Management Council conference*, ed. C. W. Slaughter. 19–23 October 1998, Boise, Idaho. *Water Resources Center Report 98*. Centers for Water and Wildlife Resources, University of California, Riverside. http://www.watershed.org/wmc/pdf/seventh_biennial_wmc_proc.pdf (2000 May 1).

Swenson, J. E., K. L. Alt, and R. L. Eng. 1986. Ecology of bald eagles in the Greater Yellowstone Ecosystem. *Wildlife Monographs* 95:1–46.

Swenson, J. E., and P. Angelstam. 1993. Habitat separation by sympatric forest grouse in Fennoscandia in relation to boreal forest succession. *Canadian Journal of Zoology* 71:1303–1310.

Swift, M. J. 1977. The ecology of wood decomposition. *Science Progress* 67:175–199.

Swift Parrot Recovery Team. 2000. *Draft swift parrot recovery plan 2001–2005*. Parks and Wildlife Service, Hobart, Tasmania, Australia.

Sydes, M. 1994. Orchids: Indicators of management success. *Victorian Naturalist* 111:213–217.

Syrjänen, K., R. Kalliola, A. Puolasmaaa, and J. Mattsson. 1994. Landscape structure and forest dynamics in subcontinental Russian European taiga. *Annales Zoologici Fennici* 31:19–34.

Szacki, J., J. Babinska-Werka, and A. Liro. 1993. The influence of landscape spatial structure on small mammal movements. *Acta Theriologica* 38:113–123.

Szaro, R. C., and M. D. Jakle. 1985. Avian use of a desert riparian island and its adjacent scrub habitat. *Condor* 87:511–519.

Szaro, R. C., and D. W. Johnson, eds. 1996. *Biodiversity in managed landscapes: Theory and practice*. New York: Oxford University Press.

Tague, C., and L. Baud. 2001. Simulating the impact of road construction and forest harvesting on hydrologic response. *Earth Surface Processes and Landforms* 26:135–151.

Tait, N. N., D. A. Briscoe, and D. M. Rowell. 1995. *Onycophora—ancient and modern radiations*. Association of Australasian Paleontologists. 18:21–30.

Tallmon, D., and L. S. Mills. 1994. Use of logs within home ranges of Californian red-backed voles on a remnant of forest. *Journal of Mammalogy* 75:97–101.

Tang, S. M., J. F. Franklin, and D. R. Montgomery. 1997. Forest harvest patterns and landscape disturbance processes. *Landscape Ecology* 12:349–363.

Tappeiner, J. C., D. Lavender, J. Walstad, R. O. Curtis, and D. S. DeBell. 1997. Silvicultural systems and regeneration methods: Current practices and new alternatives. Pp. 151–164 in *Creating a forestry for the twenty-first century: The science of ecosystem management*, ed. K. A. Kohm and J. F. Franklin. Washington, D.C.: Island Press.

Tarp, P., F. Helles, P. Holten-Andersen, J. B. Larsen, and N. Strange. 2000. Modeling near-natural silvicultural regimes for beech—an economic sensitivity analysis. *Forest Ecology and Management* 130:187–198.

Tasmanian Forestry Commission. 1987. *Forest practices codes*. Forestry Commission of Tasmania, Hobart, Tasmania, Australia.

Taulman, J. F., K. G. Smith, and R. E. Thill. 1998. Demographic and behavioral responses of southern flying squirrels to experimental logging in Arkansas. *Ecological Applications* 8:1144–1155.

Taylor, B., L. Kremsater, and R. Ellis. 1997. *Adaptive management of forests in British Columbia*. British Columbia Ministry of Forests, Forest Practices Branch. British Columbia Ministry of Forests, Victoria, British Columbia.

Taylor, C. R., S. L. Caldwell, and V. J. Rowntree. 1972. Running up and down hills: Some consequences of size. *Science* 178:1096–1097.

Taylor, P. D., L. Fahrig, K. Henein, and G. Merriam. 1993. Connectivity is a vital element of landscape structure. *Oikos* 68:571–573.

Taylor, R. 1979. How the Macquarie Island parakeet became extinct. *New Zealand Journal of Ecology* 2:42–45.

Taylor, R. J. 1990. Occurrence of log-dwelling invertebrates in regeneration and old-growth wet sclerophyll forest in southern Tasmania. *Papers and Proceedings of the Royal Society of Tasmania* 119:7–15.

———. 1991. The role of retained strips for fauna conservation in production forests in Tasmania. Pp. 265–270 in *Conservation of Australia's forest fauna*, ed. D. Lunney. Royal Zoological Society of New South Wales, Sydney, Australia.

Taylor, R. J., S. L. Bryant, D. Pemberton, and T. W. Norton. 1985. Mammals of the Upper Henty River region, Western Tasmania. *Papers and Proceedings of the Royal Society of Tasmania* 119:7–15.

Taylor, R. J., and M. Haseler. 1993. Occurrence of potential nest trees and their use by birds in sclerophyll forest in north-east Tasmania. *Australian Forestry* 56:165–171.

Taylor, W. E., and R. G. Hooper. 1991. *A modification of Copeyon's drilling technique for making artificial red-cockaded woodpecker cavities*. USDA Forest Service General Technical Report SE-72.

Telleria, J. L., and T. Santos. 1992. Spatiotemporal patterns of egg predation in forest islands: An experimental approach. *Biological Conservation* 62:29–33.

Temple, S. A. 1986. The problem of avian extinctions. *Current Ornithology* 3:453–485.

Temple, S. A., and J. R. Cary. 1988. Modeling dynamics of habitat interior bird populations in fragmented landscapes. *Conservation Biology* 2:340–347.

Temple, S. A., and J. A. Wiens. 1989. Bird populations and environmental changes: Can birds be bio-indicators? *American Birds* 43:260–270.

Terborgh, J. 1974. Preservation of natural diversity: The problem of extinction prone species. *BioScience* 24: 715–722.

———. 1986. Keystone plant resources in the tropical forest. Pp. 330–344 in *Conservation biology: The science of scarcity and diversity*, ed. M. E. Soulé. Sunderland, Mass.: Sinauer Associates.

———. 1989. *Where have all the birds gone?* Princeton, N.J.: Princeton University Press.

———. 1992. Maintenance of diversity in tropical forests. *Biotropica* 24:283–292.

———. 1999. *Requiem for nature*. Washington, D.C: Island Press.

Thiel, R. P. 1985. Relationship between road density and wolf habitat suitability in Wisconsin. *American Midland Naturalist* 113:404–407.

Thiollay, J. 1992. The influence of selective logging on bird species diversity in a Guianian rain forest. *Conservation Biology* 6:47–63.

———. 1997. Disturbance, selective logging and bird diversity: A Neotropical forest study. *Biodiversity and Conservation* 6:1155–1173.

———. 1999. Responses of avian community to rain forest degradation. *Biodiversity and Conservation* 8:513–534.

Thomas, C. D. 1990. What do real population dynamics tell us about minimum viable population sizes? *Conservation Biology* 4:324–327.

Thomas, C. D., J. A. Thomas, and M. S. Warren. 1992a. Distributions of occupied and vacant habitats in fragmented landscape. *Oecologica* 92:563–567.

Thomas, D. W. 1995. Hibernating bats are sensitive to nontactile human disturbance. *Journal of Mammalogy* 76:940–946.

Thomas, J. A., and M. G. Morris. 1995. Rates and patterns of extinction among British invertebrates. Pp. 111–130 in *Extinction rates*, ed. J. H. Lawton and R. M. May. Oxford: Oxford University Press.

Thomas, J. W. 1990. From managing a deer herd to moving a mountain—one pilgrim's progress. *Journal of Wildlife Management* 64:1–10.

———, ed. 1979. *Wildlife habitats in managed forests: The Blue Mountains of Oregon and Washington*. USDA Agricultural Handbook 553. Washington, D.C.: U.S. Government Printing Office.

Thomas, J. W., E. D. Forsmann, J. B. Lint, E. C. Meslow, B. R. Noon, and J. Verner. 1990. *A conservation strategy for the northern spotted owl*. Portland, Ore.: U.S. Government Printing Office.

Thomas, J. W., L. F. Ruggiero, R. W. Mannan, J. W. Schoen, and R. A. Lancia. 1988. Management and conservation of old-growth forests in the United States. *Wildlife Society Bulletin* 16:252–262.

Thomas, K., R. H. Norris, and G. A. Chilvers. 1992b. Litterfall in riparian and adjacent forest zones near a perennial upland stream in the Australian Capital Territory. *Australian Journal of Marine and Freshwater Research* 43:511–516.

Thompson, I. D., and P. Angelstam. 1999. Special species. Pp. 434–459 in *Maintaining biodiversity in forest ecosystems*, ed. M. L. Hunter. Cambridge: Cambridge University Press.

Thurber, J. M., R. O. Peterson, T. D. Drummer, and S. A. Thomas. 1994. Gray wolf response to refuge boundaries and roads in Alaska. *Wildlife Society Bulletin* 22:61–68.

Tickle, P., S. Hafner, R. Lesslie, D. B. Lindenmayer, C. McAlpine, B. Mackey, P. Norman, and S. Phinn. 1998. *Scoping study: Final report. Montreal Indicator 1.1e. Fragmentation of forest types—identification of research priorities*. A study prepared for the Forest and Wood Products Research and Development Corporation. Canberra, Australia.

Tietje, W. D., and R. L. Ruff, 1980. Denning behaviour of black bears in boreal forest of Alberta. *Journal of Wildlife Management* 44:858–870.

Tilghman, N. G. 1989. Impacts of white-tailed deer on forest regeneration in northwestern Pennsylvania. *Journal of Wildlife Management* 53:524–532.

Tilman, D. 1996. Biodiversity: Population versus ecosystem stability. *Ecology* 77:350–363.

Tilman, D., R. M. May, C. L. Lehman, and M. A. Nowak. 1994. Habitat destruction and the extinction debt. *Nature* 371:65–66.

Tischendorf, L. 2001. Can landscape indices predict ecological processes consistently? *Landscape Ecology* 16:235–254.

Tocher, M. D., C. Gascon, and B. L. Zimmerman. 1997. Fragmentation effects on a central American frog community: A ten-year study. Pp. 124–137 in *Tropical forest remnants: Ecology, management and conservation of fragmented communities*, ed. W. F. Laurance and R. O. Bierregaard. Chicago: University of Chicago Press.

Torsvik, V., J. Goksoyr, and F. Daae. 1990. High diversity in DNA of soil bacteria. *Applied and Environmental Microbiology* 56:782–787.

Trayler, K. M., and J. A. Davis. 1998. Forestry issues and the vertical distribution of stream invertebrates in south-western Australia. *Freshwater Biology* 4:331–342.

Trombulak, S. C., and C. A. Frissell. 2000. Review of ecological effects of roads on terrestrial and aquatic communities. *Conservation Biology* 14:18–30.

Trzcinski, M. K., L. Fahrig, and G. Merriam. 1999. Independent effects of forest cover and fragmentation on the distribution of forest breeding birds. *Ecological Applications* 9:586–593.

Tscharntke, T. 1992. Fragmentation of phragmites habitats, minimum viable population size, habitat suitability, and local extinction of moths, midges, flies, aphids and birds. *Conservation Biology* 6:530–536.

Tuchmann, E. T., K. P. Connaughton, L. E. Freedman, and C. B. Moriwaki. 1996. *The Northwest Forest Plan: A report to the president and Congress*. USDA Office of Forestry and Economic Assistance, Portland, Ore.

Tuckey, W. 2001. *Protecting our forest*. Commonwealth of Australia, Canberra.

Turner, D. P., and E. H. Franz. 1985. The influence of western hemlock and western redcedar on microbial numbers, nitrogen mineralization, and nitrification. *Plant and Soil* 88:259–267.

Turner, I. M. 1996. Species loss in fragments of tropical rain forest: A review of the evidence. *Journal of Applied Ecology* 33:200–209.

Turner, I. M., and R. T. Corlett. 1996. Conservation value of small, isolated fragments of lowland tropical rain forest. *Trends in Ecology and Evolution* 11:330–333.

Turner, M. G., W. L. Baker, C. J. Peterson, and R. K. Peet. 1998. Factors influencing succession: Lessons from large, infrequent natural disturbances. *Ecosystems* 1:511–523.

Turner, M. G., R. H. Gardner, and R. V. O'Neill. 1995. Ecological dynamics at broad scales. *Bioscience* (science and biodiversity policy supplement):29–35.

Turner, M. G., S. M. Pearson, W. H. Romme, and L. L. Wallace. 1997. Landscape heterogeneity and ungulate dynamics: What spatial scales are important? Pp. 331–348 in *Wildlife and landscape ecology: Effects of pattern and scale*, ed. J. Bissonette. New York: Springer-Verlag.

Tyndale-Biscoe, C. H., and R. F. Smith. 1969. Studies of the marsupial glider, *Schoinobates volans* (Kerr). III. Response to habitat destruction. *Journal of Animal Ecology* 38:651–659.

UHEKS-toimikunta. 1992. *Uhanalaisten elainten ja kasvien seurantatoimikunnan mietinto—Betankande av kommisionen for overvakning av hotade djur och vaxter*. Report on the monitoring of threatened animals and plants in Finland (in Finnish with Swedish and English summaries). Ymparistoministerio, Ministry of Environment. Komiteanmietinto, Committee Report. 1991.

Underwood, A. J. 1995. Ecological research, and research into environmental management. *Ecological Applications* 5:232–247.

Upton, C., and S. Bass. 1996. *The forest certification handbook*. Delray Beach, Fla.: St. Lucie Press.

Urban, D., R. V. O'Neill, and H. H. Shugart. 1987. Landscape ecology: A hierarchical perspective can help scientists understand spatial patterns. *BioScience* 37:119–127.

Urban, D. L. 2000. Using model analysis to design monitoring programs for landscape management and impact assessment. *Ecological Applications* 10:1820–1832.

Urban, D. L., and T. M. Smith. 1989. Microhabitat pattern and the structure of forest bird communities. *American Naturalist* 133:811–829.

USDA Forest Service. 2000. *Forest Service roadless area conservation final environmental impact statement*. Vol. 1. USDA Forest Service, Washington, D.C.

———. 2001a. *Sierra Nevada forest plan amendment. Final environmental impact statement*. 6 vols. USDA Forest Service, Pacific Southwest Region, San Francisco, Calif.

———. 2001b. *Sierra Nevada forest plan amendment. Final environmental impact statement. Record of Decision*. USDA Forest Service, Pacific Southwest Region, San Francisco, Calif.

USDA Forest Service and USDI Bureau of Land Management. 1994a. *Final supplemental environmental impact statement on management of habitat for late-successional and old-growth forest related species within the range of the northern spotted owl*. Vols. 1 and 2. USDA Forest Service and USDI Bureau of Land Management, Portland, Ore.

———. 1994b. *Record of decision for amendments to Forest Service and Bureau of Land Management planning documents within the range of the northern spotted owl and standards and guidelines for development of habitat for late-successional and old-growth forest related species within the range of the northern spotted owl*. USDA Forest Service Region 6, Portland, Ore.

———. 2000. *Interior Columbia Basin supplemental draft environmental impact statement*. Vol. 1. Interior Columbia Basin Ecosystem Management Project. Boise, Idaho.

USDI Fish and Wildlife Service. 1995. *Recovery plan for the Mexican spotted owl* (Strix occidentalis lucida). USDI Fish and Wildlife Service, Southwestern Region, Albuquerque, N.M.

Uuttera, J., M. Maltamo, and K. Kuusela. 1996. Impact of forest management history on the state of forests in relation to natural forest succession comparative study, North Karelia, Russian Federation. *Forest Ecology and Management* 83:71–85.

Vandermeer, J., B. Hoffman, S. Krantz-Ryan, U. Wijayratne, J. Buff, and V. Franciscus. 2001. Effect of habitat fragmentation on gypsy moth (*Mymantria dispar*) dispersal: The quality of the matrix. *American Midland Naturalist* 145:188–193.

Van der Meer, P. J., P. Dignan, and A. G. Savaneh. 1999. Effects of gap size on seedling establishment, growth and survival at three years in mountain ash (*Eucalyptus regnans* F. Muell.) forest in Victoria, Australia. *Forest Ecology and Management* 117:33–42.

Van der Zee, F. F., J. Wiertz, C. J. Ter Braak, R. C. van Apeldoorn, and J. Vink. 1992. Landscape change as a possible cause of the badger *Meles meles* decline in The Netherlands. *Biological Conservation* 61:17–22.

van Dorp, D., and P. F. Opdam. 1987. Effects of patch size, isolation and regional abundance of forest bird communities. *Landscape Ecology* 1:59–73.

Van Horne, B. 1983. Density as a misleading indicator of habitat quality. *Journal of Wildlife Management* 47: 893–901.

van Nieuwstadt, M. G., D. Sheil, and K. Kartawinata. 2001. The ecological consequences of logging in the burned forests of east Kalimantan, Indonesia. *Conservation Biology* 15:1183–1186.

Vannote, R. L., G. W. Minshall, K. W. Cummins, J. R. Sedell, and C. E. Cushing. 1980. The river continuum concept. *Canadian Journal Fisheries and Aquatic Sciences* 37:130–137.

Van Pelt, R., and J. F. Franklin. 1999. Response of understory trees to experimental gaps in old-growth Douglas-fir forests. *Ecological Applications* 9:504–512.

Van Wagner, C. E. 1978. Age-class distribution and the forest fire cycle. *Canadian Journal of Forest Research* 8:220–227.

Veblen, T. T., C. Donoso, T. Kitzberger, and A. J. Rebertus. 1996. Ecology of southern Chilean and Argentinean *Nothofagus* forests. Pp. 293–353 in *The ecology and biogeography of* Nothofagus *forests*, ed. T. T. Veblen, R. S. Hill, and J. Read. New Haven, Conn.: Yale University Press.

Veblen, T. T., T. Kitzberger, and J. Donnegan. 2000. Climatic and human influences on fire regimes in Ponderosa pine forests in the Colorado front range. *Ecological Applications* 10:1178–1195.

Vellak, K., and J. Paal. 1999. Diversity of bryophyte vegetation in some forest types in Estonia: A comparison of old unmanaged and managed forests. *Biodiversity and Conservation* 8:1595–1620.

Verboom, B., and H. Huitema. 1997. The importance of linear landscape elements for the pipistrelle *Pipistrellus pipistrellus* and the serotine bat *Eptesicus serotinus*. *Landscape Ecology* 12:117–125.

Vermeulen, R., and T. Opsteeg. 1994. Movements of some carabid beetles in road-side verges: Dispersal in a simulation programme. Pp. 393–398 in *Carabid beetles: Ecology and evolution*, ed. K. Desender. The Hague: Kluwer Academic Press.

Verner, J., K. S. McKelvey, B. R. Noon, R. J. Gutierrez, G. I. Gordon, and T. W. Beck. 1992. *The California spotted owl: A technical assessment of its current status*. USDA Forest Service General Technical Report PSW-GTR-133.

Vesely, D., and W. C. McComb. 1996. Terrestrial amphibian abundance in riparian buffer strips in the Oregon Coast Range. *Coastal Oregon Productivity Enhancement Program* 9:6–7.

Vestjens, W. J. 1973. Wildlife mortality on a road in New South Wales. *Emu* 73:107–112.

Viana, V. M., J. Ervin, R. Z. Donovan, C. Elliott, and H. Gholz. 1996. *Certification of forest products*. Washington, D.C.: Island Press.

Villard, M. A., and P. D. Taylor. 1994. Tolerance to habitat fragmentation influences the colonization of new habitat by forest birds. *Oecologia* 98:393–401.

Villard, M. A., M. K. Trzcinski, and G. Merriam. 1999. Fragmentation effects on forest birds: Relative influence of woodland cover and configuration on landscape occupancy. *Conservation Biology* 13:774–783.

Virkkala, R., A. Rajasarrkka, R. A. Vaisanen, M. Vickholm, and E. Virolainen. 1994. The significance of protected areas for the land birds of southern Finland. *Conservation Biology* 8:532–544.

Virolainen, K. M., T. Virola, J. Suhonen, M. Kuitunen, A. Lammin, and P. Siikamaki. 1999. Selecting networks of nature reserves: Methods do affect the long-term outcome. *Proceedings of the Royal Society of London. Series B* 266:1141–1146.

von Hartman, L. 1971. Population dynamics. Pp. 391–459 in *Avian Biology*, vol. 1, ed. D. S. Farner and J. R. King. London: Academic Press.

Vos, C. C., and J. P. Chardon. 1998. Effects of habitat fragmentation and road density on the distribution patterns of the moor frog *Rana arvalis*. *Journal of Applied Ecology* 35:44–56.

Wace, N. 1977. Assessment of the dispersal of plant species—the car-borne flora of Canberra. *Proceedings of the Ecological Society of Australia* 10:166–186.

Wadleigh, L., and M. J. Jenkins. 1996. Fire frequency and the vegetative mosaic of a spruce-fir forest in northern Utah. *Great Basin Naturalist* 56:28–37.

Walker, A. 1999. Examination of the barriers to movements of Tasmanian fish. Honors thesis, School of Zoology, University of Tasmania, Hobart, Australia.

Walker, B. 1998. The art and science of wildlife management. *Wildlife Research* 25:1–9.

Walker, J., F. Bullen, and B. G. Williams. 1993. Ecohydrological changes in the Murray-Darling Basin. I. The number of trees cleared over two centuries. *Journal of Applied Ecology* 30:265–273.

Walker, R. B. 1954. The ecology of serpentine soils. II. Factors affecting plant growth on serpentine soils. *Ecology* 35:259–266.

Wallis, A., D. Stokes, G. Wescott, and T. McGee. 1997. Certification and labelling as a new tool for sustainable forest management. *Australian Journal of Environmental Management* 4:224–238.

Walters, C. 1997. Adaptive policy design: Thinking at large spatial scales. Pp. 386–394 in *Wildlife and landscape ecology: Effects of pattern and scale*, ed. J. Bissonette. New York: Springer-Verlag.

Walters, C. J. 1986. *Adaptive management of renewable resources.* New York: Macmillan.

Walters, C. J., and C. S. Holling. 1990. Large scale management experiments and learning by doing. *Ecology* 71:2060–2068.

Wapstra, M., and R. J. Taylor. 1998. Use of retained trees for nesting by birds in logged eucalypt forest in north-eastern Tasmania. *Australian Forestry* 61:48–52.

Wardell-Johnson, G., and P. Horowitz. 1996. Conserving biodiversity and the recognition of heterogeneity in ancient landscapes: A case study from south-western Australia. *Forest Ecology and Management* 85:219–238.

Wardell-Johnson, G., and O. Nichols. 1991. Forest wildlife and habitat management in southwestern Australia: Knowledge, research and direction. Pp. 161–192 in *Conservation of Australia's forest fauna*, ed. D. Lunney. Chipping Norton, Australia: Surrey Beatty and Sons.

Wardell-Johnson, G., and J. D. Roberts. 1993. Biogeographic barriers in a subdued landscape: The distribution of the *Geocrinia rosea* (Anura: Myobatrachidae) complex in south-western Australia. *Journal of Biogeography* 20:95–108.

Waring, R. H., and J. Major. 1964. Some vegetation of the California coastal redwood region in relation to gradients of moisture, nutrients, light, and temperature. *Ecological Monographs* 34:167–215.

Warneke, R. M. 1962. Internal memorandum to Director (of Fisheries and Wildlife Division). 10 May. Fisheries and Wildlife Division, Melbourne, Australia.

Warren, L. M. 1993. The precautionary principle: Use with caution. Pp. 97–111 in *Environmentalism: The view from anthropology*, ed. M. Kay. New York: Routledge.

Washington State Department of Natural Resources. 1997. *Final habitat conservation plan.* Department of Natural Resources, Olympia, Wash.

Wauters, L., P. Casale, and A. Dhondt. 1994. Space use and dispersal of red squirrels in fragmented habitats. *Oikos* 69:140–146.

Wayburn, L. A., J. F. Franklin, J. C. Gordon, C. S. Binkley, D. J. Mladhoff, and N. L. Christensen. 2000. *Forest carbon in the United States: Opportunities and options for private lands.* Pacific Forest Trust, Santa Rosa, Calif.

Weathers, K. C., and G. E. Likens. 1997. Clouds in southern Chile: An important source of nitrogen to nitrogen-limited ecosystems? *Environmental Science and Technology* 31:210–213.

Weaver, J. C. 1994. Indicator species and scale of observation. *Conservation Biology* 9:939–942.

Webb, G. 1995. Habitat use and activity patterns in some southeastern Australian skinks. Pp. 23–30 in *Biology of Australasian frogs and reptiles*, ed. G. Grigg, R. Shine, and H. Ehmann. Royal Zoological Society of New South Wales, Sydney, Australia.

Webb, N. R., R. T. Clarke, and J. T. Nicholas. 1984. Invertebrate diversity on fragmented *Calluna* heathland: Effects of surrounding vegetation. *Journal of Biogeography* 11:41–46.

Webb, N. R., and P. J. Hopkins. 1984. Invertebrate diversity on fragmented *Calluna* heathland. *Journal of Applied Ecology* 21:921–933.

Webster, R. 1988. *The superb parrot: A survey of the breeding distribution and habitat requirements.* Australian National Parks and Wildlife Service, Report Series no. 12. Canberra, Australia.

Webster, R., and L. Ahern. 1992. *Management for conservation of the superb parrot (Polytelis swainsonii) in New South Wales and Victoria.* Department of Conservation and Natural Resources, Melbourne, Australia.

Wegner, J. 1994. *Ecological landscape variables for monitoring and management of forest biodiversity in Canada.* Report to Canadian Ministry of Natural Resources by GM Group, Ecological Land Management, Manotick, Canada.

Wegner, J. F., and G. Merriam. 1979. Movements by birds and small mammals between a wood and adjoining farmland habitats. *Journal of Applied Ecology* 16:349–357.

Weishampel, J. F., H. H. Shugart, and W. E. Westman. 1997. Phenetic variation in insular populations of a rainforest centipede. Pp. 111–123 in *Tropical forest remnants: Ecology, management and conservation*, ed. W. F. Laurance and R. O. Bierregaard. Chicago: University of Chicago Press.

Welsh, A. H., R. B. Cunningham, and R. L. Chambers. 2000. Methodology for estimating the abundance of rare animals: Seabird nesting on North East Herald Cay. *Biometrics* 56:22–30.

Welsh, C. J., and W. M. Healy. 1993. Effects of even-aged timber management on bird diversity and composition in northern hardwoods of New Hampshire. *Wildlife Society Bulletin* 21:143–154.

Wemple, B. C., J. A. Jones, and G. E. Grant. 1996. Channel network extension by logging roads in two basins, Western Cascades, Oregon. *Water Resources Bulletin* 32:1195–1207.

Wemple, B. C., F. J. Swanson, and J. A. Jones. 2001. Forest roads and geomorphic process interactions, Cascade Range, Oregon. *Earth Surface Processes and Landforms* 26:191–204.

Wesolowski, T. 1995. The loss of avian cavities by injury compartmentalization in a primeval European forest. *Condor* 97:256–257.

———. 1996. Natural nest sites of marsh tit (*Parus palustris*) in a primaeval forest (Bialowieza National Park, Poland). *Die Vogelwarte* 38:235–249.

Western, D., and Gichohi. 1993. Segregation effects and the impoverishment of savanna parks: The case for ecosystem viability analysis. *African Journal of Ecology* 31:269–281.

Westman, W. E. 1990. Managing for biodiversity. *BioScience* 40:26–33.

Weyerhaeuser. 2000. *Summary of the Second Year Critique Workshop on the Weyerhaeuser BC Coastal Forest Project, July 12–14, 2000.* Weyerhaeuser, Vancouver, British Columbia.

White, P. S., and S.T.A. Pickett. 1985. Natural disturbance and patch dynamics: An introduction. Pp. 3–13 in *The ecology of natural disturbance and patch dynamics*, ed. S.T.A. Pickett and P. S. White. Orlando, Fla.: Academic Press.

Whitmore, T. C. 1997. Tropical forest disturbance, disappearance and species loss. Pp. 3–12 in *Tropical forest remnants: Ecology, management and conservation of fragmented communities*, ed. W. F. Laurance and R. O. Bierregaard. Chicago: University of Chicago Press.

Whittaker, D. M., and W. A. Montevecchi. 1999. Breeding bird assemblages inhabiting riparian buffer strips in Newfoundland, Canada. *Journal of Wildlife Management* 63:167–179.

Whittaker, R. H. 1954a. The ecology of serpentine soils. I. Introduction. *Ecology* 35:258–259.

———. 1954b. The ecology of serpentine soils. IV. The vegetational response to serpentine soils. *Ecology* 35:275–288.

Whittaker, R. H., S. A. Levin, and R. B. Root. 1973. Niche, habitat and ecotope. *American Naturalist* 107:321–328.

Wiegand, J. F., R. W. Haynes, and J. L. Mikowski. 1994. *High quality forestry workshop: The idea of long rotations.* University of Washington College of Forest Resources Center for International Trade in Forest Products Special Paper 15. Seattle, Wash.

Wiens, J. 1989. Spatial scaling in ecology. *Functional Ecology* 3:385–397.

———. 1992. What is landscape ecology, really? *Landscape Ecology* 7:149–150.

———. 1994. Habitat fragmentation: Island versus landscape perspectives on bird conservation. *Ibis* 137:S97–S104.

———. 1995. Landscape mosaics and ecological theory. Pp. 1–26 in *Landscape mosaics and ecological processes*, ed. L. Hansson, L. Fahrig, and G. Merriam. London: Chapman and Hall.

———. 1997a. Wildlife in patchy environments: Metapopulations, mosaics and management. Pp. 53–84 in *Metapopulations and wildlife conservation*, ed. D. R. McCullough. Washington, D.C.: Island Press.

———. 1997b. Metapopulation dynamics and landscape ecology. Pp. 43–68 in *Metapopulation biology: Ecology, genetics and evolution*, ed. I. Hanski and M. Gilpin. San Diego: Academic Press.

———. 1999. The science and practice of landscape ecology. Pp. 37–383 in *Landscape ecological analysis: Issues and applications*, ed. J. M. Klopatek and R. H. Gardner. New York: Springer-Verlag.

Wiens, J. A., R. L. Schooley, and R. D. Weekes. 1997. Patchy landscapes and animal movements: Do beetles percolate? *Oikos* 78:257–264.

Wilcove, D. S. 1985. Nest predation in forest tracts and the decline of migratory songbirds. *Ecology* 66:1211–1214.

———. 1989. Protecting biodiversity in multiple-use lands: Lessons from the U.S. Forest Service. *Trends of Evolution and Ecology* 4:385–388.

Wilcove, D. S., C. H. McLellen, and A. P. Dobson. 1986. Habitat fragmentation in the temperate zone. Pp. 237–256 in *Conservation biology: The science of scarcity and diversity*, ed. M. E. Soulé. Sunderland, Mass.: Sinauer Associates.

Wilcox, B. A., and D. D. Murphy. 1984. Conservation strategy: The effects of fragmentation on extinction. *American Naturalist* 12:879–887.

Wilkes, J. 1982a. Stem decay in deciduous hardwoods—an overview. *Australian Forestry* 45:42–50.

———. 1982b. Pattern and process of heartrot in *Eucalyptus microcorys. Australian Forestry* 45:51–56.

Wilkinson, D. A., G. C. Grigg, and L. A. Beard. 1998. Shelter selection and home range of echidnas, *Tachyglossus aculeatus*, in the highlands of south-east Queensland. *Wildlife Research* 25:219–232.

Williams, B., and B. Marcot. 1991. Use of biodiversity indicators for analyzing and managing forest landscapes. *Transactions of the North American Wildlife and Natural Resources Conference* 56:613–627.

Williams, B. K., and F. A. Johnson. 1995. Adaptive management and the regulation of waterfowl harvest. *Wildlife Society Bulletin* 23:430–436.

Williams, J., and A. M. Gill. 1995. *The impact of fire regimes on native forests in eastern New South Wales.* Environmental Heritage Monograph Series 2. NSW National Parks and Wildlife Service. Pp. 1–68.

Williams, P. 1998. Key sites for conservation: Area-selection methods for biodiversity. Pp. 211–249 in *Conservation in a changing world*, ed. G. Mace, A. Balmford, and J. Ginsberg. Cambridge: Cambridge University Press.

Williams, P., D. Gibbons, C. R. Margules, A. Rebelo, C. Humphries, and R. Pressey. 1996. A comparison of richness hotspots, rarity hotspots and complementary areas for conserving diversity of British birds. *Conservation Biology* 10:155–174.

Wills, R., and R. Hobbs, eds. 1998. *Ecology for everyone: Communicating ecology to scientists, the public and the politicans.* Chipping Norton, Australia: Surrey Beatty and Sons.

Willson, M. F., T. L. De Samto, C. Sabag, and J. J. Armesto. 1994. Avian communities of fragmented south-temperate rainforests in Chile. *Conservation Biology* 8:508–520.

Wilson, A. M., and D. B. Lindenmayer. 1996. *The role of wildlife corridors in the conservation of biodiversity in multi-use landscapes.* Centre for Resource and Environmental Studies, Greening Australia, and The Australian Nature Conservation Agency, Canberra, Australia.

Wilson, B. A., and T. W. Clark. 1995. The Victorian Flora and Fauna Guarantee Act 1988: A five-year review of its implementation. Pp. 87–103 in *People and nature conservation: Perspectives on private land use and endangered species recovery*, ed. A. Bennett, G. Backhouse, and T. Clark. Transactions of the Royal Society of New South Wales. Chipping Norton, Australia: Surrey Beatty and Sons.

Wilson, E. O., and E. O. Willis. 1978. Applied biogeography. Pp. 522–534 in *Ecology and evolution of communities*, ed. M. L. Cody and J. M. Diamond. Cambridge: Harvard University Press, Belknap Press.

Wilson, G., N. Dexter, P. O'Brien, and M. Bomford. 1992. Pest animals in Australia. A survey of introduced wild mammals. Sydney, Australia: Bureau of Rural Resources and Kangaroo Press.

Wilson, J. 1978. Caves: Changing ecosystems? *Studies in Speleology* 3:35–42.

Wimberly, M. C., and T. A. Spies. 2001. Influences of environment and disturbance on forest patterns in coastal Oregon watersheds. *Ecology* 82:1443–1459.

Wisdom, M. J., L. R. Bright, C. G. Carey, W. W. Hines, R. J. Pedderson, D. A. Smithey, J. W. Thomas, and G. W. Witmer. 1986. *A model to evaluate elk habitat in western Oregon.* USDA Forest Service, Portland, Ore.

Wisdom, M. J., R. S. Holthausen, B. K. Wales, C. D. Hargis, V. A. Saab, D. C. Lee, W. J. Hann, T. D. Rich, M. M. Rowland, W. J. Murphy, and M. R. Eames. 2000. *Source habitats for terrestrial vertebrates of focus in the Interior Columbia River Basin: Broad-scale trends and management implications.* USDA Forest Service General Technical Report PNW-GTR-485. Portland, Ore.

With, K. A. 1994. Ontogenic shifts in how grasshoppers interact with landscape structure—an analysis of movement patterns. *Functional Ecology* 8:477–485.

———. 1997. The application of neutral landscape models in conservation biology. *Conservation Biology* 11:1069–1080.

———. 1999. Is landscape connectivity necessary and sufficient for wildlife management? Pp. 97–115 in *Forest wildlife and fragmentation: Management implications*, ed. J. Rochelle, L. A. Lehmann, and J. Wisniewski. Leiden, Germany: Brill.

With, K. A., and T. O. Crist. 1995. Critical thresholds in species' responses to landscape structure. *Ecology* 76:2446–2459.

With, K. A., and A. W. King. 1999. Extinction thresholds for species in fractal landscapes. *Conservation Biology* 13:314–326.

Witting, L., and V. Loeschcke. 1995. The optimization of biodiversity conservation. *Biological Conservation* 71:205–207.

Woinarski, J., H. F. Recher, and J. D. Majer. 1997. Vertebrates and eucalypt formations. Pp. 303–341 in *Eucalypt ecology*, ed. J. Williams and J. Woinarski. Melbourne, Australia: Cambridge University Press.

Woinarski, J. M., and J. C. Cullen. 1984. Distribution of invertebrates on foliage in forests in south-eastern Australia. *Australian Journal of Ecology* 9:207–232.

Woinarski, J. M., and S. C. Tidemann. 1991. The bird fauna of a deciduous woodland in the wet-dry tropics of northern Australia. *Wildlife Research* 18:479–500.

Woinarski, J. M., P. J. Whitehead, D. Bowman, and J. Russell-Smith. 1992. Conservation of mobile species in a variable environment: The problem of reserve design in the Northern Territory, Australia. *Global Ecology and Biogeography Letters* 2:1–10.

Wolfenbarger, D. O. 1946. Dispersion of small organisms. *American Midland Naturalist* 35:1–152.

Wolff, J. O., E. M. Schauber, and W. D. Edge. 1997. Effects of habitat loss and fragmentation in the behavior and demography of gray-tailed voles. *Conservation Biology* 11:945–956.

Wolseley, P. A., and B. Aguirre-Hudson. 1991. Lichens as indicators of environmental change in the tropical forests of Thailand. *Global Ecology and Biogeography Letters* 1:170–175.

Wondzell, S. M., and F. J. Swanson. 1999. Floods, channel change, and the hyporheic zone. *Water Resources Research* 35:555–567.

Woodgate, P., W. Peel, K. Ritman, J. E. Coram, A. Brady, A. J. Rule, and J. C. Banks. 1994. *A study of the old-growth forests of East Gippsland.* Department of Conservation and Natural Resources, Melbourne, Australia.

Woodward, F. I. 1987. *Climate and plant distribution.* Cambridge: Cambridge University Press.

Woodward, F. L., and B. G. Williams 1987. Climate and plant distribution at global and local scales. *Vegetatio* 69:189–197.

Woodwell, G. M. 1989. On causes of biotic impoverishment. *Ecology* 70:14–15.

Woolley, A., and J. B. Kirkpatrick. 1999. Factors related to condition and rare and threatened species occurrence in lowland, humid basalt remnants in northern Tasmania. *Biological Conservation* 87:131–142.

World Commission on Forests and Sustainable Development. 1999. *Our forests, our future: Report of the World Commission on Forests and Sustainable Development.* Cambridge: Cambridge University Press.

World Resources Institute. 1992. *World Resources 1992–1993.* New York: Oxford University Press.

Worldwatch Institute. 1998. *State of the world 1998.* Worldwatch Institute, Washington D.C.

World Wide Fund for Nature. 1996. *Forests for life: Timber certification in Australia and New Zealand.* Discussion Paper. September. Sydney, Australia: World Wide Fund for Nature.

Wright, R. G. 1999. Wildlife management in national parks: Questions in search of answers. *Ecological Applications* 9:30–36.

Wright, S. 1969. *Evolution and the genetics of populations.* Vol. 2. *The Theory of Gene Frequencies.* Chicago: University of Chicago Press.

Yaffee, S. L. 1994. *The wisdom of the spotted owl: Policy lessons for a new century.* Washington, D.C.: Island Press.

Yahner, R. H. 1983. Seasonal dynamics, habitat relationships and management of avifauna in farmstead shelterbelts. *Journal of Wildlife Management* 47:85–104.

———. 1988. Changes in wildlife communities near edges. *Conservation Biology* 2:333–339.

Yoakum, J. D. 1978. Pronghorn. Pp. 103–121 in *Big game of North America*, ed. J. L. Schmidt and G. L. Douglas. Harrisburg, Penn.: Stackpole Books.

York, A. 2000. Long-term effects of frequent low-intensity burning on ant communities in coastal blackbutt forests of southeastern Australia. *Austral Ecology* 25:83–98.

Young, A., T. Boyle, and A. Brown. 1996. The population genetic consequences of habitat fragmentation for plants. *Trends in Ecology and Evolution* 11:413–418.

Young, M., and B. Howard. 1996. Can Australia afford a representative reserve network by 2000? *Search* 27:22–26.

Zackrisson, O. 1977. Influence of forest fires on the north Swedish boreal forest. *Oikos* 29:22–32.

Zanette, L., P. Doyle, and S. M. Trémont. 2000. Food shortage in small fragments: Evidence from an area-sensitive passerine. *Ecology* 81:1654–1666.

Zanuncio, J. C., J. A. Mezzomo, R. N. Guedes, and A. C. Oliveria. 1997. Influence of strips of native vegetation on Lepidoptera associated with *Eucalyptus cloezianna* in Brazil. *Forest Ecology and Management* 108:85–90.

Zeide, B. 1994. Big projects, big problems. *Environmental Monitoring and Assessment* 33:115–133.

Zenner, E. K. 2000. Do residual trees increase structural complexity in Pacific Northwest coniferous forests? *Ecological Applications* 10:800–810.

Zielinski, W. J., and S. T. Gellman. 1999. Bat use of remnant old-growth redwood stands. *Conservation Biology* 13:160–167.

Zimmerman, B. L., and R. O. Bierregaard. 1986. Relevance of equilibrium theory of island biogeography and species-area relations to conservation with a case study from Amazonia. *Journal of Biogeography* 13:133–143.

Zinke, P. J. 1962. The pattern of influence of individual trees on soil properties. *Ecology* 43:130–133.

Zinke, P. J., and R. L. Crocker. 1962. The influence of giant sequoia on soil properties. *Forest Science* 8:2–11.

Zobel, D. B., A. McKee, G. M. Hawker, and C. T. Dyrness. 1976. Relationships of environment to composition, structure and diversity of forest communities. *Ecological Monographs* 46:135–156.

Zobel, D. B., L. F. Roth, and G. M. Hawk. 1985. *Ecology, pathology, and management of Port Orford cedar.* USDA Forest Service General Technical Report GTR-PNW-184.

Zuidema, P. A., J. Sayer, and W. Dijkman. 1996. Forest fragmentation and biodiversity: The case for intermediate-sized reserves. *Environmental Conservation* 2:290–297.

ABOUT THE AUTHORS

David B. Lindenmayer is an associate professor at the Centre for Resource and Environmental Studies at The Australian National University in Canberra (Australia). He has worked on forestry-related projects for twenty years and has published widely on many aspects of forest ecology, wildlife biology, and conservation biology. He maintains active research programs in the Central Highlands of Victoria and in southern New South Wales from which he has published over 260 articles and books on logging impacts on biodiversity, landscape change and modification, metapopulation dynamics, the habitat requirements of vertebrates, plantation design, and numerous other topics.

Jerry F. Franklin has been professionally involved with forests for nearly fifty years. For thirty-five years he was a scientist with the U.S. Forest Service Pacific Northwest Forest Experiment Station in Corvallis, Oregon, except for occasional duties in other parts of the world, such as a year-long loan to the Japanese Forest Experiment Station and two years as ecosystem studies program officer in the National Science Foundation in Washington, D.C. Since 1986 he has been a professor of ecosystem science at the University of Washington in Seattle. Franklin has participated in several important policy analyses, including the Clayoquot Sound science panel (British Columbia), President Clinton's Forest Ecosystem Management Assessment Team, and the Sierra Nevada Ecosystem Project. His major research interests are structure and function of natural (especially older) forests, forest responses to disturbance (heightened by the eruption of Mount St. Helens in western Washington), and application of ecosystem knowledge in forest management. He has authored or coauthored over 300 articles and books on forests, ecosystem science, and conservation.

INDEX